Biotechnology in Invertebrate Pathology and Cell Culture

Biotechnology in Invertebrate Pathology and Cell Culture

Edited by

KARL MARAMOROSCH

Department of Entomology
Rutgers University
New Brunswick, New Jersey

ACADEMIC PRESS, INC.
Harcourt Brace Jovanovich, Publishers
San Diego New York Berkeley Boston
London Sydney Tokyo Toronto

ACADEMIC PRESS, INC.
1250 Sixth Avenue, San Diego, California 92101

United Kingdom Edition published by
ACADEMIC PRESS INC. (LONDON) LTD.
24–28 Oval Road, London NW1 7DX

Library of Congress Cataloging in Publication Data

Biotechnology in invertebrate pathology and cell culture.

Includes index.
1. Microbial insecticides. 2. Viral insecticides.
3. Microbial insecticides industry. 4. Viral insecticides
industry. 5. Microbial genetics. I. Maramorosch, Karl.
[DNLM: 1. Cells, Cultured–drug effects. 2. Genetic
Intervention. 3. Genetics, Microbial. 4. Insecticides–
pharmacodynamics. 5. Insects–genetics. QW 51 B616]
TP248.I57B56 1987 668'.651 87–1206
ISBN 0–12–470255–4 (alk. paper)

PRINTED IN THE UNITED STATES OF AMERICA

87 88 89 90 9 8 7 6 5 4 3 2 1

Contents

I. GENETIC MANIPULATION OF MICROBIAL INSECTICIDES

8. *Bacillus thuringiensis* Isolate with Activity against Coleoptera

Corinna Herrnstadt, Frank Gaertner, Wendy Gelernter, and David L. Edwards

9. Parasporal Body of Mosquitocidal Subspecies of *Bacillus thuringiensis*

Brian A. Federici, Jorge E. Ibarra, Leodegario E. Padua, Niels J. Galjart, and Natarajan Sivasubramanian

10. Current Status of the Microbial Larvicide *Bacillus sphaericus*

Samuel Singer

II. MASS PRODUCTION OF MICROBIAL AND VIRAL INSECTICIDES

III. GENE MANIPULATION AND CELL CULTURE

14. Expression of Human Interferon α in Silkworms with a Baculovirus Vector

S. Maeda

15. Biologically Active Influenza Virus Hemagglutinin Expressed in Insect Cells by a Baculovirus Vector

Kazumichi Kuroda, Charlotte Hauser, Rudolf Rott, Hans-Dieter Klenk, and Walter Doerfler

16. Transfection of *Drosophila melanogaster* Transposable Elements into the *Drosophila hydei* Cell Line

Tadashi Miyake, Naomi Mae, Tadayoshi Shiba, Shunzo Kondo, Manabu Takahisa, and Ryu Ueda

17. FP Mutation of Nuclear Polyhedrosis Viruses: A Novel System for the Study of Transposon-Mediated Mutagenesis

Malcolm J. Fraser

18. Expression of Foreign Genes in Insect Cells

Lois K. Miller

19. Genotypic Variants in Wild-Type Populations of Baculoviruses

A. H. McIntosh, W. C. Rice, and C. M. Ignoffo

20. Biotechnological Application of Invertebrate Cell Culture to the Development of Microsporidian Insecticides

Timothy J. Kurtti and Ulrike G. Munderloh

21. Grasshopper and Locust Control Using Microsporidian Insecticides

S. K. Raina, M. M. Rai, and A. M. Khurad

22. Establishment of Embryonic Cell Lines from the Brown Ear Tick *Rhipicephalus appendiculatus* and Their Immunogenicity in Rabbits

M. Nyindo, L. R. S. Awiti, and T. S. Dhadialla

23. Establishment of an Ovarian Cell Line in the Cotton Bollworm *Heliothis armigera* and *in Vitro* Replication of Its Cytoplasmic Polyhedrosis Virus

De-Ming Su, Zhong-Jian Shen, and Yun-Xian Yue

IV. CELL FUSION

24. Fusion of Insect Cells

Jun Mitsuhashi

25. Protoplast Fusion of Insect Pathogenic Fungi

S. Shimizu

V. FUTURE PERSPECTIVES

26. University–Industry Perspectives

George M. Gould

27. Control of Invertebrate Pests through the Chitin Pathway

H. M. Mazzone

Contributors

Numbers in parentheses indicate the pages on which the authors' contributions begin.

MICHAEL J. ADANG (85), Agrigenetics Advanced Science Company, Madison, Wisconsin 53716

KEIO AIZAWA (3), Institute of Biological Control, Faculty of Agriculture, Kyushu University, Fukuoka 812, Japan

JOJI AOKI (183), Faculty of Agriculture, Tokyo University of Agriculture and Technology, Fuchu, Tokyo 183, Japan

L. R. S. AWITI (367), The International Centre of Insect Physiology and Ecology, Nairobi, Kenya

H. de BARJAC (63), Institut Pasteur, 75–724 Paris, Cedex 15, France

ANJA C. G. DERKSEN (167), Boyce Thompson Institute, Cornell University, Ithaca, New York 14853

T. S. DHADIALLA (367), The International Centre of Insect Physiology and Ecology, Nairobi, Kenya

WALTER DOERFLER (235), Institute of Genetics, University of Cologne, Cologne, Federal Republic of Germany

KATHLEEN G. DWYER (167), Boyce Thompson Institute, Cornell University, Ithaca, New York 14853

DAVID L. EDWARDS (101), Mycogen Corp., San Diego, California 92121

BRIAN A. FEDERICI (115), Division of Biological Control, Department of Entomology, University of California, Riverside, California 92521

MALCOLM J. FRASER (265), Department of Biological Sciences, University of Notre Dame, Notre Dame, Indiana 46556

FRANK GAERTNER (101), Mycogen Corp., San Diego, California 92121

NIELS J. GALJART (115), Division of Biological Control, Department of Entomology, University of California, Riverside, California 92521

R. C. GARBER (197), Department of Plant Pathology, Cornell University, Ithaca, New York 14853

RANDY GAUGLER (457), Department of Entomology, Rutgers University, New Brunswick, New Jersey 08903

WENDY GELERNTER (101), Mycogen Corp., San Diego, California 92121

GEORGE M. GOULD (417), Hoffmann-La Roche Inc., Nutley, New Jersey 07110

ROBERT R. GRANADOS (167), Boyce Thompson Institute, Cornell University, Ithaca, New York 14853

I. HARPAZ[1] (451), Department of Entomology, Hebrew University of Jerusalem, Rehovot 76100, Israel

CHARLOTTE HAUSER (235), Institute of Genetics, University of Cologne, Cologne, Federal Republic of Germany

CORINNA HERRNSTADT (101), Mycogen Corp., San Diego, California 92121

MICHIO HIMENO (29), Department of Polymer Engineering, Faculty of Technology, Tokyo University of Agriculture and Technology, Koganei-shi, Tokyo 184, Japan

JORGE E. IBARRA (115), Division of Biological Control, Department of Entomology, University of California, Riverside, California 92521

KEN F. IDLER (85), Agrigenetics Advanced Science Company, Madison, Wisconsin 53716

C. M. IGNOFFO (305), Biological Control of Insects Laboratory, U.S. Department of Agriculture, Columbia, Missouri 65201

DONOVAN E. JOHNSON (45), U.S. Grain Marketing Research Laboratory, Agricultural Research Service, USDA, Manhattan, Kansas 66502

KOHZO KANDA (75), Institute of Biological Control, Faculty of Agriculture, Kyushu University, Fukuoka 812, Japan

A. M. KHURAD (345), Biological Pest Management, Department of Zoology, Nagpur University, Nagpur 440 010, India

HANS-DIETER KLENK (235), Institut für Virologie der Justus, Liebig Universität Giessen, Giessen, Federal Republic of Germany

SHUNZO KONDO (251), Technical Laboratory, Mitsubishi-Kasei Institute of Life Sciences, Tokyo 194, Japan

KAZUMICHI KURODA (235), Institut für Virologie der Justus, Liebig Universität Giessen, Giessen, Federal Republic of Germany

TIMOTHY J. KURTTI (327), Department of Entomology, University of Minnesota, St. Paul, Minnesota 55108

J. C. LARA (13), Department of Microbiology and Immunology, University of Washington, Seattle, Washington 98195

NAOMI MAE (251), Laboratory of Cell Biology, Mitsubishi-Kasei Institute of Life Sciences, Tokyo 194, Japan

[1]Deceased.

S. MAEDA (221), Zoecon Research Institute, Palo Alto, California 94304
KARL MARAMOROSCH (485), Department of Entomology, Rutgers University, New Brunswick, New Jersey 08903
H. M. MAZZONE (439), U.S. Department of Agriculture, Forest Service, Hamden, Connecticut 06514
A. H. McINTOSH (305), Biological Control of Insects Laboratory, U.S. Department of Agriculture, Columbia, Missouri 65201
LOIS K. MILLER (295), Departments of Entomology and Genetics, The University of Georgia, Athens, Georgia 30602
JUN MITSUHASHI (387), Division of Forest Protection, Forestry and Forest Products Research Institute, Kukizaki, Inashiki, Ibaraki 305, Japan
TADASHI MIYAKE (251), Laboratory of Cell Biology, Mitsubishi-Kasei Institute of Life Sciences, Tokyo 194, Japan
ULRIKE G. MUNDERLOH (327), Department of Entomology, University of Minnesota, St. Paul, Minnesota 55108
M. NYINDO (367), The International Centre of Insect Physiology and Ecology, Nairobi, Kenya
LEODEGARIO E. PADUA (115), National Institute of Biotechnology and Applied Microbiology, University of the Philippines at Los Baños, Laguna, Philippines
M. M. RAI (345), Biological Pest Management, Department of Zoology, Nagpur University, Nagpur 440 010, India
S. K. RAINA (345), Biological Pest Management, Department of Zoology, Nagpur University, Nagpur 440 010, India
W. C. RICE (305), Biological Control of Insects Laboratory, U.S. Department of Agriculture, Columbia, Missouri 65201
THOMAS A. ROCHELEAU (85), Agrigenetics Advanced Science Company, Madison, Wisconsin 53716
RUDOLF ROTT (235), Institut für Virologie der Justus, Liebig Universität Giessen, Giessen, Federal Republic of Germany
H. E. SCHNEPF (13), Department of Microbiology and Immunology, University of Washington, Seattle, Washington 98195
ZHONG-JIAN SHEN (375), Biology Department, Fudan University, Shanghai, People's Republic of China
TADAYOSHI SHIBA (251), Laboratory of Cell Biology, Mitsubishi-Kasei Institute of Life Scienses, Tokyo 194, Japan
S. SHIMIZU[2] (401), Fukushima-Ken Sericulture Experiment Station, Yanagawa-machi, Date-gun, Fukushima 960-07, Japan

[2]Present address: Faculty of Textile Science, Kyoto Institute of Technology, Matsugasaki, Sakyo-Ku, Kyoto 606, Japan.

SAMUEL SINGER (133), Western Illinois University, Macomb, Illinois 61455

NATARAJAN SIVASUBRAMANIAN (115), Division of Biological Control, Department of Entomology, University of California, Riverside, California 92521

DE-MING SU (375), Biology Department, Fudan University, Shanghai, People's Republic of China

MANABU TAKAHISA (251), Laboratory of Cell Biology, Mitsubishi-Kasei Institute of Life Sciences, Tokyo 194, Japan

K. TOMCZAK (13), Department of Microbiology and Immunology, University of Washington, Seattle, Washington 98195

B. G. TURGEON (197), Department of Plant Pathology, Cornell University, Ithaca, New York 14853

RYU UEDA (251), Laboratory of Cell Biology, Mitsubishi-Kasei Institute of Life Sciences, Tokyo 194, Japan

H. D. VANETTEN (197), Department of Plant Pathology, Cornell University, Ithaca, New York 14853

K. WELTRING (197), Department of Plant Pathology, Cornell University, Ithaca, New York 14853

H. R. WHITELEY (13), Department of Microbiology and Immunology, University of Washington, Seattle, Washington 98195

O. C. YODER (197), Department of Plant Pathology, Cornell University, Ithaca, New York 14853

YUN-XIAN YUE (375), Biology Department, Fudan University, Shanghai, People's Republic of China

Preface

Severe effects from the repeated use of toxic chemical insecticides demand the investigation of alternatives. Microbial control of pests and vectors of disease agents provides a safe alternative to the use of chemical insecticides by avoiding pollution of the environment and by sparing beneficial insects. Successful use of several microbial preparations applied against target pests in both agriculture and medicine has prompted active research on the enhancement of epizootics for long-term pest management strategies. Genetically engineered microbial organisms and viruses have been obtained, and new mass production technology has been developed in laboratories in several countries.

Questions often asked by scientists and the general public concern the safety of genetically engineered viruses and microorganisms used for pest and vector control. Among the risks that have to be considered is the possibility of excessive reproduction of genetically engineered biocontrol agents. It can be expected that pests will eventually become resistant to the newly created control agents, and current research is, therefore, aimed at the proper manipulation of these agents. The anticipated risks have to be weighed against the benefits, and government regulations for controlling the release of genetically manipulated microorganisms have to be tailored accordingly. Hazards also have to be evaluated because new biocontrol agents may have unexpected effects on the ecosystem. Prior to their release, newly created agents must be extensively tested in the laboratory. After evaluation of the data and the establishment of the genetic stability of these agents and their ability to persist in the ecosystem, a complete risk assessment study must be carried out. These studies represent a new era in environmental research. The concerns surrounding the use of biological control agents in the environment are controversial. The issue is not whether or not they should be used, but how they should be regulated to avoid the environmental and health problems that were encountered when new technologies were previously introduced.

The manipulation of invertebrate pathogens and cells using recombinant DNA and cell fusion technology to improve crop production and human health holds considerable promise for the future. This is not a narrow field of biological interest, but has remarkably far-ranging implications. Applications have already transcended the boundaries of the agricultural sciences and have entered advanced frontiers of medicine.

The aims of this volume are to collate existing information in one source, to present important basic and applied advances, and, above all, to provide a stimulating forum for the discussion of new ideas and observations on genetically manipulated microbial and viral agents and thus benefit all who are interested in the development and uses of pathogens of invertebrates.

It was my intent to provide readers interested in one particular aspect of this subject with a self-contained chapter on their topic of choice. As a result, it was sometimes impossible to avoid overlap of information among some chapters.

The contributors are well known in their fields. Each has prepared a thoughtful treatment of experimental data and recent literature, and has included as yet unpublished data, personal interpretations, and conclusions.

Fusion of invertebrate cells, safety of viral insecticides, potential hazards of biocontrol agents, biochemical and genetic mechanisms of gene variation and transfer, transposable elements, virus-mediated transfer, applications of genetically engineered organisms to biological pest control, and new, intriguing medical applications through the use of invertebrate cell culture and baculoviruses are among the subjects covered in the six sections of this treatise.

Microbiologists, entomologists, virologists, parasitologists, geneticists, medical researchers, biocontrol specialists working in the field, and graduate students in related fields of biomedical research will find this work of interest.

I wish to pay special tribute to all the authors for their excellent contributions and to the staff of Academic Press for its part in producing this treatise.

Karl Maramorosch

I

Genetic Manipulation of Microbial Insecticides

1

Strain Improvement of Insect Pathogens

KEIO AIZAWA

The development and success of microbial insecticides have been dependent on the selection of effective pathogenic microorganisms for the control of injurious insects. However, not only selection but also improvement of effective strains has been emphasized (Aizawa, 1971a, 1982).

In this chapter, we will discuss our investigations on the selection of effective strains and our attempts at strain improvement of insect pathogenic microorganisms focused on microbial control in sericultural countries (Aizawa, 1971b).

I. *Bacillus thuringiensis* PREPARATIONS AND SERICULTURE

When *B. thuringiensis* preparations were marketed in the United States and France, they were also imported into Japan. Although the

BIOTECHNOLOGY IN INVERTEBRATE PATHOLOGY AND CELL CULTURE

effectiveness of *B. thuringiensis* preparations against injurious insects was recognized, the importation of *B. thuringiensis* strains isolated in foreign countries and their preparations had been forbidden through plant quarantine owing to their toxicity against the silkworm, *Bombyx mori*, except for experimental uses approved by the Ministry of Agriculture and Forestry (later, Ministry of Agriculture, Forestry and Fisheries). At that time, spread of the disease caused by *B. thuringiensis* in sericultural farms was feared, since *B. thuringiensis* had been isolated for the first time in Japan by Ishiwata (1901) as a causative agent of "a severe flacherie" (sotto disease).

The effects of *B. thuringiensis* preparations on silkworm rearing and sericulture were investigated in Japan, focusing on the following topics: toxicity of *B. thuringiensis* strains and *B. thuringiensis* preparations to the silkworm (Aizawa and Sato, 1961; Aizawa and Fujiyoshi, 1968); susceptibility of silkworm strains to *B. thuringiensis* strains and preparations (Aizawa and Sato, 1961; Aizawa *et al.*, 1962); rearing of the silkworm on mulberry leaves sprayed with sublethal doses of *B. thuringiensis* preparation (Aizawa and Sato, 1961); survey of the relationship between the incidence of silkworm diseases and the existence of *B. thuringiensis* spores in sericultural farms (Aizawa *et al.*, 1961); isolation of *B. thuringiensis* from litters from sericultural farms (distribution of *B. thuringiensis* serotypes in Japan) (Aizawa *et al.*, 1961; Ohba and Aizawa, 1978; Ohba *et al.*, 1979; Dulmage and Aizawa, 1982); multiplication of *B. thuringiensis* in dead larvae of the silkworm (Aizawa *et al.*, 1962; Aizawa and Fujiyoshi, 1964); multiplication of *B. thuringiensis* in the soil (Akiba *et al.*, 1979, 1980); drift experiment of *B. thuringiensis* preparations in the mulberry plantation; disinfection of silkworm rearing tools and rearing rooms with formalin (Aizawa and Sato, 1961); and utilization of a *B. thuringiensis* strain which exerts low toxicity to the silkworm but is highly toxic to injurious insects (Aizawa and Fujiyoshi, 1968; Aizawa *et al.*, 1975; Aizawa, 1976, 1982)

Based on these experiments, foreign *B. thuringiensis* preparations were made exempt from plant quarantine in January 1971 in Japan. In 1972 the Japan Plant Protection Association set up the Study Committee on *Bacillus thuringiensis* Preparations, which comprised five sections, for fundamental research, effectiveness, safety, sericulture, and apiculture.

The study committee investigated the data on the efficacy of *B. thuringiensis* preparations against injurious insects and selected target insects for the utilization of *B. thuringiensis* preparations. The study committee also investigated the methods of evaluation of the potency of *B. thuringiensis* preparations and adopted the following procedures for regulation of the potency of preparations.

The test insect is the silkworm, *B. mori*, strain Shunrei × strain Shogetsu, which is reared on an artificial diet at 24–25°C. Larvae of the 2nd day of the 3rd instar are used for bioassay. The *B. thuringiensis* preparation is diluted $1\frac{1}{3}$–2 times and and 10 g of artificial diet and 0.5 ml of diluted sample are mixed in a Petri dish. Thirty larvae (10 larvae × 3 Petri dishes) are used for each dilution of the sample. Larvae are reared for 3 days on a diet incorporating the sample and for 2 more days on a normal diet. Mortalities are recorded, and if more than 2 of 30 control larvae die the results are discarded.

Each formulator should prepare a "self-standard" and deposit it with the Agricultural Chemicals Inspection Station, Ministry of Agriculture, Forestry and Fisheries. Each self-standard has 1000 BmU (*Bombyx mori* units)/mg, irrespective of the toxicity to the silkworm. The potency of the sample is calculated from the formula

$$\text{Potency of sample} = \frac{LC_{50}\ (\mu g/ml)\ \text{self-standard}}{LC_{50}\ (\mu g/ml)\ \text{sample}} \times 1000\ \text{BmU/mg}$$

The potency should remain between 85 and 200% of the indicated BmU value of the sample. Each self-standard will be expressed as international units per milligram, using E-61 (1000 IU/mg) or U.S. standard (HD-1-S-1971: 18,000 IU/mg), if necessary.

This procedure is used only for regulation of the potency of preparations, and the recommended amount in the preparation for control of the target insects should be determined by efficacy tests (Aizawa *et al.*, 1975; Aizawa, 1976; 1982).

Based on investigations of the effect of *B. thuringiensis* preparations on silkworm rearing and sericulture, these preparations can be disseminated in sericultural environments, without causing damage to sericulture if the preparation is handled carefully. Also, the cause of any particular incidence of silkworm disease has to be determined to ascertain whether it was due to the dissemination of *B. thuringiensis* preparations.

In Japan, one spore-killed *B. thuringiensis* preparation was registered in 1981 and living spore–crystal mixture preparations (four native preparations and three foreign preparations) were registered in 1982.

II. SELECTION OF EFFECTIVE *Bacillus thuringiensis* STRAINS

In sericultural countries, particularly Japan, it is highly desirable to have microorganisms with low pathogenicity to the silkworm and at the same time high pathogenicity to injurious insects. Based on these criteria we have carried out experiments on the selection and breeding

TABLE I

Potencies of *Bacillus thuringiensis* Preparations[a]

Preparation	Subspecies	Spore count/g	IU[b]/mg		
			Silkworm (*Bombyx mori*)	Fall webworm (*Hyphantria cunea*)	Tobacco cutworm (*Spodoptera litura*)
U.S. standard (HD-1-S-1971)	*kurstaki*	3.0×10^{10}	18,000	18,000	18,000
AY self-standard (1000 BmU/mg)[c]	*aizawai*	2.1×10^{10}	28,000	6,800	38,500
AF 101 self-standard (1000 BmU/mg)[c]	*sotto*	2.6×10^{10}	95	12,600	0

[a]From Aizawa, 1982.
[b]IU, International units.
[c]BmU, *Bombyx mori* (silkworm) units.

of *B. thuringiensis* strains with low toxicity to the silkworm. One strain, designated AF 101 and belonging to serotype 4a : 4b, was selected. This strain is toxic to the common cabbageworm, *Pieris rapae crucivora*, the fall webworm, *Hyphantria cunea*, and the diamondback moth, *Plutella xylostella*. Physical, chemical, and biological treatments of *B. thuringiensis* strains, including lysogenic phage, were applied to produce this selection; however, the real cause of the appearance of this particular strain has not yet been verified.

The tobacco cutworm, *Spodoptera litura*, is an important injurious insect of vegetables and pastures in Japan and is resistant to ordinary *B. thuringiensis* strains. However, strains belonging to subsp. *aizawai* (serotype 7) and HD-1 strain (subsp. *kurstaki*, serotype 3a : 3b) showed toxicity against the tobacco cutworm. Strains belonging to subsp. *aizawai* were selected and a strain designated AY was obtained.

The potencies of AF 101 and AY preparations against the silkworm, the fall webworm, and the tobacco cutworm are indicated in Table I. The AF 101 preparation is applied in sericultural environments and the AY preparation in nonsericultural environments.

III. STRAIN IMPROVEMENT OF *Bacillus thuringiensis*

A. Temperature-Sensitive Mutant on Spore Formation

Whenever germination and multiplication of *B. thuringiensis* spores occur easily in the soils of mulberry plantations, the phenomena correlated with the reproduction of spores (S) and toxic crystals (C) will cause damage to sericulture. Taking into consideration the effect of living spores contained in *B. thuringiensis* preparations, a spore-killed preparation obtained by chemical treatment was produced and was registered in Japan.

Bacillus thuringiensis sporeless mutants were obtained by treatment with *N*-methyl-*N′*-nitro-*N*-nitrosoguanidine (NTG). However, the maintenance of such strains presented a problem. In addition to NTG treatment, we carried out selection of sporeless mutants by the combination of resistance to rifampicin and temperature-sensitive (*ts*) mutants. We obtained a strain ($S^{ts}C^{+}$) which forms spores and toxic crystals at 27°C, but forms only toxic crystals at 37°C (Aizawa, 1976).

This investigation contributed to the understanding of the nature of *B. thuringiensis* preparations and also to the application of the preparations in sericultural environments.

In connection with the multiplication of *B. thuringiensis* in the soil, which has been extensively investigated, other experimental results are included here. Cell numbers decreased gradully at different rates, depending on the nature of the soils and *B. thuringiensis* strains. However, no accumulation of *B. thuringiensis* cells on repeated applications to soils was observed (Akiba *et al.*, 1979, 1980).

B. Selecting a *Bacillus thuringiensis* Strain Which Produces Chitinase

If a *B. thuringiensis* strain produces enzymes for the digestion of substrates in the cuticle or gut tissues, the target insect will be killed by invasion of spores through the epidermis or through the gut tissues. Selection of *B. thuringiensis* strains which produce chitinase was attempted (Aizawa, 1975).

Commercial chitin was dissolved in concentrated HCl and washed with distilled water, and the extracted chitin was lyophilized. A solid medium was prepared with the lyophilized chitin (0.5 g) as the only organic substrate, inorganic substances ($Na_2HPO_4 \cdot 12H_20$, 0.8 g; NaCl, 0.1 g; $MgSO_4 \cdot 7H_2O$, 0.05 g; KH_2PO_4, 0.27 g), agar (2.5 g), and distilled water (100 ml) (pH 7.0).

Thoroughly washed *B. thuringiensis* spores were plated on the surface of this chitin medium and incubated. Colonies with clear zones showing digestion of chitin were selected. Sometimes the selected colonies were cultured in broth and the vegetative cells were UV-irradiated. With repeated cultures on the chitin medium, colonies which form large clear zones appeared.

Although the strains which form clear zones consistently have not yet been selected, spore suspensions were smeared on the body surface of the silkworm and the rice stem borer, *Chilo suppressalis*. However, invasion of spores into the insect body through the cuticle has not been observed. Peroral ingestion tests were also done, but no increase in pathogenicity was observed.

C. Selecting a *Bacillus thuringiensis* Strain Which Multiplies in an Insect Gut Juice

Ordinary *B. thuringiensis* strains do not multiply in the gut juice (pH greater than 10) of the silkworm, since the gut juice contains an antibacterial substance(s) which inhibits the growth of *B. thuringiensis*. The nature of the antibacterial substance(s) in the silkworm gut juice was investigated, and selection of a *B. thuringiensis* strain which multiplies in the gut juice was attempted (Aizawa, 1975).

Gut juice of the silkworm was added at a concentration of 10–90% to a broth. Growth of *B. thuringiensis* T84A1 strain (serotype 4a : 4b) was observed in cultures containing up to 50% gut juice but not in those containing more than 60%.

Gut juice of the silkworm was dialyzed against distilled water at 4°C for 57 hr. Growth of *B. thuringiensis* was observed in the dialyzed gut juice 2 days after the inoculation. However, the growth was inhibited by addition of the dialyzate of the gut juice. Thus, the inhibitory effect has been shown to be due to the dialyzable and thermostable substance(s) of the gut juice.

Selection of a *B. thuringiensis* strain which multiplies in the gut juice was attempted. Ordinary *B. thuringiensis* strains did not multiply in media containing more than 50% gut juice and 25% broth, but a *B. thuringiensis* strain which multiplied in a medium containing 90% gut juice and 25% broth was obtained. However, the selected strain did not multiply in 80 and 90% gut juice diluted with H_2O or in 100% gut juice.

IV. USE OF AN ATTENUATED INSECT VIRUS FOR CONTROL OF SILKWORM NUCLEAR POLYHEDROSIS

The rice stem borer, *C. suppressalis*, is susceptible to the nuclear polyhedrosis virus of the silkworm and that of the greater wax moth, *Galleria mellonella*. The rice stem borer was infected with the virus of *G. mellonella*-adapted *Bombyx mori* nuclear polyhedrosis virus (BG virus) (Aizawa, 1962) and serial passages of BG virus through the rice stem borer by injection were carried out for more than 30 generations. The infectivity titer after injection of the supernatant (3000 rpm for 15 min) of the larval homogenates increased markedly after the 18th generation. The infectivity to the silkworm of *C. suppressalis*-adapted BG virus (BGC virus) after serial passages through the rice stem borer decreased, but the BGC virus was highly infectious to the rice stem borer.

After passage of the BG virus through *G. mellonella* for more than 100 generations, the infectivity due to peroral inoculation of polyhedra of the BG virus in the newly hatched larvae of the silkworm was observed to decrease. A decrease of infectivity in the silkworm pupae after injection was observed in certain silkworm strains.

Newly hatched larvae of the silkworm were fed with nuclear polyhedra of the BG virus (10^3–10^5 polyhedra/ml) and after 34 hr were challenged with nuclear polyhedra of the silkworm. A defense reaction was observed in the larvae previously fed with polyhedra of the BG virus, particularly at a concentration of 10^3 polyhedra/ml. The basis for this phenomenon is not clear but it will be included in the category of interference.

TABLE II

Research Requirements for Developing Microbial Insecticides

1. Target insects
   Choice of target insects
   Mass rearing of target insects
2. Selection and strain improvement of microorganisms
   Selection of effective microorganisms
   Screening of effective strains
   Microbial breeding and genetic manipulation for strain improvement
3. Effectiveness
   Laboratory test of microorganisms
   Field test for insect pest control
4. Safety of microorganisms
5. Effects of microorganisms on environment
   Effect of microorganisms on nontarget organisms
   Microbial ecology
   Check of development of resistance to microorganisms in insects
6. Mass production of microorganisms
7. Formulation and quality control of microbial insecticides
   Formulation
   Bioassay for measurement of potency
   Stability
8. Application technology
   Rationalization of application
   Countermeasures for any problems caused by microbial insecticides

V. RESEARCH REQUIREMENTS FOR DEVELOPMENT OF MICROBIAL INSECTICIDES

Selection of effective strains of insect pathogenic microorganisms and microbial breeding and genetic manipulation for strain improvement are playing an important role in the development of microbial insecticides. These research subjects belong to categories indicated in Table II. Also, inspection of microbial insecticides from the point of view of fermentation technology is necessary and should be emphasized. Innovation in and expansion of these subjects will definitely contribute to the development of microbial insecticides (Aizawa, 1982).

REFERENCES

Aizawa, K. (1962). Infection of the greater wax moth, *Galleria mellonella* (Linnaeus), with the nuclear polyhedrosis virus of the silkworm, *Bombyx mori* (Linnaeus). *J. Insect Pathol.* **4,** 122–127.

Aizawa, K. (1971a). Strain improvement and preservation of virulence of pathogens. *In* "Microbial Control of Insects and Mites" (H. D. Burges and N. W. Hussey, eds.), pp. 655–672. Academic Press, New York.
Aizawa, K. (1971b). Present status of investigations on microbial control in Japan. *In* "Entomological Essays to Commemorate the Retirement of Professor K. Yasumatsu" (S. Asahina, J. L. Gressitt, Z. Hidaka, T. Nishida, and K. Nomura, eds.), pp. 381–389. Hokuryukan Publ. Co., Ltd. Tokyo.
Aizawa, K. (1975). Selection and strain improvement of insect pathogenic micro-organisms for microbial control. *JIBP Synth.* **7,** 99–105.
Aizawa, K. (1976). Recent development in the production and utilization of microbial insecticides in Japan. *Proc. Int. Colloq. Invertebr. Pathol., 1st, and 9th Annu. Meet., Soc. Invertebr. Pathol.,* 1976, pp. 59–63.
Aizawa, K. (1982). Microbial control of insect pests. *In* "Advances in Agricultural Microbiology" (N. S. Subba Rao, ed.), pp. 397–417. Oxford & IBH Publ. Co., New Delhi, Bombay, and Calcutta.
Aizawa, K., and Fujiyoshi, N. (1964). The growth of *Bacillus thuringiensis* in dead larvae of the silkworm, *Bombyx mori. J. Seric. Sci. Jpn.* **33,** 399–402 (in Japanese, with English summary).
Aizawa, K., and Fujiyoshi, N. (1968). Selection and breeding of bacteria for control of insect pests in the sericultural countries. *Proc. Jt. U.S.-Japan Semin. Microbial Control Insect Pests, 1967,* pp. 79–83.
Aizawa, K., and Sato, F. (1961). Effect of bacterial insecticides upon the silkworm, *Bombyx mori. Sanshi Kenkyu* **39,** 38–43 (in Japanese, with English summary).
Aizawa, K., Takasu, T., and Kurata, K. (1961). Isolation of *Bacillus thuringiensis* from the dust of silkworm rearing houses of farmers. *J. Seric. Sci. Jpn.* **30,** 451–455 (in Japanese, with English summary).
Aizawa, K., Kawarabata, T., and Sato, F. (1962). Response of the silkworm, *Bombyx mori,* to *Bacillus thuringiensis* Berliner. *J. Seric. Sci. Jpn.* **31,** 253–257 (in Japanese, with English summary).
Aizawa, K., Fujiyoshi, N., Ohba, M., and Yoshikawa, N. (1975). Selection and utilization of *Bacillus thuringiensis* strains for microbial control. *Proc. Intersect. Congr. IAMS, 1st,* 1974, Vol. 2, pp. 597–606.
Akiba, Y., Sekijima, Y., Aizawa, K., and Fujiyoshi, N. (1979). Microbial ecological studies on *Bacillus thuringiensis.* III. Effect of pH on the growth of *Bacillus thuringiensis* in soil extracts. *Jpn. J. Appl. Entomol. Zool.* **23,** 220–223 (in Japanese, with English summary).
Akiba, Y., Sekijima, Y., Aizawa, K., and Fujiyoshi, N. (1980). Microbial ecological studies on *Bacillus thuringiensis.* IV. The growth of *Bacillus thuringiensis* in soils of mulberry plantations. *Jpn. J. Appl. Entomol. Zool.* **24,** 13–17 (in Japanese, with English summary).
Dulmage, H. T., and Aizawa, K. (1982). Distribution of *Bacillus thuringiensis* in nature. *In* "Microbial and Viral Pesticides" (E. Kurstak, ed.), pp. 209–237. Dekker, New York.
Ishiwata, S. (1901). On a kind of severe flacherie (sotto disease). I. *Dainihon Sanshi Kaiho* **114,** 1–5 (in Japanese).
Ohba, M., and Aizawa, K. (1978). Serological identification of *Bacillus thuringiensis* and related bacteria isolated in Japan. *J. Invertebr. Pathol.* **32,** 303–309.
Ohba, M., Aizawa, K., and Furusawa, T. (1979). Distribution of *Bacillus thuringiensis* serotypes in Ehime Prefecture, Japan. *Appl. Entomol. Zool.* **14,** 340–345.

2

Structure and Regulation of the Crystal Protein Gene of Bacillus thuringiensis

H. R. WHITELEY, H. E. SCHNEPF, K. TOMCZAK, AND
J. C. LARA

I. INTRODUCTION

Over 20 different subspecies or varieties of *Bacillus thuringiensis* have been distinguished on the basis of flagellar serotypes (de Barjac and Bonnefoi, 1962) and interesting new varieties are still being discovered (e.g., Herrnstadt *et al.* this volume). By far the greatest number of subspecies are toxic to lepidopteran larvae. These subspecies have been investigated more thoroughly than the more recently discovered subspecies, which are toxic to dipteran or coleopteran larvae. The insecticidal activity of all *B. thuringiensis* strains is due largely to the crystal proteins which are synthesized during sporulation. In the subspecies lethal to lepidopterans, it is known that the crystal protein is a protoxin (M_r 130,000 to 160,000) which can be processed *in vitro* or in the insect gut to a toxic peptide of M_r ca. 68,000 (Tyrell *et al.*, 1981). However, electrophoretic analyses of crystals of different subspecies (Calabrese *et al.*, 1980) indicate that the crystals of some subspecies contain not one but several large peptides and some contain only smaller peptides. Considerable diversity has been reported (Dulmage and Cooperators, 1981) with regard to the spectrum of insects susceptible to different *B. thuringiensis* strains and the reasons for such differential toxicity are not understood. The differences in crystal composition and toxicity of different subspecies, as well as the sporulation-dependent production of the crystal proteins, raise obvious questions concerning the number of crystal protein genes, their properties, and how their synthesis is regulated. An attempt has been made to study some of these questions using recombinant DNA techniques to isolate individual crystal protein genes.

A. Number of Crystal Protein Genes

The first crystal protein gene was cloned (Schnepf and Whiteley, 1981) into *Escherichia coli* from a very large plasmid present in *B. thuringiensis* subsp. *kurstaki* strain HD-1-Dipel (a variant of strain HD-1 isolated from a commercial insecticide named Dipel). Extracts of the recombinant *E. coli* strain were toxic to tobacco hornworm larvae and contained an M_r 135,000 peptide which reacted with antibodies elicited to a crystal protein peptide. Subsequently, crystal protein genes were cloned from both pasmids and chromosome of *B. thuringiensis* subsp. *kurstaki* HD-1 (Held *et al.*, 1982) and subsp. *thuringiensis* (Klier *et al.*, 1982) and from plasmids from several other subspecies (Kronstad and Whiteley, 1984; Klier *et al.*, 1985; McLinden *et al.*, 1985; Shibano *et al.*, 1985; Adang *et al.*, 1985). Mosquitocidal genes from *B. thuringiensis* subsp. *israelensis* have also been cloned (Ward *et al.*, 1984; Waalwijk *et al.*, 1985; Walfield et al., 1986).

A survey of a number of *B. thuringiensis* subspecies (Kronstad *et al.*, 1983; Carlton and Gonzalez, 1985) showed that all strains contained at least one plasmid and most contained several (up to 17). Hybridization studies using a probe derived from the first cloned gene (Kronstad *et al.*, 1983) identified plasmids bearing crystal protein genes. Unexpectedly, in approximately half of the strains examined, more than one plasmid was hybridized, indicating the presence of either homologous genes or multiple copies of the same gene. Hybridization to restriction digests and restriction enzyme mapping supported the former explanation—the existence of homologous genes—especially in different strains of *B. thuringiensis* subsp. *kurstaki* and *thuringiensis*. Table I

TABLE I

***Hind*III Restriction Fragments and Plasmids That Hybridize with an Intragenic Crystal Protein Probe**[a,b]

B. thuringiensis strain	*Hind*III fragments (kb)	Plasmids (MDa)
kurstaki		
HD-1	6.6, 5.3, 4.5	~150,* 44
HD-1-Dipel	6.6, 4.5	~150*
HD-73	6.6	50
thuringiensis		
berliner 1715	5.3, 4.5, 2.3	~150, 60, 55
HD-290	5.3, 4.5,	~150, 57
HD-2	5.3	57*

[a]From Whiteley *et al.*, 1984.
[b]The cloned genes and the plasmids from which they were derived have been underlined; * indicates a doublet.

shows three "classes" of homologous genes distinguished by the sizes of the *Hind*III fragments which hybridized with the gene-specific probe.

To compare the three classes of genes, we cloned one "6.6 kb class gene" from *B. thuringiensis* subsp. *kurstaki* HD-73 and two "5.3 kb class genes" (one from *B. thuringiensis* subsp. *kurstaki* HD-1 and the other from *B. thuringiensis* subsp. *thuringiensis* HD-2). The gene cloned from strain HD-1-Dipel was a "4.5 kb class gene." Analysis of extracts of each recombinant *E. coli* strain by immunoblotting disclosed that the protoxins produced from the 4.5 and 6.6 kb class genes were both about M_r 134,000 in size, whereas the 5.3 kb class gene product was about M_r 130,000. Studies of solubilized crystals isolated from strains shown in Table I were in accord with the pattern of peptides expected from the distribution of homologous genes in the plasmid of each subspecies, suggesting that each homologous gene is expressed. Thus, the finding of more than one large peptide in crystals of some subspecies probably indicates the expression of different genes rather than the degradation of a single gene product or contamination by other unrelated peptides.

The finding of homologous genes in the *kurstaki* strains has been confirmed by Wilcox *et al.* (1986) and by Shivakumar *et al.* (1986). Wilcox *et al.* reported that preliminary tests for toxicity showed that a strain containing three of the homologous genes was measurably more toxic than a strain containing only two of these genes and that the two strains showed differences in their activities toward *Trichoplusia ni* and *Spodoptera exigua*. These observations suggest that individual crystal proteins may vary in their toxicity to different insects and that the insecticidal activity of a given subspecies depends on the total complement of expressed crystal protein genes. Availability of individual cloned genes should greatly facilitate the elucidation of the contribution of each crystal protein to the toxicity of a given strain.

B. Location of Crystal Protein Genes

A number of techniques (curing experiments, hybridization, cloning, and transconjugation experiments) indicate that crystal protein genes are located on different plasmids in different strains (reviewed by Whiteley and Schnepf, 1986). Furthermore, experiments by Carlton and co-workers (reviewed by Carlton and Gonzalez, 1985) suggested that plasmid-borne crystal protein genes could transfer to the chromosome and/or to other plasmids. The discovery of two repeated elements, IR1750 and IR2250 (Kronstad and Whiteley, 1984) could provide a mechanism to account for the diverse location of the genes. These repeated elements were found on virtually all plasmids which carry a crystal protein gene and they were also found on the chromosome in many strains. In three strains studied in detail (Kronstad and Whiteley), 1986), several copies of these elements were located in direct and inverted orientation around the crystal protein gene. These elements could, therefore, provide a means for DNA rearrangement via recombination events or through transposition. Evidence for transposition came from studies by Lereclus *et al.* (1983, 1984), who demonstrated transfer of *B. thuringiensis* DNA to a *Streptococcus faecalis* plasmid. More convincingly, analysis of the DNA sequence of IR1750 (Mahillon *et al.*, 1985) showed a DNA arrangement typical of a transposable element with an open reading frame; the deduced amino acid sequence of the latter showed homology with an *E. coli* transposase.

II. DNA SEQUENCE ANALYSIS

A. Promoter Structure

At present, two crystal protein genes have been sequenced completely: the 4.5 kb class gene from *B. thuringiensis* subsp. *kurstaki* HD-1-Dipel (Schnepf *et al.*, 1985) and the 6.6 kb class gene from *B. thuringiensis* subsp. *kurstaki* HD-73 (Adang *et al.*, 1985). A third gene, isolated from *B. thuringiensis* subsp. *sotto*, has been partially sequenced (Shibano *et al.*, 1985). All three have identical DNA sequences

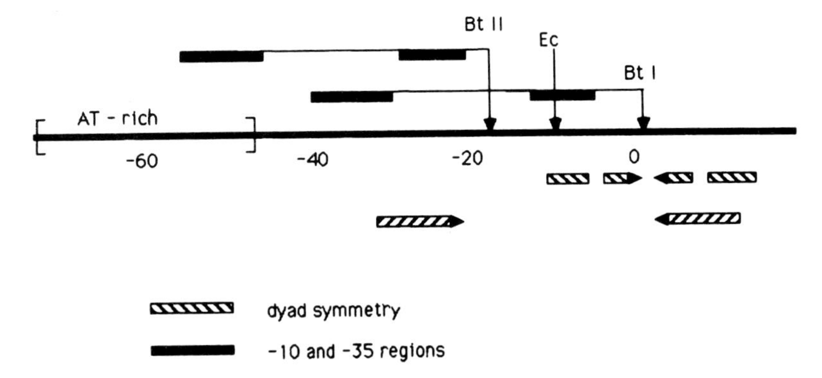

Fig. 1. Crystal protein gene promoter. The diagram of the promoter shows the following features relative to the position of the BtI transcriptional start site: the positions of the −10 and −35 regions (solid bars) of the BtI and BtII promoters, the transcriptional start sites (arrows) for BtI, BtII, and *E. coli* (Ec), the regions of dyad symmetry (striped arrows), and the AT-rich region (in brackets).

in the promoter region (Whiteley and Schnepf, 1986). Wong *et al.* (1983) found that the crystal protein gene is transcribed from two temporally regulated, overlapping promoters (BtI and BtII), shown schematically in Fig. 1. Transcription in *E. coli* is initiated from a start site (Ec) located between the two Bt start sites. Comparisons of the DNA sequence of the putative −10 and −35 regions of these promoters showed little or no homology to promoter sequences recognized by *Bacillus subtilis* RNA polymerase containing sigma-43 (vegetative polymerase), sigma-32, sigma-37, or sigma-29 (Losick and Youngman, 1984). Some homology was found with the promoter of the "0.3" sporulation gene; transcription of this gene *in vitro* has not yet been reported. On the basis of these observations, it can be predicted that two different sigma factors will be involved in the transcription of the crystal protein gene. Recently, we have isolated an RNA polymerase from sporulating *B. thuringiensis* cells which contains a new sigma subunit (M_r 35,000). This polymerase can initiate RNA synthesis *in vitro* from the BtI start site (Brown and Whiteley, 1987).

B. Coding Region

The two completely sequenced genes contain open reading frames of 1176 and 1178 amino acids, respectively, encoding peptides of M_r 133,500 (the 4.5 kb class gene from HD-1-Dipel) and M_r 133,300 (the 6.6 kb class gene from HD-73). Analyses for predicted hydropathy and secondary structure indicate that the N-terminal half of each molecule

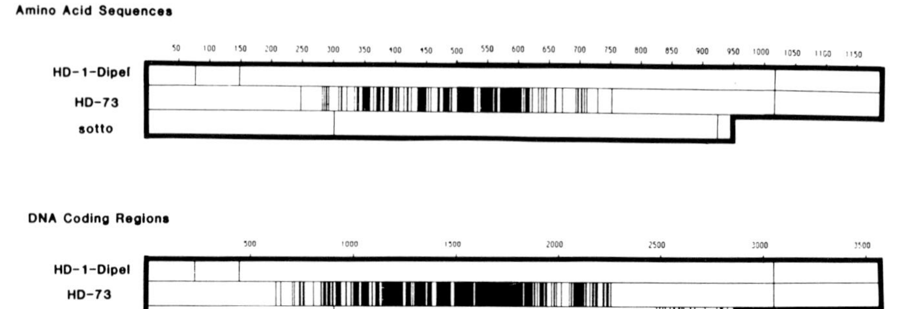

Fig. 2. Comparison of the nucleotide and amino acid sequences of three crystal protein genes. The diagram shows the positions of divergent amino acids or nucleotides (bars) relative to the beginning of the coding region.

is different from the C-terminal half. The N-terminal region contains more hydrophobic segments, very few cysteine residues, and more predicted β structure. The C-terminal region contains more predicted hydrophilic segments, most of the cysteine and lysine residues, and more predicted α-helical structure.

Figure 2 presents a comparison of the nucleotide sequences and deduced amino acid sequences of the coding portions of the three crystal protein genes analyzed to date. It should be noted that earlier studies (Dulmage and Cooperators, 1981) established that HD-1 and HD-73 crystals differ in their toxicity to *Trichopleusia ni* and *Heliothis virescens;* the toxicity of *sotto* crystals to these insects was not reported. Figure 2 shows that the three genes are remarkably similar in amino acid sequence in the N-terminal region (up to codon 290 or 300) and in the C-terminal portion beyond codon 750. However, the 6.6 kb class gene from HD-73 differs greatly in amino acid sequence from the HD-1-Dipel and *sotto* genes in the region between codons 290 and 750; as indicated below, this divergent region is included in the toxic portion of the molecule.

C. Location of the Toxin-Encoding Fragment

Two general approaches have been used to delineate the toxic portion of the protoxin molecule: studies of proteolytic fragments generated from crystals and studies of deleted cloned genes. A thorough study of the former type was presented by Nagamatsu *et al.* (1984). These investigators found that amino acid residues 29–32 of the protoxin were present on a toxic fragment of ca. 516 amino acids isolated after tryptic digestion of crystal protein from *B. thuringiensis* subsp. *dendrolimus*. We analyzed peptides produced by recominant *E. coli* strains having deletions and fusions of the protoxin gene (Schnepf and Whiteley, 1985). Analyses of peptides produced by strains carrying 5′ end deletions showed that the N-terminal 55% of the protoxin was sufficient for toxicity; fusions to *lacZ* located the N-terminus of the toxic fragment between codons 10 and 50. Deletions at the 3′ end up to codon 645 but not codon 603 allowed synthesis of the toxic peptide, thus giving a peptide with a minimal estimated size of about M_r 73,000. Using similar methodology, Adang *et al.* (1985) reported a toxic fragment of about M_r 68,000 for the HD-73 gene. It is noteworthy that the N-terminus of the toxic fragment contains the most hydrophobic segment of the protoxin molecule, suggesting that this region may be involved in the interaction of the toxin with the insect gut epithelial cells. Based on the comparison presented in Fig. 2, it may be speculated that differences in insect toxicity are determined by sequences located in the C-terminal portion of the toxic fragment, although clearly additional data are needed to identify the important sequences.

III. EXPRESSION OF THE CRYSTAL PROTEIN IN *Escherichia coli*

A. Properties of an Overproducer Strain

Tn5 insertional analysis was used in our earlier experiments (Wong *et al.*, 1983) to determine the position of the crystal protein gene on the recombinant plasmid. In the course of these studies, we found one Tn5 insertion strain, designated B8, which produced significantly larger amounts of the protoxin than the parent recombinant strain.

To compare expression in the overproducer and parent strains, we fused the *lacZ* gene to fragments of *B. thuringiensis* DNA containing the first 10 codons of the crystal protein gene and several hundred base pairs of DNA upstream from the *E. coli* initiation site (i.e., the transcriptional and translational start sites were provided by the *B. thuringiensis* DNA). Measurements of β-galactosidase activities showed that the overproducer, which had Tn5 inserted at position −145, had three to five times more activity than the parent strain or mutants with Tn5 inserted at position −377 or −600. S1 mapping experiments demonstrated that RNA synthesis in the overproducer strain was initiated at the same site as in the parent strain, indicating that the higher activity in the overproducer was not due to transcription from a promoter located in Tn5.

Deletion of sequences upstream from the promoter region also yielded higher β-galactosidase activities. Specifically, two to three times higher values were found in recombinant strains lacking sequences upstream from positions −48, −87, and −176, whereas strains with deletions to positions −258 and −432 had the same β-galactosidase activities as the parent recombinant strain and a strain with a deletion to position −25 (i.e., into the promoter region) had very little activity. These observations suggest that the high β-galactosidase activity of the recombinant *E. coli* overproducer strain results from the interruption of a region of negative regulation located approximately between positions −87 and −258.

A number of different *E. coli* strains were transformed with the B8 plasmid and the amount of protoxin synthesized was estimated by scanning Coomassie blue-stained polyacrylamide gels after electrophoresis. Two *E. coli* strains, CS412 and MC1000, displayed slightly higher relative amounts of the M_r 135,000 peptide band (ca. 6–10% of the total protein) than the HB101 strain used in cloning (ca. 2–5% of the total protein). These differences in expression may be related to differences in the copy numbers of the recombinant plasmid in the different *E. coli* strains or to genotypic differences between the hosts.

B. Properties of the Crystal Protein Made in *Escherichia coli*

1. Morphology of Inclusions

Examination of the *E. coli* overproducer strain by phase microscopy showed that many cells contained granules. Electron microscopy showed that some of these cells had regularly shaped crystal-like inclusions (Fig. 3A). These inclusions appeared to be more variable in size and were smaller than the crystals seen in sporulating cultures of *B. thuringiensis* subsp. *kurstaki* HD-1-Dipel (Fig. 3B), the source of the cloned crystal protein gene. However, the appearance of the crystal was dependent on the temperature at which the culture was grown. In *E. coli* cultures grown at 32°C the crystals usually had sharp edges (Fig. 3C) but typically were not regularly bypyramidal in shape; a well-defined crystalline lattice could be seen in some sections. Inclusions from cells grown at 38°C were rounded or lobed aggregates (Fig. 3D) rather than bipyramidal, and usually only some of the subunits of the inclusions were organized into a crystalline array. The reasons for the

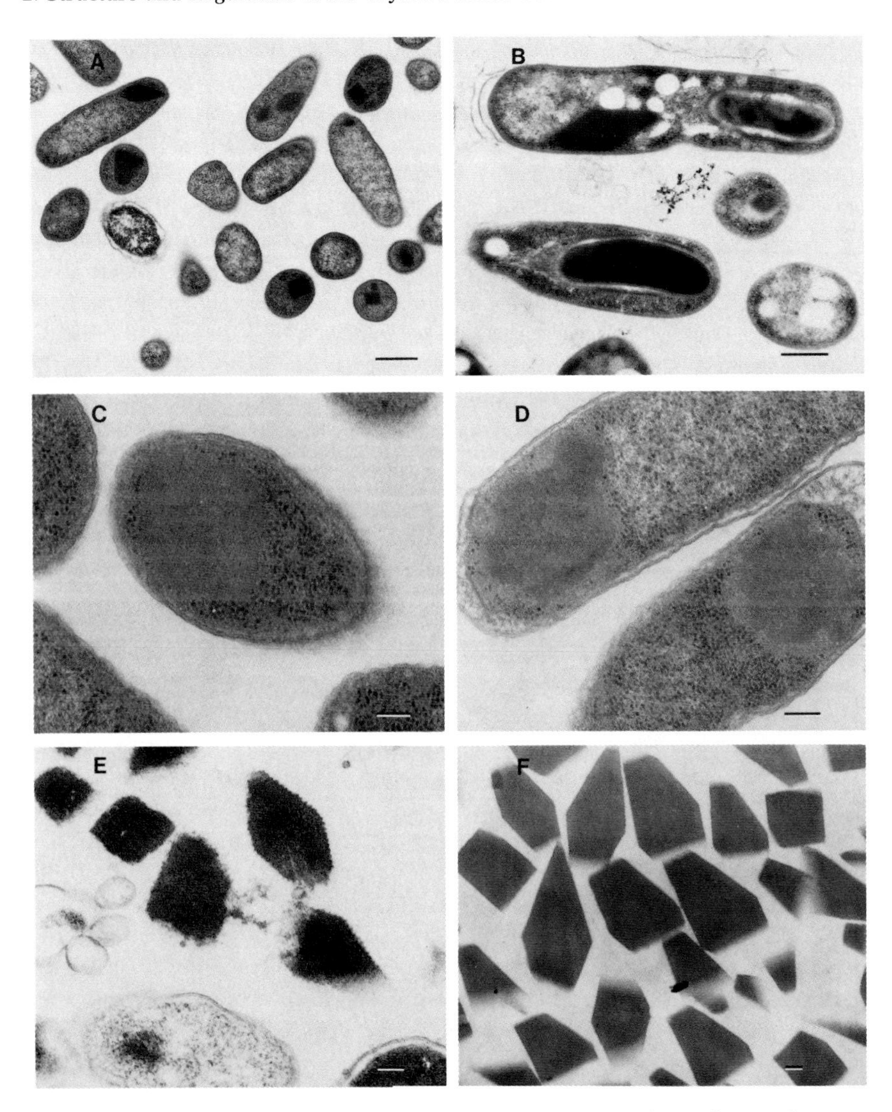

Fig. 3. Electron micrographs of *B. thuringiensis* crystals and *E. coli* crystal protein inclusions. Thin sections of: (A) *E. coli* overproducer strain containing crystal-like inclusions; (B) sporulating cells of *B. thuringiensis* subsp. *kurstaki* HD-1-Dipel; (c) *E. coli* overproducer grown at 32°C (note bipyramidal crystal with lattice structure); (D) *E. coli* overproducer grown at 38°C (note irregular lobed crystal with a patch of lattice structure in center); (E) crystal inclusions isolated from *E. coli* overproducer; (F) crystals isolated from *B. thuringiensis*. Bar = 0.5 μm (A, B); 0.1 μm (C, D, E, F).

differences in the shapes of the inclusions are not known. Measurements of the latticelike structure showed a periodicity of 4.7 × 11.8 nm in the repeated spacing. Similar values were reported for the spacing of the lattice in *B. thuringiensis* crystals (Holmes and Monro, 1965) and in crystalline inclusions (presumably consisting of spore coat protein) occasionally found in sporulating *B. subtilis* (Kaneko and Matsushima, 1975). We also observed inclusions in recombinant strains bearing a *lacZ* promoter fused to (a) the intact BT coding region, (b) codon 10 of the crystal protein gene, (c) codon 50 of the gene, or (d) a segment containing the N-terminus and up to codon 645 followed by the *B. thuringiensis* terminator. Most of the inclusions found in these strains consisted of rounded or irregularly shaped, poorly defined aggregates (data not shown).

2. Biochemistry and Toxicity of the Inclusions

The inclusions from the overproducing *E. coli* could be applied to Renografin gradients and recovered from a well-defined band in the density gradient in a manner similar to that used to purify *B. thuringiensis* crystals. However, unlike preparations isolated from *B. thuringiensis* cultures (Fig. 3F), little purification of the crystal protein was achieved: inclusions from *E. coli* were about 50% pure as judged by electron microscopy (Fig. 3E) and by analysis of stained protein gels. Attempts to solubilize the *E. coli* inclusions with the same relatively mild alkaline reducing conditions used to solubilize *B. thuringiensis* crystals (0.1 *M* Na_2CO_3, 2% α-mercaptoethanol, pH 9.5) demonstrated that only about 10% of the crystal protein could be dissolved. Complete solubilization could be achieved only with addition of urea (4 *M*) or guanidine (3 *M*) to the solubilization buffer, or through use of an ionic detergent (1% sodium dodecyl sulfate) and a reducing agent. This is in contrast to our earlier findings with the parent recombinant strain (Schnepf and Whiteley, 1981), which indicated that the lower amounts of crystal protein produced by this strain could largely be solubilized using mild alkaline reducing conditions. Many other proteins have been observed to form insoluble aggregates when overproduced in *E. coli*, and this phenomenon has been attributed in part to improper disulfide band formation (Williams *et al.*, 1982).

As stated earlier, the protein isolated from *B. thuringiensis* crystals is a protoxin which can be cleaved to a toxic, protease-resistant fragment either *in vitro* or in the insect midgut. Using conditions for trypsin cleavage described by Chestukhina *et al.* (1977), we were able to digest solubilized crystal proteins produced in both *E. coli* and *B. thuringien-*

sis to produce a protease-resistant fragment of M_r 60,000. The trypsin-resistant fragment could be obtained from crystal preparations from *B. thuringiensis* which had been solubilized by using 3 *M* guanidine and mercaptoethanol or the alkaline reducing conditions described above. A trypsin-resistant fragment from the *E. coli* crystal protein inclusions was obtained reproducibly only from the inclusions which were solublized by alkaline reducing conditions. When the *E. coli* inclusions were solubilized in 3 *M* guanidine and mercaptoethanol, precipitates were frequently formed during the dialysis step preceding trypsinization, and proteolysis was difficult to control. Presumably the presence of some insoluble *E. coli* proteins in the inclusion preparations contributed to the difficulty in controlling the digestion of this material. As shown in Fig. 4, the trypsin fragments of the crystal proteins synthesized in *E. coli* and *B. thuringiensis* have very similar sizes (ca. M_r 60,000); however, the peptide derived from the *E. coli* inclusions consistently had a slightly higher apparent molecular weight.

The inclusions from *E. coli* containing plasmid pHES50, which encodes a truncated crystal protein gene consisting of the first 645 codons of the gene fused to the final 74 codons and the transcriptional terminator (Schnepf and Whiteley, 1985), were also analyzed. Although these inclusions could be solubilized in a solution containing 1% sodium dodecyl sulfate (SDS) and 2% 2-mercaptoethanol, they were not soluble under the alkaline reducing conditions noted above even in the presence of high levels of chaotropic agents (e.g., 8 *M* urea). Analysis of the inclusions by SDS-polyacrylamide gel electrophoresis revealed that the truncated crystal protein was the predominant protein in the inclusions but that it accounted for less than half of the total protein in the inclusions (data not shown). The insolubility of these inclusions and the relative impurity of the truncated crystal protein contained in them have hampered our ability to further analyze the truncated products of this crystal protein gene.

Preliminary quantitative toxicity tests comparing the crystal protein from *B. thuringiensis* subsp. *kurstaki* HD-1-Dipel, the trypsin fragment from the *B. thuringiensis* crystal protein, and the crystal protein from the inclusions produced in *E. coli* indicate that all three proteins have LC_{50} values in the range of 2–20 ng/cm^2 of food for larvae of *Manduca sexta*. Approximate LC_{50} data have not yet been obtained for the trypsin fragment derived from the inclusions produced in *E. coli* or for the truncated crystal protein; however, the tests which have been performed indicate that the truncated protein, assayed in the form of inclusions, may be less toxic by a factor of 10.

In conclusion, the current study has revealed a number of similarities

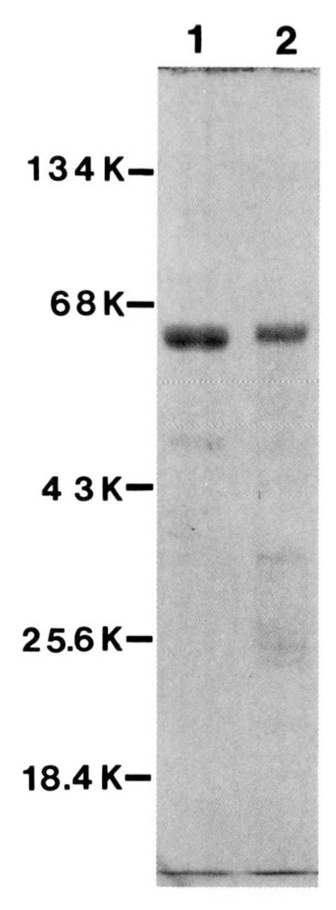

Fig. 4. Comparison of trypsin fragments obtained from crystal proteins synthesized in *B. thuringiensis* subsp. *kurstaki* HD-1-Dipel (lane 1) and recombinant *E. coli* (lane 2). Photograph of a Coomassie blue-stained 10% polyacrylamide gel. The mobilities of molecular weight standards, as indicated with bars, are: *B. thuringiensis* crystal protein, M_r 134,000; bovine serum albumin, M_r 68,000; ovalbumin, M_r 43,000; α-chymotrypsinogen, M_r 25,600; β-lactoglobulin, M_r, 18,400.

and differences in the ultrastructure and biochemistry of crystal protein inclusions produced in *E. coli* and in *B. thuringiensis*. Production of the crystal protein in large amounts in *E. coli* is dependent on enhanced transcription achieved by interruption or removal of DNA upstream from the promoter and is influenced by the particular *E. coli* host strain. We are currently investigating the effect of the upstream region on transcription of the crystal protein gene promoter in *Bacillus*.

The crystalline inclusions produced in *E. coli* generally have a less regular shape than those produced in *B. thuringiensis* and a relatively small proportion have a well-defined crystalline lattice structure. The

E. coli inclusions are somewhat less soluble than crystals produced in *B. thuringiensis*, although they have roughly equivalent toxicity to *M. sexta*. More disturbing is the observation that the product of a truncated crystal protein gene formed very insoluble inclusions when over-produced in *E. coli*, indicating that further genetic dissection of the toxicologic properties of this protein may require the analysis of mutations which are less disruptive than large deletions.

ACKNOWLEDGMENTS

This research was supported by U.S. Public Health Service Grant GM-20784 from the National Institute of General Medical Sciences, by Grant PCM-8315859 from the National Science Foundation, and by a grant from Cetus Corporation. H. R. W. is a recipient of Research Career Award K6-GM-442 from the U.S. Public Health Service, National Institute of General Medical Sciences.

REFERENCES

Adang, M. J., Staver, M. J., Rocheleau, T. A., Leighton, J., Barker, R. F., and Thomson, D. V. (1985). Characterized full-length and truncated plasmid clones of the crystal protein of *Bacillus thuringiensis* subsp. *kurstaki* HD-73 and their toxicity to *Manduca sexta*. *Gene* **36,** 289–300.

Brown, K. L., and Whiteley, H. R. (1987). In preparation.

Calabrese, D. M., Nickerson, K. W., and Lane, L. C. (1980). A comparison of protein crystal subunit sizes in *Bacillus thuringiensis*. *Can. J. Microbiol.* **26,** 1006–1010.

Carlton, B. C., and Gonzalez, J. M., Jr. (1985). The genetics and molecular biology of *Bacillus thuringiensis*. *In* "The Molecular Biology of the Bacilli" (D. A. Dubnau, ed.), Vol. 2, pp. 211-249. Academic Press, New York.

Chestukhina, G. G., Kostina, L. I., Zalunin, I. A., Kotova, T. S., Katruhka, S. P., Kuznetsov, Y. S., and Stepanov, V. W. (1977). Proteins of *Bacillus thuringiensis* delta-endotoxin crystals. *Biokhimiya (Moscow)* **42,** 1660–1667.

de Barjac, H., and Bonnefoi, A. (1962). Essai de classification biochimique et sérologique de 24 sources de bacillus du type *B. thuringiensis*. *Entomophaga* **8,** 5–31.

Dulmage, H. T., and Cooperators (1981). Insecticidal activity of isolates of *Bacillus thuringiensis* and their potential for pest control. *In* "Microbial Control of Pests and Plant Diseases 1970–1980" (H. D. Burges, ed.), pp. 193–222. Academic Press, London.

Held, G. A., Bulla, L. A., Jr., Farrari, E., Hoch, J., Aronson, A. I., and Minnich, S. A. (1982). Cloning and localization of the lepidopteran protoxin gene of *Bacillus thuringiensis* subsp. *kurstaki*. *Proc. Natl. Acad. Sci. U.S.A.* **79,** 6065–6059.

Holmes, K. C., and Monro, R. E. (1965). Studies on the structure of parasporal inclusions from *Bacillus thuringiensis*. *J. Mol. Biol.* **14,** 572–581.

Kaneko, I., and Matsushima, H. (1975). Crystalline inclusions in sporulating *Bacillus subtilis* cells. *In* "Spores VI" (P. Gerhardt, R. N. Costilow, and H. L. Sadoff, eds.), pp. 580–585. Am. Soc. Microbiol., Washington, D. C.

Klier, A., Fargette, F., Ribier, J., and Rapoport, G. (1982). Cloning and expression of the crystal protein genes from *Bacillus thuringiensis* strain *berliner* 1715. *EMBO J.* **1,** 791–799.

Klier, A., Lereclus, G., Ribier, D., Bourgouin, C., Menou, G., Lecadet, M.-M., and Rapoport, G. (1985). Cloning and expression in *Escherichia coli* of the crystal protein gene from *Bacillus thuringiensis* strain *aizawa* 7-29 and comparison of the structural organization of genes from different serotypes. *In* "Molecular Biology of Microbial Differentiation" (J. A. Hoch and P. Setlow, eds.), pp. 217–224. American Society for Microbiology, Washington, D. C.

Kronstad, J. W., and Whiteley, H. R. (1984). Inverted repeat sequences flank a *Bacillus thuringiensis* crystal protein gene. *J. Bacteriol.* **160,** 95–102.

Kronstad, J. W., and Whiteley, H. R. (1986). Three classes of homologous *Bacillus* thuringiensis crystal protein genes. *Gene* **43,** 41–50.

Kronstad, J. W., Schnepf, H. E., and Whiteley, H. R. (1983). Diversity of locations for the *Bacillus thuringiensis* crystal protein gene. *J. Bacteriol.* **154,** 419–428.

Lereclus, D., Menou, G., and Lecadet, M.-M. (1983). Isolation of a DNA sequence related to several plasmids from *Bacillus thuringiensis* after a mating involving the *Streptococcus faecalis* plasmid pAMB1. *Mol. Gen. Genet.* **191,** 307–313.

Lereclus, D., Ribier, J., Klier, A., Menou, G., and Lecadet, M.-M (1984). A transposon-like structure related to the δ-endotoxin gene of *Bacillus thuringiensis. EMBO J.* **3,** 2561–2567.

Losick, R., and Youngman, P. (1984). Endospore formation in *Bacillus. In* "Microbial Development" (R. Losick and L. Shapiro, eds.), pp. 63–88. Cold Spring Harbor Lab., Cold Spring Harbor, New York.

McLinden, J. H., Sabourin, J. R., Clark, B. D., Gensler, D. R., Workman, W. E., and Dean, D. H. (1985). Cloning and expression of an insecticidal k-73 type crystal protein gene from *Bacillus thuringiensis* var. *kurstaki* into *Escherichia coli. Appl. Environ. Microbiol.* **50,** 623–628.

Mahillon, J., Seurinck, J., Rompuy, L. V., Delcour, J., and Zabeau, M. (1985). Nucleotide sequence and structural organization of an insertion sequence element (IS231) from *Bacillus thuringiensis* strain *berliner* 1715. *EMBO J.* **4,** 3895–3899.

Nagamatsu, Y., Itai, Y., Hatanaka, C., Fumatsu, G., and Hayashi, K. (1984). A toxic fragment from the entomocidal crystal protein of *Bacillus thuringiensis. Agric. Biol. Chem.* **48,** 611–619.

Schnepf, H. E., and Whiteley, H. R. (1981). Cloning and expression of the *Bacillus thuringiensis* crystal protein gene in *Escherichia coli. Proc. Natl. Acad. Sci. U.S.A.* **78,** 2893–2897.

Schnepf, H. E., and Whiteley, H. R. (1985). Delineation of a toxin-encoding segment of a *Bacillus thuringiensis* crystal protein gene. *J. Biol. Chem.* **260,** 6273–6280.

Schnepf, H. E., Wong, H. C., and Whiteley, H. R. (1985). The amino acid sequence of a crystal protein from *Bacillus thuringiensis* deduced from the DNA base sequence. *J. Biol. Chem.* **160,** 6264–6272.

Shibano, Y., Yamagata, A., Nakamura, N., Iizuka, T., Sugisaki, H., and Takanami, M. (1985). Nucleotide sequence coding for the insecticidal fragment of the *Bacillus thuringiensis* crystal protein. *Gene* **34,** 243–251.

Shivakumar, A. G., Grundling, G. J., Benson, T. A., Casuto, D., Miller, M. F., and Spear, B. B. (1985). Vegetative expression of the δ-endotoxin genes of *Bacillus thuringiensis* subsp. *kurstaki* in *Bacillus subtilis. J. Bacteriol.* **166,** 194–204.

Tyrell, D. J., Bulla, L. A., Jr., Andrews, R. E., Jr., Kramer, K. J., Davidson, L. I., and Nordin, P. (1981). Comparative biochemistry of entomocidal parasporal crystals of selected *Bacillus thuringiensis* strains. *J. Bacteriol.* **145,** 1052–1062.

Waalwijk, C., Dullemans, A. M., van Workum, M. E. S., and Visser, B. (1985). Molecular cloning and nucleotide sequence of the M_r 28,000 crystal protein gene of *Bacillus thuringiensis*. *Nucleic Acids Res.* **13,** 8207–8217.

Walfield, A. M., Garduno, F., Thorne, L., Zounes, M., Decker, D. J., Wild, M. A., and Pollock, T. J. (1986). Cloning of a gene that codes for a mosquitocidal toxin from *Bacillus thuringiensis* var. *israelensis*. *In* "Bacillus Molecular Genetics and Biotechnology Applications" (A. T. Ganesan and J. A. Hoch, eds), pp. 321–333. Academic Press, Orlando, Florida.

Ward, E. S., Ellar, D. J., and Todd, J. A. (1984). Cloning and expression in *Escherichia coli* of the insecticidal δ-endotoxin of *Bacillus thuringiensis* var. *israelensis*. *FEBS Lett.* **175,** 377–3821.

Whiteley, H. R., and Schnepf, H. E. (1986). The molecular biology of parasporal crystal body formation in *Bacillus thuringiensis*. *Annu. Rev. Microbiol.* **40,** 549–576.

Whiteley, H. R., Schnepf, H. E., Kronstad, J. W., and Wong, H. C. (1984). Structural and regulatory analysis of a cloned *Bacillus thuringiensis* crystal protein gene. *In* "Genetics and Biotechnology of Bacilli" (A. T. Ganesan and J. A. Hoch, eds.), pp. 375–386. Academic Press, Orlando, Florida.

Wilcox, D. R., Shivakumar, A. G., Melin, B. E., Miller, M. F., Benson, T. A., Schopp, C. W., Casuto, D., Gundling, G. J., Bolling, T. J., Spear, B. B., and Fox, J. L. (1986). Genetic engineering of bioinsecticides. *In* "Protein Engineering: Applications in Science, Industry and Medicine" (M. Inouye and R. Sarma, eds.), pp. 395–413. Academic Press, Orlando.

Williams, D. C., Van Frank, R. M., Muth, W. L., and Burnett, J. P. (1982). Cytoplasmic inclusion bodies in *Escherichia coli* producing biosynthetic human insulin proteins. *Science* **215,** 687–688.

Wong, H. C., Schnepf, H. E., and Whiteley, H. R. (1983). Transcriptional and translational start sites for the *Bacillus thuringiensis* crystal protein gene. *J. Biol. Chem.* **258,** 1960–1967.

3

Mechanism of *Bacillus thuringiensis* Insecticidal δ-Endotoxin Action on Insect Cells in Vitro

MICHIO HIMENO

I. INTRODUCTION

Bacillus thuringiensis is a gram-positive insect pathogenic bacterium which produce a crystalline protein body (called crystal) in the cell during spore formation. The protein body is a strong toxin to lepidopteran or dipteran larvae and is called δ-endotoxin. When these protein bodies are ingested by the larvae, they are dissolved by the digestive

BIOTECHNOLOGY IN INVERTEBRATE
PATHOLOGY AND CELL CULTURE

juice in the larval midgut and they attack the columnar cells in the midgut epithelium. The cells swell and are released from the basement membrane and finally burst (Endo and Nishiitsutsuji-Uwo, 1980). Studies using cultured insect cells to determine the mechanism of action by lepidopteran-specific δ-endotoxin showed that the cultured cells swelled and burst on treatment with the dissolved δ-endotoxin as shown in the midgut cells (Murphy *et al.*, 1976; Nishiitsutsuji-Uwo *et al.*, 1979). When insects susceptible to δ-endotoxin ingested the crystals, the pH of their hemolymph was elevated and the K^+ concentration in the hemolymph increased (Nishiitsutsuji-Uwo and Endo, 1981). Harvey and Wolfersberger (1979) showed that influx of potassium from the blood to the lumen was unaffected, but flux in the reverse direction was tripled in an experiment on the δ-endotoxin treatment of isolated midgut (*Manduca secta* larvae). The cytotoxic response induced by the dissolved δ-endotoxin seemed to be specific on cultured insect cells but was not observed on mammalian cells (Nishiitsutsuji-Uwo *et al.*, 1980). The dissolved toxin from *B. thuringiensis* var. *israelensis* caused rapid cytolytic action in cultured insect cells and mammalian cells. In addition, the dissolved toxin caused hemolysis of rat, mouse, sheep, and human erythrocytes (Thomas and Ellar, 1983).

This chapter deals with the mechanism of action of lepidopteran-specific δ-endotoxin on cultured insect cells, TN-368 cells. The Na^+ and K^+ concentrations in cells treated with δ-endotoxin from *B. thuringiensis* var. *aizawai* were determined (Himeno *et al.*, 1985). The effects of these ions on the cells treated by the δ-endotoxin are discussed.

II. MATERIALS AND METHODS

A. Cells

TN-368 cells were grown in TC 199-MK medium (McIntosh *et al.*, 1973) containing 5% fetal bovine serum and 50 μg/ml kanamycin sulfate at 27°C. The cells were cultured in plastic vessels (Falcon No. 3013 tissue culture flasks) to determine Na^+ or K^+ in the cells treated with δ-endotoxin and on a Falcon Micro-Test plate 3040 to determine toxic activity.

B. Purification of Crystals

Crystals from *B. thuringiensis* var. *aizawai* I45 (Nishiitsutsuji-Uwo and Eda, 1975) were purified by the biphasic separation technique described by Nishiitsutsuji-Uwo *et al.* (1979) and the flotation procedure of Yamamoto (1983).

C. Dissolution of Crystals

The purified crystals were dissolved by partially purified proteolytic enzyme from silkworm gut juice (Nishiitsutsuji-Uwo *et al.*, 1979). The dissolved crystal toxin was dialyzed against water in a plastic vessel. The precipitate in the toxin solution was removed by centrifugation. The δ-endotoxin solution was stored in a freezer until it was used.

D. Assay for δ-Endotoxin Activities

Confluent monolayers of TN-368 cells were obtained 3–4 days after subculture and were washed twice with Hanks', phosphate-buffered saline (PBS) or isotonic solution (150 m*M* NaCl, 230 m*M* KCl, or 190 m*M* sucrose solution). The dissolved δ-endotoxin (5–12 μg/ml protein) was applied to the monolayers at room temperature. The reagents for assay of the activity were dissolved in the isotonic solution and were applied to the cultures with or without δ-endotoxin. The cell response was recorded continuously in photographs with a phase-contrast microscope. Swollen cells were counted as a positive response.

E. Concentration of Na^+ or K^+ in TN-368 Cells

The cells were subcultured in plastic vessels (Falcon No. 3013 treated with 0.1 *N* HNO_3) at 4-day intervals. Confluent monolayers were washed with the isotonic solution and treated with the toxin solution. After treatment with the toxin for a suitable period the cells were washed with the sucrose isotonic solution. Na^+ or K^+ ions in the monolayer cells were extracted with 0.1 *N* HNO_3 for 72 hr. The cells were dislodged and disrupted with an ultrasonic oscillator. The optical density of the suspension was determined at 260 nm in order to determine the relative amounts of Na^+ and K^+ (concentration of Na^+ or K^+ per optical density at 610 nm). The suspension was centrifuged at 3000 rpm for 15 min and the amounts of Na^+ or K^+ in the supernatants were determined with a Shimazu atomic absorption/flame emission spectrometer (Himeno *et al.*, 1985).

III. RESULTS

A. Cell Swelling and Ion Concentration

When the confluent monolayer cells were treated with the ion-free δ-endotoxin in the sucrose isotonic solution, the cells did not swell.

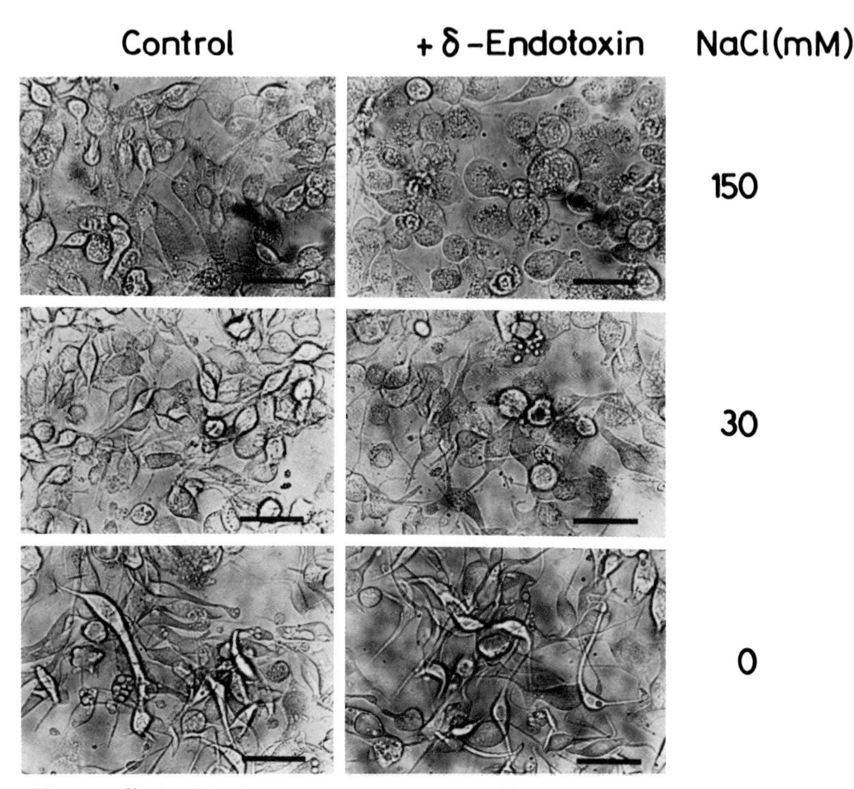

Fig. 1. Effects of NaCl concentration on cell swelling caused by δ-endotoxin in NaCl–sucrose isotonic solution. The TN-368 cells were treated with 7 μg/ml δ-endotoxin (+δ-endotoxin) or without (control) in NaCl–sucrose isotonic solution at various NaCl concentrations. The concentration of NaCl is shown at the margin of the photograph. The cells were incubated at room temperature for 90 min. The bar represents 50 μm.

However, when cells in an isotonic solution containing Na^+ or K^+ were treated with the δ-endotoxin, the cells swelled (Figs. 1 and 2). When the sucrose isotonic solution was gradually replaced by NaCl (Fig. 1) or KCl (Fig. 2) isotonic solution, cell swelling was observed. NaCl or KCl is an essential factor in cell swelling due to δ-endotoxin and the degree of the swelling depended on the ion concentration. The degree of cell swelling is shown in Fig. 3 as the ratio of length to width of the cells. The ratio of cells swollen like a ballon was 1, but the maximum ratio of cells responding to the toxin in the NaCl isotonic solution was 0.8 for 90-min incubation at 25°C. The swollen cells in the culture were counted (Fig. 4). In the NaCl isotonic solution all the cells swelled, but in the KCl isotonic solution 40% of cells swelled 90 min after addition of the toxin. The time course for cell response to δ-endotoxin was also observed in NaCl, KCl, and sucrose isotonic solutions (Fig. 5).

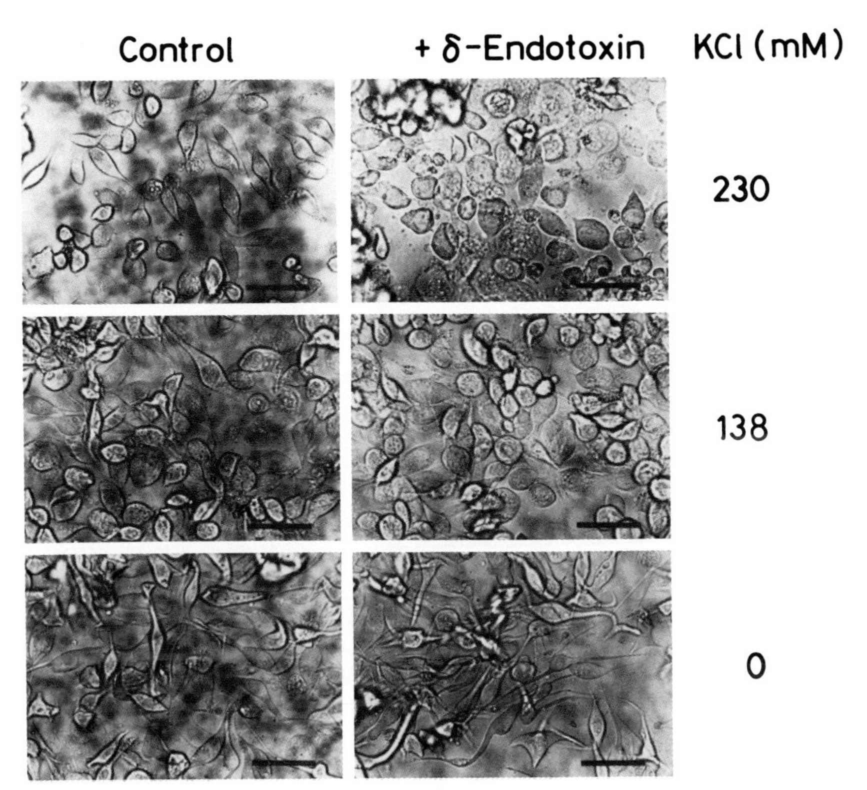

Fig. 2. Effects of KCl concentration on cell swelling of δ-endotoxin in KCl–sucrose isotonic solution. The detailed procedure is described in the discussion of Fig. 1.

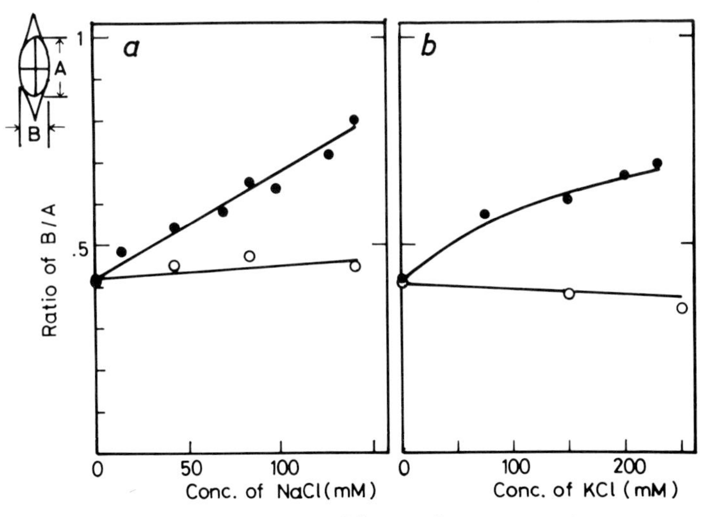

Fig. 3. Degree of cell swelling caused by δ-endotoxin in various concentrations of (a) NaCl or (b) KCl–sucrose isotonic solution. Cells were treated with (●) or without (○) 7 μm/ml δ-endotoxin. Because 250 m*M* KCl solution (slightly hypertonic) was used as an isotonic solution, the cells shrank slightly.

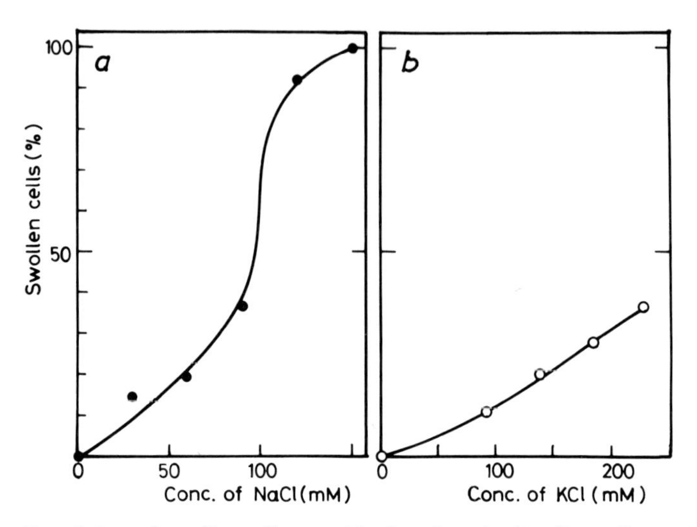

Fig. 4. Population of swollen cell caused by δ-endotoxin. Swollen cells caused by 7 μg/ml δ-endotoxin were counted in (a) NaCl or (b) KCl–sucrose isotonic solutions having various ion concentrations.

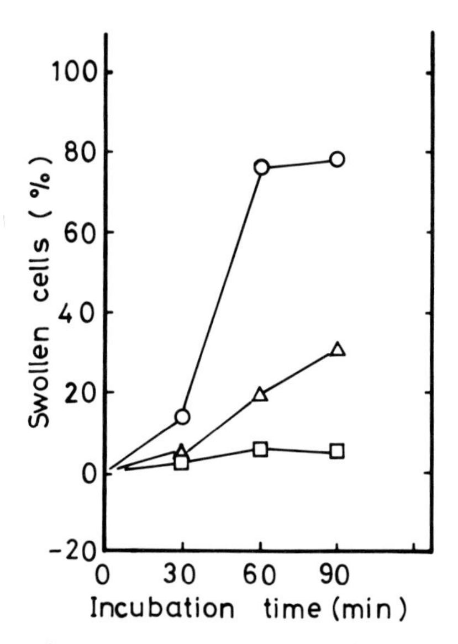

Fig. 5. Time course of cytotoxic response to δ-endotoxin in NaCl, KCl, or sucrose isotonic solution. The TN-368 cells were treated with 7 μg/ml δ-endotoxin in NaCl (○), KCl (△), or sucrose (□) isotonic solutions at room temperature.

B. Amount of Na^+ and K^+ in Cells Treated with δ-Endotoxin

Cells in the NaCl isotonic solution were treated with δ-endotoxin and the concentrations of Na^+ and K^+ in the cells were determined at various times. Five minutes after addition of the toxin the relative amount of Na^+ increased 2-fold, but that of K^+ decreased to 30% at 5 min after the addition and finally down to 10% (Fig. 6). Uptake of Na^+ into the cells and release of K^+ from the cells were dependent on the concentration of δ-endotoxin added (Fig. 7) and the incubation period (Fig. 6). When the cells treated with the toxin released K^+, an intracellular substance was also released simultaneously.

C. Inhibition of Cell Swelling by Tetrodotoxin

Tetrodotoxin (TTX) is a well-known Na^+ channel blocker for nerve (Narahashi *et al.*, 1964) or muscle (Munson *et al.*, 1979) cells. TTX was

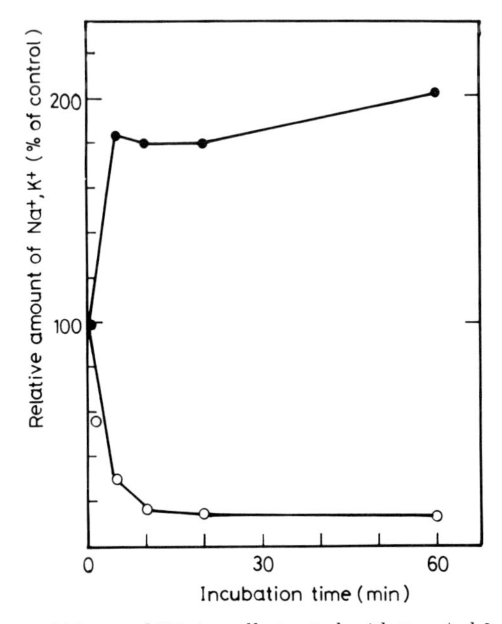

Fig. 6. Amounts of Na^+ and K^+ in cells treated with 7 μg/ml δ-endotoxin in NaCl isotonic solution at 27°C. The amounts of Na^+ (●) and K^+ (○) were determined. The detailed procedure is described in the text. The amounts of Na^+ and K^+ in normal cells were estimated as 100% and plotted at zero time (Himeno *et al.*, 1985).

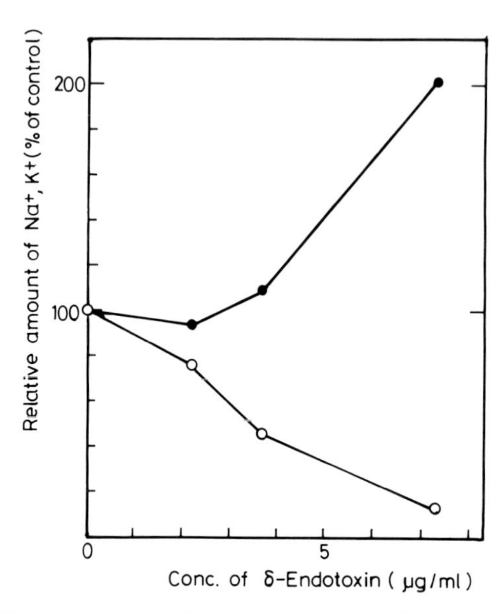

Fig. 7. Effect of δ-endotoxin on the amount of intracellular Na^+ or K^+. The cells were treated with various concentrations of δ-endotoxin in 150 m*M* NaCl solution for 60 min and were carefully washed with only 190 m*M* sucrose solution. Intracellular Na^+ and K^+ were extracted with 0.1 *N* HNO_3. The amounts of Na^+ (●) and K^+ (○) were determined. The amounts of normal cells were estimated as 100% (Himeno *et al.*, 1985).

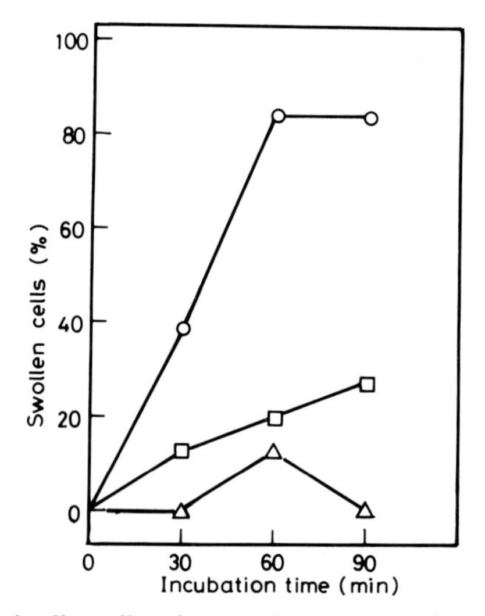

Fig. 8. Inhibition of cell swelling by tetrodotoxin (TTX). Cell swelling caused by δ-endotoxin was observed at various times in the NaCl isotonic solution after addition of 11.6 μg/ml δ-endotoxin (○), 157 μ*M* TTX (△), and both TTX and δ-endotoxin (□).

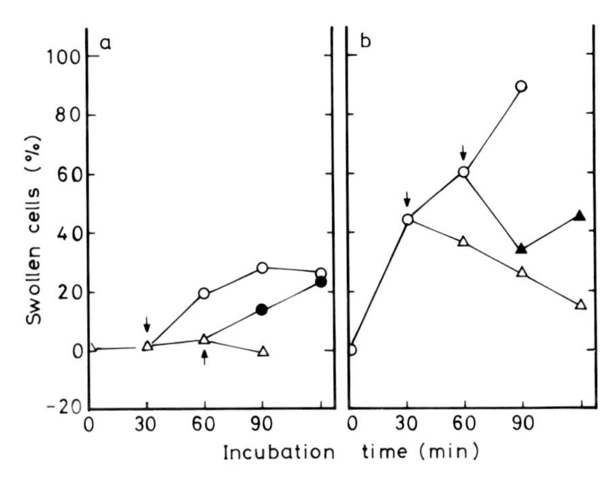

Fig. 9. Effect of TTX on cell swelling before and after treatment with δ-endotoxin. (a) After treatment of the cells with 1.57 μ*M* TTX (△) in the NaCl isotonic solution, TTX was removed at 30 (○) or 60 (●) min, as shown by the arrows. The cells were washed with the isotonic solution and incubated with 8.3 μg/ml δ-endotoxin in the isotonic solution. (b) After treatment of the cells with δ-endotoxin (○), the toxin was removed at 30 (△) or 60 (▲) min, as shown by the arrows. The cells were washed with the isotonic solution and incubated with TTX (Himeno *et al.*, 1985).

added to the insect cells in the presence or absence of δ-endotoxin in NaCl or KCl (data not shown) isotonic solution. The cell swelling caused by the toxin was inhibited by the addition of TTX (Fig. 8). The cell swelling caused by δ-endotoxin in the KCl isotonic solution was also inhibited by the addition of TTX (Himeno *et al.*, 1985). The inhibition of TTX could be reversed up to 30 min after addition of δ-endotoxin. The swollen cells could return to their normal shape up to 30 min after addition of δ-endotoxin, but not after 60 min after the addition (Fig. 9b). When TTX was added together with δ-endotoxin in the isotonic ion solutions, the cell swelling due to δ-endotoxin was inhibited by TTX. After remove of TTX, cell swelling was induced by δ-endotoxin (Fig. 9a).

D. Stimulation of Cell Swelling by 4-Aminopyridine

4-Aminopyrine (4-AP) is a selective blocker of K^+ channels (Yeh *et al.*, 1976) that inhibits influx and efflux of K^+; on the other hand, tetraethylammomium chloride (TEA) is effective only on the efflux of K^+ from inside the cell. The swelling caused by δ-endotoxin was stimulated by the addition of 4-AP in the KCl isotonic solution (Fig.10b) but not in the NaCl solution (Fig. 10a). Addition of 4-AP alone in the NaCl

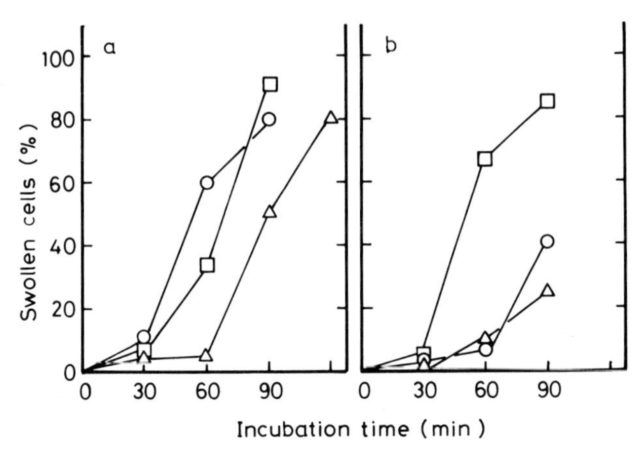

Fig. 10, Stimulation of cell swelling by 4-aminopyridine (4-AP). δ-Endotoxin (5.2 μg/ml) was added to the cells in the presence (□) or absence (○) of 500 μM 4-AP in (a) NaCl or (b) KCl isotonic solution. As the control, 4-AP alone (△) was added to cells (Himeno *et al.*, 1985).

isotonic solution induced cell swelling with δ-endotoxin (Fig. 10a) and in the KCl isotonic solution it slightly induced swelling without the toxin (Fig. 10b). TEA did not affect the cell swelling caused by the toxin because TEA could not enter into the cell. This evidence suggested that 4-AP blocked the efflux of K^+ induced by δ-endotoxin but did not block the uptake of Na^+ and K^+.

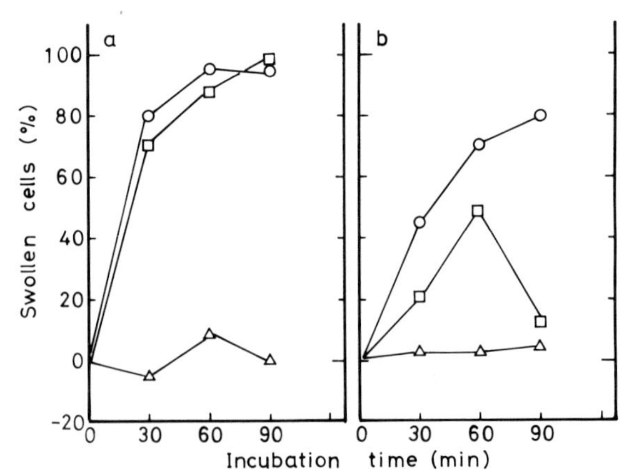

Fig. 11. Inhibition of the cell swelling by ouabain. δ-Endotoxin (11.6 μg/ml) was added to the cells in the presence (□) or absence (○) of 686 μg/ml ouabain in (a) NaCl or (b) KCl isotonic solution. As the control, ouabain alone (△) was added to the cells (Himeno *et al.*, 1985).

E. Inhibition of Cell Swelling by Ouabain

When ouabain, an inhibitor of Na^+,K^+-ATPase (Post *et al.*, 1960), was added to the cells in presence of δ-endotoxin, the cell swelling caused by the toxin was inhibited in the KCl isotonic solution (Fig. 11b) but not in the NaCl isotonic solution (Fig. 11a). Valinomycin, which is an ionophore, did not compete with the action of the endotoxin. Valinomycin stimulated cell swelling caused by the toxin in the KCl isotonic solution (Nishiitsutsuji-Uwo *et al.*, 1979) but not in the NaCl isotonic solution (data not shown).

F. Effect of Various Nucleotide Derivatives on δ-Endotoxin Activity

δ-Endotoxin was mixed with various nucleotide derivatives in the NaCl isotonic solution and swelling activities were determined (Fig. 12 and 13). Cell swelling caused by δ-endotoxin was inhibited by ATP, AMP, cAMP, NAD, GTP, and CTP. The time course of inhibition by cAMP in the NaCl isotonic solution was determined (Figs. 14a and

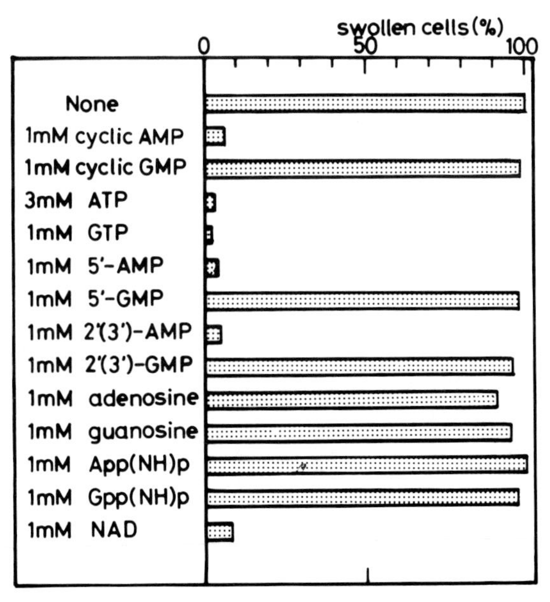

Fig. 12. Effects of nucleotide derivatives on cell swelling caused by δ-endotoxin. The cells were treated with 7.1 μg/ml δ-endotoxin and the indicated nucleotide derivatives in NaCl isotonic solution for 60 min at room temperature. As the control, δ-endotoxin alone was added to the cells ("None").

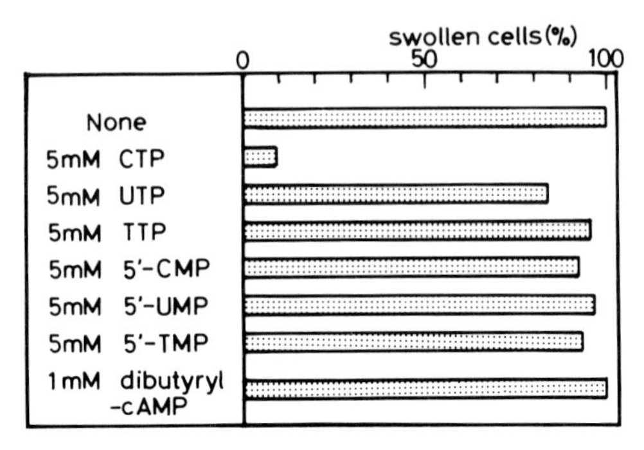

Fig. 13. Effect of pyrimidine nucleotide derivatives and dibutyryl-cAMP. The procedure is described in the legend of Fig. 12.

14b). cAMP is a competitive inhibitor of a compound induced by δ-endotoxin in the cells and therefore cell swelling resumed with the addition of cAMP (Fig. 14). Cell swelling was caused by addition of App(NH)p (5′-adenylylimidodiphosphate), Gpp(NH)p, and 2′(3′)-GMP without δ-endotoxin, and these drugs stimulated the swelling by δ-endotoxin (Table I). Dibutyryl-cAMP did not affect cell swelling (Fig. 13 and Table I). Sodium nitroprusside (Fig. 15), arginine, and phosphatidylserine as activators (Schultz *et al.*, 1977) of guanylate cyclase

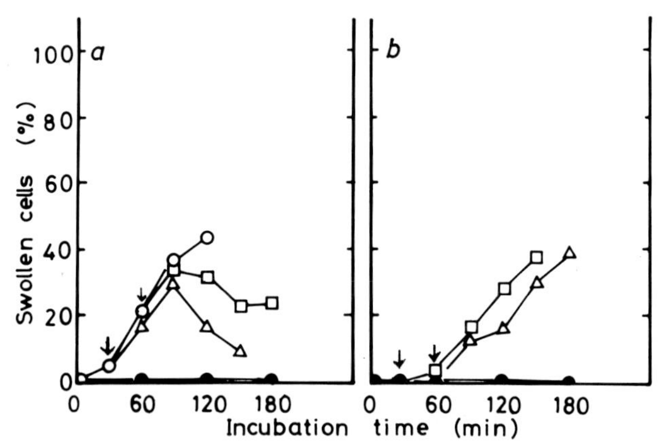

Fig. 14. Inhibition of cell swelling by cAMP. (a) After treatment of the cells with 7.1 μg/ml δ-endotoxin in the NaCl isotonic solution, the toxin was removed at 30 min (△) or 60 min (□), as shown by the arrows. The cells were washed in the isotonic solution and incubated with 1 m*M* cAMP. As a control, cells were treated with δ-endotoxin (○) or cAMP (●). (b) After treatment of the cells with cAMP, the cAMP was removed at 30 (□) or 60 (△) min. The cells were washed with the isotonic solution and incubated with δ-endotoxin. As the control, cells were treated with cAMP only (●).

TABLE I
Swelling Activity of Various Nucleotide Derivatives on TN-368 Cells[a]

Nucleotide derivatives	Concentration (m*M*)	Swollen cells (%)	
		60 min	180 min
None	—	0	0
2′(3′)-GMP	1	6	77
	5	9	90
App(NH)p	1	4	80
	5	18	83
Gpp(NH)p	1	0	10
	5	21	80
Dibutyryl-cAMP	5	4	0

[a]The cells were incubated with the indicated drug for 60 or 180 min at a concentration of 1 or 5 m*M* in NaCl isotonic solution. As a control, cells were incubated in the isotonic solution ("None").

stimulated cell swelling by δ-endotoxin. The cGMP content of the cells treated with δ-endotoxin was higher than that of the normal cells (data not shown).

IV. DISCUSSION

Na^+ or K^+ is an essential component for cell swelling caused by δ-endotoxin from *B. thuringiensis* var. *aizawai*. Swelling is dependent on

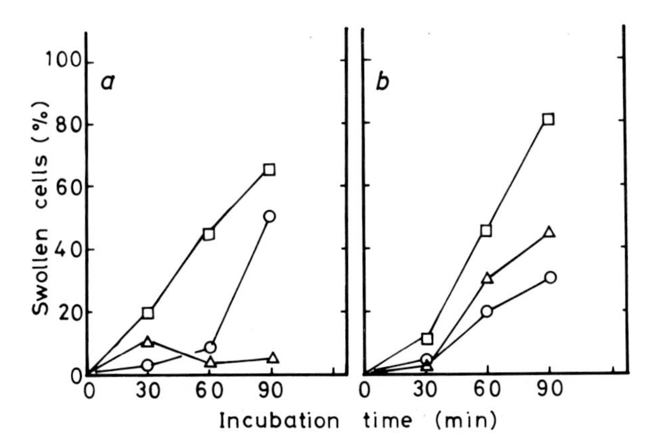

Fig. 15. Stimulation of cell swelling by nitroprusside (NP). δ-Endotoxin (5.2 μg/ml) was added to the cells in the presence (□) or absence (○) of NP at (a) 1 m*M* in the NaCl isotonic solution or (b) 5 m*M* in the KCl isotonic solution. As the control, NP alone (△) was added to the cells.

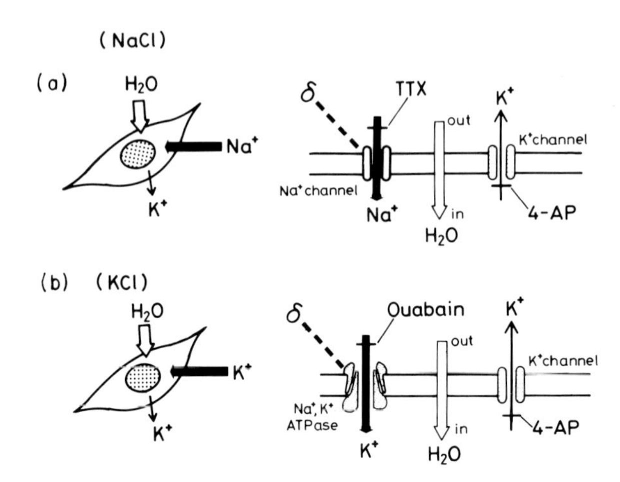

Fig. 16. Model of cell swelling caused by δ-endotoxin. (a) The sodium ion channel in the cell membrane is indirectly activated by δ-endotoxin and then Na^+ influx is stimulated in the NaCl isotonic solution. This stimulation is inhibited by TTX. The potassium leak channel is also inhibited by 4-AP and K^+ efflux is blocked. (b) Na^+,K^+-ATPase is activated by δ-endotoxin and K^+ influx is stimulated in the KCl isotonic solution.

the concentration of NaCl, KCl, and δ-endotoxin. Swelling caused by δ-endotoxin may be due to the stimulation of Na^+ influx through Na^+ channels, because the swelling is inhibited by addition of TTX to the NaCl isotonic solution. Also, the swelling induced by 4-AP may be caused by blocking K^+ efflux through K^+ leak channels. Ouabain, a blocker of Na^+,K^+-ATPase (Na^+,K^+ pump), inhibits cell swelling caused by δ-endotoxin in the KCl isotonic solution but not in the NaCl isotonic solution.

These results are illustrated in Fig. 16. δ-endotoxin in the NaCl isotonic solution stimulates influx of the Na^+ channel and the K^+ leak channel passive, but in the KCl isotonic solution the toxin stimulates Na^+,K^+-ATPase and the leak channel of K^+. Therefore a large volume of water comes into the cells, and the cells swell. It was suggested that a cell membrane perturbation was induced by the endotoxin because cell sap was released from the cytoplasm due to the high osmotic pressure induced by the toxin which caused cells to burst.

Nucleotide derivatives, especially nucleotide compounds containing the AMP moiety, inhibited cell swelling, but adenosine did not. cAMP, which is a competitor of cGMP in the cells, was the most effective inhibitor of swelling caused by the toxin. Sodium nitroprusside, App(NH)p, arginine, and phosphatidylserine are activators of guanylate cyclase in cell membrane. The presence of these compounds suggests that δ-endotoxin stimulated guanylate cyclase and enhanced the intracellular cGMP content. Then this high level of cGMP directly or indirectly activated the Na^+ channel and Na^+,K^+-ATPase.

REFERENCES

Endo, Y., and Nishiitsutsuji-Uwo, J. (1980). Mode of action of *Bacillus thuringiensis* δ-endotoxin: Histopathological change in the silkworm midgut. *J. Invertebr. Pathol.* **36,** 90–103.

Harvey, W., and Wolfersberger, M. G. (1979). Mechanism of inhibition of active potassium transport in isolated midgut of *Manduca sexta* by *Bacillus thuringiensis* endotoxin. *J. Exp. Biol.* **83,** 293–308.

Himeno, M., Koyama, N., Funato, J., and Komano, T. (1985). Mechanism of action of *Bacillus thuringiensis* insecticidal delta-endotoxin on insect cells *in vitro*. *Agric. Biol. Chem.* **49,** 1461–1468.

McIntosh, A., Maramorosch, K., and Rechtoris, C. (1973). Adaptation of an insect cell line (*Agallia constricta*) in a mammalian cell culture medium. *In Vitro* **8,** 375–378.

Munson, R., Jr., Westermark, B., and Glaser, L. (1979). Tetrodotoxin-sensitive sodium channels in normal human fibroblasts and normal human glia-like cells. *Proc. Natl. Acad. Sci. U.S.A.* **76,** 6425–6429.

Murphy, D. W., Sohi, S. S., and Fast, P. G. (1976). *Bacillus thuringiensis* enzyme-digested delta-endotoxin: Effect on cultured insect cells. *Science* **194,** 954–956.

Narahashi, T., Moore, J. W., and Scott, W. R. (1964). Tetrodotoxin blockage of sodium conductance increase in lobster giant axons. *J. Gen. Physiol.* **47,** 965–974.

Nishiitsutsuji-Uwo, J., and Eda, M. (1975). Sporeless mutant of *Bacillus thuringiensis*. II. Mutants derived from var. *thuringiensis* and *sotto*. *Experientia* **31,** 1285–1287.

Nishiitsutsuji-Uwo, J., and Endo, Y. (1981). Mode of action of *Bacillus thuringiensis* δ-endotoxin: Changes in hemolymph pH and ions of *Pieris, Lymantria* and *Ephestia* larvae. *Appl. Entomol. Zool.* **16,** 225–230.

Nishiitsutsuji-Uwo, J., Endo, Y., and Himeno, M. (1979). Mode of action of *Bacillus thuringiensis* δ-endotoxin: Effect on TN-368 cells. *J. Invertebr. Pathol.* **34,** 267–275.

Nishiitsutsuji-Uwo, J., Endo, Y., and Himeno, M. (1980). Effects of *Bacillus thuringiensis* δ-endotoxin on insect and mammalian cells *in vitro*. *Appl. Entomol. Zool.* **15,** 133–139.

Post, R. L., Meritt, C. R., Kinsolving, C. R., and Albright, C. D. (1960). Membrane adenosine triphosphatase as a participant in the active transport of sodium and potassium in the human erythrocyte. *J. Biol. Chem.* **235,** 1795–1802.

Schultz, K. D., Schultz, K., and Schultz, G. (1977). Sodium nitroprusside and other smooth muscle-relaxants increase cyclic GMP levels in rat ductus deferens. *Nature (London)* **265,** 750–751.

Thomas, W. E., and Ellar, D. (1983). *Bacillus thuringiensis* var. *israelensis* crystal δ-endotoxin: Effect on insect and mammalian cells *in vitro* and *in vivo*. *J. Cell Sci.* **60,** 181–197.

Yamamoto, T. (1983). Identification of entomocidal toxins of *Bacillus thuringiensis* by high performance liquid chromatography. *J. Gen. Microbiol.* **129,** 2595–2603.

Yeh, J. Z., Oxford, G. S., Wu, C. H., and Narahashi, T. (1976). Dynamics of aminopyridine block of potassium channel in squid axon membrane. *J Gen. Physiol.* **68,** 519–535.

4

Entomocidal Activity of Crystal Proteins from Bacillus thuringiensis toward Cultured Insect Cells

DONOVAN E. JOHNSON

I. INTRODUCTION

Of the 28 known subspecies of *Bacillus thuringiensis* (DeLucca, 1984), two have gained prominence due to their specific insect toxicities and consequent usefulness as natural insect control agents. They are *B. thuringiensis* subsp. *kurstaki*, which is primarily effective against lepidopteran insects, and *B. thuringiensis* subsp. *israelensis*, whose toxicity is limited to dipteran insects, including mosquitos. Both subspecies produce one or more inclusion bodies (crystals) concomitant with sporulation, which produce a series of ion-transport alterations within the epithelial cells lining the midgut when ingested by the feeding insect larvae. The usual consequence is that the insect stops feeding and ultimately starves to death.

BIOTECHNOLOGY IN INVERTEBRATE
PATHOLOGY AND CELL CULTURE

The entomocidal protein contained in the inclusion body of *B. thuringiensis* subsp. *kurstaki* is not toxic in its native (insoluble) form (Heimpel and Angus, 1960). On ingestion by a susceptible larva, the particulate proteinaceous crystal is rapidly hydrolyzed by a combination of factors in the midgut, resulting in the release of soluble toxic fragments which quickly block the absorption of nutrients and lead to the eventual destruction of epithelial cells lining the midgut.

Histopathological changes occur at the level of cell organelles as well as macroscopically. Microvilli swell and become distorted, vacuoles appear within the membranous network of the rough endoplasmic reticulum, and mitochondria become swollen, distorted, and condensed (Endo and Nishiitsutsuji-Uwo, 1980). The morphological changes at the gut epithelium occur within 10–30 min, depending on the toxic protein concentration. The swollen cells often burst or whole agglomerates are detached from the epithelium. The events within the gut cause larvae to stop feeding within as little as 2 min after ingestion of the toxin (Heimpel and Angus, 1959). This feeding inhibition is a universal response to the toxin among susceptible larvae and is a major factor in the successful application of *B. thuringiensis* for crop protection.

An alternative method of assaying inclusion body protein has been developed that involves cultured insect tissue (Murphy *et al.*, 1976; Johnson *et al.*, 1980; Johnson, 1981; Johnson and Davidson, 1984). The *in vitro* technique has several advantages, including specificity, time of bioassay, and ease of manipulation. The method combines a cellular tissue system with a bioluminescent ATP bioassay procedure, which provides an efficient quantitative measurement of toxin activity on host tissue. It avoids the more common subjective observations employing visual evidence of cytolysis, including vital staining. It also provides an ideal cellular system that can be used for the investigation of toxin-mediated tissue damage. Histopathological reactions seen in the columnar midgut cells can also be shown with tissue culture derived from *Choristoneura fumiferana* (Spruce budworm) (Ebersold *et al.*, 1979); Johnson, 1981) and from *Trichoplusia ni* (cabbage looper) (Nishiitsutsuji-Uwo *et al.*, 1979). Vacuole formation, swelling, and membrane disintegration are clearly seen in treated tissue cells. Symptoms develop in tissue culture only if proteolytically activated crystal protein is used, which parallels the natural progression of toxigenesis in the intact feeding larvae.

I have sought to characterize the toxic response of cultured insect cells to crystal protein from *B. thuringiensis* according to specificity and relationship to larval sensitivity. Within lepidopteran and dipteran species, host specificity is apparent with crystals from *B. thuringiensis* subsp. *kurstaki* toxic toward lepidopteran larvae but not toward dipteran (mosquito) larvae (Tyrell *et al.*, 1981; Ignoffo *et al.*, 1981). Conversely, inclusion bodies from *B. thuringiensis* subsp. *israelensis* are significantly more toxic toward mosquito larvae than they are toward lepidopteran larvae. The two toxins differ biochemically by subunit molecular wight and amino acid composition (Tyrell *et al.*, 1981), as well as in their response to certain cations (such as calcium) and cell surface modulators (lectins, cholesterol, etc.). Despite these differences, the specificity of larval response to crystal protein is conserved in most cell lines observed to date, which strengthens the argument that *in vitro* tissue culture sensitivity to *B. thuringiensis* entomocidal protein is related to actual larval sensitivity.

II. CELL CULTURE

The cell lines used in this study were obtained as follows: Spruce budworm (*Choristoneura fumiferana* Clemens), IPRI-CF1, a suspended culture started from minced neonate larvae by Sohi (1973); tobacco hornworm (*Manduca sexta* L.), MRRL-CHE-20, an attached cell line established from embryonic tissue by Eide *et al.* (1975); Indianmeal moth (*Plodia interpunctella*), IAL-PID2, a line initiated from imaginal wing disc tissue by Lynn and Oberlander (1981); and two mosquito cell lines established from 1st instar larvae and embryonic tissue [*Anopheles gambiae* (Pudney and Varma), and *Aedes aegypti* (Peleg), respectively], which were kindly provided by I. Schneider, Walter Reed Army Institute of Research, Washington, D.C. The spruce budworm and tobacco hornworm cells were grown at 28°C, and the mosquito and Indianmeal moth cell lines were cultured at 25°C. Growth medium composition was as follows: IPRI-CF1 was grown on Grace's insect tissue culture medium supplemented with 15% inactivated fetal calf serum (iFCS); MRRL-CHE-20 and IAL-PID2 were grown on modified Grace's medium (Yunker *et al.*, 1967) containing 15 and 10% iFCS, respectively. *Aedes aegypti* was grown on Singh's (1967) adaptation of Mitsuhashi and Maramorosch (1964) (M & M) medium with 15% iFCS, and *Anopheles gambiae* was grown on Singh's (1967) medium, also with 15% iFCS. No antibiotics were used in the culture media for any of the lines cultivated. Cells were counted before transfer and prior to bioassay with a Model ZF Coulter Counter (Coulter Electronics, Hialeah, Fla.) equipped with a 100-μm aperture tube.

All procedures for growth of *B. thuringiensis* subsp. *kurstaki* (HD-1) and *B. thuringiensis* subsp. *israelensis* (HD-500), as well as harvesting,

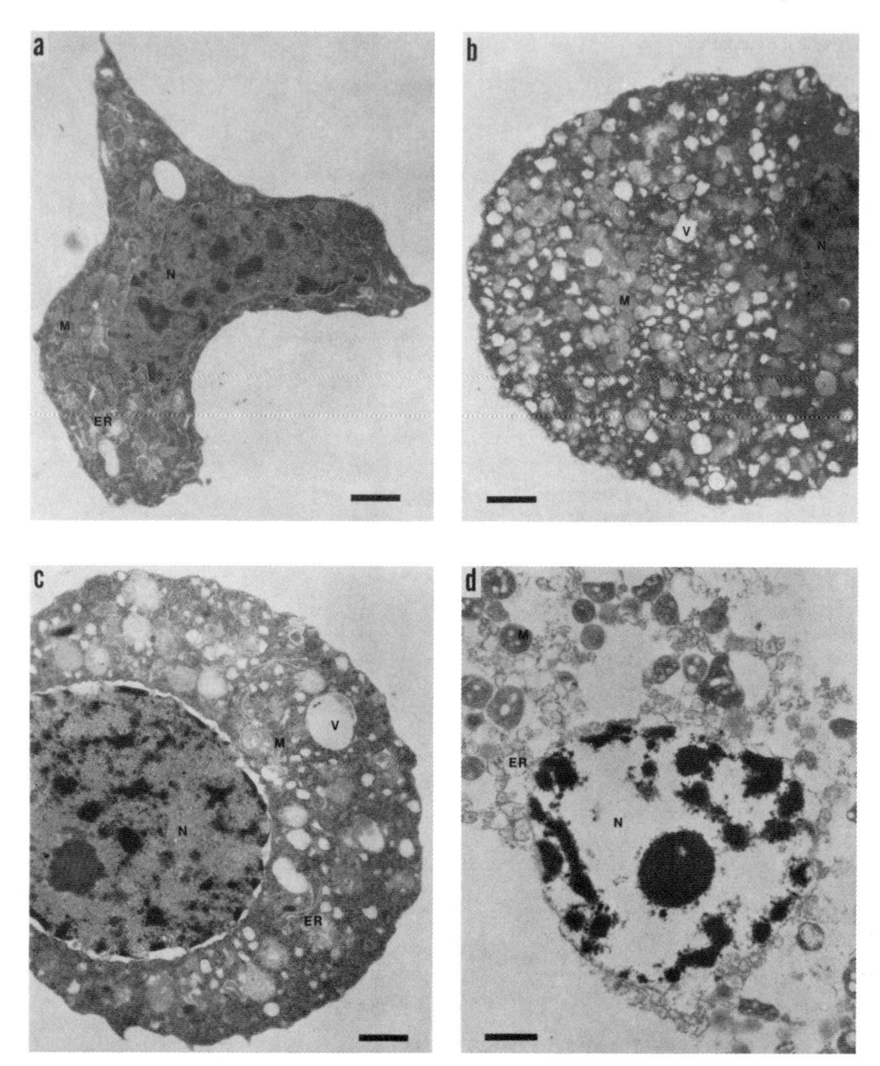

Fig. 1. Ultrastructure of IPRI-CF1 cells following treatment with an LC_{50} dose of activated δ-endotoxin. A normal untreated cell is shown in (a), followed by representative cells treated for 30 (b), 60 (c), and 120 (d) min. The nucleus (N), numerous mitochondria (M), vacuoles (V), and the rough endoplasmic reticulum (ER) are labeled in each sequence. Bar = 1 μm. Reprinted from Johnson (1981) with permission of the *Journal of Invertebrate Pathology*.

purification, dissolution, and activation of crystals, have been reported elsewhere (Johnson, 1981; Bulla *et al.*, 1981).

Cytological Response

The toxicity of protein isolated from inclusion bodies of *B. thuringiensis* subsp. *kurstaki* for cultured insect cells of the spruce budworm (*C. fumiferana* Clemens) is well established (Murphy *et al.*, 1976; Johnson, 1981). The cytological response of the suspended IPRI-CF1 cell line to activated crystal protein is progressive with time and eventually leads to cellular lysis (Fig. 1). Electron micrographs of tissue cells treated for 30 min with an amount of activated toxin equivalent to the LC_{50} showed that cells appeared to swell, the cytoplasm became granular, and the mitochondrial cristae dissolved. At 120 min posttreatment, cells burst completely, but the nuclear membrane (although swollen) appeared to remain intact. Mitochondria darkened and became enlarged, and many vacuoles appeared in the rough endoplasmic reticulum. Ribosomes arranged along the latter became clearly visible during the later stages of rupture. Untreated control cells remained unchanged during the treatment interval.

III. TOXIN BIOASSAY

The bioluminescent reaction can be used to detect ATP in living cells, and this forms the basis for entomocidal activity measurement in insect tissue culture (Fig. 2). ATP is rapidly lost in cells damaged by exposure to a cytolytic toxin, resulting in a reduced bioluminescent response. Thus, there is an inverse relationship between the amount of toxin used and residual cellular ATP content. Assay response is linear for ATP concentrations of 10^{-6} *M* or less, and the cell/substrate is adjusted to provide ca. 2–4 × 10^5 cells/ml (approximately 10^{-7} *M* ATP). Control of these parameters combined with a multiple sampling routine helps to ensure a reproducible and accurate *in vitro* measurement of toxic response in susceptible insect tissue.

The reaction employs the principles of firefly bioluminescense, in which ATP reacts with an enzyme (luciferase)-bound substrate (luciferin) in the presence of Mg^{2+}, producing a luciferyl adenylate complex. The enzyme-bound complex is oxidized by molecular oxygen to produce a molecularly excited intermediate, which quickly reverts to the ground state, producing a photon of light, CO_2, AMP, and the product oxyluciferin (Deluca, 1976). Under optimum conditions, one pho-

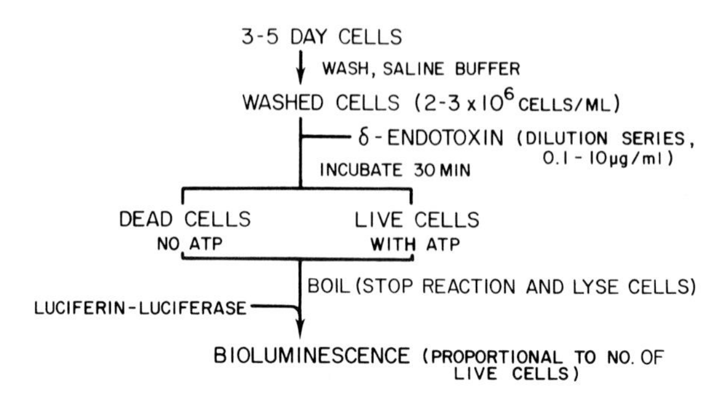

Fig. 2. ATP–bioluminescence analysis method used to measure toxicity of crystal proteins from *B. thuringiensis* to cultured insect cells.

ton of light is produced for each molecule of ATP, permitting a maximum sensitivity of approximately 10^{-18} mole of ATP.

Light (photons) produced by the cyclic reaction of ATP with luciferin–luciferase was measured with a bioluminescent photometer. A Lumac Biocounter Model M2010 (3M Medical Products, St. Paul, Minn.) interfaced with a Commodore Model 4032 computer (Commodore Business Machines, Norristown, Pa.) provided sample measurement and data analysis. All data were processed according to a program especially written to provide a least-squares fit of the experimental data based on log protein per milliliter versus percent toxicity.

Statistical analysis was performed using a computer-generated least-squares fit of raw data from the response of tissue cells to inclusion body protein. The LC_{50} was extrapolated from the least-squares best fit at the point where 50% of the cells underwent cytolysis. This method of determining LC_{50} was employed throughout this study.

A. Crystal Protein Activation and Purification

Enzymatic activation of dissolved crystal protein from *B. thuringiensis* subsp. *kurstaki* is necessary before toxicity is apparent with cultured lepidopteran insect cells. Alkali-dissolved crystal protein in the presence of 5 m*M* phenylmethylsulfonyl fluoride (PMSF, an inhibitor of alkaline protease) yields primarily high molecular weight protoxin (135,000), which is only slightly active *in vitro* (LC_{50} = 479.8 μg protein/ml) (Fig. 3). Some endogenous proteolysis occurs during dissolution in the absence of PMSF (LC_{50} = 199.1 μg protein/ml). Addition of an alkaline protease (such as chymotrypsin) converts the inactive protoxin from *B. thuringiensis* subsp. *kurstaki* to an active toxin of molecular weight of approximately $60{,}000_{(P\text{-}60)}$ which is effective against

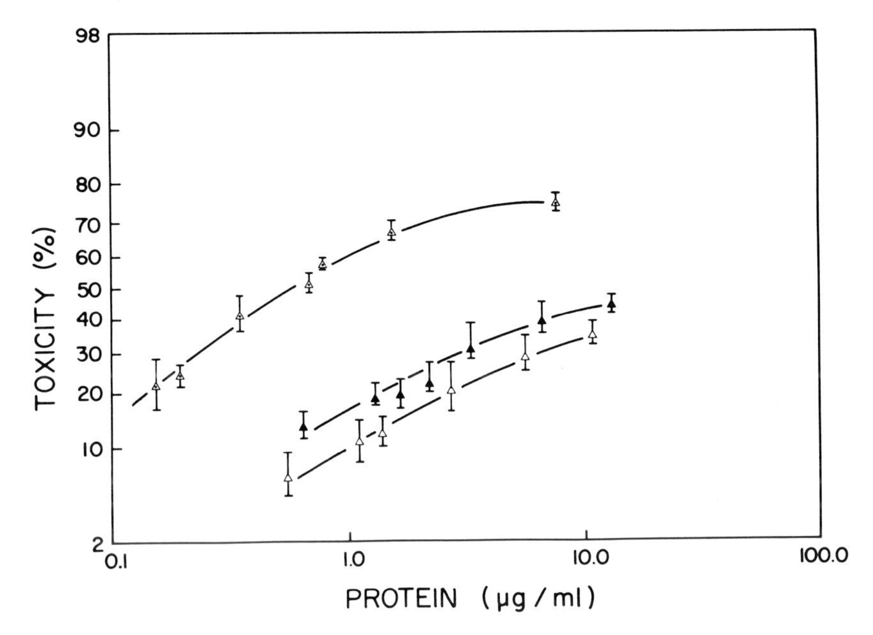

Fig. 3. Toxicity of *B. thuringiensis* subsp. *kurstaki* crystal protein toward cultured cells of the spruce budworm during various stages of dissolution and activation. Crystal protein, dissolved either in the presence of 5 m*M* PMSF (△) or without PMSF (▲); dissolved crystal protein followed by activation with chymotrypsin (⚠). Revised figure reprinted from Johnson and Davidson (1984) with permission of *In Vitro*.

lepidopteran insect tissue cells (LC_{50} = 5.76 μg protein/ml) as well as larvae. The active mosquito toxin from *B. thuringiensis* subsp. *israelensis* crystal protein is substantially smaller ($M_r \approx 26{,}000)_{(P\text{-}26)}$ and does not require additional protease treatment after alkaline dissolution in order to generate a toxic response from mosquito cells *in vitro*.

Table I lists the various steps taken during purification of *B. thuringiensis* subsp. *kurstaki* crystal protein and the degree of purity obtained with each technique. It should be noted that units of toxicity are taken as a reciprocal of the LC_{50} since mortality is inversely related to concentration. Specific activity increased over 1800-fold based on the unactivated crystal protein, or 79-fold if one uses the appearance of P-60 (activation step) as a basis for comparison. The molecular sieving step with Sephadex was not necessary, as comparable purity could be achieved with protein passed twice through DEAE-BioGel A without prior passage through G-100. However, recovery was improved with inclusion of the molecular sieve column. A specific activity of 9.07 units/ml was achieved with *B. thuringiensis* subsp. *kurstaki* crystal

TABLE I

Purification of Entomocidal Protein from Crystals of *B. thuringiensis* subsp. *kurstaki*

Step	Volume (ml)	Protein (mg/ml)	Total protein	LC_{50}[a]	Units/ml[b]	Specific activity[c]	Increase in purity	Recovery
1. 0.0135 N NaOH	5.2	3.970	20.62	54.21	0.0184	0.005	—	100
2. Protease activation[d]	5.2	3.990	20.72	2.17	0.4608	0.115	23	100
3. Sephadex G-100SF	38.6	0.264	10.19	1.95	0.5128	1.942	388.4	49.4
4. DEAE-BioGel A	64.0	0.140	6.65	1.35	0.7407	5.291	1058.2	32.3
5. DEAE (second pass)	29.0	0.104	4.06	1.06	0.9434	9.071	1814.2	20.1

[a] 50% lethal concentration, in micrograms per milliliter.
[b] Reciprocal of LC_{50}.
[c] Units per milligram.
[d] α-Chymotrypsin, 10 μg/mg soluble crystal protein.

protein, and separate experiments have yielded specific activities as high as 28–30 units/ml. A smiliar purification scheme for *B. thuringiensis* subsp. *israelensis* crystal toxin was followed (not shown), with the single exception of the chymotrypsin activation step. Specific activity of P-26 purified on the DEAE-BioGel A column was 9.53 units/ml.

B. Response Time

The response of cultured insect cells to purified *B. thuringiensis* crystal protein from both subspecies was rapid and effective, reaching 50% of maximal response within 5 min of exposure to the toxin (Fig. 4). This finding is consistent with previous work involving uptake of radiolabeled glucose in toxin-treated larvae, which increased within 1 min after exposure but ceased altogether after 10 min (Fast and Donaghue, 1971). Cell age is an important factor in the reliability of bioassay results, and cell cultures of the spruce budworm were most sensitive to toxin dose only after 5 days in culture (Johnson, 1981). Consequently, results were most consistent when cultures to be used for bioassay were harvested at the same age on a routine basis.

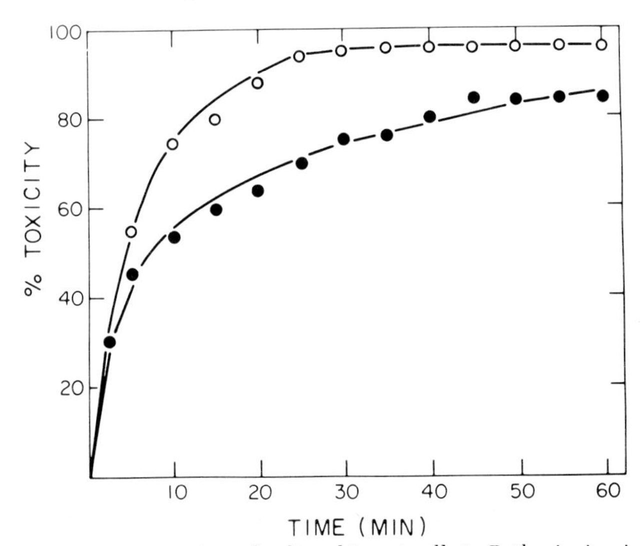

Fig. 4. Bioassay response time of cultured insect cells to *B. thuringiensis* δ-endotoxin. Percent toxicity was based on ATP content of treated versus nontreated (control) cells incubated for the indicated amount of time. Quantity of toxin used was twice the LC_{50} for each species. Mosquito (*A. gambiae*) cells were used to measure *B. thuringiensis* subsp. *israelensis* toxin response (○), and spruce budworm (*C. fumiferana*) cells were used to measure *B. thuringiensis* subsp. *kurstaki* toxin response (●).

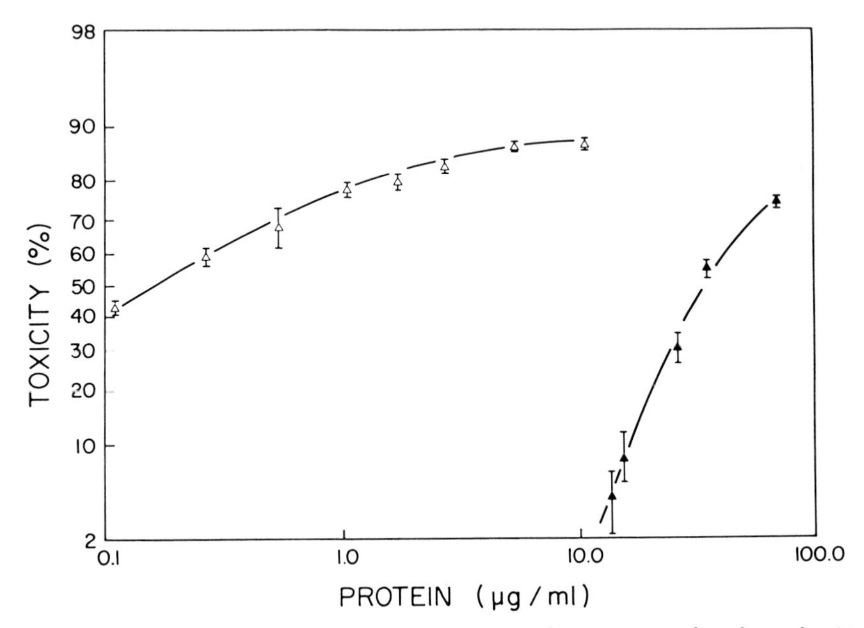

Fig. 5 Toxicity of activated crystal protein from *B. thuringiensis* subsp. *kurstaki* (△) and *B. thuringiensis* subsp. *israelensis* (▲) to cultured cells of the spruce budworm (*C. fumiferana*). Revised figure reprinted from Johnson and Davidson (1984) with permission of *In Vitro*.

C. Specificity

The sensitivity of spruce budworm (IPRI-CF1) cells to purified crystal protein from *B. thuringiensis* subsp. *kurstaki* and subsp. *israelensis* is shown in Fig. 5. Not only was there a great variation in cellular sensitivity to the toxins from the two *B. thuringiensis* subspecies (LC_{50} = 0.46 μg *B. thuringiensis* subsp. *kurstaki* protein/ml vs. 4.99 μg *B. thuringiensis* subsp. *israelensis* protein/ml), but also the spruce budworm cells responded differentially to toxin dose. The IPRI-CF1 cell line responds linearly to activated *B. thuringiensis* subsp. *kurstaki* δ-endotoxin protein only at low doses (protein concentrations ranging up to approximately the LC_{50}). Beyond this concentration, the response curve flattens, rarely exceeding 70–80% sensitivity. The remaining cells (20–30%) apparently represent the resistant part of a mixed population that appears genetically stable (Johnson, 1984). However, there was no evidence of resistance to *B. thuringiensis* subsp. *israelensis* crystal protein among IPRI-CF1 cells, leading to a linear dose–response relationship even at high concentrations of crystal pro-

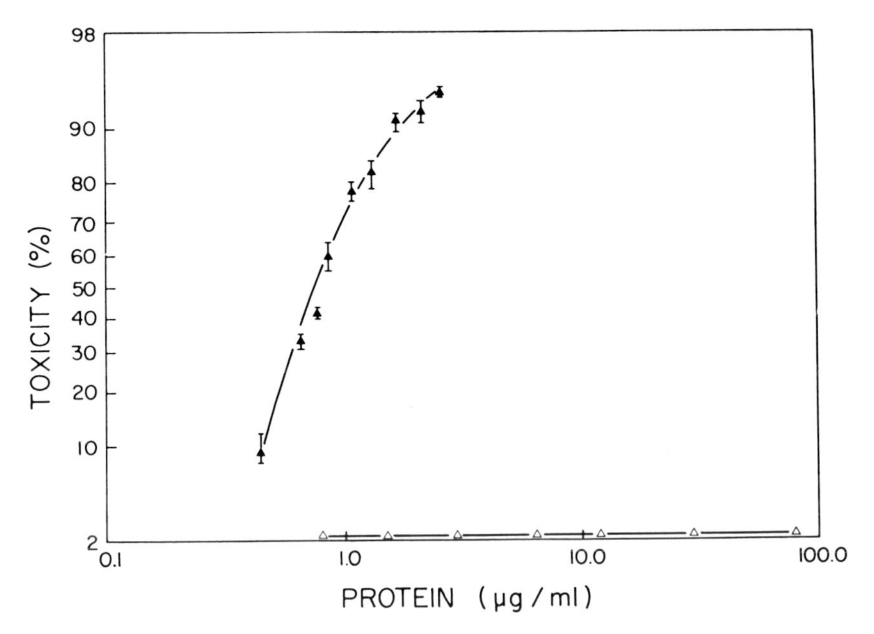

Fig. 6. Toxicity of activated crystal protein from *B. thuringiensis* subsp. *kurstaki* (△) and *B. thuringiensis* subsp. *israelensis* (▲) to cultured cells of the mosquito (*A. gambiae*). Revised figure reprinted from Johnson and Davidson (1984) with permission of *In Vitro*.

tein. The effective LC_{50} for lepidopteran cell lines treated with toxin from *B. thuringiensis* subsp. *israelensis,* however, was sevenfold worse than the LC_{50} for crystal protein from *B. thuringiensis* subsp. *kurstaki.*

The response of cultured mosquito cells (*A. gambiae*) to purified crystal protein from both *B. thuringiensis* subspecies is shown in Fig. 6. Sensitivity to *B. thuringiensis* subsp. *israelensis* crystal toxin was good (LC_{50} = 0.78 μg protein/ml) and the slope of the response curve was steep and uniform. Conversely, mosquito cells were totally insensitive to crystal protein from *B. thuringiensis* subsp. *kurstaki,* even at concentrations as high as 68 μg protein/ml. Cultured tissue cells from *A. aegypti* were almost as sensitive to crystal protein from *B. thuringiensis* subsp. *israelensis* (LC_{50} = 1.28 μg protein/ml) as *Anopheles* cells. No resistance to this toxic protein was observed in either mosquito cell line. This is in contrast to the resistance of spruce budworm cells to *B. thuringiensis* subsp. *kurstaki* crystal protein described above.

A compilation of toxicity data for activated crystal protein isolated from *B. thuringiensis* subsp. *kurstaki* and subsp. *israelensis* assayed against a variety of lepidopteran and dipteran cell lines is shown in

TABLE II

Toxicity of *B. thuringiensis* Purified Crystal Protein to Cultured Insect Tissue from Lepidopteran and Dipteran Sources[a]

		LC_{50}[b]	
Cell line	Tissue origin	*B. thuringiensis* subsp. *kurstaki*	*B. thuringiensis* subsp. *israelensis*
Choristoneura fumiferana	Neonate larvae	0.46 ± 0.19	4.99 ± 1.43
Manduca sexta	Embryo	0.21 ± 0.36	55.20 ± 9.85
Plodia interpunctella	Imaginal wing disc	NT[c]	NT
Aedes aegypti	Embryo	NT	1.28 ± 0.92
Anopheles gambiae	Neonate larvae	NT	0.78 ± 0.27

[a]Revised table reprinted from Johnson and Davidson (1984) with permission of *In Vitro*.

[b]Micrograms of activated δ-endotoxin protein per milliliter, the amount of toxin that is lethal for 50% of the cells under normal assay conditions.

[c]No measurable cellular toxicity observed at a maximum crystal protein level of 82 μg/ml.

Table II. The sensitivity of the lepidopteran cell lines tested (with the exception of IAL-PID2) was restricted primarily to crystal protein isolated from *B. thuringiensis* subsp. *kurstaki*, whereas the dipteran cell lines were sensitive only to crystal protein originating from *B. thuringiensis* subsp. *israelensis*. The single exception to these findings was the *P. interpunctella* cell line, which was totally insensitive to both toxins. The insensitivity of this cell line to crystal protein may be partly due to the origin of this line, which was established from the imaginal wing disc of the Indianmeal moth. The activated protein from crystals of *B. thuringiensis* subsp. *israelensis* possessed slight toxicity toward spruce budworm and tobacco hornworm cell lines, in partial agreement with larval sensitivity to these two crystal types as reported by Ignoffo *et al.* (1981). However, the activated protein from crystals of *B. thuringiensis* subsp. *kurstaki* exhibited no toxicity to dipteran cell lines. Based on the results shown in Table II, we believe that the toxic action of activated crystal protein from *B. thuringiensis* subsp. *kurstaki* and subsp. *israelensis* on embryonic insect cells cultured *in vitro* is specific and may reflect the actual physiological response generated in larval midgut tissue.

IV. MODIFIERS OF CELL SURFACE ACTIVITY

Certain molecules suppress or alter crystal protein toxicity, apparently through a mechanism involving interference with toxin–membrane binding. Cholesterol, a major lipid component of the insect tissue cell membrane and a likely site for toxin binding, interferes with the toxic response *in vitro* when present in the reaction mixture but exerts no effect when preincubated with the toxin. Certain phospholipids (especially phosphatidylcholine and phosphatidylserine) are reported to affect *B. thuringiensis* subsp. *israelensis* activity in mammalian and insect cells *in vitro* (Thomas and Ellar, 1983a). However, we have been unable to duplicate this response with ATP-measured activity.

The influence of various cations on toxicity of *B. thuringiensis* crystal proteins is especially revealing. The buffer used in all bioassays reported here contained 10 m*M* morpholinopropanesulfonic acid (MOPS), 89 m*M* KCl, 2.25 m*M* each $MgCl_2$ and $MgSO_4$, 0.9 m*M* $CaCl_2$, and 22.4 m*M* NaCl. If, however, the KCl–NaCl levels were imbalanced by addition of NaCl to 85 m*M* concomitant with a decrease in KCl to 2 m*M*, a significant (73%) improvement occurred in the LC_{50} of *B. thuringiensis* subsp. *kurstaki* crystal protein (the LC_{50} was reduced from 0.804 to 0.011 μg/ml). An equivalent response was not obtained with *B. thuringiensis* subsp. *israelensis* protein, as toxicity to *Anopheles* cells was unchanged in either buffer system. These findings are consistent with the data of Himeno *et al.* (1985), who described increased swelling and subsequent cytotoxicity of cabbage looper tissue cells as a result of exposure to *B. thuringiensis* subsp. *aizawai* crystal protein under similar conditions.

Other cations influence toxicity as well. Calcium ion at concentrations of 2–10 m*M* exerted a major inhibitory effect on toxic protein from *B. thuringiensis* subsp. *kurstaki* but failed to affect *B. thuringiensis* subsp. *israelensis* activity (Fig. 7). Ouabain, an inhibitor of Na^+,K^+-ATPase in eukaryotic membranes, was ineffective at concentrations as high as 570 μ*M* toward either mosquito or lepidopteran cells during the bioassay interval with their respective *B. thuringiensis* toxin proteins.

Two conclusions can be drawn from this work: (1) Certain cations modify toxicity of *B. thuringiensis* subsp. *kurstaki* crystal protein, apparently through interaction at the membrane/cellular surface, and (2) there is a discrete difference in the activity patterns of toxic crystal protein from *B. thuringiensis* subsp. *kurstaki* and subsp. *israelensis*. The lepidopteran toxin appears to interact with a membrane-associated receptor molecule. When insect cells were treated before bioassay with pronase (3–10 μg/ml)), different results were observed when they were later assayed with δ-endotoxin. Activity of *B. thuringiensis* subsp.

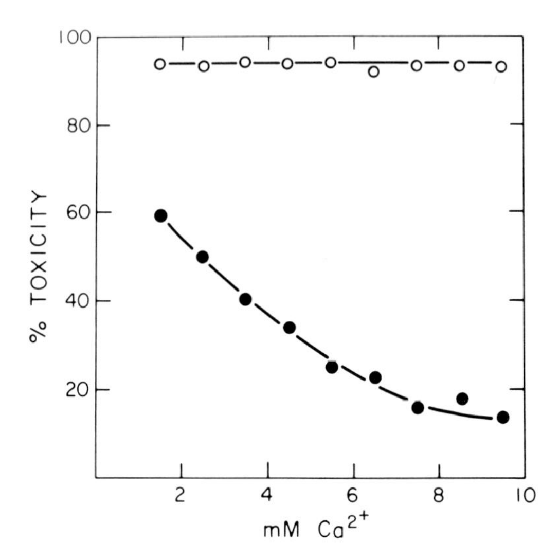

Fig. 7. Effect of calcium on toxicity of *B. thuringiensis* δ-endotoxin. Calcium (as $CaCl_2$) was added to bioassay tubes in appropriately diluted stock solutions to yield final desired concentrations. Bioassay volume, toxin concentration (equivalent to twice the LC_{50}), and buffer composition were uniform throughout the dilution series. Toxins from *B. thuringiensis* subsp. *kurstaki* (●) and subsp. *israelensis* (○) were measured with spruce budworm cells (*C. fumiferana*) and mosquito cells (*A. gambiae*), respectively.

kurstaki toxin toward spruce budworm cells pretreated with pronase was reduced (5- to 7-fold) (not shown), whereas pronase treatment had no effect on *B. thuringiensis* subsp. *israelensis* toxin activity with mosquito cells. Consequently, receptor site composition in spruce budworm cells must reflect a proteinaceous character necessary for proper binding of *B. thuringiensis* subsp. *kurstaki* toxin. Knowles *et al.* (1984) found that certain monosaccharides (*N*-acetylgalactosamine and *N*-acetylneuraminic acid) inactivated *B. thuringiensis* subsp. *kurstaki* toxin activity toward spruce budworm cells. According to their report, various lectins (wheat germ agglutinin and soybean agglutinin) provided partial protection from cell lysis by *B. thuringiensis* subsp. *kurstaki* toxin. However, a monosaccharide concentration of 125 m*M* was necessary to obtain complete inactivation of δ-endotoxin. Using the more sophisticated method of data quantification by ATP measurement, our results do not confirm the work of Knowles *et al.* (1984), but reveal only minor inactivation of δ-endotoxin even at higher concentrations of *N*-acetylgalactosamine (up to 300 m*M*). Nevertheless, firm binding of toxin to the surface of the tissue cell membrane can be

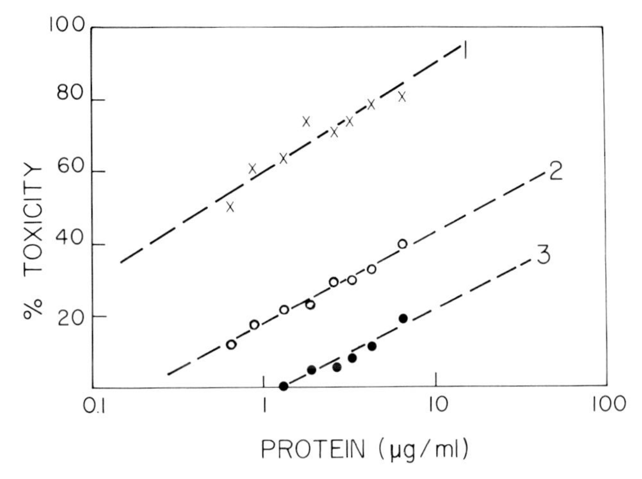

Fig. 8. Binding (adsorption) of entomocidal protein from crystals of *B. thuringiensis* to cultured insect cells. A quantity of toxin protein was used to treat three equal aliquots of washed insect cells sequentially, so that the toxic effects of the same preparation could be followed with successive trials of fresh cells. Controls were taken at each step to account for any increase in the quantity of ATP released into solution from each aliquot of cells and were subsequently subtracted from bioassay data for the following round. First (×), second (○), and third (●) treatments.

demonstrated. In both cases, toxins from *B. thuringiensis* subsp. *kurstaki* and subsp. *israelensis* are bound, perhaps irreversibly, to their respective assay cells, as evidenced by loss of activity from a preparation recycled for a secondary and tertiary treatment. Toxin remained bound to treated cells and membrane residues, so that the supernatant contained proportionately less toxic activity when bioassayed with a second and third batch of fresh cells (Fig. 8).

V. RESISTANCE

Normal populations of IPRI-CF1 contain cells which are resistant to concentrations of crystal protein from *B. thuringiensis* subsp. *kurstaki* in excess of 10 times the usual LC_{50}. These cells persist in the cultivation of the cell line and may overcome the sensitive population with time. The proportion of IPRI-CF1 cells resistant to high concentrations of crystal protein in a 4- to 5-day-old population of cells ranged from 20 to 30% (see Figs. 3 and 5). Spruce budworm cells in culture are spindle-shaped and uniform in size, and no morphological distinction is evident between resistant and sensitive cells. The resistant cells can, however, readily be selected from a mixed cell population by cultivation in the presence of crystal protein (Johnson, 1984). The activity of a cell line treated in this way diminished 79-fold, to an extrapolated LC_{50} of approximately 412 μg protein/ml. However, the resistant cell strain (designated CF1-R) was not stable but slowly lost resistance when returned to normal culture medium (lacking toxin). Eventually, the cell strain returned to a stable but mixed population of cells that exhibited a sensitivity ratio to toxin similar to that of the original cell line (Johnson, 1984). It is not known whether resistnace to crystal protein from *B. thuringiensis* subsp. *israelensis* could be acquired in a similar fashion by mosquito cells.

Similar instances of acquired resistance have been reported in several cell lines. Constant exposure of a tobacco hornworm (*M. sexta*) cell line to β-ecdysone produced a resistant cell population which later diminished after removal of the hormone from the growth medium (Marks and Holman, 1979). Likewise, Couregeon (1972) was able to obtain a subline of *Drosophila* cells that no longer responded to β-ecdysone by growing the cells in the continuous presence of the hormone for several passages. Data presented by Johnson (1984) indicate that spruce budworm cells undergo a similar physiological adaptation, leading to nearly complete resistance to activated crystal protein. The resistance was not hereditary, however, since removal of the toxin during subsequent cultivation resulted in reversion to the original level of sensitivity. Attempts to establish clones of the sensitive and the resistant cells in the IPRI-CF1 cell line have been unsuccessful.

VI. CONCLUSIONS

Insect tissue cells are a valid model system for studying the toxic response to activated protein from crystals of *B. thuringiensis*. In general, the toxic response of cultured cells of various lepidopteran and dipteran insects paralleled the larval sensitivity to protein isolated from crystals of *B. thuringiensis* subsp. *kurstaki* and subsp. *israelensis*. This work contrasts with that of Nishiitsutsuji-Uwo *et al.* (1980), who concluded that cellular swelling and lysis appeared to be a universal response of insect tissue cells cultured *in vitro* on exposure to *B. thuringiensis* crystal protein. These investigators used insect tissue cells from *Trichoplusia ni* (Lepidoptera) and *Culex molestus* (Diptera) treated with gut juice-activated δ-endotoxin protein from dissolved crystals of *B. thuringiensis* subsp. *aizawai*. They observed cytologic changes in cells of *Trichoplusia* and *Culex* on exposure to 50 μg toxin protein/ml, whereas we were unable to detect any response from dipteran cells when we used *B. thuringiensis* subsp. *kurstaki* crystal protein at even greater concentrations. The *Culex* cell line used by these investigators was ovarian, in contrast to the embryonic origin of the two mosquito cell lines used in the work described in this chapter. This observation may help explain the absence of any response from the *Plodia* cell line on exposure to activated crystal protein in these studies, since it was also nonembryonic. The data reported here are supported by the work of Thomas and Ellar (1983b), who reported specific toxic responses from lepidopteran and dipteran cell lines to solubilized crystal protein *B. thuringiensis* subsp. *kurstaki* and subsp. *israelensis*. All of their cell lines were ovarian, however, with the exception of *C. fumiferana* (the same line used in this study). Nevertheless, the tissue source origin may be an important determinant of tissue cell sensitivity to soluble δ-endotoxin protein derived from crystals of various *B. thuringiensis* subspecies.

ACKNOWLEDGMENT

I am very grateful for the expert technical assistance of Loren I. Davidson and Brian D. Barnett.

REFERENCES

Bulla, L. A., Jr., Kramer, K. J., Cox, D. J., Jones, B. L., Davidson, L. I., and Lookhart, G. L. (1981). Purification and characterization of the entomocidal protoxin of *Bacillus thuringiensis*. *J. Biol. Chem.* **256,** 3000–3004.

Couregeon, A. M. (1972). Action of insect hormones at the cellular level. *Exp. Cell Res.* **74,** 327–336.

Deluca, M. (1976). Firefly luciferase. *Adv. Enzymol.* **44,** 37–68.

DeLucca, A. J. (1984). Lectin grouping of *Bacillus thuringiensis* serovars. *Can. J. Microbiol.* **30,** 1100–1104.

Ebersold, H. R., Luethy, P., and Huber, H. E. (1979). Membrane damaging effect of the δ-endotoxin of *Bacillus thuringiensis*. *Experientia* **36,** 495.

Eide, P. E., Caldwell, J. M., and Marks, E. P. (1975). Establishment of two cell lines from embryonic tissue of the tobacco hornworm, *Manduca sexta* (L.). *In Vitro* **11,** 395–399.

Endo, Y., and Nishiitsutsuji-Uwo, J. (1980). Mode of action of *Bacillus thuringiensis* δ-endotoxin: Histopathological changes in the silkworm midgut. *J. Invertebr. Pathol.* **36,** 90–103.

Fast, P. G., and Donaghue, T. P. (1971). The δ-endotoxin of *Bacillus thuringiensis*. II. On the mode of action. *J. Invertebr. Pathol.* **18,** 135–138.

Heimpel, A. M., and Angus, T. A. (1959). The site of action of crystalliferous bacteria in Lepidoptera larvae. *J. Insect Pathol.* **1,** 152–170.

Heimpel, A. M., and Angus, T. A. (1960). Bacterial insecticides. *Bacteriol. Rev.* **24,** 266–288.

Himeno, M., Koyama, N., Funato, T., and Komano, T. (1985). Mechanism of action of *Bacillus thuringiensis* insecticidal delta-endotoxin on insect cells *in vitro*. *Agric. Biol. Chem.* **49,** 1461–1468.

Ignoffo, C. M., Couch, T. L., Garcia, C., and Kroha, M. J. (1981). Relative activity of *Bacillus thuringiensis* var. *kurstaki* and *B. thuringiensis* var. *israelensis* against larvae of *Aedes aegypti, Culex quinquefasciatus, Trichoplusia ni, Heliothis zea,* and *Heliothis virescens.* J. Econ. Entomol. **74,** 218–222.

Johnson, D. E. (1981). Toxicity of *Bacillus thuringiensis* entomocidal protein toward cultured insect tissue. *J. Invertebr. Pathol.* **38,** 94–101.

Johnson, D. E. (1984). Selection for resistance to *Bacillus thuringiensis* δ-endotoxin in an insect cell line (*Choristoneura fumiferana*). *Experientia* **40,** 274–275.

Johnson, D. E., and Davidson, L. I. (1984). Specificity of cultured insect tissue cells for bioassay of entomocidal protein from *Bacillus thuringiensis*. *In Vitro* **20,** 66–70.

Johnson, D. E., Niezgodski, D. M., and Twaddle, G. M. (1980). Parasporal crystals produced by oligosporogenous mutants of *Bacillus thuringiensis* ($Spo^{-}Cr^{+}$). *Can. J. Microbiol.* **26,** 486–491.

Knowles, B. H., Thomas, W. E., and Ellar, D. J. (1984). Lectin-like binding of *Bacillus thuringiensis* var. *kurstaki* lepidopteran-specific toxin is an initial step in insecticidal action. *FEBS Lett.* **168,** 197–202.

Lynn, D. E., and Oberlander, H. (1981). Development of cell lines from imaginal wing discs of Lepidoptera. *In Vitro* **17,** 208.

Marks, E. P., and Holman, G. M. (1979). Ecdysone action on insect cell lines. *In Vitro* **15,** 300–307.

Mitsuhashi, J., and Maramorosch, K. (1964). Leaf hopper tissue culture: Embryonic, nymphal and imaginal tissues from aseptic insects. *Contrib. Boyce Thompson Inst.* **22,** 435–460.

Murphy, D. W., Sohi, S. S., and Fast, P. G. (1976). *Bacillus thuringiensis* enzyme-digested delta endotoxin: Effect on cultured insect cells. *Science* **194,** 954–956.

Nishiitsutsuji-Uwo, J., Endo, Y., and Himeno, M. (1979). Mode of action of *Bacillus thuringiensis* δ-endotoxin: Effect on TN-368 cells. *J. Invertebr. Pathol.* **34,** 267–275.

Nishiitsutsuji-Uwo, J., Endo, Y., and Himeno, M. (1980). Effects of *Bacillus thuringiensis* δ-endotoxin on insect and mammalian cells *in vitro*. *Appl. Entomol. Zool.* **15,** 133–139.

Singh, K. R. P. (1967). Cell cultures from larvae of *Aedes albopictus* (Skuse) and *Aedes aegypti* (L.). *Curr. Sci.* **36,** 506–508.

Sohi, S. S. (1973). *In vitro* cultivation of larval tissues of *Choristoneura fumiferana* (Clemens) (Lepidoptera: Torticidae). *Proc. Int. Colloq. Invertebr. Tissue Cult. 3rd, 1971,* pp. 75–92.

Thomas, W. E., and Ellar, D. J. (1983a). Mechanism of action of *Bacillus thuringiensis* var. *israelensis* insecticidal δ-endotoxin. *FEBS Lett.* **154,** 362–368.

Thomas, W. E., and Ellar, D. J. (1983b). *Bacillus thuringiensis* var. *israelensis* crystal δ-endotoxin: Effects on insect and mammalian cells *in vitro* and *in vivo*. *J. Cell Sci.* **60,** 181–197.

Tyrell, D. J., Bulla, L. A., Jr., Andrews, R. E., Jr., Kramer, K. J., Davidson, L. I., and Nordin, P. (1981). Comparative biochemistry of entomocidal parasporal crystals of selected *Bacillus thuringiensis* strains. *J. Bacteriol.* **145,** 1052–1062.

Yunker, C. E., Vaughn, J. L., and Cory, J. (1967). Adaptation of a cell line (Grace's *Antherea* cells) to a medium free of insect hemolymph. *Science* **155,** 1565–1566.

5

Operational Bacterial Insecticides and Their Potential for Future Improvement

H. DE BARJAC

Although the exact origin of biological control of insects is still controversial, the phenomenon is very old. Its widespread commercial application especially with regard to insect vectors is fairly recent. The strict requirements necessary for a successful biocontrol agent—specificity, high speed of action, high potency, good stability and economic feasibility—explain why there are such a limited number used at present.

The inconvenience of using chemical insecticides and the modern awareness of environmental protection have helped to promote biological control agents. A major advantage of these bioinsecticides is their safety, a consequence of their specificity, as well as their biodegradability.

The development of insect resistance to the most commonly used chemical insecticides has also favored the use of biological controls. The complex mechanism for the action of numerous biocontrol agents is unlikely to initiate insect resistance.

Several classes of biological control agents exist. They are found among microorganisms such as bacteria, fungi, viruses, protozoa, and among larger organisms such as nematodes and other insects which act as parasites or predators.

Apart from their common features of specificity and safety, these control agents have very different characteristics and target insects. Accordingly, these target insects can be divided into two groups: insect pests and insect vectors. A list of the corresponding biological insecticides is given in Tables I and II. A fairly high number of biocontrol agents are presently known, but only a few of them have been commercially marketed (trademark names given in Tables I and II).

Among these commercial bioinsecticides, one bacterium has achieved the highest priority: *Bacillus thuringiensis*. It is probably the best known and most widely used bioinsecticide at present, as well as one of the most promising ones. It has the ability to attack both kinds of target insects, agricultural pests and disease vectors, a property which is not shared by most bioinsecticides.

Known since the early 1900s, *B. thuringiensis* is currently being applied to agricultural crops and many trees for the control of Lepidoptera larvae. Serotype H3a,3b (*B. thuringiensis* subsp. *kurstaki*) is most frequently used, along with serotype H7.

Literature relevant to *B. thuringiensis* is so enormous that one can only mention some main characteristics of this species with emphasis on new developments. The references given here are very limited, but many reviews exist including a very recent one by Aronson *et al.* (1986).

In 1977 Goldberg and Margalit discovered a *Bacillus* strain that killed mosquito and blackfly larvae. It was further characterized as being the H14 serotype of *B. thuringiensis* and was named *B. thuringiensis* subsp. (or serovar) *israelensis* (de Barjac, 1978a,b). This fact completely ruled out the theory of the strict specificity of *B. thuringiensis* for Lepidoptera larvae, which had prevailed for more than 30 years. Now a more flexible approach should be used when discussing *B. thuringiensis* pathogenicity.

The 24 existing H serotypes and the corresponding 33 varieties present a rather broad spectrum of action. For instance, activity againt Coleoptera larvae has been reported for a *B. thuringiensis* strain (256–82) belonging to serotype H8a,8b (Krieg *et al.*, 1983).

Until now, pathogenicity for Coleoptera larvae was restricted to *Bacillus popilliae;* the commercial product is being used effectively for the control of the Japanese beetle *Popillia japonica*. Unfortunately, the quasi-impossibility of obtaining infectious spores by classical *in vitro* cultures has greatly hampered new research developments.

Mosquitocidal activity has also been reported in *B. thuringiensis* serotypes other than H14. A strain of *B. thuringiensis* of serotype H8a,8b (PG14) has been described with a high mosquitocidal activity

TABLE I
Biological Control of Agricultural and Forestry Insect Pests

Pathogenic microorganism or type	Trade name/Source
Bacteria	
Bacillaceae	Bactospéine/Belgium, France
Bacillus thuringiensis	Dipel, Thuricide, Certan/United States
(H3a,3b, H7, H1, . . .)	Entobakterine, Dendrobacilline, Bitoxibacilline/USSR
	Toaqosei BT/Japan
Bacillus popilliae	Doom/United States
Viruses	
Baculovirus: NPV	Biocontrol 1, San 404, Elcar, Gypcheck/United States
Reovirus: CPV	Matsukmin/Japan
Fungi	
Hyphomycetes	
(Deuteromycetes)	Biotrol FBB/United States, Beauverine/USSR
Beauveria spp.	
Hirsutella thompsoni	Mycar/United States
Verticillium lecanii	Vertek, Vertalec/Great Britain
Metarrhizium anisopliae	Brazil
Nomuraea (= Spicaria) rileyi	United States
Zygomycetes	
Entomophtora spp.	
Protozoa	
Microsporidia	Noloc/United States
Nosema locustae	
Parasites	
Nematodes	
Neoplectana spp.	
Insects	
Trichogramma	
Predators	
Insects	
Hymenoptera, Coleoptera (Coccinellidae)	
Genetic control	
Male sterility, cytoplasmic incompatibility, chromosome translocations	
Biochemical control	
Growth regulators	
Pheromones	Muscamone, Bag A, Bug, Gossyplure

TABLE II
Biological Control of Vector Insects (Mosquitoes, Blackflies) and Other Diptera (Flies)

Pathogenic microorganism or type	Trade name/Source
Bacteria	
Bacillaceae	Bactimos/France, Vectobac, Teknar/ United States
	Bactocoulitside/USSR
Bacillus thuringiensis H14	Skeetal/Great Britain
Bacillus thuringiensis H1	Muscabac/Finland, Bitoxibacilline/USSR
Bacillus sphaericus (H5a,5b and H25)	
Fungi	
Hyphomycetes (Deuteromycetes)	
Culicinomyces clavisporus	
Tolypocladium cylindrosporum	
Metarhizium spp.	
Phycomycetes	
Lagenidium giganteum	
Coelomomyces (iliensis)	
Leptolegna spp.	
Zygomycetes	
Entomophtora spp.	
Protozoa	
Microsporidia	
Nosema algerae	
Vavraia (= *Pleistophora*) culicis	
Viruses	
Baculovirus	
Densonucleovirus	
Parasites	
Nematodes	
Mermithidae: *Romanomermis culicivorax*	Q-licide/United States
Predators	
Insects	
Toxorhynchites (brevipalpis)	
Fishes	
Gambusia affinis	
Aplocheilus blochi	
Poecilia reticulata	
Plants	
Utricularia spp.	
Genetic control	
Male or hybrids sterility, cytoplasmic incompatibility, chromosome translocations, lethal genes	
Biochemical control	
Growth regulators	Altosid/United States
Pheromones	

(Padua *et al.*, 1984) and two strains of serotype H10 (73 E 10-16 and 10-2), with a significant, even if lower, toxicity for mosquito and blackfly larvae, have been reported (Padua *et al.*, 1980; Finney and Harding, 1982).

Bacillus thuringiensis subsp. *israelensis* has also been found useful as a larvicide for sandflies (*Phlebotomus* spp.), horn flies (*Haematobia irritans*), and sciarid flies (*Lycoriella mali*). But its main application remains the control of mosquitoes and blackfly larvae, this last usage making an inestimable contribution to the fight against river blindness (onchocerciasis) in West Africa. Tons of commercial *B. thuringiensis* subsp. *israelensis* products are applied each year to control *Culex*, *Anopheles*, and *Aedes* larvae, thus helping to prevent tropical diseases such as filariasis, malaria, dengue, various types of encephalitis, and yellow fever.

Bacillus thuringiensis acts mainly by means of a proteinaceous δ-endotoxin, which is produced at the sporulation stage in the form of a crystallized inclusion (called a crystal). This crystal, which is readily visible under phase-contrast microscopy, represents a protoxin which is activated into a toxin, after ingestion, by the alkaline gut proteases of the insect (Lecadet, 1965). This toxin has stimulated a considerable amount of research and is still attracting more studies, especially with the development of genetic engineering.

It is believed that most *B. thuringiensis* crystals are multimeric structures made of polypeptide subunits or dimers of about 230 kDa each. At alkaline pH in the presence of reducing agents, a basic protoxin subunit is produced, a monomer of about 130 kDa. Activation by insect gut proteases leads to the final δ-endotoxin, which is about 68 kDa. In fact, there are discrepancies among authors concerning the size of the toxic unit, or units, especially in the case of *B. thuringiensis* subsp. *israelensis*, where special solubilization conditions lead to another smaller peptide of about 26 kDa having cytolytic and hemolytic activity. A synergism between the 26 and 65 kDa proteins *B. thuringiensis* subsp. *israelensis* has been suggested by different authors (Wu and Chang, 1985; Ibarra and Federici, 1986).

In fact, with progressing studies, the complexity appears to increase and several different toxins produced by the same *B. thuringiensis* strain are likely to be detected.

Bacillus thuringiensis crystal toxins induce gut paralysis or general paralysis of the susceptible insect larvae shortly after ingestion. The insect quickly stops feeding. Cytopathological events are represented by extensive damage of midgut epithelial cells, which usually swell before undergoing lysis (de Barjac, 1978c). Mitochondria are also modified (Charles and de Barjac, 1983).

The active site of the *B. thuringiensis* δ-endotoxin and its molecular mode of action are still unknown. Apparently this toxin interacts with the epithelial cell membrance. A study of the *B. thuringiensis* serotype H3a,3b has recently suggested recognition by the toxin of a specific glycoconjugate receptor with a terminal *N*-acetylgalactosamine residue (Knowles *et al.*, 1984). Other studies of the *B. thuringiensis* subsp. *israelensis* serotype H14 have suggested specific binding of the toxin to membrane sphingolipids, leading to cytolysis by disarrangement of the lipid bilayer (Thomas and Ellar, 1983).

Increased interest in *B. thuringiensis* crystal toxin has been generated by the discovery of its plasmid regulation and the cloning of the genes responsible. Important breakthroughs might represent the first steps in creating a new generation of insecticides.

With the cloning and expression in *Escherichia coli* and *Bacillus subtilis* of the genes coding for the crystal toxins of *B. thuringiensis* serotypes H3a,3b and H1, respectively, in 1981 (Schnepf and Whiteley) and 1982 (Klier *et al.*), the simultaneous discovery of a conjugationlike system and the transfer of toxin plasmids in *B. thuringiensis* (Gonzalez *et al.*, 1982) confirmed the localization of the toxin-coding genes, but also triggered further developments by genetic manipulation.

The organization of the crystal gene and the surrounding sequences within a composite transposon, as suggested by Lereclus *et al.* (1984), would account for the multiple location of the genes in one strain as well as for exchanges between different strains. The complete amino acid sequence of the toxin has been determined in *B. thuringiensis* serotype H3a,3b with the delineation of the toxin-coding fragment of the crystal gene. (Schnepf *et al.*, 1985; Schnepf and Whiteley, 1985). Further similar studies of other serotypes have also been made.

Recently, molecular cloning of the crystal protein genes of *B. thuringiensis* subsp. *israelensis* or *B. thuringiensis* serotype H14 has been reported in *E. coli* or *B. subtilis* (Ward *et al.*, 1984; Waalwijk *et al.*, 1985; Thorne *et al.*, 1986), and in *B. megaterium* and *B. subtilis* (Sekar and Carlton, 1985). Concommitantly, numerous new approaches to the study of bioinsecticides have been made, e.g., the cloning of *B. thuringiensis* H3a,3b toxin gene in *Pseudomonas* sp. and its possible use as a seed coating for the protection of corn roots against *Agrotis ypsilon*.

Another example comes from Belgium (van Montagu), where the toxin-coding gene of the same *B. thuringiensis* was introduced in a tobacco plant to give the plant the ability to produce the crystal toxins and hence ensure its self-defense against the tobacco hornworm. Along the same line, researchers are now trying to transfer the toxin gene of *B. thuringiensis* serotype H14 into a blue-green alga (*cyanobacteria* such as *Anacystis* or *Anabaena*) in order to establish this toxin permanently in the environment and natural food source of mosquito larvae.

Other workers are trying to relate the specificity of different *B. thuringiensis* crystal toxins to one or another amino acid in the sequence of the toxic peptide units.

All these trends in genetic engineering point toward considerable progress in the development of future biopesticides.

Besides the crystal toxins of *B. thuringiensis,* mention should be made of another thermostable and nucleotide-like toxin excreted in the culture medium by a few strains in serotypes such as H1, H9, H10: the β-exotoxin. Known for more than 20 years (de Barjac and Dedonder, 1965), its use has been restricted until now because of its nucleotide analog nature, teratogenic effects on insects (Burgerjon and Biache, 1967), and toxicity when injected into mammals (de Barjac and Riou, 1969). However, its broad spectrum of action has permitted its application for the control of housefly larvae (and in the USSR, various cattle fly larvae). β-Exotoxin is now used experimentally in the United States for the control of the Colorado potato beetle. It appears that small dosages of β-exotoxin are no less safe than most chemical insecticides; therefore its future use could be increased.

Information obtained on *B. thuringiensis* stimulates research on another cadidate for the control of mosquito larvae: *Bacillus sphaericus.* A complete review of the relevant literature will not be given here (see, for instance, Yousten, 1984; Davidson, 1984 and the chapter by S. Singer in this volume.)

The first really potent strains of *B. sphaericus* identified included strain 1593 (Singer, 1977), strain 2297 (Wrickemesinghe and Mendis, 1980), and strain 2362 (Weiser, 1984).

Of the 45 H serotypes of *B. sphaericus* actually characterized, only five have been described as containing mosquitocidal strains, which have different levels of toxicity (de Barjac *et al.*, 1985). The most studied and the most promising strains belong to serotypes H5a,5b (main strains 1593 and 2362) and H25 (strain 2297). In 1986 a strain of serotype H6 was characterized as being one of the most toxic (de Barjac *et al.*, 1987).

Some analogies exist between *B. sphaericus* and *B. thuringiensis* H14 if we consider that they both have crystallized protein toxins which appear at the sporulation stage (de Barjac and Charles, 1983). "Activation" of such toxins by gut proteases of mosquito larvae has been suggested for both species and different peptide subunits have been characterized. In *B. sphaericus* a toxic unit of 43 kDa has been found (Baumann *et al.*, 1985). Comparable pathological effects on midgut epithelial cells are also known.

However, fundamental differences exist. *Bacillus sphaericus* strains are likely to harbor toxins other than the crystallized ones, and the action and specificity of *B. thuringiensis* subsp. *israelensis* and *B. sphaericus* are different. The former is extremely toxic to *Simulium* larvae and is more potent against *Aedes* and *Culex* larvae than against *Anopheles* larvae. The known *B. sphaericus* strains are not toxic to blackfly larvae, but are extremely to very toxic to *Culex* and *Anopheles* larvae, and have little or not toxicity to *Aedes* larvae.

Consequently, in applications for mosquito control, *B. thuringiensis* subsp. *israelensis* and *B. sphaericus* can complement each other. It has also been shown that *B. sphaericus* is more efficient than *B. thuringiensis* subsp. *israelensis* in polluted water and its action is more lasting. In both cases, the quality of the formulation is a key to the success of the application. *Bacillus thuringiensis* and *B. sphaericus* act as food poisons therefore the requirement for best efficacy in field studies consists in an adequately long-lasting availability of active material in the feeding zones of the target larvae.

There may be special problems associated with environmental conditions, for instance, dense vegetal cover in mosquito breeding sites or high-speed running water in blackfly breeding sites. That is why industry already offers a variety of *B. thuringiensis* formulations, from powders or liquid suspensions to granules, pellets, or briquets (for *B. thuringiensis* subsp. *israelensis*).

Progress in formulations is only one of the avenues to further developments in the study of microbial insecticides. Another includes genetic manipulation of *Bacillus* strains, as mentioned above for *B. thuringiensis*. Here, the possibilities are many, covering not only the creation of strains with a new or broader spectrum of action or higher toxicity, but also the creation of strains that excrete their toxins into the culture medium, and the transfer of toxin-coding genes into other organisms.

Research on *B. thuringiensis* is more advanced than on *B. sphaericus*. Transfer of *B. thuringiensis* toxin gene into plants and strains with different crystal toxins has been reported. Cloning of a toxin gene from *B. sphaericus* has been reported but needs further confirmation (Ganesan *et al.*, 1983; Louis *et al.*, 1984).

Once these different bacterial toxins are completely characterized, the biochemical, or strictly chemical, synthesis *in vitro* of molecules derived from these models can be attempted.

However, the behavior, efficacy, and fate of such toxins under field conditions needs to be determined. The same holds true for genetically modified *Bacillus* strains.

Another development using bacterial insecticides is the creation of asporogenous or oligosporogenous toxic mutants, free of the disadvantages attributed to the spores. Even if they have been proved quite safe for the environment and nontarget fauna, spore applications should be scrutinized from time to time because of possible environmental contamination, especially of water. For this reason asporogenous *B. thuringiensis* subsp. *israelensis* is now available to treat drinking water in urban zones against mosquito larvae.

The use of spore-free preparations for the control of agricultural and forestry insect pests is somewhat questionable because the killing of some insect species (e.g., *Anagasta kuhniella*) may be dependent on or enhanced by the *Bacillus* spores. One exception exists in areas in India and Japan with cottage silkworm industry. The use of spore-free products is recommended in order to prevent contamination and death of silkworms. Irradiated preparations (γ- or X-rays) with nonviable spores are available commercially.

With more ecological research on insects and environments, discovery of other entomopathogenic bacteria, especially *Bacillus* spp., is highly probable.

When looking for improvements in bioinsecticides, we ought to keep in mind that we do not yet understand all of nature's "tricks" for balancing insect populations and that, besides laboratory manipulations, nature is one of the main sources of new biocontrol agents. The search for new strains, especially in tropical or developing countries, should be actively pursued.

REFERENCES

Aronson, A. A., Beckman, W., and Dunn, P. (1986). *B. thuringiensis* and related insect pathogens. *Microbiol. Rev.* **50,** 1–24.

Baumann, P., Unterman, B. M., Baumann, L., Broadwell, A. H., Abbene, S. J., and Bowditch, R. D. (1985). Purification of the larvicidal toxin of *Bacillus sphaericus* and evidence for high-molecular-weight precursors. *J. Bacteriol.* **163**(2), 738–747.

Burgerjon, A., and Biache, G. (1967). Effets tératologiques chez les nymphes et les adultes d'insectes, dont les larves ont ingéré des doses subléthales de toxine thermostable de *Bacillus thuringiensis* Berliner. *C. R. Hebd. Seances Acad. Sci.* **264,** 2423–2425.

Charles, J. F., and de Barjac, H. (1983). Action des critaux de *B. thuringiensis* var. *israelensis* sur l'intestin moyen des larves de *Aedes aegypti* L. en microscopie électronique. *Ann. Microbiol. (Paris),* **134A,** 197–218.

Davidson, E. W. (1984). Microbiology, pathology and genetics of *Bacillus sphaericus:* Biological aspects which are important to field use. *Mosq. News* **44**(2), 147–152.

de Barjac, H. (1978a). Une nouvelle variété de *Bacillus thuringiensis* très toxique pour les moustiques: *B. thuringiensis* var. *israelensis* sérotype H14. *C. R. Hebd. Seances Acad. Sci., Ser. D* **286,** 797–800.

de Barjac, H. (1978b). Toxicité de *Bacillus thuringiensis* var. *israelensis* pour les larves d'*Aedes aegypti* et d'*Anopheles stephensi*. *C. R. Hebd. Seances Acad. Sci., Ser. D* **286,** 1175–1178.

de Barjac, H. (1978c). Etude cytologique de l'action de *Bacillus thuringiensis* var. *israelensis* sur larves de moustiques. *C. R. Hebd. Seances Acad. Sci., Ser. D* **286,** 1629–1632.

de Barjac, H., and Charles, J. F. (1983). Une nouvelle toxine active sur les moustiques, présente dans des inclusions cristallines produites par *Bacillus sphaericus. C. R. Seances Acad. Sci., Ser. 3* **297,** 905–910.

de Barjac, H., and Dedonder, R. (1965). Isolement d'un nucéotide identifiable à la "toxine thermostable" de *Bacillus thuringiensis* berliner. *C. R. Hebd. Seances Acad. Sci.* **260,** 7050–7051.

de Barjac, H., and Riou, J.-Y. (1969). Action de la toxine thermostable de *B. thuringiensis* var. *thuringiensis* administrée à des souris. *Rev. Pathol. Comp. Med. Exp.* **6**(805), 367–374.

de Barjac, H., Larget-Thiery, I., Cosmao Dumanoir, V., and Ripouteau, H. (1985). Serological classification of *Bacillus sphaericus* strains in relation with toxicity to mosquito larvae. *Appl. Microbiol. Biotechnol.* **21,** 85–90.

de Barjac, H. *et al.* (1987). In press.

Finney, J. R., and Harding, J. B. (1982). The susceptibility of *Simulium verecundum* (Diptera: Simuliidae) to three isolates of *Bacillus thuringiensis* serotype 10 (*darmstadiensis*). *Mosq. News* **43**(3), 434–435.

Ganesan, S., Kamdar, H., Jayaraman, K., and Szulmajster, J. (1983). Cloning and expression in *Escherichia coli* of a DNA fragment from *Bacillus sphaericus* coding for biocidal activity against mosquito larvae. *Mol. Gen. Genet.* **189,** 181–183.

Goldberg, L. J., and Margalit, J. (1977). A bacterial spore demonstrating rapid larvicidal activity against *Anopheles sergentii, Uratrotaenia unguiculata, Culex univitattus, Aedes aegypti* and *Culex pipiens. Mosq. News* **37,** 355–358.

Gonzalez, J. M., Brown, B. S., and Carlton, B. C. (1982). Transfer of *Bacillus thuringiensis* plasmids coding for δ-endotoxin among strains of *B. thuringiensis* and *B. cereus. Proc. Natl. Acad. Sci. U.S.A.* **79,** 6951–6955.

Ibarra, J. E., and Federici, B. A. (1986). Isolation of a relatively non-toxic 65 kDa protein inclusion from the parasporal body of *Bacillus thuringiensis* subsp. *israelensis. J. Bacteriol.* **165,** 527–533.

Klier, A., Fargette, F., Ribier, J., and Rapoport, G. (1982). Cloning and expression of the crystal protein genes from *Bacillus thuringiensis* strain berliner 1715. *EMBO J.* **1,** 791–799.

Knowles, B. H., Thomas, W. E., and Ellar, D. J. (1984). Lectin-like binding of *Bacillus thuringiensis* var. *kurstaki* lepidopteran-specific toxin is an initial step in insecticidal action. *FEBS Lett.* **168,** 197–202.

Krieg, A., Huger, A. M., Langenbruch, G. A., and Schnetter, W. (1983). *Bacillus thuringiensis* var. *tenebrionis,* a new pathotype effective against larvae of Coleoptera. *Z. Angew. Entomol.* **96,** 500–508.

Lecadet, M. M. (1965). Isolement et caractérisation de deux protéases des chenilles de *Pieris brassicae* et étude de leur action sur l'inclusion parasporale de *B. thuringiensis.* Thèse Doc. Etat Fac. Sci., Paris.

Lereclus, D., Ribier, J., Klier, A., and Lecadet, M. M. (1984). A transposon-like structure related to the δ-endotoxin gene of *Bacillus thuringiensis. EMBO J.* **3,** 2561–2567.

Louis, J., Jayaraman, G., and Szulmajster, J. (1984). Biocide gene(s) and biocidal activity in different strains of *B. sphaericus.* Expression of the gene(s) in *E. coli* maxi cells. *Mol. Gen. Genet.* **195,** 23–28.

Padua, L. E., Ohba, M., and Aizawa, K. (1980). The isolates of *Bacillus thuringiensis* serotype H10 with a high preferential toxicity to mosquito larvae. *J. Invertebr. Pathol.* **36,** 180–186.

Padua, L. E., Ohba, M., and Aizawa, K. (1984). Isolation of a *Bacillus thuringiensis* strain (serotype 8a8b) highly and selectively toxic against mosquito larvae. *J. Invertebr. Pathol.* **44,** 12–17.

Schnepf, H. E., and Whiteley, H. R. (1981). Cloning and expression of the *Bacillus thuringiensis* crystal protein gene in *Escherichia coli*. *Proc. Natl. Acad. Sci. U.S.A.* **78,** 2893–2897.

Schnepf, H. E., and Whiteley, H. R. (1985). Delineation of a toxin-encoding segment of a *Bacillus thuringiensis* crystal protein gene. *J. Biol. Chem.* **260,** 6273–6280.

Schnepf, H. E., Wong, H. C., and Whiteley, H. R. (1985). The amino acid sequence of a crystal protein from *Bacillus thuringiensis* deduced from the DNA base sequence. *J. Biol. Chem.* **260,** 6264–6272.

Sekar, V., and Carlton, B. C. (1985). Molecular cloning of the δ-endotoxin gene of *Bacillus thuringiensis* var. *israelensis*. *Gene* **33,** 151–158.

Singer, S. (1977). Isolation and development of bacterial pathogens of vectors *in* Biological regulation of vectors. *DHEW Publ. NIH (U.S.)* **NIH-77-1180,** 3–18.

Thomas, W. E., and Ellar, D. J. (1983). Mechanism of action of *Bacillus thuringiensis* var. *israelensis* insecticidal δ-endotoxin. *FEBS Lett.* **154,** 362–368.

Thorne, L., Garduno, F., Thompson, T., Decker, D., Zounes, M., Wild, M., Walfield, A., and Pollock, T. (1986). Structural similarities between the Lepidoptera and Diptera specific insecticidal endotoxin genes of *Bacillus thuringiensis* subsp. *kurstaki* and *israelensis*. *J. Bacteriol.* **166,** 801–811.

Waalwijk, C., Dullenmans, A. M., van Workum, M. E. S., and Visser, B. (1985). Molecular cloning and the nuclotide sequence of the M^r 28000 crystal protein gene of *Bacillus thuringiensis* subsp. *israelensis*. *Nucleic Acids Res.* **13,** 8207–8217.

Ward, E. S., Ellar, D. J., and Todd, J. A. (1984). Cloning and expression in *Escherichia coli* of the insecticidal δ-endotoxin gene of *Bacillus thuringiensis* var. *israelensis*. *FEBS Lett.* **175,** 377–382.

Weiser, J. (1984). A mosquito-virulent *Bacillus sphaericus* in adult *Simulium damnosum* from northern Nigeria. *Zentralbl. Mikrobiol.* **139,** 57–60.

Wrickemesinghe, R. S. B., and Mendis, C. L. (1980). *Bacillus sphaericus* spore from Sri Lanka demonstrating rapid larvicidal activity on *Culex quinquefasciatus*. *Mosq. News* **40,** 387–389.

Wu, D., and Chang, F. N. (1985). Synergism in mosquitocidal activity of 26 and 65 kDa proteins from *Bacillus thuringiensis* subsp. *israelensis* crystal. *FEBS Lett.* **190,** 232–236.

Yousten, A. A. (1984). *Bacillus sphaericus:* Microbiological factors related to its potential as a mosquito larvicide. *Adv. Biotechnol. Processes* **3,** 315–343.

6

Expression of δ-Endotoxin Gene of Bacillus thuringiensis

KOHZO KANDA

I. INTRODUCTION

Bacillus thuringiensis is well known for the production of δ-endotoxin (crystal) highly toxic to lepidopterous and dipterous insects and has become one of the agents with the greatest potential for insect pest control as an alternative to chemical control methods. Investigation of the crystal production in *B. thuringiensis* contributes not only to the development of the utilization of this bacterium as a microbial control agent but also to the fundamental knowledge of spore-forming bacteria. Crystal production in *B. thuringiensis* is not stable. The ability to produce crystal can be eliminated easily by treatments such as continuous subculture, acridine orange, and high-temperature cultivation; however, the lost ability cannot be regained. This evidence suggests that the crystal is coded by a plasmid. It has been shown by gene cloning experiments that the crystal gene of *B. thuringiensis* is located on a plasmid in some subspecies of this bacterium (Schnepf and Whiteley, 1981; Held *et al.*, 1982; Klier *et al.*, 1982; Ward *et al.*, 1984; Shibano *et al.*,

BIOTECHNOLOGY IN INVERTEBRATE
PATHOLOGY AND CELL CULTURE

1985; Adang *et al.*, 1985), and it was suggested by experiments using cloned crystal gene that transcriptional regulation occurred in the crystal production (Wong *et al.*, 1983).

This chapter attempts to reveal the possible crystal gene in *B. thuringiensis* subsp. *alesti* by transformation of *Bacillus subtilis* with plasmids of either a crystalliferous or an acrystalliferous strain of *B. thuringiensis* subsp. *alesti*. The expression of the gene was investigated with *in vivo* and *in vitro* synthesis of crystal protein.

II. EXPERIMENTAL RESULTS

A. Plasmid Analysis of *Bacillus thuringiensis* subsp. *alesti*

Plasmid analysis of crystalliferous and acrystalliferous strains of *B. thuringiensis* was attempted to investigate the possible location of crystal gene on plasmid. A crystalliferous (cry^+) strain of *B. thuringiensis* subsp. *alesti* provided by Pasteur Institute was used in this experiment. An acrystalliferous strain (cry^-) was spontaneously obtained from the crystalliferous strain by subculturing. Plasmids were extracted from each strain (cry^+ and cry^-) by using lysozyme and Sarkosyl NL97 treatment. Cells for plasmid extraction were cultivated for 4 hr at 27°C with shaking, after the inoculation of 2 ml of cells cultured for 16 hr into 200 ml of LB broth. Cultured cells were harvested by centrifugation and resuspended in 3 ml of TES buffer (0.02 *M* Tris-HCl, pH 7.9, 0.15 *M* EDTA, and 15% sucrose) containing 4 mg/ml of lysozyme (Wako Pure Chemical Industries, Ltd.). The cell suspension was incubated for 1 hr at 37°C, and cleared lysate was prepared by adding 1 ml of Proteinase K (2.5 mg/ml) and 1 ml of Sarkosyl NL97 (4%) to the reaction mixture with further incubation for 1 hr at 37°C. Plasmid was then purified from cleared lysate treated with water-saturated phenol by ethidium bromide–CsCl ultracentrifugation. Plasmid patterns of the strains (cry^+ and cry^-) were compared by using 1% agarose gel electrophoresis with TAE buffer (0.04 *M* Tris-acetate, pH 8.0, 2 m*M* EDTA).

The comparison of plasmid patterns showed that the smallest molecule of plasmids in the crystalliferous strain was not present in the acrystalliferous strain (Fig. 1).

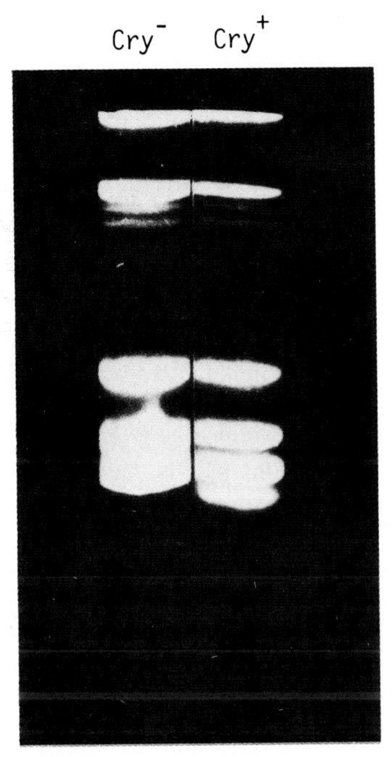

Fig. 1. Plasmids pattern of *B. thuringiensis* subsp. *alesti*. Cry$^+$, plasmids from the type strain of subsp. *alesti*; Cry$^-$, plasmids from an acrystalliferous strain isolated from the type strain of subsp. *alesti*.

B. Transformation of *Bacillus subtilis* by *Bacillus thuringiensis* DNA

The possibility of a crystal gene located on a plasmid was then examined biologically by transformation. In this experiment, *B. subtilis* Y12S, a derivative from Marburg 168 strain, was used as the recipient strain. Streptomycin-resistant strain ApSm2 was prepared from *B. thuringiensis* subsp. *alesti* by treatment with *N*-methyl-*N*′-nitro-*N*-nitrosoguanidine and used as the donor strain of DNA for transformation. Total DNA from ApSm2 strain was prepared by the method of Saito and Miura (1963). Transformation of *B. subtilis* Y12S strain was done according to Anagnostpouls and Spizizen (1961), and transformants were selected by the genetic marker for streptomycin resistance coded by donor DNA.

Transformation of *B. subtilis* Y12S strain was measured by total DNA of ApSm2 strain with low frequency (Table I). Transformants selected by streptomycin resistance were then observed under a phase-contrast

TABLE I

Transformation of *B. subtilis* Y12S Strain as Determined by Total DNA of *B. thuringiensis* subsp. *alesti*

DNA[a]	Viable cells/ml	Sm^r transformants/ml	Number of crystal-forming colonies/Sm^r colonies
ApSm2 DNA	1.3×10^9	1.3×10^2	2/126
W23 DNA	1.4×10^8	1.7×10^3	0/168
Control	0	0	—

[a]ApSm2 DNA is DNA from a streptomycin-resistant mutant of the crystalliferous strain of *B. thuringiensis* subsp. *alesti*. W23 DNA is from *B. subtilis* W23 strain (Sm^r), and 1 × SSC is used as the control of transformation.

microscope, and two transformants which produced crystals were obtained (Fig. 2). These transformants did not share the flagellar antigen of any known H serotypes of *B. thuringiensis*, and the auxotrophy of cells was identical to that of *B. subtilis* Y12S strain.

The results mentioned above suggest two possibilities. One is that the crystal gene is located on chromosomal DNA which is very near the

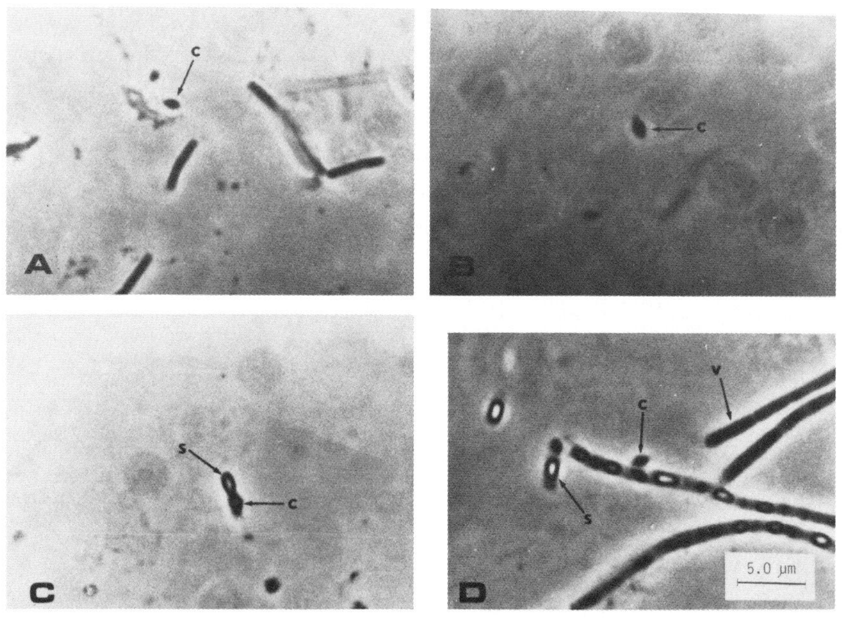

Fig. 2. Crystal formation in transformants of *B. subtilis* Y12S strain by total DNA of *B. thuringiensis* subsp. *alesti*. (A–C) Transformants; (D) type strain of *B. thuringiensis* subsp. *alesti*. Crystals are indicated by "c."

TABLE II
Cotransformation of *B. subtilis* Y12S Strain

DNA	Viable cells/ml	Sm^r transformants/ml	Number of crystal-forming colonies/Sm^r colonies
W23 DNA			
Cry^+ plasmids	1.5×10^8	3.5×10^4	2/154
Cry^- plasmids	2.0×10^8	3.2×10^4	0/108

streptomycin-resistant gene, and the other is that the crystal gene and streptomycin-resistant gene are located on plasmid and on chromosomal DNA independently. To test those possibilities, cotransformation of *B. subtilis* Y12S strain was attempted. Chromosomal DNA of *B. subtilis* W23 strain whose genetic marker was streptomycin resistant and plasmids extracted from cry^+ and cry^- strains of *B. thuringiensis* subsp. *alesti* were used as donor DNA for the cotransformation. Although DNA from *B. subtilis* W23 strain was used as the selective genetic marker, two transformants were found to produce crystals by using plasmids from *B. thuringiensis* cry^+ strain (Table II). The results strongly suggest that the crystal production is coded by plasmid.

C. Protoplast Transformation of *Bacillus subtilis* by Plasmid of *Bacillus thuringiensis*.

Protoplast transformation was reported to be an effective method for the analysis of plasmid which did not code any selective genetic marker such as drug resistance, because of its high frequency (Chang and Cohen, 1979). *Bacillus subtilis* Y12S strain was transformed by plasmids from cry^+ and cry^-, respectively, according to the method of Chang and Cohen. (1979).

In the protoplast transformation, the frequency of transformation of *B. subtilis* Y12S by pUB110 was more than 30%. All transformants which produced crystals were obtained when plasmids extracted from the cry^+ strain of *B. thuringiensis* subsp. *alesti* were used as donor DNA (Table III).

D. Expression of Crystal Gene

The expression of crystal gene in *B. thuringiensis* was examined serologically by *in vivo* and *in vitro* synthesis of crystal protein. Crystals were purified by 50–80% Urografin 76 density gradient ultra-

TABLE III
Protoplast Transformation of *B. subtilis* Y12S Strain by Plasmids from *B. thuringiensis* subsp. *alesti*

Plasmid	Viable cells/ml	Transformants/ml	Number of crystal-forming colonies/viable cells
Cry^+ plasmids	9.3×10^6	1.0×10^6[a]	7/63
Cry^- plasmids	1.2×10^7	0[a]	0/101
pUB 110	4.6×10^6	2.7×10^6[b]	0/105

[a]Crystal-forming cells.
[b]Neomycin-resistant (5 μg/ml) cells.

centrifugation and were dissolved with 0.05 *N* NaOH at 37°C for 1 hr. Crystal antigen was prepared from the crystal solution dialyzed against 0.01 *M* carbonate buffer, pH 9.6, at 4°C for 16 hr with several changes of buffer. Rabbits were immunized with alkali-solubilized crystals and anticrystal rabbit IgG was prepared.

In vivo synthesis of crystal protein in the type strain of *B. thuringiensis* subsp. *alesti* was investigated by the detection of crystal antigen production in growing cells. The direct method of ELISA was used in this experiment with alkaline phosphatase–conjugated anticrystal rabbit IgG according to the method of Voller *et al.* (1976). The concentration of crystal antigen detected by this method was 30 ng/ml. Cells cultivated at 27°C in nutrient broth containing 0.1% NaCl were harvested at 1-hr intervals and dissolved with 0.05NaOH at 37°C for 1 hr. Cell debris was removed by filtration on Millex-HA (Millipore Corporation). Filtrate was dialyzed against 0.01 *M* carbonate buffer, pH 9.6 at 4°C for 16 hr and used as the crystal antigen for ELISA. The crystal antigen was synthesized not only during the sporulating phase but also during the logarithmic phase of bacterial cells (Fig. 3). Furthermore, it was demonstrated that the crystal antigen was synthesized at the T_{-2} stage of the logarithmic phase (Fig. 4). These results indicate that the crystal protein was mainly synthesized during sporulation, but a small amount of crystal protein was also synthesized during the logarithmic phase.

In vivo synthesis of crystal protein produced the same results as *in vitro* synthesis of crystal protein in *B. thuringiensis* subsp. *kurstaki* HD-1 strain. *In vitro* translation of mRNA of crystal protein was carried out according to a previous report (Andrews *et al.*, 1982). RNA extracted from cultivated cells at 27°C at 1-hr intervals was translated *in vitro* with ^{14}C-labeled amino acids (Amersham International plc). Crystal protein

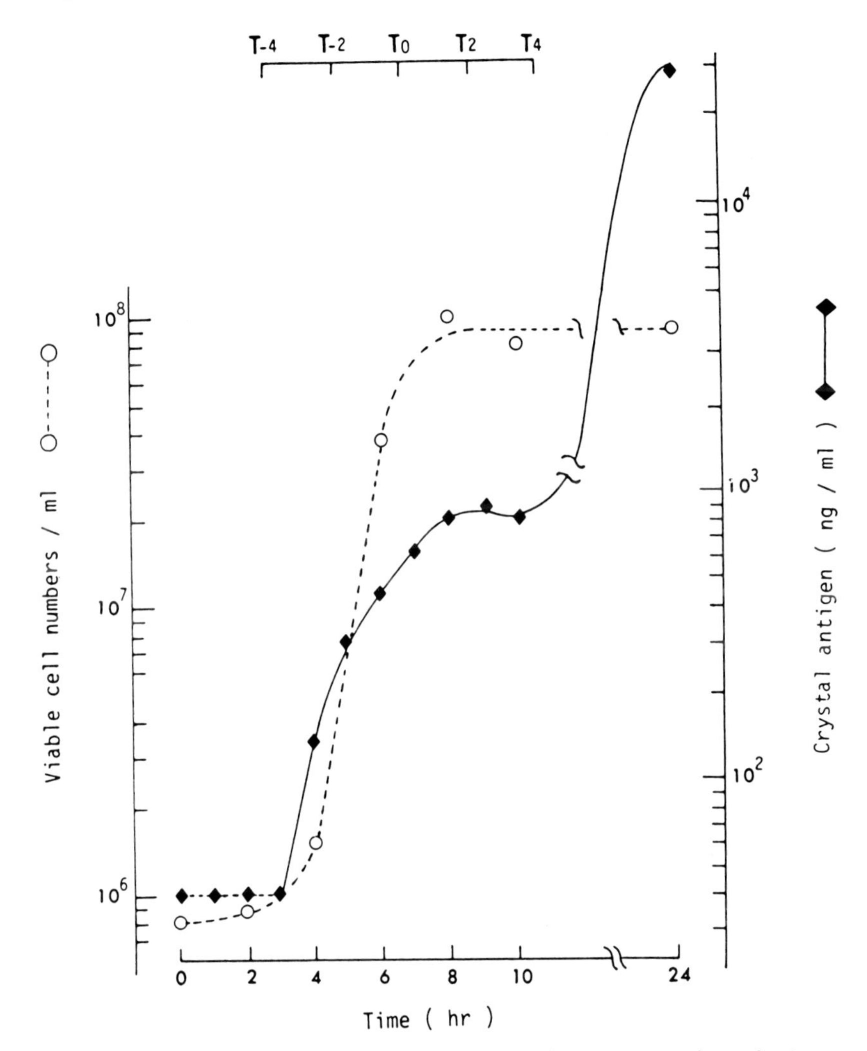

Fig. 3. *In vivo* synthesis of crystal protein in *B. thuringiensis* subsp. *alesti*.

translated from each stage of mRNA was precipitated by anticrystal rabbit IgG, and the radioactivity of each precipitate was counted.

In these experiments, the crystal-specific mRNA was transcribed mainly during the sporulating phase (T_3) but also during the logarithmic phase (T_{-1}) of cells (Fig. 5). However, the transcriptional amount of crystal-specific mRNA at the T_{-1} stage was about 17% of that at the T_3 stage.

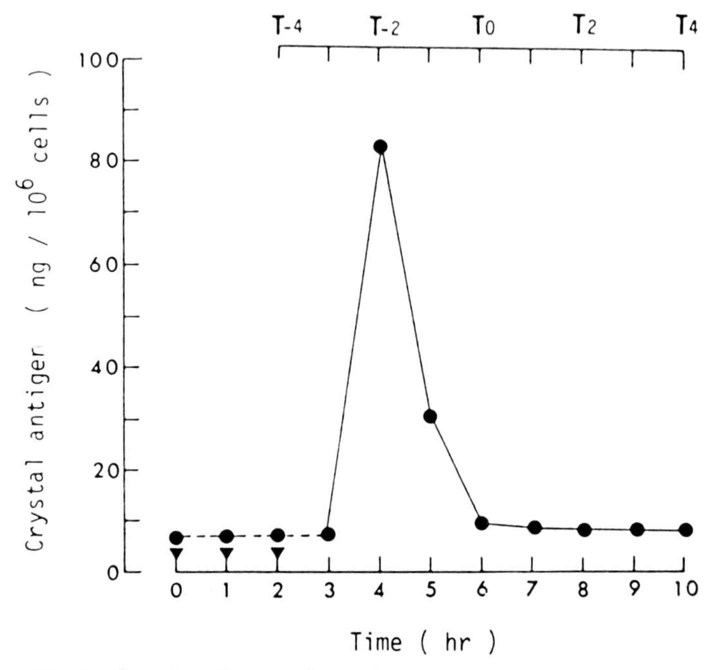

Fig. 4. Crystal antigen synthesized in *B. thuringiensis* subsp. *alesti*.

III. CONCLUSION

Plasmid analysis and transformation of *B. subtilis* suggest that the crystal gene (or genes) of the type strain of *B. thuringiensis* subsp. *alesti* is located on plasmid. The crystal gene is expressed not only during the sporulating phase but also during the logarithmic phase of cells. *In vivo* synthesis of crystal protein in *B. thringiensis* subsp. *alesti* investigated by ELISA detected the production of crystal antigen during stages T_{-4} to T_6 in cultivated cells. About 0.8% of the total amount of crystal antigen produced during the sporulating phase is already synthesized during the logarithmic phase (T_{-2}). Moreover, *in vitro* translation of mRNA in *B. thuringiensis* subsp. *kurstaki* HD-1 strain shows that most of the mRNA which coded crystal antigen was transcribed at T_3 during the sporulating phase, but a small amount of crystal-specific mRNA was also transcribed at T_{-1} during the logarithmic phase.

In *B. subtilis* the sigma factor of RNA polymerase, which has promoter specificity during the logarithmic phase, changes molecular structure during the sporulating phase. During the logarithmic phase and the sporulating phase each sigma factor is able to recognize different types of gene promoters. Modification of RNA polymerase occurs in *B. thuringiensis* strain Berliner 1715 (Lecadet *et al.*, 1974), and the cloned

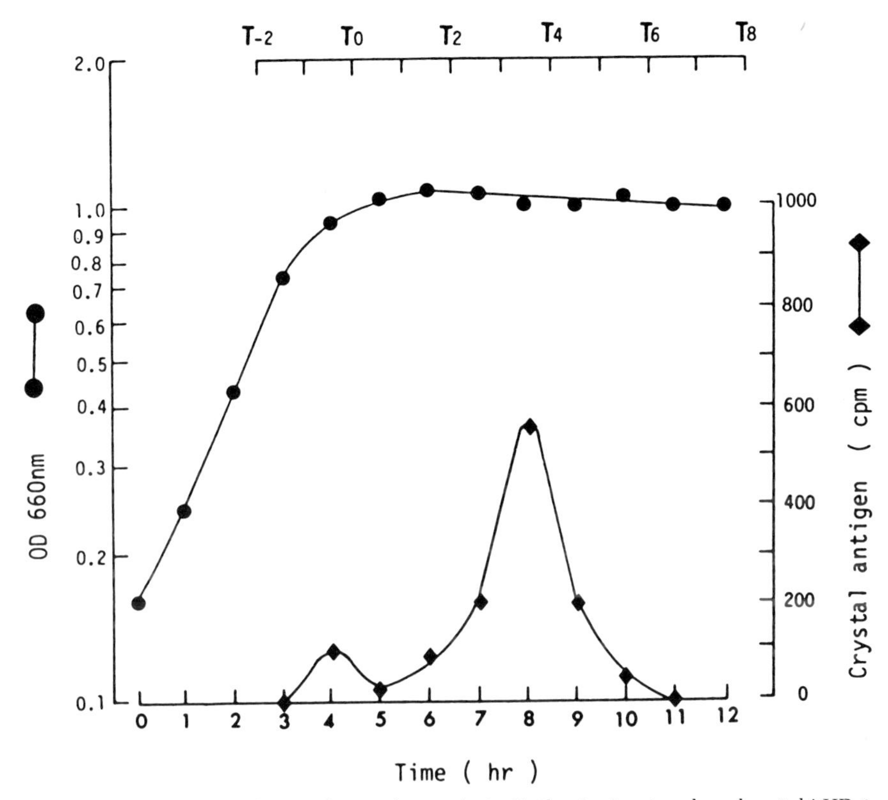

Fig. 5. *In vitro* synthesis of crystal protein in *B. thuringiensis* subsp. *kurstaki* HD-1 strain.

gene of the crystal protein in *B. thuringiensis* subsp *kurstaki* HD-1 strain has two types of promoter (Wong *et al.*, 1983).

Based on the evidence presented above, it is suggested that the crystal gene has two types of promoters that can be recognized by different types of RNA polymerase during the logarithmic phase and the sporulating phase; i.e., each crystal gene has different types of promoters in *B. thuringiensis*. It was also shown that crystal gene expression during the logarithmic phase is regulated at the transcriptional level; however, the regulation system for the expression of crystal gene during this period is still obscure.

REFERENCES

Adang, M. J., Staver, M. J., Rocheleau, T. A., Leighton, J., Barker, R. F., and Thompson, D. V. (1985). Characterized full-length and truncated plasmid clones of the crystal protein of *Bacillus thuringiensis* subsp. *kurstaki* HD-73 and their toxicity to *Manduca sexta*. *Gene* **36,** 289–300.

Anagnostpouls, C., and Spizizen, J. (1961). Requirements for transformation in *Bacillus subtilis. J. Bacteriol.* **81,** 741–746.

Andrews, R. E., Kanda, K., and Bulla, L. A., Jr. (1982). *In vitro* translation of entomocidal toxin of *B. thuringiensis. In* "Molecular Cloning and Gene Regulation of Bacilli" (A. T. Ganesan, S. Chang, and J. A. Hoch, eds.), pp. 121–130. Academic Press, New York.

Chang, S., and Cohen, S. N. (1979). High frequency transformation of *Bacillus subtilis* protoplast by plasmid DNA. *Mol. Gen. Genet.* **168,** 111–115.

Held, G. A., Bulla, L. A., Jr., Ferrari, E., Hoch, J., Aronson, A. I., and Minich, S. A. (1982). Cloning and localization of lepidopteran protoxin gene of *Bacillus thuringiensis* subsp. *kurstaki. Proc. Natl. Acad. Sci. U.S.A.* **79,** 6065–6069.

Klier, A., Fargette, F., Ribier, J., and Rapoport, G. (1982). Cloning and expression of crystal protein genes from *Bacillus thuringiensis* strain berliner 1715. *EMBO J.* **1,** 791–799.

Lecadet, M.-M., Klier, A. F., and Ribier, J. (1974). Isolation and characterization of two asporogenous rifampicin resistant mutants of *B. thuringiensis. Biochimie* **56,** 1471–1479.

Saito, H., and Miura, K. (1963). Preparation of transforming deoxyribonucleic acid by phenol treatment. *Biochim. Biophys. Acta* **72,** 619–639.

Schnepf, H. E., and Whitely, H. R. (1981). Cloning and expression of *Bacillus thuringiensis* crystal protein gene in *Escherichia coli. Proc. Natl. Acad. Sci. U.S.A.* **78,** 2893–2897.

Shibano, Y., Yamagata, A., Nakamura, N., Iizuka, T., Sugisaki, H., and Takanami, M. (1985). Nucleotide sequence coding for the insecticidal fragment of *Bacillus thuringiensis* protein. *Gene* **34,** 243–251.

Voller, A., Bidwell, D. E., and Bartlett, A. (1976). Enzyme immunoassay in diagnostic medicine. *Bull. W. H. O.* **53,** 55–65.

Ward, E. S., Ellar, D. J., and Todd, J. A. (1984). Cloning and expression in *Escherichia coli* of the insecticidal δ-endotoxin gene of *Bacillus thuringiensis* var. *israelensis. FEBS Lett.* **175,** 377–382.

Wong, H. C., Schnepf, H. E., and Whiteley, H. R. (1983). Transcriptional and translational start sites for the *Bacillus thuringiensis* crystal protein gene. *J. Biol. Chem.* **258,** 1960–1967.

7

Structural and Antigenic Relationships among Three Insecticidal Crystal Proteins of Bacillus thuringiensis subsp. *kurstaki*

MICHAEL J. ADANG, KEN F. IDLER,
AND THOMAS A. ROCHELEAU

I. INTRODUCTION

Bacillus thuringiensis produces a characteristic crystal during sporulation. Depending on the isolate, these crystals are toxic to Lepidoptera, Diptera, or Coleoptera. Because of this spectrum of activity,

BIOTECHNOLOGY IN INVERTEBRATE
PATHOLOGY AND CELL CULTURE

several *B. thuringiensis* strains are produced commercially for the control of pest insects.

The diversity of *B. thuringiensis* strains is due to the protein subunits of their crystals. Lepidopteran-active crystals contain subunit peptides of M_r 130,000–140,000, designated the protoxin or insecticidal crystal protein (ICP). Some crystals contain several distinct peptides (Calabrese *et al.*, 1980), which can have different toxic ranges (Jarrett, 1985). Examples of strains that harbor multiple plasmids encoding several ICP genes are subspecies *kurstaki* HD-1 and HD-1-Dipel (Kronstad *et al.*, 1983). DNA hybridization experiments by Kronstad *et al.* (1983) showed that strain HD-1 harbors three ICP genes characterized by hybridizing 6.6, 5.3, and 4.5 kb *Hin*dIII fragments. This provides a useful terminology because the respective coding regions can be designated as 6.6, 5.3, and 4.5 types. The 4.5-type gene was cloned and the nucleotide sequence determined (Schnepf and Whiteley, 1981; Schnepf *et al.*, 1985). This chapter describes the cloning of the 6.6 and 5.3 ICP genes from strain HD-1. The 6.6 gene is very similar to the 6.6 subsp. *kurstaki* HD-73 gene (Adang *et al.*, 1985; McLinden *et al.*, 1985) and the 5.3 is a unique gene type. The nucleotide sequence of the 5.3 gene is presented and the open reading frame (ORF) compared with the two previously published sequences. These open reading frames can be divided into four regions based on their comparative homologies. Using a monoclonal antibody, the position of a unique antigenic site is located in the 6.6-type gene.

II. MATERIALS AND METHODS

A. Bacterial Strains and Plasmids

Escherichia coli strains MC1061 (Casadaban and Cohen, 1980) and HB101 (Boyer and Roulland-Dussoix, 1969) were used for transformations. *Escherichia coli* pES1 (Schnepf and Whiteley, 1981) was obtained from the American Type Culture Collection (ATCC No. 31995); *E. coli* pBT73-16 from our laboratory was described previously (Adang *et al.*, 1985). *Bacillus thuringiensis* subsp. *kurstaki* HD-73 was obtained from the Bacillus Genetics Stock Collection (Dean; Ohio State University) and subsp. *kurstaki* HD-1 was isolated from a commercial formulation (Dipel; Abbott Laboratories) in 1983. *Escherichia coli* plasmid DNA was prepared by an alkaline lysis method (Birnboim and Doly, 1979); *B. thuringiensis* plasmid DNAs, 30 MDa and larger, were prepared by the procedure of Kronstad *et al.* (1983).

B. Growth Media

L-broth (Miller, 1972) was the standard *E. coli* growth medium. *Bacillus thuringiensis* for plasmid isolation was grown in SPY (Spizizen, 1958) with 0.1% yeast extract and 0.1% glucose. Cultures for crystal isolation were grown in modified G medium (Aronson *et al.*, 1971).

C. Antibody Preparation and ICP Gene Cloning

Crystals were purified from sporulated *B. thuringiensis* cultures by three passes in Hypaque 76 (Winthrop) gradients (Meenakshi and Jayaraman, 1979), washed with 1 *M* NaCl and then with deionized water, and lyophilized. Antigen was prepared by dissolving crystals in 2% 2-mercaptoethanol, pH 10. Rabbit antiserum was made against solubilized HD-73 crystals. MAb-1 prepared against solubilized HD-73 crystals was described previously (Adang *et al.*, 1985).

A *Sau*3A partial digest of HD-1 plasmid DNA was ligated into pUC18 vector and transformed into *E. coli* MC1061. Colonies were hybridized to a nick-translated 3.7 kb DNA fragment containing the complete HD-73 gene from plasmid pBT73-16. Hybridizing transformants were further screened for ICP antigen. Bacterial colonies on nitrocellulose filters were lysed in $CHCl_3$ vapor followed by DNase and lysozyme treatment (Helfman *et al.*, 1983). Filters were treated with rabbit anti-ICP serum or MAb-1, and ICP antigen was detected with an alkaline phosphatase ELISA system (Blake *et al.*, 1984). Immunodetection of peptides on protein blots was performed according to Towbin *et al.* (1979) with slight modifications (Adang *et al.*, 1985).

D. DNA Sequencing Reactions

All the sequencing reactions were done according to the method of Maxam and Gilbert (1980) with modifications described by Barker *et al.* (1983). Further details are described in Adang *et al.* (1985). Computer analysis of the sequence data was performed with programs made available by O. Smithies, J. Deveraux, and F. Blattner (University of Wisconsin, Madison) and by H. Martinez, B. Katzung, and T. Farrah (University of California, San Francisco).

E. Insect Bioassays

Bioassays were done as described by Schesser *et al.* (1977) against larvae of tobacco hornworm, *Manduca sexta* (Carolina Biological Supply). Tests used either sonicated *E. coli* extracts in 10 m*M* NaCl, 10 m*M* Tris-HCl, pH 8.0, and 1 m*M* EDTA or purified *B. thuringiensis* crystals.

III. RESULTS

A. Molecular Cloning

Previously, Kronstad *et al.* (1983) demonstrated by DNA hybridization that *B. thuringiensis* HD-1 plasmids contain three regions of homology to an ICP gene. This suggested the presence of multiple crystal protein genes that can be distinguished by the distance between the first internal *Hin*dIII site and the 5′ flanking *Hin*dIII site. These distances were 6.6, 5.3, and 4.5 kb. To clone these genes, plasmid DNA from strain HD-1 was partially digested with *Sau*3A, ligated into *Bam*HI-digested pUC18, and transformed in *E. coli* MC1061. Individual colonies were selected that hybridized with the previously cloned HD-73 ORF from pBT73-16. The 676 hybridizing colonies were screened for crystal protein expression by a colony immunoblot method using MAb-1 and rabbit anti-HD-73 ICP serum. MAb-1 recognized antigen produced by three colonies on the nitrocellulose filters. Preliminary restriction enzyme mapping of insert DNA in these recombinant clones showed that each harbored a 6.6-type gene. One of these recombinants, *E. coli* pBT1-89A (Fig. 1A) was mapped more thoroughly. It has the same *Eco*RI, *Xba*I, *Eco*RV, *Bgl*II, *Sac*I, *Pst*I, *Ava*I, *Xho*I, *Kpn*I, *Hin*dIII, and *Pvu*II sites in its ORF as pBT73-16. This series of restriction digests indicates that the 6.6 gene of HD-1 is very similar and possibly identical to the 6.6 HD-73 gene.

The subject of this study is the gene harbored in *E. coli* pBT1-106A (Fig. 1B). Restriction enzyme analysis indicated that it differed from the 6.6 gene in pBT1-89A (Fig. 1A) and from the 4.5 type in pES1 (Schnepf *et al.*, 1985). Based on these differences, the pBT1-106A ICP gene can be designated a 5.3-type gene. To compare the expressed HD-1 gene products with pES1 and pBT73-16, they were subcloned into the same vector and host. The pBT1-89A gene was subcloned on a 5.1 kb *Sph*I fragment, and the pBT1-106A gene on a 5.4 kb *Pst*I fragment into the matching sites in pBR322 and transformed into *E. coli* HB101. The resulting strains are designated *E. coli* pBT1-89B and pBT1-106B. Both the original and derived strains were lethal in bioassays against tobacco hornworm larvae.

B. Antigenicity of *Escherichia coli*–Expressed ICP

Immunoblots were prepared from *E. coli* extracts containing ICP for each of the three gene types. Parallel blots were reacted with either

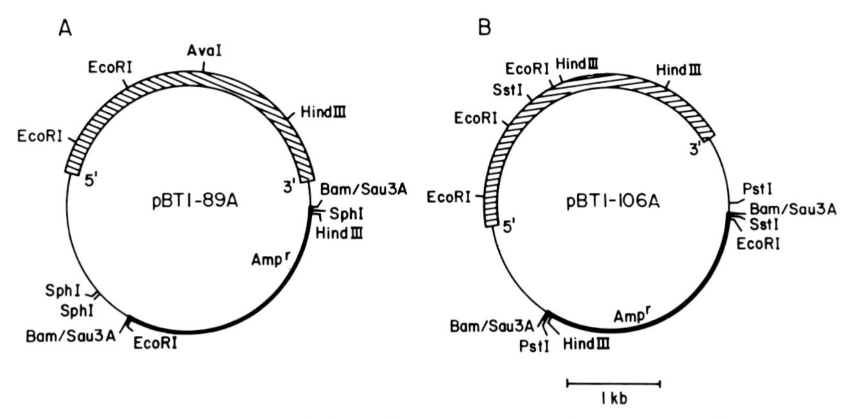

Fig. 1. Restriction maps of plasmids containing a *B. thuringiensis* subsp. *kurstaki* strain HD-1 insecticidal crystal protein gene: (A) pBT1-89A and (B) pBT1-106A. Shaded areas indicate crystal protein coding DNA and thin lines *B. thuringiensis* flanking DNA. Thick lines indicate pUC18 vector.

polyclonal sera (Fig. 2A) or a monoclonal antibody (Fig. 2B) to solubilized HD-73 crystals. The polyclonal serum shows that each strain harboring an ICP gene produces an immunoreactive peptide of M_r 130,000–135,000 and several smaller molecular weight peptides down to approximately M_r 68,000. The 6.6 gene products of pBT1-89B and pBT73-16 react more strongly with the HD-73 antiserum than extracts of pES1 (4.5 type) or pBT1-106 (5.3 type).

The immunoblot treated with MAb-1 shows binding only to the 6.6-type ICP genes (Fig. 2B). Proteins expressed in *E. coli* strains pBT1-106B (5.3 type) and *E. coli* pES1 (4.5 type) did not react with this monoclonal antibody.

C. Nucleotide Sequence of a Unique HD-1 Crystal Protein Gene

Figure 3 shows the restriction enzyme map and the strategy used to sequence the 5.3-type ICP gene in pBT1-106A. Figure 4 shows the complete nucleotide sequence with the ORF beginning with an ATG at nucleotide position 1 and terminating with TAA at position 3468 encoding a peptide of 1155 amino acids of M_r 130,641.

D. Codon Usage of the Gene

The codon usage preference of the pBT1-106A gene is very similar to that of the other two ICP genes. Because of the relatively high A + T content (61.15%) there is a preference for A or T at the third codon

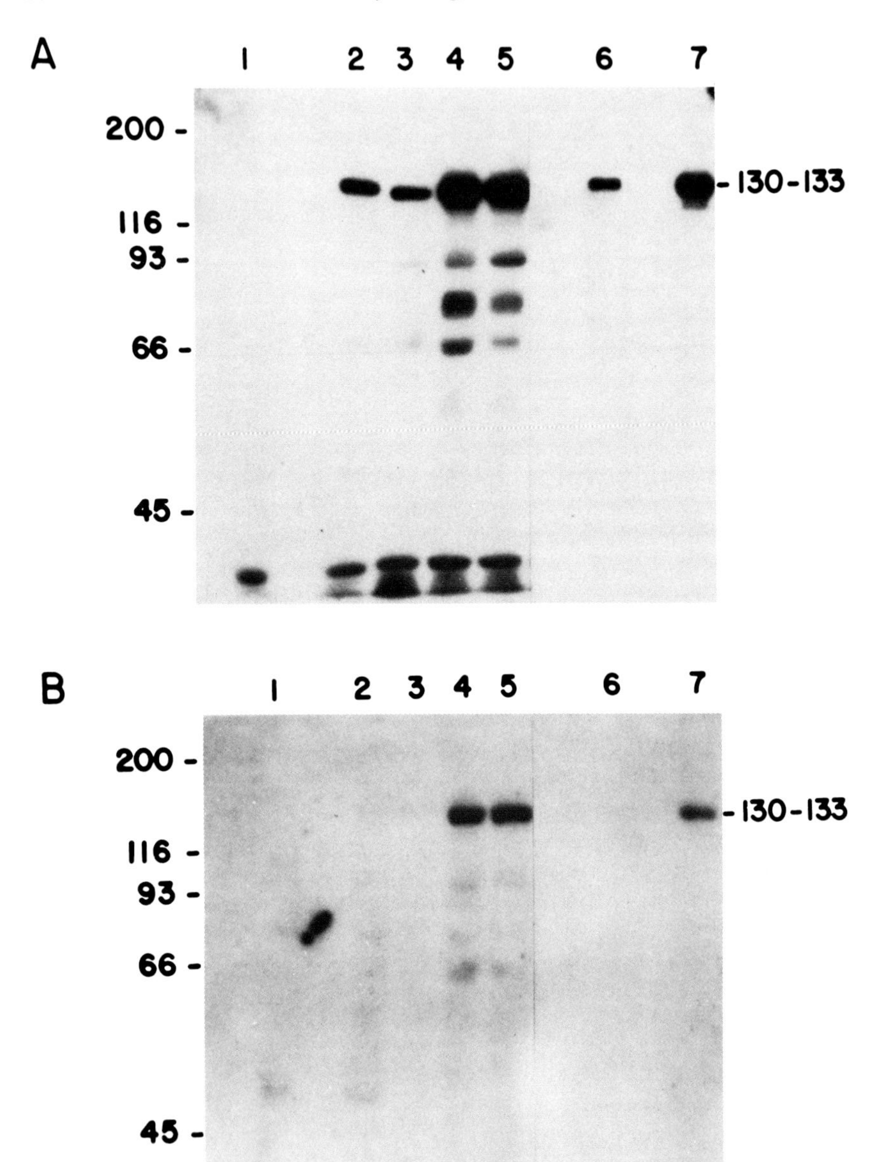

Fig. 2. Immunoblots of *E. coli*–expressed ICP reacted with (A) polyclonal antiserum or (B) MAb-1 to HD-73 protoxin after SDS-10% PAGE. Lanes 1–5 each contain 100 μg of the following *E. coli* extracts: lane 1, *E. coli* pBR322; lane 2, *E. coli* pES1; lane 3, *E. coli* pBT1-106B; lane 4, *E. coli* pBT1-89B; lane 5, *E. coli* pBT73-16. Lane 6 contains 0.1 μg of HD-73 crystal protein and lane 7, 1 μg of HD-73 crystal protein. Protein standards with their sizes (kDa) were myosin (200), δ-galactoside (116), phosphorylase (93), bovine serum albumin (66), and ovalbumin (45). The apparent protoxin sizes are indicated on the right. These autoradiograms were exposed to film for 6 hr.

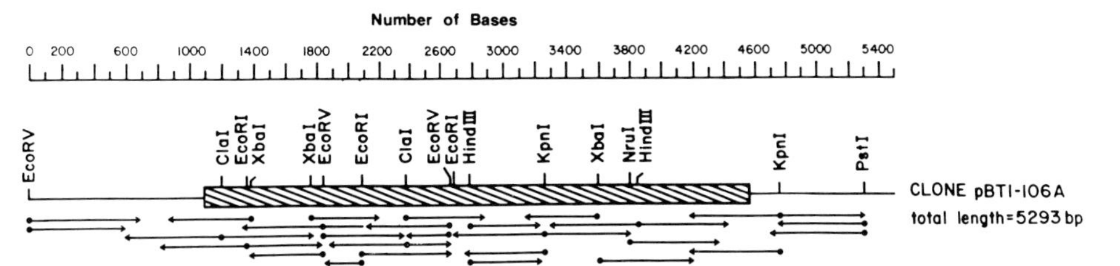

Fig. 3. Restriction endonuclease map and the sequencing strategy employed to sequence the *B. thuringiensis* subsp. *kurstaki* HD-1 insecticidal crystal protein gene in plasmid pBT1-106A. The dots indicate the position of the 5′-end labeling and the arrows indicate the direction and extent of sequencing. The shaded area indicates ICP coding DNA and the lines *B. thuringiensis* flanking DNA.

position. The ratios for A : G and T : C are 2.54 : 1 and 2.66 : 1, respectively. This bias was also reported recently for the 28 kd gene from *B. thuringiensis* var. *israelensis* (Waaljik *et al.*, 1985).

E. Comparison of Three ICP Genes

Figure 5 shows the deduced amino acid sequences of the pBT73-16 (6.6 type), pBT1-106A (5.3 type), and pES1 (4.5 type) ICP genes aligned for optimal homology. Individual coding sequence comparisons are shown as graphs in Fig. 6, where mismatches are indicated by vertical bars. Based on their comparative homologies, these genes can be divided into four regions. Region 1, extending from the N-terminal methionine to amino acid 282, is 98% conserved for the three coding regiions. In region 2 (amino acids 283–466) the 6.6 and 5.3 genes are conserved with 98% homology but are only 63% homologous with the 4.5 gene. This pattern changes in region 3 (amino acids 467–788), where the 5.3 and 4.5 genes are identical but only 63% homologous with the 6.6 gene. In the carboxyl terminal 367 residues of region 4, the 6.6 and 4.5 genes differ by two amino acids and the 5.3 gene is 88% homologous. The predominant change is a 26-amino-acid deletion in the 5.3 ORF. Overall, the protoxin structures are conserved in the N- and C-terminal thirds of the molecules while the central portion, designated regions 2 and 3, varies for each of the genes studied.

IV. DISCUSSION

The toxicity of each *B. thuringiensis* strain to Lepidoptera is determined by its crystal composition, which for some strains is complex.

```
   1 ATGGATAACAATCCGAACATCAATGAATGCATTCCTTATAATTGTTTAAGTAACCCTGAAGTAGAAGTATTAGGTGGAGAAAGAATAGAA 90
     MetAspAsnAsnProAsnIleAsnGluCysIleProTyrAsnCysLeuSerAsnProGluValGluValLeuGlyGlyGluArgIleGlu

  91 ACTGGTTACACCCCAATCGATATTTCCTTGTCGCTAACGCAATTTCTTTTGAGTGAATTTGTTCCCGGTGCTGGATTTGTGTTAGGACTA 180
     ThrGlyTyrThrProIleAspIleSerLeuSerLeuThrGlnPheLeuLeuSerGluPheValProGlyAlaGlyPheValLeuGlyLeu

 181 GTTGATATAATATGGGGAATTTTTGGTCCCTCTCAATGGGACGCATTTCTTGTACAAATTGAACAGTTAATTAACCAAAGAATAGAAGAA 270
     ValAspIleIleTrpGlyIlePheGlyProSerGlnTrpAspAlaPheLeuValGlnIleGluGlnLeuIleAsnGlnArgIleGluGlu

 271 TTCGCTAGGAACCAAGCCATTTCTAGATTAGAAGGACTAAGCAATCTTTATCAAATTTACGCAGAATCTTTTAGAGAGTGGGAAGCAGAT 360
     PheAlaArgAsnGlnAlaIleSerArgLeuGluGlyLeuSerAsnLeuTyrGlnIleTyrAlaGluSerPheArgGluTrpGluAlaAsp

 361 CCTACTAATCCAGCATTAAGAGAAGAGATGCGTATTCAATTCAATGACATGAACAGTGCCCTTACAACCGCTATTCCTCTTTTTGCAGTT 450
     ProThrAsnProAlaLeuArgGluGluMetArgIleGlnPheAsnAspMetAsnSerAlaLeuThrThrAlaIleProLeuPheAlaVal

 451 CAAAATTATCAAGTTCCTCTTTTATCAGTATATGTTCAAGCTGCAAATTTACATTTATCAGTTTTGAGAGATGTTTCAGTGTTTGGACAA 540
     GlnAsnTyrGlnValProLeuLeuSerValTyrValGlnAlaAlaAsnLeuHisLeuSerValLeuArgAspValSerValPheGlyGln

 541 AGGTGGGGATTTGATGCCGCGACTATCAATAGTCGTTATAATGATTTAACTAGGCTTATTGGCAACTATACAGATCATGCTGTACGCTGG 630
     ArgTrpGlyPheAspAlaAlaThrIleAsnSerArgTyrAsnAspLeuThrArgLeuIleGlyAsnTyrThrAspHisAlaValArgTrp

 631 TACAATACGGGATTAGAGCGTGTATGGGGACCGGATTCTAGAGATTGGATAAGATATAATCAATTTAGAAGAGAATTAACACTAACTGTA 720
     TyrAsnThrGlyLeuGluArgValTrpGlyProAspSerArgAspTrpIleArgTyrAsnGlnPheArgArgGluLeuThrLeuThrVal

 721 TTAGATATCGTTTCTCTATTTCCGAACTATGATAGTAGAACGTATCCAATTCGAACAGTTTCCCAATTAACAAGAGAAATTTATACAAAC 810
     LeuAspIleValSerLeuPheProAsnTyrAspSerArgThrTyrProIleArgThrValSerGlnLeuThrArgGluIleTyrThrAsn

 811 CCAGTATTAGAAAATTTTGATGGTAGTTTTCGAGGCTCGGCTCAGGGCATAGAAGGAAGTATTAGGAGTCCACATTTGATGGATATACTT 900
     ProValLeuGluAsnPheAspGlySerPheArgGlySerAlaGlnGlyIleGluGlySerIleArgSerProHisLeuMetAspIleLeu

 901 AACAGTATAACCATCTATACGGATGCTCATAGAGGAGAATATTATTGGTCAGGGCATCAAATAATGGCTTCTCCTGTAGGGTTTTCGGGG 990
     AsnSerIleThrIleTyrThrAspAlaHisArgGlyGluTyrTyrTrpSerGlyHisGlnIleMetAlaSerProValGlyPheSerGly

 991 CCAGAATTCACTTTTCCGCTATATGGAACTATGGGAAATGCAGCTCCACAACAACGTATTGTTGCTCAACTAGGTCAGGGCGTGTATAGA 1080
     ProGluPheThrPheProLeuTyrGlyThrMetGlyAsnAlaAlaProGlnGlnArgIleValAlaGlnLeuGlyGlnGlyValTyrArg

1081 ACATTATCGTCCACTTTATATAGAAGACCTTTTAATATAGGGATAAATAATCAACAACTATCTGTTCTTGACGGGACAGAATTTGCTTAT 1170
     ThrLeuSerSerThrLeuTyrArgArgProPheAsnIleGlyIleAsnAsnGlnGlnLeuSerValLeuAspGlyThrGluPheAlaTyr

1171 GGAACCTCCTCAAATTTGCCATCCGCTGTATACAGAAAAAGCGGAACGGTAGATTCGCTGGATGAAATACCGCCACAGAATAACAACGTG 1260
     GlyThrSerSerAsnLeuProSerAlaValTyrArgLysSerGlyThrValAspSerLeuAspGluIleProProGlnAsnAsnAsnVal

1261 CCACCTAGGCAAGGATTTAGTCATCGATTAAGCCATGTTTCAATGTTTCGTTCAGGCTTTAGTAATAGTAGTGTAAGTATAATAAGAGCT 1350
     ProProArgGlnGlyPheSerHisArgLeuSerHisValSerMetPheArgSerGlyPheSerAsnSerSerValSerIleIleArgAla

1351 CCTATGTTCTCTTGGATACATCGTAGTGCTGAATTTAATAATATAATTCCTTCATCACAAATTACACAAATACCTTTAACAAAATCTACT 1440
     ProMetPheSerTrpIleHisArgSerAlaGluPheAsnAsnIleIleProSerSerGlnIleThrGlnIleProLeuThrLysSerThr

1441 AATCTTGGCTCTGGAACTTCTGTCGTTAAAGGACCAGGATTTACAGGAGGAGATATTCTTCGAAGAACTTCACCTGGCCAGATTTCAACC 1530
     AsnLeuGlySerGlyThrSerValValLysGlyProGlyPheThrGlyGlyAspIleLeuArgArgThrSerProGlyGlnIleSerThr

1531 TTAAGAGTAAATATTACTGCACCATTATCACAAAGATATCGGGTAAGAATTCGCTACGCTTCTACCACAAATTTACAATTCCATACATCA 1620
     LeuArgValAsnIleThrAlaProLeuSerGlnArgTyrArgValArgIleArgTyrAlaSerThrThrAsnLeuGlnPheHisThrSer

1621 ATTGACGGAAGACCTATTAATCAGGGGAATTTTTCAGCAACTATGAGTAGTGGGAGTAATTTACAGTCCGGAAGCTTTAGGACTGTAGGT 1710
     IleAspGlyArgProIleAsnGlnGlyAsnPheSerAlaThrMetSerSerGlySerAsnLeuGlnSerGlySerPheArgThrValGly

1711 TTTACTACTCCGTTTAACTTTTCAAATGGATCAAGTGTATTTACGTTAAGTGCTCATGTCTTCAATTCAGGCAATGAAGTTTATATAGAT 1800
     PheThrThrProPheAsnPheSerAsnGlySerSerValPheThrLeuSerAlaHisValPheAsnSerGlyAsnGluValTyrIleAsp
```

Fig. 4. Complete nucleotide sequence of the *B. thuringiensis* subsp. *kurstaki* HD-1 insecticidal crystal protein gene in pBT1-106A. The derived amino acid sequence is given below.

Bacillus thuringiensis subsp. *kurstaki* HD-1 harbors multiple ICP genes whose products accumulate in its crystals (Kronstad *et al.*, 1983). Schnepf and Whiteley (1981) cloned an ICP gene from subsp. *kurstaki* strain HD-1-Dipel. In this study we have cloned two different ICP genes from the related *kurstaki* strain, HD-1. Colony hybridization with the 6.6-type HD-73 gene identified a number of recombinant *E. coli* that contained a partial or complete ICP gene. Colonies expressing ICP antigen were selected using an immunoassay technique. *Escherichia coli*

```
1801 CGAATTGAATTTGTTCCGGCAGAAGTAACCTTTGAGGCAGAATATGATTTAGAAAGAGCACAAAAGGCGGTGAATGAGCTGTTTACTTCT 1890
     ArgIleGluPheValProAlaGluValThrPheGluAlaGluTyrAspLeuGluArgAlaGlnLysAlaValAsnGluLeuPheThrSer

1891 TCCAATCAAATCGGGTTAAAAACAGATGTGACGGATTATCATATTGATCAAGTATCCAATTTAGTTGAGTGTTTATCTGATGAATTTTGT 1980
     SerAsnGlnIleGlyLeuLysThrAspValThrAspTyrHisIleAspGlnValSerAsnLeuValGluCysLeuSerAspGluPheCys

1981 CTGGATGAAAAAAAAGAATTGTCCGAGAAAGTCAAACATGCGAAGCGACTTAGTGATGAGCGGAATTTACTTCAAGATCCAAACTTTAGA 2070
     LeuAspGluLysLysGluLeuSerGluLysValLysHisAlaLysArgLeuSerAspGluArgAsnLeuLeuGlnAspProAsnPheArg

2071 GGGATCAATAGACAACTAGACCGTGGCTGGAGAGGAAGTACGGATATTACCATCCAAGGAGGCGATGACGTATTCAAAGAGAATTACGTT 2160
     GlyIleAsnArgGlnLeuAspArgGlyTrpArgGlySerThrAspIleThrIleGlnGlyGlyAspAspValPheLysGluAsnTyrVal

2161 ACGCTATTGGGTACCTTTGATGAGTGCTATCCAACGTATTTATATCAAAAAATAGATGAGTCGAAATTAAAAGCCTATACCCGTTACCAA 2250
     ThrLeuLeuGlyThrPheAspGluCysTyrProThrTyrLeuTyrGlnLysIleAspGluSerLysLeuLysAlaTyrThrArgTyrGln

2251 TTAAGAGGGTATATCGAAGATAGTCAAGACTTAGAAATCTATTTAATTCGCTACAATGCCAAACACGAAACAGTAAATGTGCCAGGTACG 2340
     LeuArgGlyTyrIleGluAspSerGlnAspLeuGluIleTyrLeuIleArgTyrAsnAlaLysHisGluThrValAsnValProGlyThr

2341 GGTTCCTTATGGCCGCTTTCAGCCCCAAGTCCAATCGGAAAATGTGCCCATCATTCCCATCATTTCTCCTTGGACATTGATGTTGGATGT 2430
     GlySerLeuTrpProLeuSerAlaProSerProIleGlyLysCysAlaHisHisSerHisHisPheSerLeuAspIleAspValGlyCys

2431 ACAGACTTAAATGAGGACTTAGGTGTATGGGTGATATTCAAGATTAAGACGCAAGATGGCCATGCAAGACTAGGAAATCTAGAATTTCTC 2520
     ThrAspLeuAsnGluAspLeuGlyValTrpValIlePheLysIleLysThrGlnAspGlyHisAlaArgLeuGlyAsnLeuGluPheLeu

2521 GAAGAGAAACCATTAGTAGGAGAAGCACTAGCTCGTGTGAAAAGAGCGGAGAAAAAATGGAGAGACAAACGTGAAAAATTGGAATGGGAA 2610
     GluGluLysProLeuValGlyGluAlaLeuAlaArgValLysArgAlaGluLysLysTrpArgAspLysArgGluLysLeuGluTrpGlu

2611 ACAAATATTGTTTATAAAGAGGCAAAAGAATCTGTAGATGCTTTATTTGTAAACTCTCAATATGATAGATTACAAGCGGATACCAACATC 2700
     ThrAsnIleValTyrLysGluAlaLysGluSerValAspAlaLeuPheValAsnSerGlnTyrAspArgLeuGlnAlaAspThrAsnIle

2701 GCGATGATTCATGCGGCAGATAAACGCGTTCATAGCATTCGAGAAGCTTATCTGCCTGAGCTGTCTGTGATTCCGGGTGTCAATGCGGCT 2790
     AlaMetIleHisAlaAlaAspLysArgValHisSerIleArgGluAlaTyrLeuProGluLeuSerValIleProGlyValAsnAlaAla

2791 ATTTTTGAAGAATTAGAAGGGCGTATTTTCACTGCATTCTCCCTATATGATGCGAGAAATGTCATTAAAAATGGTGATTTTAATAATGGC 2880
     IlePheGluGluLeuGluGlyArgIlePheThrAlaPheSerLeuTyrAspAlaArgAsnValIleLysAsnGlyAspPheAsnAsnGly

2881 TTATCCTGCTGGAACGTGAAAGGGCATGTAGATGTAGAAGAACAAAACAACCACCGTTCGGTCCTTGTTGTTCCGGAATGGGAAGCAGAA 2970
     LeuSerCysTrpAsnValLysGlyHisValAspValGluGluGlnAsnAsnHisArgSerValLeuValValProGluTrpGluAlaGlu

2971 GTGTCACAAGAAGTTCGTGTCTGTCCGGGTCGTGGCTATATCCTTCGTGTCACAGCGTACAAGGAGGGATATGGAGAAGGTTGCGTAACC 3060
     ValSerGlnGluValArgValCysProGlyArgGlyTyrIleLeuArgValThrAlaTyrLysGluGlyTyrGlyGluGlyCysValThr

3061 ATTCATGAGATCGAGAACAATACAGACGAACTGAAGTTTAGCAACTGTGTAGAAGAGGAAGTATATCCAAACAACACGGTAACGTGTAAT 3150
     IleHisGluIleGluAsnAsnThrAspGluLeuLysPheSerAsnCysValGluGluGluValTyrProAsnAsnThrValThrCysAsn

3151 GATTATACTGCGACTCAAGAAGAATATGAGGGTACGTACACTTCTCGTAATCGAGGATATGACGGAGCCTATGAAAGCAATTCTTCTGTA 3240
     AspTyrThrAlaThrGlnGluGluTyrGluGlyThrTyrThrSerArgAsnArgGlyTyrAspGlyAlaTyrGluSerAsnSerSerVal

3241 CCAGCTGATTATGCATCAGCCTATGAAGAAAAAGCATATACAGATGGACGAAGAGACAATCCTTGTGAATCTAACAGAGGATATGGGGAT 3330
     ProAlaAspTyrAlaSerAlaTyrGluGluLysAlaTyrThrAspGlyArgArgAspAsnProCysGluSerAsnArgGlyTyrGlyAsp

3331 TACACACCACTACCAGCTGGCTATGTGACAAAAGAATTAGAGTACTTCCCAGAAACCGATAAGGTATGGATTGAGATCGGAGAAACGGAA 3420
     TyrThrProLeuProAlaGlyTyrValThrLysGluLeuGluTyrPheProGluThrAspLysValTrpIleGluIleGlyGluThrGlu

3421 GGAACATTCATCGTGGACAGCGTGGAATTACTTCTTATGGAGGAATAA3468
     GlyThrPheIleValAspSerValGluLeuLeuLeuMetGluGluEnd
```

Fig. 4. *(Continued)*

pBT1-89A reacted strongly with a monoclonal antibody prepared to HD-73 protoxin. Extensive restriction enzyme analysis showed no differences between the pBT1-89A 6.6-type gene and the HD-73 6.6-type gene. A second gene type was selected using polyclonal anti-HD-73 serum. The 5.3 gene, contained in *E. coli* pBT1-106A, is distinct from the HD-73 gene and the previously cloned HD-1-Dipel gene harbored in *E. coli* pES1. Southern blots of *Hin*dIII-cut HD-1 plasmid DNA indicated that this strain has a pES1-type gene (data not shown), but no clones of this type were identified in the pUC18 library. The above two gene types and the presence of a pES1-type gene correspond to the number of protoxin genes determined by Southern hybridization for

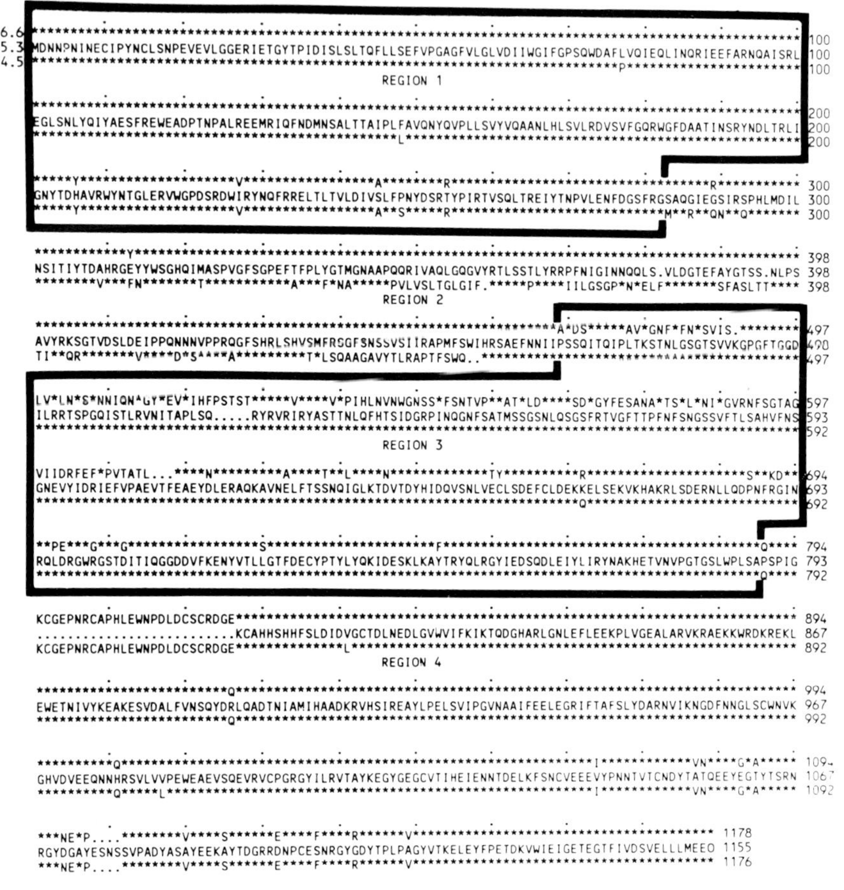

Fig. 5. Alignment of *B. thuringiensis* subsp. *kurstaki* insecticidal crystal proteins at the amino acid level. The pBT73-16 (6.6 type), pBT1-106A (5.3 type), and pES1 (4.5 type) ICP amino acid sequences were compared and gaps inserted to compute the optimal alignment. Numbering is from the N-terminal methionine. An asterisk (*) indicates identity with the pBT1-106A sequence; a dot (·) indicates an amino acid gap. Substitutions are shown by the single-letter code for the amino acid. Regions 1 and 3 are boxed. Sequence data for pBT73-16 are from Adang *et al.* (1985) and for pES1 from Schnepf *et al.* (1985).

strain HD-1 (Kronstad *et al.*, 1983). The three hybridizing *Hin*dIII fragments detected by Kronstad *et al.* represent three distinct gene structures and not polymorphism in 5′ flanking restriction enzyme sites.

Escherichia coli strains harboring these ICP genes produce immunoreactive peptides with the expected molecular weights of 130,000–135,000. Figure 2 demonstrates the differential binding of these pep-

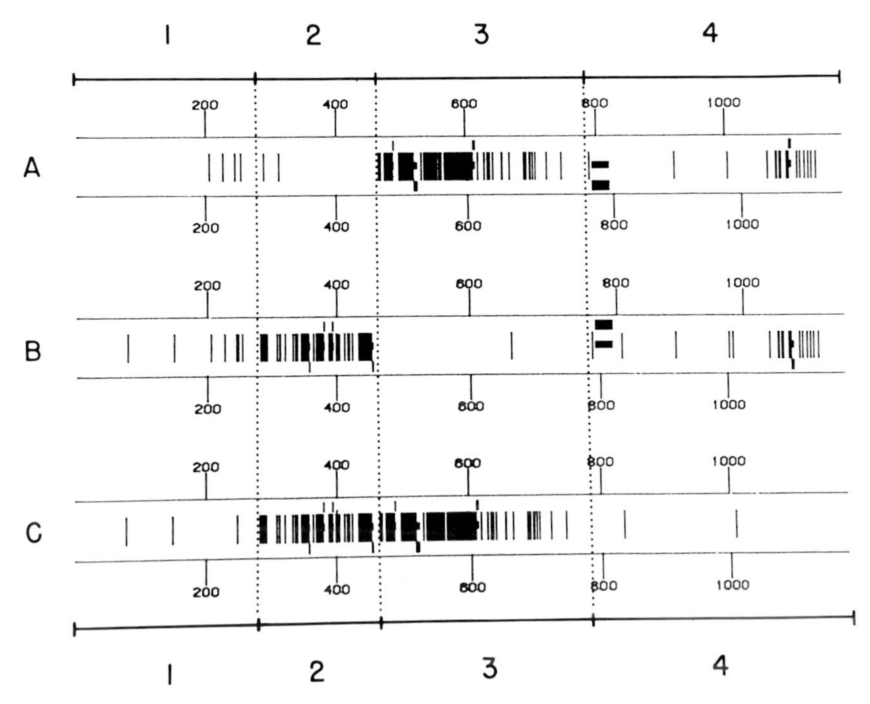

Fig. 6. Graphic alignment of the *B. thuringiensis* subsp. *kurstaki* ICP amino acid sequence. The alignments are the same as in Fig. 5 and sequences are displayed one above the other. (A) pBT73-16 gene above pBT1-106A; (B) pBT1-106A above pES1; (C) pBT73-16 above pES1. Vertical bars (|) represent amino acid changes. Gaps are shown as blocks above or below the expanded sequence. Regions 1–4 are labeled at the bottom of the figure.

tides with rabbit antiserum and MAb-1. *Escherichia coli* strains pBT1-89B and pBT73-16 harboring 6.6-type genes react more strongly with the homologous polyclonal serum than the other *E. coli* strains tested. One explanation for the higher avidity for the 6.6-type products may be that some antigenic sites vary between the 6.6-type protoxin and the 5.3- and 4.5-type protoxins. This differential antigenicity correlates with the divergence, discussed below, of the 6.6-type genes from the 5.3 and 4.5 genes. Also, these strains produced ICP reactive with MAb-1, whereas 5.3- and 4.5-type peptides are not recognized by this antibody. To our knowledge, this is the first use of a monoclonal antibody to differentiate protoxins. Previously, several authors used polyclonal antisera to distinguish *B. thuringiensis* crystals. Strain HD-73 crystals are classified as serotype K-73 and strain HD-1 as serotype K-1 based on their differential reaction in Ouchterlony immunodiffusion (Krywienczyk *et al.*, 1978) and microcomplement fixation tests (Lynch and Baumann, 1985). Monoclonal antibodies such as MAb-1 that react with a particular protoxin can be used as probes to identify crystal subunits. They are more specific than polyclonal antibodies, so they can supplement antisera in the identification and classification of *B. thuringiensis* insecticidal crystals.

The primary nucleotide structure of the 5.3-type gene in pBT1-106A (Fig. 4) contains an ORF encoding a peptide with a molecular weight of 130,641. This size is smaller than the M_r 133,330 6.6-type or the M_r 133,500 4.5-type ICP. This difference in mass is due to a 26-amino-acid deletion near the carboxyl end of the peptide.

Figure 5 aligns the deduced amino acid sequences of the pBT73-16 (6.6 type), pBT1-106A (5.3 type), and pES1 (4.5 type) coding regions. This alignment is show graphically in Fig. 6, where mismatched residues are represented by vertical bars. Based on the amount of homology, the sequences can be divided into four regions. The N-terminal 282 amino acids of region 1 are 98% conserved for each sequence. The conservation in this region is not surprising because this area is important in processing the inactive protoxin to an active toxin. Nagamatsu *et al.* (1984) found that the protoxin is preferentially cleaved by trypsin between residues 28 and 29 during this step. Schnepf and Whiteley (1985) used DNA deletions and fusion peptides to demonstrate the importance of this N-terminal region. Fusions to the 10th codon produced toxic peptides while fusions to the 50th codon did not. Region 2 includes the next 184 amino acids, which are characterized by the divergence of the 4.5 from the 6.6 and 5.3 sequences. This pattern changes for the 322 residues of region 3, where the 5.3 and 4.5 genes are identical and the 6.6 gene is 63% homologous. Figure 6 shows that most of these substitutions occur from residue 467 to 616. Apparently, this portion of the gene is necessary for toxicity to insects. Adang *et al.* (1985) deleted the 3′ end of the HD-73 protoxoin sequence to the 612th codon and the expressed polypeptide was toxic. A deletion to the 476th codon resulted in an inactive gene product. Schnepf and Whitely (1985) constructed similar 3′ deletions with the pES1 gene and found that deletions to the 645th codon allowed the synthesis of toxic peptides, whereas deletions to the 603rd codon yielded a nontoxic peptide. The carboxyl terminus of the trypsin-resistant 516-residue toxic fragment from *B. thuringiensis* subsp. *dendrolimus* is located in region 3 (Nagamatsu *et al.*, 1984). These data show that important determinants of toxicity are located in region 3. In addition, the 6.6 protoxin has a unique antigenic determinant in region 3 (Fig. 2B). MAb-1 recognizes an epitope between residues 476 and 613 of the 6.6 protoxin (Adang *et al.*, 1985) in a segment that is only 27% homologous with the 5.3 and 4.5 protoxins. Region 4 includes the C-terminal 32% (367 residues) of the protoxins. It starts where the pBT1-106A (5.3 type) ORF diverges from the 6.6 and 4.5 ORFs. The most significant change is a 26-amino-acid deletion in the 5.3 ORF. It is not clear what effect, if any, differences in region 4 have on the biological activity of the protoxins. Dividing the ICP genes into homologous regions assists in their comparison. As seen in Fig. 6, most changes are clustered between residues 340 and 617. This portion of the protoxin gene is required for insect toxicity and may also be involved in its activity spectrum.

We have no direct evidence to prove that the two protoxin genes we have cloned are expressed in HD-1, but the conservation of the 5′ and 3′ regulatory sequences identified previously for the pES1 gene (Wong *et al.*, 1983) suggests that each gene type is under the same transcriptional control. Therefore, these polypeptides are probably components of HD-1 crystals. In support of this, Calabrese *et al.* (1980) resolved HD-1 crystals into subunit peptides with apparent molecular weights of 143,000 and 155,000 on sodium dodecyl sulfate (SDS) gels. These estimated sizes are larger than those deduced from the nucleotide sequences, but the smaller peptide may be the pBT1-106A product and the larger a composite of the other two ICP gene products. This contrasts with the HD-1-Dipel protoxin, which migrates as a single peptide of M_r 134,000 (Bulla *et al.*, 1981). The SDS gel data of Bulla *et al.* and the Southern hybridization data showing the lack of a hybridizing 5.3 kb *Hin*dIII fragment (Kronstad *et al.*, 1983) indicate that their HD-1-Dipel does not contain a 5.3-type ICP gene. The absence of an ICP gene from a strain is not surprising, since there are extensive data demonstrating the instability of *B. thuringiensis* plasmids containing these genes (González *et al.*, 1981). Transposonlike structures flanking ICP genes may also account for this instability (Lereclus *et al.*, 1984; Kronstad and Whiteley, 1984). Strains HD-1 and HD-1-Dipel produce serotype k1 crystals. Members of this group are noted for their heterogeneity in toxicity to particular insect species (Dulmage and Cooperators, 1981). An explanation for this variation in k1 strains is that crystals with differing biological activity contain different amounts of these ICP gene products. Insect bioassays are needed with *E. coli* and *B. thuringiensis* strains expressing defined ICP genes to measure the toxicity of related protoxins against different insect species.

Through cloning and sequencing ICP genes, we can compare the structures of these proteins. Variation in the three subspecies *kurstaki* protoxins studied is clustered in regions of the polypeptides. Information of this type, combined with biological studies, will provide new tools to develop more efficacious *B. thurigiensis* strains.

ACKNOWLEDGMENTS

We are grateful to Ms. Jo Adang for graphics and to Jeff Lotzer for computer data analysis. This is Agrigenetics Advanced Science Company Publication No. 59.

REFERENCES

Adang, M. J., Staver, M. J., Rocheleau, T. A., Leighton, J., Barker, R. F., and Thompson, D. V. (1985). Characterized full-length and truncated plasmid clones of the crystal protein of *Bacillus thuringiensis* subsp. *kurstaki* HD-73 and their toxicity to *Manduca sexta*. *Gene* **36**, 289–300.

Aronson, A. I., Angelo, N., and Holt, S. C. (1971). Regulation of extracellular protease production in *Bacillus cereus* T: Characterization of mutants producing altered amounts of protease. *J. Bacteriol.* **106**, 1016–1025.

Barker, R. F., Idler, K. B., Thompson, D. V., and Kemp, J. D. (1983). Nucleotide sequence of the T-DNA region from the *Agrobacterium tumefaciens* octopine Ti plasmid pTi15955. *Plant Mol. Biol.* **2**, 335–350.

Birnboim, H. C., and Doly, J. (1979). A rapid alkaline extraction procedure for screening recombinant plasmid DNA. *Nucleic Acids Res.* **7**, 1513–1523.

Blake, M. S., Johnston, K. H., Russell-Jones, G. J., and Gotschlich, E. C. (1984). A rapid, sensitive method for detection of alkaline phosphatase-conjugated antibody on Western blots. *Anal. Biochem.* **136**, 175–179.

Boyer, H. W., and Roulland-Dussoix, D. (1969). A complementation analysis of the restriction and modification of DNA in *Escherichia coli*. *J. Mol. Biol.* **41**, 459–472.

Bulla, L. A., Kramer, K. J., Cox, D. J., Jones, B. L., Davidson, L. I., and Lookhart, G. L. (1981). Purification and characterization of the entomocidal protoxin of *Bacillus thuringiensis*. *J. Biol. Chem.* **256**, 3000–3004.

Calabrese, D. M., Nickerson, K. W., and Lane, L. C. (1980). A comparison of protein crystal subunit sizes in *Bacillus thuringiensis*. *Can. J. Microbiol.* **26**, 1006–1010.

Casadaban, M. J., and Cohen, S. N. (1980). Analysis of gene control signals by DNA fusion and cloning *Escherichia coli*. *J. Mol. Biol.* **138**, 179–207.

Dulmage, H. T., and Cooperators (1981). Insecticidal activity of isolates of *Bacillus thuringiensis* and their potential for pest control. *In* "Microbial Control of Pests and Plant Diseases 1970–1980" (H. D. Burges, ed.), pp. 223–247. Academic Press, London.

González, J. M., Dulmage, H. T., and Carlton, B. C. (1981). Correlation between specific plasmids and δ-endotoxin production in *Bacillus thuringiensis*. *Plasmid* **5**, 351–365.

Helfman, D. M., Feramisco, J. R., Fiddes, J. C., Thomas, G. P., and Hughes, S. H. (1983). Identification of clones that encode chicken tropomyosin by direct immunological screening of a cDNA expression library. *Proc. Natl. Acad. Sci. U.S.A.* **80**, 31–35.

Jarrett, P. (1985). Potency factors in the delta-endotoxin of *Bacillus thuringiensis* var.

aizawai and the significance of plasmids in their control. *J. Appl. Bacteriol.* **58,** 437–448.

Kronstad, J. W., and Whiteley, H. R. (1984). Inverted repeat sequences flank a *Bacillus thuringiensis* crystal protein gene. *J. Bacteriol.* **160,** 95–102.

Kronstad, J. W., Schnepf, H. E., and Whiteley, H. R. (1983). Diversity of locations for *Bacillus thuringiensis* crystal protein genes. *J. Bacteriol.* **154,** 419–428.

Krywienczyk, J., Dulmage, H. T., and Fast, P. G. (1978). Occurrence of two serologically distinct groups within *Bacillus thuringiensis* serotype 3 ab var. *kurstaki. J. Invertebr. Pathol.* **31,** 372–375.

Lereclus, D., Ribier, J., Klier, A., Menou, G., and Lecadet, M.-M (1984). A transposon-like structure related to the δ-endotoxin gene of *Bacillus thuringiensis. EMBO J.* **3,** 2561–2567.

Lynch, M. J., and Baumann, P. (1985). Immunological comparisons of the crystal protein from strains of *Bacillus thuringiensis. J. Invertebr. Pathol.* **46,** 47–57.

McLinden, J. H., Savourin, J. R., Clark, B. D., Gensler, D. R., Workman, W. E., and Dean, D. H. (1985). Cloning and expression of an insecticidal k-73 type crystal protein gene from *Bacillus thuringiensis* var. *kurstaki* into *Escherichia coli. Appl. Environ. Microbiol.* **50,** 623–628.

Maxam, A. M., and Gilbert, W. (1980). Sequencing end-labeled DNA with base-specific chemical cleavages. *In* "Methods in Enzymology" (L. Grossman and K. Moldave, eds.), Vol. 65, pp. 499–560. Academic Press, New York.

Meenakshi, K., and Jayaraman, K. (1979). On the formation of crystal proteins during sporulation in *Bacillus thuringiensis* var. *thuringiensis. Arch. Microbiol.* **120,** 9–14.

Miller, J. H. (1972). "Experiments in Molecular Genetics." Cold Spring Harbor Lab., Cold Spring Harbor, New York.

Nagamatsu, Y., Itai, Y., Hatanaka, C., Funatsu, G., and Hayashi, K. (1984). A toxic fragment from the entomocidal crystal protein of *Bacillus thuringiensis. Agric. Biol. Chem.* **48,** 611–619.

Schesser, J. H., Kramer, K. J., and Bulla, L. A. (1977). Bioassay for homogeneous parasporal crystal of *Bacillus thuringiensis* suing the tobacco hornworm, *Manduca sexta. Appl. Environ. Microbiol.* **33,** 878–880.

Schnepf, H. E., and Whiteley, H. R. (1981). Cloning and expression of the *Bacillus thuringiensis* crystal protein gene in *Escherichia coli. Proc. Natl. Acad. Sci. U.S.A.* **78,** 2893–2897.

Schnepf, H. E., and Whiteley, H. R. (1985). Delineation of a toxin-encoding segment of a *Bacillus thuringiensis* crystal protein gene. *J. Biol. Chem.* **260,** 6273–6280.

Schnepf, H. E., Wong, H. C., and Whiteley, H. R. (1985). The amino acid sequence of a crystal protein from *Bacillus thuringiensis* deduced from the DNA base sequence. *J. Biol. Chem.* **260,** 6264–6272.

Spizizen, J. (1958). Transformation of biochemically deficient strains of *Bacillus subtilis* by deoxyribonuclease. *Proc. Natl. Acad. Sci. U.S.A.* **44,** 1072–1078.

Towbin, H., Staehelin, T., and Gordon, J. (1979). Electrophoretic transfer of proteins from polyacrylamide gels to nitrocellulose sheets: Procedure and some applications. *Proc. Natl. Acad. Sci. U.S.A.* **76,** 4350–4354.

Waalwijk, C., Dullemans, A. M., van Workum, M. E. S., and Visser, B. (1985). Molecular cloning and the nucleotide sequence of the M_r 28 000 crystal protein gene of *Bacillus thuringiensis* subsp. *israelensis. Nucleic Acids Res.* **13,** 8207–8217.

Wong, H. C., Schnepf, H. E., and Whiteley, H. R. (1983). Transcriptional and translational start sites for the *Bacillus thuringiensis* crystal protein gene. *J. Biol. Chem.* **258,** 1960–1967.

8

Bacillus thuringiensis Isolate with Activity against Coleoptera

CORINNA HERRNSTADT, FRANK GAERTNER,
WENDY GELERNTER, and DAVID L. EDWARDS

I. INTRODUCTION

Bacillus thuringiensis produces a polypeptide that is toxic to a wide variety of insects. This toxin, called δ-endotoxin, is of great interest because it is a safe, effective method of controlling insects without harmful side effects. The toxin from a given strain is generally specific for a particular pest and is not harmful to nontarget organisms (Aronson *et al.*, 1986).

A large number of varietal strains of *B. thuringiensis* have been isolated (Dulmage, 1981). Most of these have activity against lepidopteran insects (caterpillars, moths), while some have activity against Diptera (flies, mosquitoes). Bioassays against lepidopteran hosts have shown that different varietal strains of *B. thuringiensis* have different relative toxicities when tested against a panel of insects (Dulmage *et al.*, 1981).

A number of genes coding for *B. thuringiensis* toxins have now been cloned (Schnepf and Whiteley, 1981; Klier *et al.*, 1982) and sequence data are beginning to become available (Adang *et al.*, 1985; Schnepf *et*

BIOTECHNOLOGY IN INVERTEBRATE
PATHOLOGY AND CELL CULTURE

al., 1985; Shibano *et al.*, 1985; Thorne *et al.*, 1986). These data show that different toxins have different (primary) amino acid sequences.

While the presently isolated strains of *B. thuringiensis* have good activity against Lepidoptera and Diptera, several economically important pests are found in the order Coleoptera (beetles); these are not affected by the current array of *B. thuringiensis* isolates. An isolate with activity against some species of Coleoptera has been reported by Krieg *et al.* (1983, 1984). We have isolated a strain of *B. thuringiensis* that has no insecticidal activity against any lepidopteran or dipteran host that we have studied. This strain has good activity against several coleopteran insects including several economically important pests (Herrnstadt *et al.*, 1986). We have named this strain *Bacillus thuringiensis* var. *san diego*.

II. MATERIALS AND METHODS

Bacillus thuringiensis var. *san diego* cells were maintained and toxin crystals were prepared as described previously (Herrnstadt *et al.*, 1986).

Bioassays against *Diabrotica undecimpunctata undecimpunctata*, *Hypera brunipennis*, *Leptinotarsa decemlineata*, and *Pyrrhalta luteola* were carried out by spraying spore/crystal preparations, purified crystals, or purified toxin from recombinant cells onto leaf discs from appropriate host plants. Assays against *Aedes egypti* were performed by adding spore/crystal preparations to larvae in water. All other assays were carried out by incorporating spore/crystal preparations or purified crystals into an appropriate diet mixture.

Electron microscopy was carried out on *B. thuringiensis* var. *san diego* cells grown for 21 hr at 28°C, harvested by centrifugation, and fixed with 1.5% glutaraldehyde and 2% paraformaldehyde in 0.1 *M* sodium cacodylate buffer, pH 7.3. The fixed cells were embedded in Epon, sectioned, and stained with lead citrate and uranyl acetate.

Western blots were carried out as described by Burnette (1981). Probit analysis was done as described by Finney (1971).

III. HOST RANGE

Preliminary screening of *B. thuringiensis* var. *san diego* showed that this isolate had activity against a number of coleopteran insects but was not active against any lepidopteran or dipteran species that we tested. These results are shown in Table I. *Bacillus thuringiensis* var. *san*

TABLE I

Insects Evaluated for Susceptibility to *B. thuringiensis* var. *san diego*

Order	Family	Species	Common name	Stage[a]	Activity[b]
Coleoptera	Bruchidae	*Callosobruchus maculatus*	Cowpea weevil	A,L	+
	Chrysomelidae	*Diabrotica balteata*	Banded cucumber beetle	A,L	+
		Diabrotica undecimpunctata	White spotted cucumber beetle	A,L	+
		Haltica tombacina	Flea beetle spp.	A,L	+++
		Leptinotarsa decemlineata	Colorado potato beetle	A	–
				L	++++
		Pyrrhalta luteola	Elm leaf beetle	A,L	++++
	Curculionidae	*Anthonomus grandis*	Boll weevil	A,L	++++
		Otiorhynchus sulcatus	Black vine weevil	A	–
				L	++
		Hypera brunneipennis	Egyptian alfalfa weevil	A	–
				L	+++
	Dermestidae	*Attagenus megatoma*	Black carpet beetle	L	–
	Nitidulidae	*Carpophilus hemipterus*	Dried fruit beetle	L	–
	Ptinidae	*Gibbium psylloides*	Spider beetle spp.	A	–
	Tenebrionidae	*Tenebrio molitor*	Yellow mealworm	L	++
		Tribolium castaneum	Red flour beetle	A,L	–
Diptera	Culicidae	*Aedes aegypti*	Yellow fever mosquito	L	–
	Muscidae	*Musca domestica*	House fly	L	–
Lepidoptera	Noctuidae	*Spodoptera exigua*	Beet army worm	L	–
		Trichoplusia ni	Cabbage looper	L	–

[a]A, Adult; L, larva.

[b]Insecticidal activities were arbitrarily classified from weak (+) to very strong (++++).

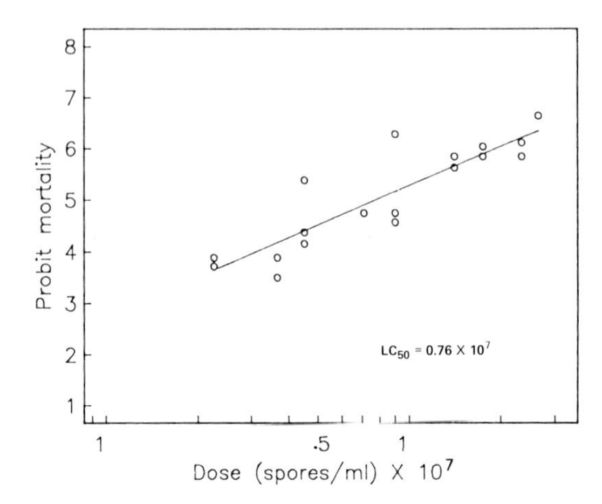

Fig. 1. Bioassay of *B. thuringiensis* var. *san diego* spore/crystal preparation against Colorado potato beetle larvae. Spore/crystal preparations of various concentrations were prepared. Potato leaves were dipped into these solutions and air-dried for 1 hr. The leaves were then exposed to 80 sec–instar Colorado potato beetle larvae. Mortality was assessed after 96 hr. Data were analyzed by probit analysis (Finney, 1971).

diego is toxic to the Egyptian alfalfa weevil (*H. brunipennis*), the Colorado potato beetle (*L. decemlineata*), and the cotton boll weevil (*Anthonomus grandis*). These are all commercially important pests. We have carried out more detailed bioassays on the Colorado potato beetle and the boll weevil. These results are shown in Figs. 1 and 2. The data

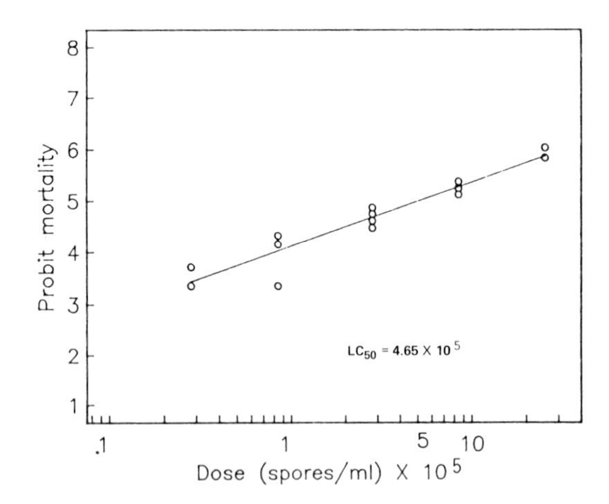

Fig. 2. Bioassay of *B. thuringiensis* var. *san diego* spore/crystal preparations against boll weevil larvae. Spore/crystal preparations were prepared and incorporated into standard synthetic diet. Assays were carried out against 80 sec–instar boll weevil larvae. Mortality was assessed after 96 hr. Data were analyzed by probit analysis (Finney, 1971).

show that *B. thuringiensis* var. *san diego* has good activity against both of these insects.

IV. BIOCHEMISTRY

Cells of *B. thuringiensis* var. *san diego* produce large, rectangular, wafer-shaped crystals. Transmission electron micrographs of these cells are shown in Fig. 3. Figures 3a and 3b show different orientations of toxin crystals within a cell. Figure 3c shows a toxin crystal outside a cell. Figure 4 shows a scanning electron micrograph of purified toxin crystals.

Figure 5 shows a sodium dodecyl sulfate (SDS)-polyacrylamide gel of purified *B. thuringiensis* var. *san diego* toxin crystals. The toxin protein migrates with a molecular weight of 64,000. This is in contrast to crystal protein from the lepidopteran-specific strain HD-73, which migrates with a molecular weight of 134,000. Treatment of HD-73 crystal protein with gut juice from susceptible insects or with a protease results in the production of an active toxin fragment of molecular weight 65,000. Toxin from *B. thuringiensis* var. *san diego* does not appear to be processed in this manner. We were not able to detect a smaller molecular weight toxic fragment when *B. thuringiensis* var. *san diego* crystal protein was treated with trypsin, papain, or gut juice from *Tenebrio molitor*.

Purified toxin from *B. thuringiensis* var. *san diego* did not cross-react immunologically with similar material from strain HD-73. In order to study the protein composition of this toxin with respect to that of HD-73 (the 65,000 molecular weight active form), both proteins were subjected to complete hydrolysis with trypsin and the resulting peptides were separated by high-performance liquid chromatography (HPLC) and compared. Elution profiles of peptides from both of these strains are shown in Fig. 6. It can be seen that the elution profiles of these two proteins are entirely different.

V. MOLECULAR GENETICS

The gene coding for the *B. thuringiensis* var. *san diego* toxin has been cloned into pBR322 as a 5.8 kb *Bam*HI fragment and has been expressed in *Escherichia coli*. Toxin material from extracts of these cells was purified using an antibody affinity column. This material was toxic to the elm leaf beetle *P. luteola*. Because we are unable to maintain colonies of Colorado potato beetle or boll weevil in California, we have

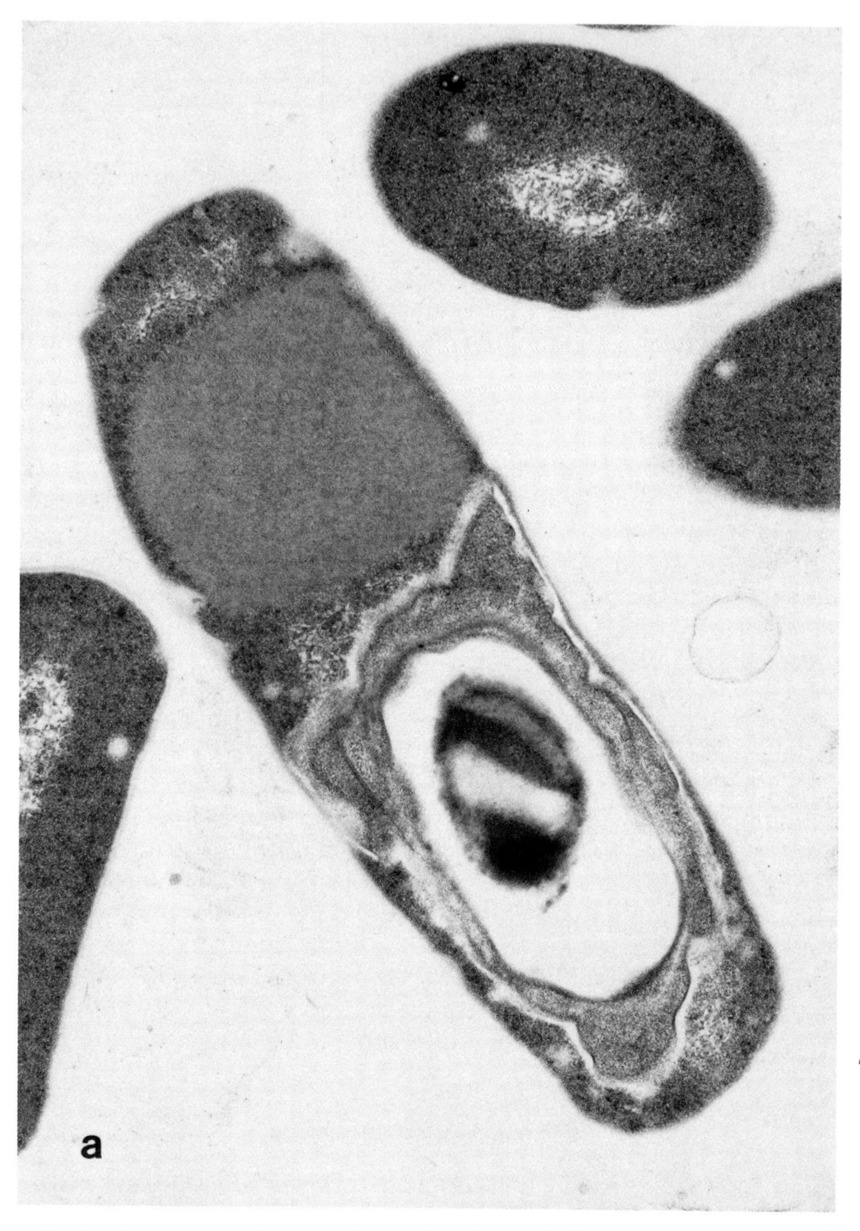

Fig. 3. Transmission electron micrographs of *B. thuringiensis* var. *san diego* cells. (a and b) Different orientations of crystals; (c) an isolated crystal. (From Herrnstadt *et al.*, 1986. © 1986, *Bio/Technology*. Used by permission.)

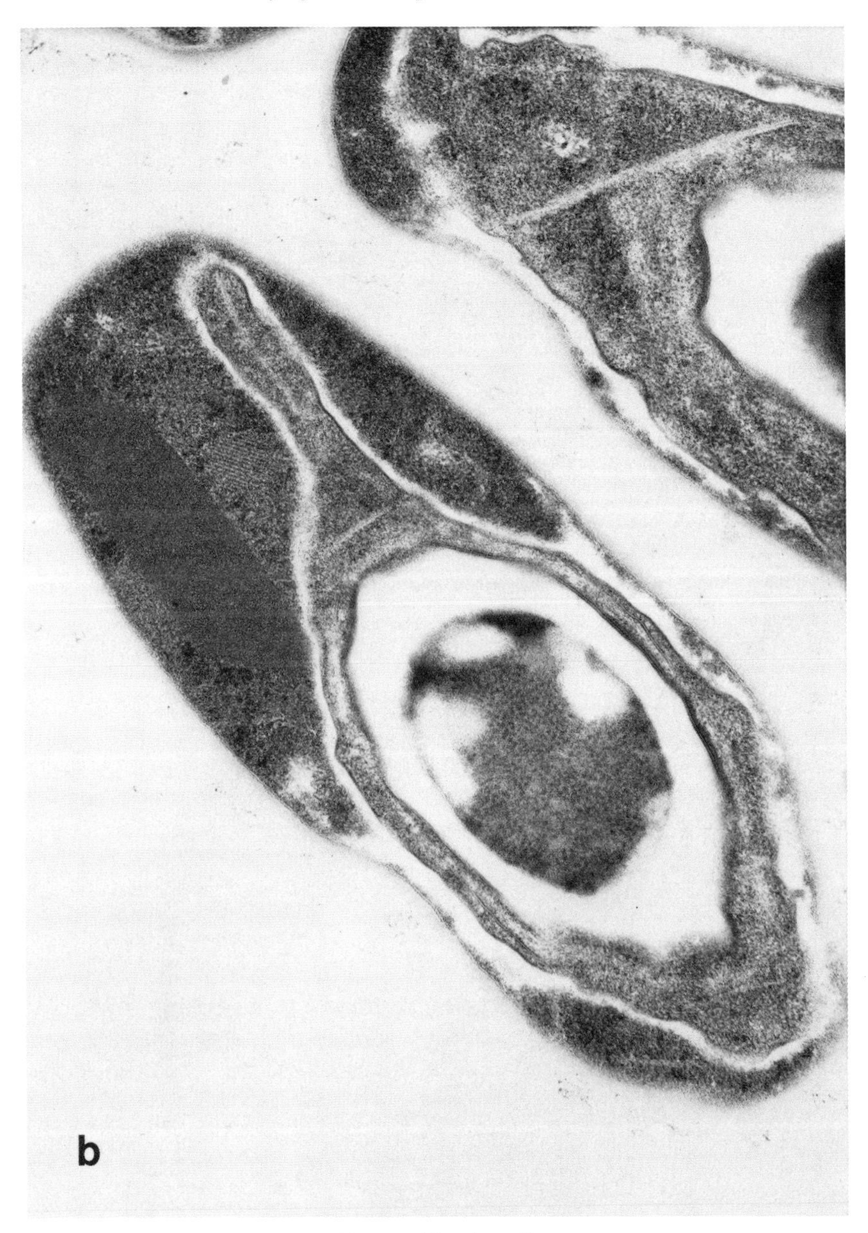

Fig. 3. (*Continued*)

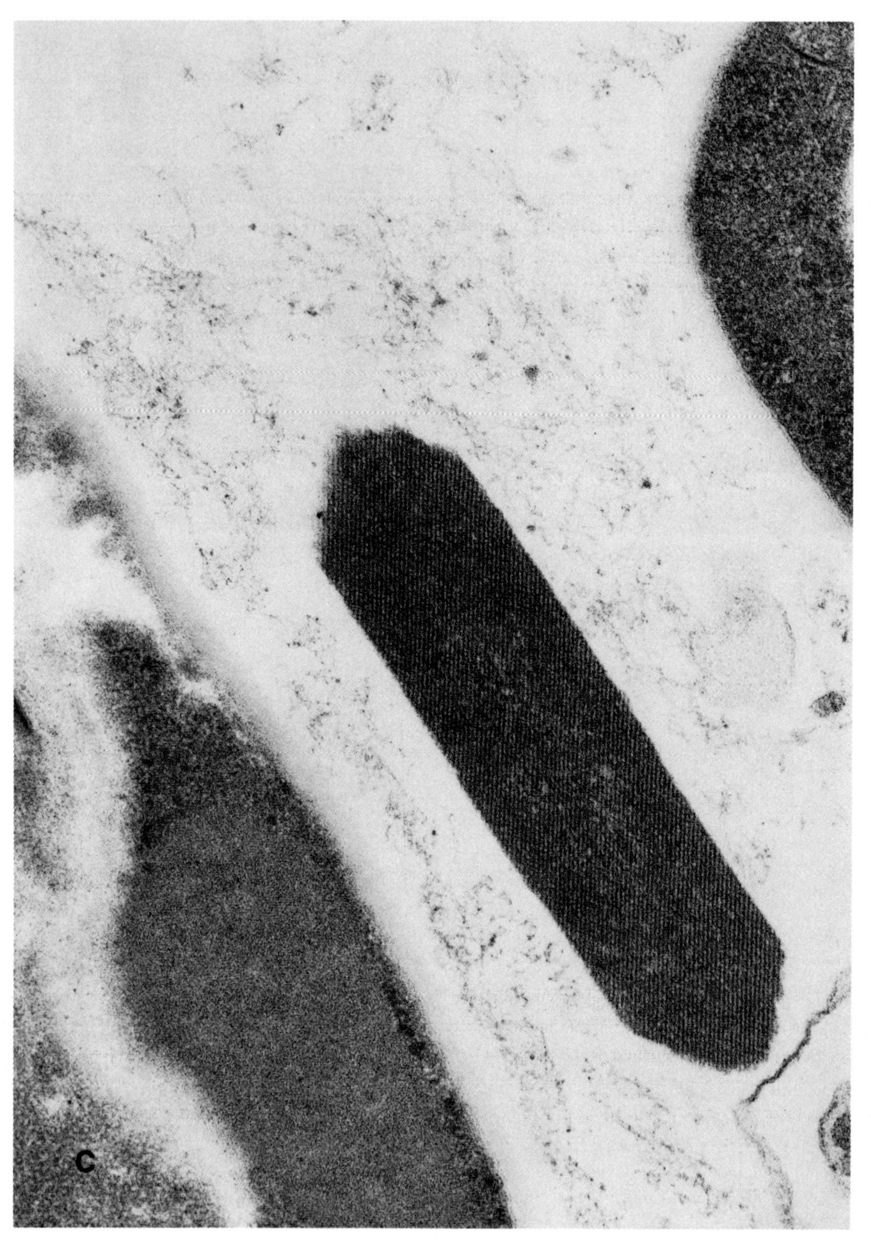

Fig. 3. (*Continued*)

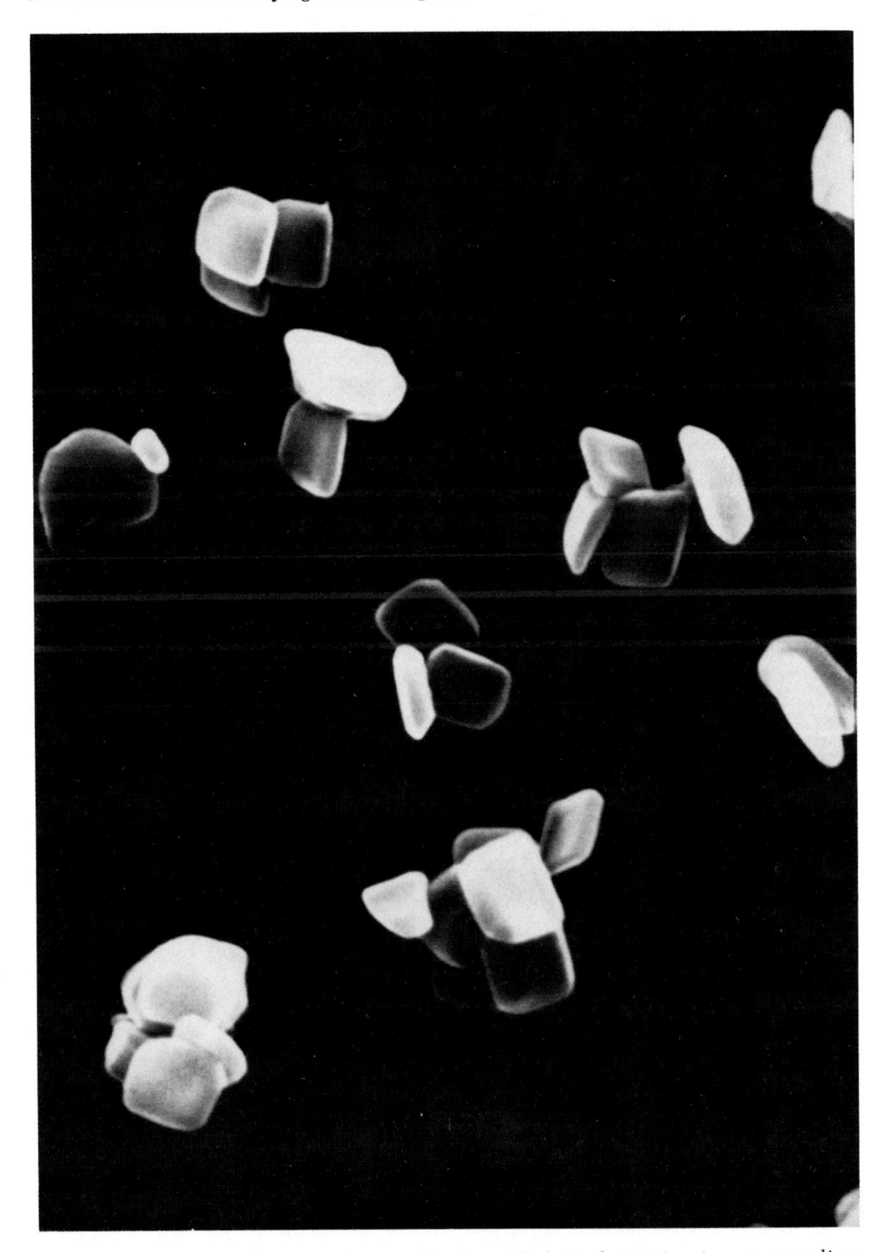

Fig. 4. Scanning electron micrograph of purified *B. thuringiensis* var. *san diego* crystals. (From Herrnstadt *et al.*, 1986. © 1986, *Bio/Technology*. Used by permission.)

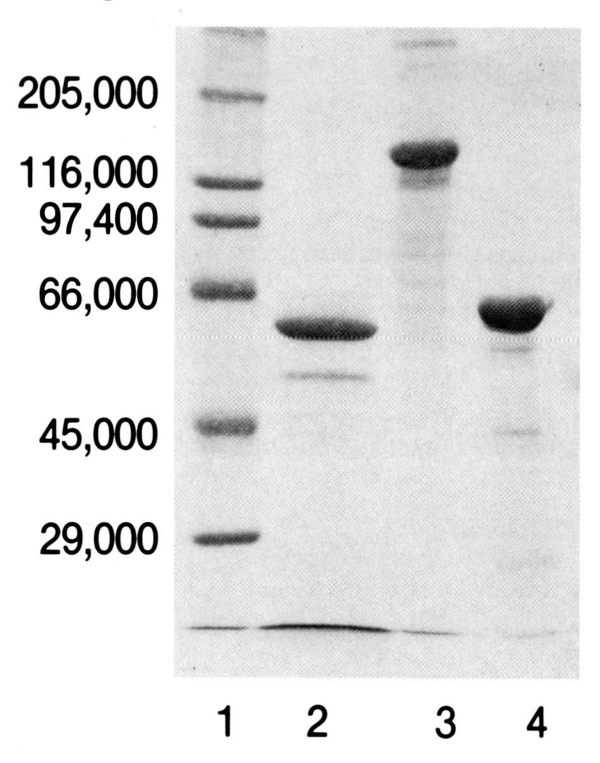

Fig. 5. SDS-polyacrylamide gel of *B. thuringiensis* toxins. Lane 1, protein standards; lane 2, HD-1 toxin after treatment of HD-1 toxin crystals with alkali (pH 10.0) and papain; lane 3, HD-1 crystals treated directly with alkali and SDS; lane 4, *B. thuringiensis* var. *san diego* toxin crystals treated directly with alkali and SDS. (From Herrnstadt *et al.*, 1986. © 1986, *Bio/Technology*. Used by permission.)

developed a laboratory bioassay for *B. thuringiensis* var. *san diego* toxin based on the elm leaf beetle. A Western blot of an *E. coli* extract containing the toxin is shown in Fig. 7. Material identified by anti-*B. thuringiensis* var. *san diego* antibodies migrates with a molecular weight of 74,400. A protein of this size is produced regardless of the orientation of the *Bam*HI fragment in the plasmid. This result suggests that the *B. thuringiensis* var. *san diego* toxin may be synthesized as a larger precursor protein that is processed before the material is deposited as crystals in the cell.

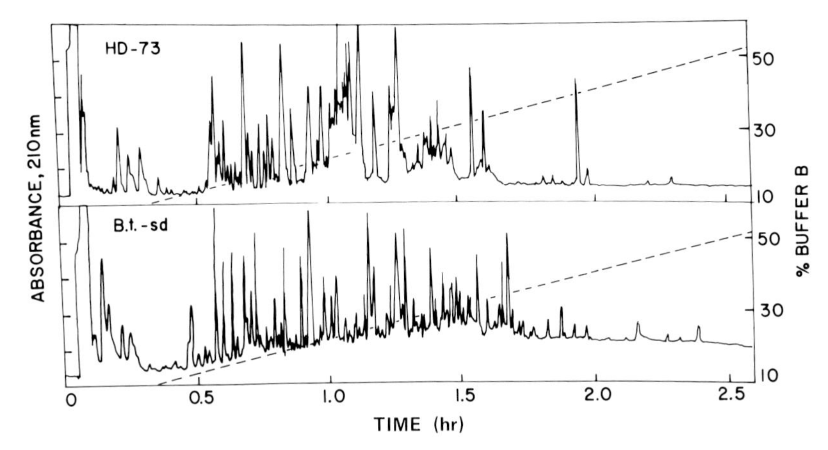

Fig. 6. Tryptic peptide elution profiles of toxins from HD-73 and *B. thuringiensis* var. *san diego*. The HD-73 toxin was prepared by solubilizing toxin crystals with alkali (pH 10.0) followed by digestion with papain and purification of the toxin by HPLC, using a DEAE ion-exchange column (Waters, Milford, Mass.). The toxin was prepared by solubilization of toxin crxstals in alkali followed by HPLC chromatography on DEAE. Tryptic peptides were prepared as described (Herrnstadt *et al.*, 1986). Peptides were dissolved in 1% trifluoroacetic acid and injected onto a C_4 reversed-phase HPLC column. Elution was with a 9–52% linear gradient of acetonitrile in 1% trifluoroacetic acid. The scale at the right gives the percent acetonitrile in the solution. (From Herrnstadt *et al.*, 1986. © 1986, *Bio/Technology*. Used by permission.)

VI. DISCUSSION

The *B. thuringiensis* var. *san diego* isolate described here has novel biological activity. This organism is of pathotype "C" as described by Krieg *et al.* (1983) and has activity only against coleopteran insects. Insecticidal activities against the alfalfa weevil, the boll weevil, and the Colorado potato beetle are indicators of the commercial importance of this strain.

The *B. thuringiensis* var. *san diego* toxin appears to be entirely different from the lepidopteran-specific toxins. It does not cross-react immunologically and the peptide profile of tryptic fragments is entirely different. The crystal protein found in the toxin crystals does not appear to be processed to a smaller molecular weight active form.

The identification of the *B. thuringiensis* var. *san diego* strain also provides encouragement that additional strains of *B. thuringiensis* or other insecticidal microorganisms with activity against economically important pests can be identified and developed as safe, effective biocontrol agents.

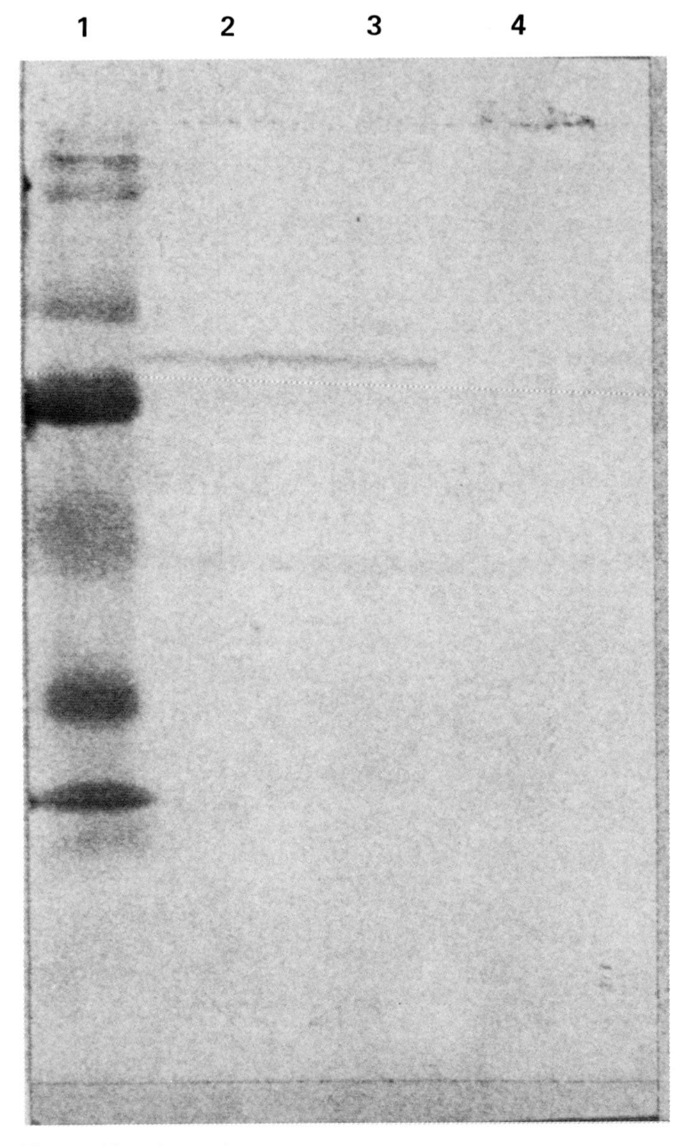

Fig. 7. Western blot of recombinant *E. coli* containing *B. thuringiensis* var. *san diego* toxin. SDS gels of recombinant cell were assayed with antibodies to the toxin as described by Burnette (1981). Lane 1, protein standards; lane 2, recombinant *E. coli* with 5.8 kb *Bam*HI insert in one orientation; lane 3, recombinant *E. coli* with 5.8 kb *Bam*HI fragment in the opposite orientation; lane 4, *E. coli* control.

ACKNOWLEDGMENTS

We thank Vanessa Lefevre, Steve McElfresh, Jewel Payne, Leonard Wittwer, and Siu-Yin Wong for excellent technical assistance.

REFERENCES

Adang, M., Staver, M., Rocheleau, T., Leighton, J., Barker, R., and Thompson, D. (1985). Characterization of full length and truncated plasmid clones of the crystal protein of *Bacillus thuringiensis* subsp. *kurstaki* HD-73 and their toxicity to *Manduca sexta*. *Gene* **36,** 289–300.

Aronson, A., Beckman, W., and Dunn, P. (1986). *Bacillus thuringiensis* and related insect pathogens. *Microbiol. Rev.* **50,** 1–24.

Burnette, W. (1981). "Western blotting": Electrolytic transfer of proteins from sodium dodecyl sulfate-acrylamide gels to unmodified nitrocellulose and radiographic detection with antibody and radioiodinated protein A. *Anal. Biochem.* **112,** 195–203.

Dulmage, H. (1981). Insecticidal activity of isolates of *Bacillus thuringiensis* and their potential for pest control. *In* "Microbial Control of Pests and Plant Diseases 1970–1980" (H. D. Burges, ed.), pp. 193–222. Academic Press, London.

Dulmage, H., de Barjac, H., Reich, D., Donaldson, G., and Krywienczyk, J. (1981). "*Bacillus thuringiensis* Cultures." Available from the U.S. Department of Agriculture, Agricultural Research Service, New Orleans, Louisiana.

Finney, D. J. (1971). "Probit Analysis," 3rd ed. Cambridge Univ. Press, London and New York.

Herrnstadt, C., Soares, G., Wilcox, E., and Edwards, D. (1986). A new strain of *Bacillus thuringiensis* with activity against coleopteran insects. *Bio/Technology* **4,** 305–308.

Klier, A., Fargette, F., Ribier, J., and Rapoport, G. (1982). Cloning and expression of the crystal protein genes from *Bacillus thuringiensis* strain *berliner* 1715. *EMBO J.* **1,** 791–799.

Krieg, A., Huger, A., Langenbruch, G., and Schnetter, W. (1983). *Bacillus thuringiensis* var. *tenebrionis:* Ein neuer gegenüber Larven von Coleopteren wirksamer Pathotyp. *Z. Angew. Entomol.* **96,** 500–508.

Krieg, A., Huger, A., Langenbruch, G., and Schnetter, W. (1984). Neue Ergebnisse über *Bacillus thuringiensis* var. *tenebrionis* unter besonderer Berücksichtigung seiner Wirkung auf den Kartoffelkäfer (*Leptinotarsa decemlineata*). *Angew. Schaedlingskd, Pflanzenschutz, Umweltschutz.* **57,** 145–150.

Schnepf, H., and Whiteley, H. (1981). Cloning and expression of the *Bacillus thuringiensis* crystal protein gene in *Escherichia coli*. *Proc. Natl. Acad. Sci. U.S.A.* **78,** 2893–2897.

Schnepf, H., Wong, H., and Whiteley, H. (1985). The amino acid sequence of a crystal protein from *Bacillus thuringiensis* deduced from the DNA base sequence. *J. Biol. Chem.* **260,** 6264–6257.

Shibano, Y., Yamagata, A., Nakamura, N., Izuka, Y., Sugisaki, H., and Takanami, M. (1985). Nucleotide sequence coding for the insecticidal fragment of the *Bacillus thuringiensis* crystal protein. *Gene* **34,** 243–251.

Thorne, L., Garduno, F., Thompson, T., Decker, D., Zounes, M., Wild, M., Walfield, A. M., and Pollock, T. (1986). Structural similarity between the Lepidoptera- and Diptera-specific insecticidal endotoxin genes of *Bacillus thuringiensis* subsp. "kurstaki" and "israelensis." *J. Bacteriol.* **166,** 801–811.

9

Parasporal Body of Mosquitocidal Subspecies of Bacillus thuringiensis

BRIAN A. FEDERICI, JORGE E. IBARRA,
LEODEGARIO E. PADUA, NIELS J. GALJART,
AND NATARAJAN SIVASUBRAMANIAN

I. INTRODUCTION

For most of the time since its discovery around the turn of this century, the insecticidal bacterium *Bacillus thuringiensis* has been considered primarily a pathogen of lepidopterous larvae (Heimpel, 1967; Faust and Bulla, 1982). Owing largely to interest in its insecticidal properties, literally hundreds of isolates of this species have been col-

BIOTECHNOLOGY IN INVERTEBRATE
PATHOLOGY AND CELL CULTURE

lected from around the world and grouped into different serotypes and varieties or subspecies based primarily on the immunological characteristics of flagellar antigens (de Barjac, 1981). Studies of several different serotypes have demonstrated that the toxicity of most isolates, with the exception of those that produce the broad-spectrum β-exotoxin, is due to a protein of approximately 135 kDa which is synthesized during sporulation and assembled into a bipyramidal crystalline parasporal body (Huber and Luethy, 1981). This protein is actually a protoxin which after ingestion by larvae is cleaved by midgut proteases, yielding an active toxin of about 67 kDa, referred to as the δ-endotoxin (Lecadet and Dedonder, 1967). By 1976, thirteen serotypes of *B. thuringiensis* were recognized, none of which produced an endotoxin with significant toxicity to anything other than lepidopterous larvae. Then in 1977, Goldberg and Margalit reported the isolation of a spore-forming bacterium from Israel that was highly toxic to mosquito larvae. This bacterium was shown by de Barjac (1978) to be a new serotype, H-14, of *B. thuringiensis* and designated variety/subspecies *israelensis*. The general mode of action of this subspecies was shown to be similar to that of other subspecies in that death resulted from ingestion of the parasporal body, which destroyed the midgut epithelium of susceptible insects, in this case the larvae of nematocerous dipterans such as mosquitoes and black flies (Undeen and Nagel, 1978; Tyrell *et al.*, 1979; Charles and de Barjac, 1983). The parasporal body of this subspecies differed, however, from those of other subspecies in that it was basically spherical, contained several different inclusions and proteins (Tyrell *et al.*, 1981), and when solubilized in alkali was cytolytic for a variety of invertebrate and vertebrate cells, including erythrocytes, and toxic to mice on injection (Thomas and Ellar, 1983a).

The initial reports of the unique characteristics of *B. thuringiensis* subsp. *israelensis*, and its rapid development as an operational larvicide, attracted considerable interest and were followed by reports of other mosquitocidal isolates of *B. thuringiensis,* including isolates of subspecies *darmstadiensis* (Padua *et al.*, 1980) and *morrisoni* (Padua *et al.*, 1984). Although the parasporal bodies of these isolates have received relatively little study to date, when compared with the parasporal body of *B. thuringiensis* subsp. *israelensis,* interesting and potentially important similarities and differences are already apparent. In this brief presentation, we will summarize the pertinent information regarding the structure, composition, and toxicity of the mosquitocidal parasporal bodies of *B. thuringiensis,* with emphasis on subspecies *israelensis* and the PG-14 isolate of subspecies *morrisoni.*

II. PARASPORAL BODY OF *Bacillus thuringiensis* subsp. *israelensis*

The first isolate of this subspecies was obtained from a *Culex pipiens* breeding site in the Negev desert of Israel (Goldberg and Margalit, 1977; Margalit and Dean, 1985). This isolate was originally labeled ONR 60A, but numerous derivatives of it have been made and given different codes and numbers, e.g., 1884, 1897, and HD-567. Although it cannot be stated with certainty, based on plasmid patterns, it appears that most of the strains used for laboratory studies or in commercial formulations are closely related to the original ONR 60A isolate.

Numerous studies beginning with those of Goldberg and Margalit (1977) have documented the high toxicity of this subspecies to mosquitoes and blackflies (for reviews see Lacey, 1985; Lacey and Undeen, 1986). It does not produce detectable levels of β-exotoxin (de Barjac, 1978), and thus this toxicity is due primarily to the protein inclusions present in the parasporal body. Here we will describe the structure and protein composition of the parasporal body and then discuss its toxicity in terms of these components.

A. Structure

The parasporal body of *B. thuringiensis* subsp. *israelensis* is spherical, 0.7–1.2 μm in diameter, and composed of at least three different types of inclusions (Fig. 1). These inclusions differ from one another to some extent in size, shape, and electron density, and each is enveloped in one or more layers of a meshlike envelope, which also surrounds the entire parasporal body (Huber and Luethy, 1981; Tyrell *et al.*, 1981; Charles and de Barjac, 1982; Mikkola *et al.*, 1982; Insell and Fitz-James, 1985; Lee *et al.*, 1985; Ibarra and Federici, 1986a). The largest inclusion is typically rounded to angular and the least electron dense of the three inclusion types (Fig. 1b). It is estimated that this inclusion represents 30–50% of the protein within the parasporal body, and, based on the relative abundance of major parasporal body proteins detected by sodium dodecyl sulfate-polyacrylamide gel electrophoresis (SDS-PAGE; Fig. 2), this inclusion is thought to contain the cytolytic 28-kDa protein (Insell and Fitz-James, 1985; Ibarra and Federici, 1986a). The two other inclusion types readily observed within the parasporal body are a moderately dense polyhedral inclusion, which often appears bar-shaped in transverse sections, and a dense, rounded to spherical inclusion (Fig. 1b). Again, based on the relative abundance of parasporal body proteins

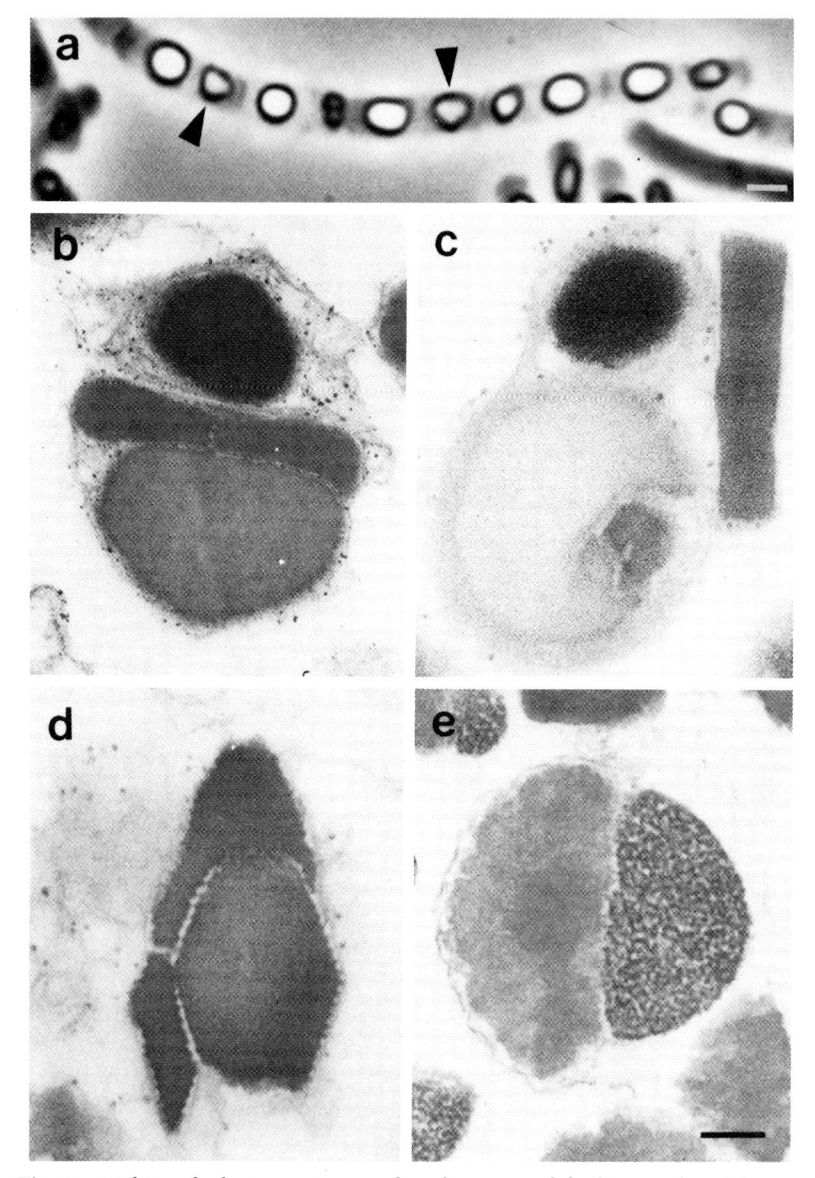

Fig. 1. Light and electron micrographs of parasporal bodies produced by mosquitocidal subspecies of *B. thuringiensis*. (a) Wet mount of a sporulated culture of subspecies *darmstadiensis* (73-E-10-2) illustrating spherical to irregular parasporal bodies (arrowheads) adjacent to spores. (b–3) Ultrathin sections through parasporal bodies of (b) subspecies *israelensis*, (c and d) subspecies *morrisoni* (PG-14), and (e) subspecies *darmstadiensis*. Note the different shapes and densities of the inclusions that make up the parasporal body. The quasi-bipyramidal inclusion from PG-14 (d) is typically found enveloped within the parasporal body. Bar in (a) = 500 nm. Micrographs in (b)–(e) are approximately the same magnification; bar in (e) = 200 nm.

observed by SDS-PAGE, these latter inclusions are thought to contain, respectively, proteins of 65 and 135 kDa. In regard to the densest inclusion type, Mikkola *et al.* (1982) have shown that two inclusions of this type can occur within the parasporal body. This is an interesting observation because there is some evidence that the parasporal body of *B. thuringiensis* subsp. *israelensis* actually contains two high-molecular-weight proteins (Pfannenstiel *et al.*, 1984; Ibarra and Federici, 1986b), with masses of 126 and 135 kDa, and thus each of these may be present in a separate inclusion.

B. Protein Composition

The intact parasporal body of *B. thuringiensis* subsp. *israelensis* is composed of a series of proteins ranging from 28 to around 140 kDa, as determined by SDS-PAGE and column chromatography (Tyrell *et al.*, 1981; Thomas and Ellar, 1983a; Pfannenstiel *et al.*, 1984; Armstrong *et al.*, 1985; Chestukina *et al.*, 1985; Hurley *et al.*, 1985; Insell and Fitz-James, 1985; Sriram *et al.*, 1985). The number of proteins reported varies and is dependent largely on the method of parasporal body preparation and degree of degradation prior to analysis. If on the basis of relative abundance only major proteins are considered, the parasporal body contains only three or four (Fig. 2). Allowing for some variation in size estimates for the same protein by different investigators, the major proteins are those of 28, 65, and 130 kDa. In several studies the latter protein has been resolved by SDS-PAGE into two with sizes of about 126 and 135 kDa, respectively. The 28-kDa protein has recently been shown, through analysis of the gene encoding it, to consist of 249 amino acids with a mass of 27,340 Da (Waalwijk *et al.*, 1985).

In addition to these proteins, several minor proteins with sizes of about 25, 36–40, 53, and 67 kDa are also observed when intact parasporal bodies are analyzed by SDS-PAGE. If parasporal bodies are solubilized in alkali, these minor proteins increase in amount concomitantly with a decrease in the amounts of the major proteins. Thus, most of the minor proteins are thought to result from cleavage of the major proteins by alkaline proteases associated with the parasporal body (Chilcott *et al.*, 1983). A similar proteolysis most likely occurs in the insect gut and, as in other species of *B. thuringiensis*, is probably important in activating toxins in the *B. thuringiensis* subsp. *israelensis* parasporal body. At this point, there is evidence in many of the studies cited above that the 25-kDa protein results from cleavage of the 28-kDa protein, the 35- to 40-kDa proteins from the 65-kDa protein, and the 53- and 67-kDa proteins from the two proteins of around 130 kDa.

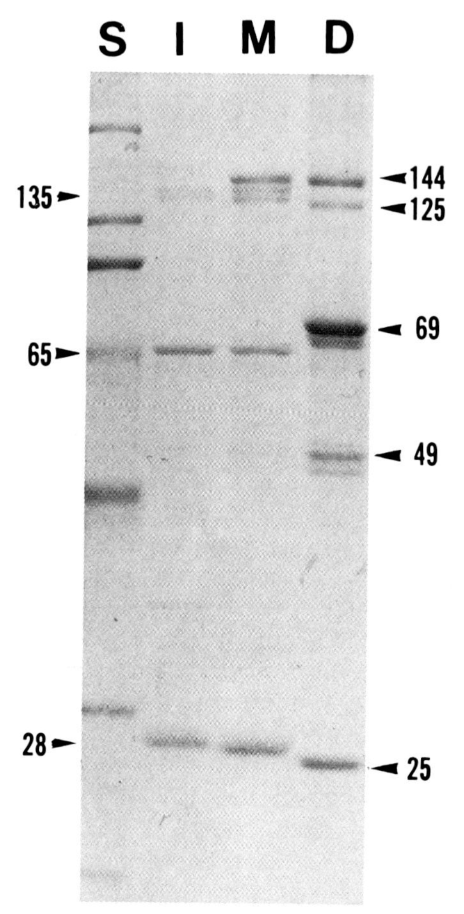

Fig. 2. Comparison of the parasporal body protein composition of mosquitocidal subspecies of *Bacillus thuringiensis*. S, Molecular weight standards; I, subspecies *israelensis;* M, subspecies *morrisoni* (PG-14); D, subspecies *darmstadiensis* (73-E-10-2).

C. Toxicity

Although there is general agreement that the toxicity of *B. thuringiensis* subsp. *israelensis* is due to the parasporal body, considerable disagreement exists in regard to which of the parasporal body proteins is essential to mosquitocidal activity. Since Thomas and Ellar (1983a) demonstrated that the alkali-solubilized parasporal body was mosquitocidal and cytolytic, there has been a trend to assign mosquitocidal activity to only one protein, thereby excluding others from such a role. For example, several initial as well as a few more recent studies reported that the 25/28-kDa protein is the mosquitocidal toxin and is also the protein responsible for the broad cytolytic activity of this subspecies (Thomas and Ellar, 1983a,b; Yamamoto *et al.*, 1983; Davidson and Yamamoto, 1984; Armstrong *et al.*, 1985; Insell and Fitz-James, 1985; Sriram *et al.*, 1985). In other studies it has been reported that the 65-kDa protein (Hurley *et al.*, 1985; Lee *et al.*, 1985) or the 130-kDa protein (Sekar, 1986; Visser *et al.*, 1986), or even a protein of around 31–35 kDa (Cheung and Hammock, 1985), is the mosquitocidal toxin. The broad cytolytic activity of the 25/28-kDa protein has been documented in many of these studies, and the controversy at present is over the degree to which this protein and each of the others within the parasporal body is mosquitocidal. A detailed discussion of the data regarding the properties of each protein is not appropriate here, but the following summary derived from the studies cited above may provide some insight into which proteins contribute to the parasporal body's toxicity.

When assayed against first or second instars of *Aedes aegypti*, the LC_{50} for the intact parasporal body is generally in the range of <0.05 to 2 ng/ml. Against third or fourth instars, the LC_{50} is four- to fivefold this amount. Toxicity of the parasporal body decreases by tenfold, or often more, once it is solubilized. Because mosquitoes feed on particulate matter, this loss of toxicity is thought to be due to lower rates of toxin ingestion. The latter creates problems in assessing the toxicity of purified proteins because typically they are assayed in a soluble form. This problem can be partially overcome by precipitating the protein or attaching it to latex beads. In any case, the toxicity of most proteins is significantly less than that of the intact parasporal body, and, which is important, also much less than that of the solubilized parasporal body. Typically, the LC_{50} values reported for purified solubilized proteins range from several hundred to well over 1000 ng/ml, indicating that these proteins are from 20 to more than 100 times less toxic than the solubilized parasporal body. A similar loss in toxicity is found when values for purified proteins bound to latex beads or precipitated are compared with values for intact parasporal bodies. Thus, it appears that on an individual basis no one of the proteins, even in particulate form, will prove to be as toxic as the intact parasporal body. The above studies do, however, provide evidence that the major proteins differ in toxicity to mosquitoes, and at this point the data indicate that the 65- and 130-kDa proteins are more toxic than the 28-kDa protein.

An important implication from these studies is that the high toxicity of the parasporal body may be due to a synergistic interaction of two or more proteins. Several recent studies have provided data which support such an interaction of parasporal body proteins. For example, Wu and Chang (1985) found that the toxicity of the 26-, 65-, or 130-kDa proteins tested alone was very low in comparison to mixtures of the 26- and 65-kDa, 26- and 130-kDa, or 26-, 65-, and 130-kDa proteins. In another study, Ibarra and Federici (1986a) showed that the bar-shaped polyhedral inclusion, which contains a 65-kDa protein, was much less toxic than the intact parasporal body, but that contamination with small amounts of the 28-kDa protein significantly increased this inclusion's toxicity.

In summary, data currently available suggest that the mosquitocidal properties of the parasporal body are not due to a single protein but rather to an interaction, perhaps synergistic, of two or more proteins.

III. PARASPORAL BODY OF *Bacillus thuringiensis* subsp. *morrisoni* (PG-14)

The PG-14 isolate of *B. thuringiensis* subsp. *morrisoni* (H-8a : 8b) was obtained from a soil sample collected from a canal in Cebu City, the Philippines (Padua *et al.*, 1981). Studies by Padua *et al.* (1984) demonstrated that this isolate, in contrast to the reference strain, produces a spherical or irregular parasporal body, is negative for β-exotoxin production, and is highly toxic to mosquitoes, but not to larvae of *Bombyx mori*. In assays of the spore–parasporal body complex against larvae of *A. aegypti* and *Culex molestus*, the PG-14 isolate was as toxic as *B. thuringiensis* subsp. *israelensis* (Table I). More recent studies by Ibarra and Federici (1986b) and Gill *et al.* (1987) have shown that the parasporal body of PG-14 is similar but not identical to the parasporal body of *B. thuringiensis* subsp. *israelensis*.

A. Structure

Like *B. thuringiensis* subsp. *israelensis*, the parasporal body of PG-14 is essentially spherical, about 1 μm in diameter, and consists of several different types of inclusions (Figs. 1c and 1d). Each inclusion is surrounded by a meshlike envelope, which also surrounds the entire parasporal body, holding the inclusions together. In PG-14 there are four different types of inclusions, three of which are the same as the three types that occur in the parasporal body of *B. thuringiensis* subsp. *israelensis*. These are (1) a large, angular to rounded inclusion of low electron density, (2) a bar-shaped inclusion of moderate electron density, and (3) a hemispherical to spherical inclusion of high electron density. The fourth inclusion type is a quasi-bipyramidal inclusion, generally of high electron density. This latter type, which does not occur in

TABLE I
Comparative Toxicities of Spore–Parasporal Body Complexes of *B. thuringiensis* subsp. *israelensis* and the PG-14 Isolate of *B. thuringiensis* subsp. *morrisoni*[a]

	LC_{50} (mg/liter)	
Species	*B. thuringiensis* subsp. *morrisoni* PG-14 (serotype 8a : 8b)	*B. thuringiensis* subsp. *israelensis* (serotype 14)
Aedes aegypti	0.0204[b]	0.0212[b]
	0.0436[c]	0.0516[c]
Culex molestus	0.0176[b]	0.0186[b]
	0.0181[c]	0.0329[c]
Bombyx mori	>100	>100

[a]From Padua *et al.* (1984).
[b]Two-day-old larvae.
[c]Fourth instars.

B. thuringiensis subsp. *israelensis*, is usually bipartite, consisting of a central polyhedral core and outer layer (Fig. 1a).

B. Protein Composition

When analyzed by SDS-PAGE, the intact parasporal body of PG-14 exhibits five major proteins with sizes, respectively, of about 28, 65, 126, 135, and 144 kDa (Fig. 2). The 28-, 65-, 126-, and 135-kDa proteins essentially comigrate with the respective proteins of similar size in *B. thuringiensis* subsp. *israelensis*, although the 28-kDa protein of PG-14 usually migrates slightly ahead of the homologous protein of *B. thuringiensis* subsp. *israelensis*, indicating that the former is somewhat smaller. The correspondence of each protein with a specific inclusion body type is thought to be the same as in *B. thuringiensis* subsp. *israelensis* (see Sections II,A and II,B). Extending the correlation of individual proteins with specific inclusions, the 144-kDa protein of PG-14 is thought to occur in the quasi-bipyramidal inclusion as this protein and inclusion type occur only in this isolate.

The similarity in the size of proteins shared by *B. thuringiensis* subsp. *israelensis* and subsp. *morrisoni* PG-14 is only circumstantial evidence that these proteins are indeed similar or have the same function. However, more convincing evidence for homology among proteins of similar size from these two isolates has been provided by Gill *et al.*

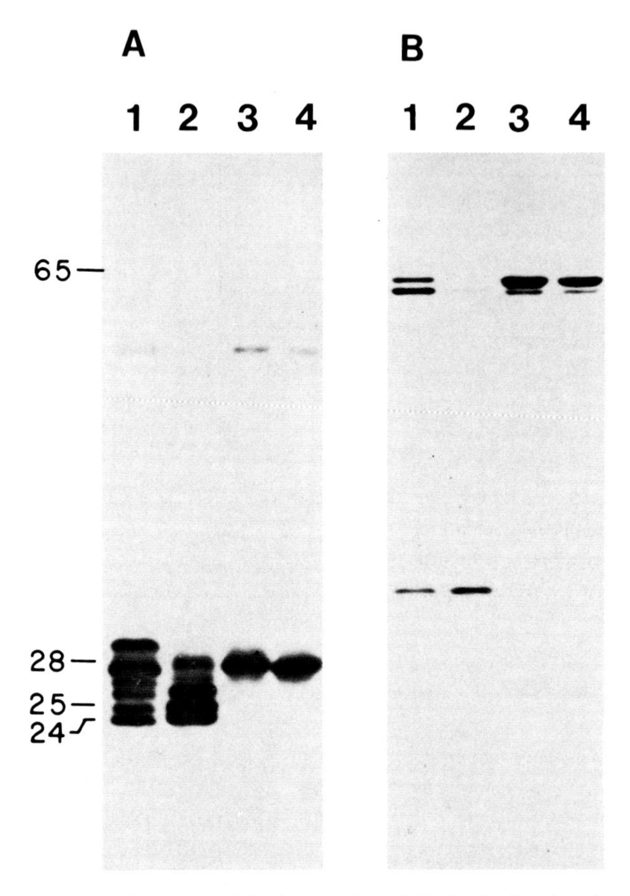

Fig. 3. Homology of parasporal body proteins of *B. thuringiensis* subsp. *israelensis* and *morrisoni* (PG-14) determined by immunoblotting. A. Probed with antibodies raised in rabbits against the 25-kDa protein of subsp. *israelensis*. B. Probed with antibodies raised against the 65-kDa protein of subsp. *israelensis*. For both A and B, the lanes are as follows: lane 1, parasporal bodies of subsp. *morrisoni* solubilized in Na_2CO_3·HCl; lane 2, parasporal bodies of subsp. *israelensis* solubilized in Na_2CO_3·HCl; lane 3, parasporal bodies of subsp. *israelensis* subjected to electrophoresis without prior solubilization in Na_2CO_3·HCl; lane 4, parasporal bodies of subsp. *morrisoni* subjected to electrophoresis without prior solubilization in Na_2CO_3·HCL. See Gill *et al.* (1987) for details of experimental methods.

(1987). They found that polyclonal antibodies raised against the 25- or 65-kDa proteins of *B. thuringiensis* subsp. *israelensis* cross-reacted extensively with the proteins of similar size from PG-14 (Fig. 3). In addition, a partial amino acid sequence we determined for the N-terminal end of the 65-kDa protein of PG-14 is virtually identical to the sequence

TABLE II

N-Terminal Amino Acid Sequences of the 65-kDa Proteins from *B. thuringiensis* subsp. *israelensis*[a] and the PG-14[b] Isolate of *B. thuringiensis* subsp. *morrisoni*

Subsp. *israelensis*	Met-Glu-Asn-Xaa-Pro-Leu-Asp-Thr-Leu-Ser-iLeu-Val-Asn-Glu-Thr
PG-14	Met-Glu-Asn-Ser -Ser-Leu-Asp-Thr-Leu-Ser-iLeu-Val-Asn-Glu-Thr

[a]From Chestukhina *et al.* (1985).
[b]Sequence was determined using a highly purified preparation of the bar-shaped polyhedral inclusion isolated from parasporal bodies (see Fig. 1). This inclusion consists almost exclusively of a 65-kDa protein.

for this region that Chestukhina *et al.* (1985) determined for a protein of similar size from *B. thuringiensis* subsp. *israelensis* (Table II). But perhaps the best evidence that the 27-, 65-, 126-, and 135-kDa proteins of PG-14 are essentially homologous to the proteins of a corresponding size in *B. thuringiensis* subsp. *israelensis* comes from the nucleotide sequence we recently determined for the gene encoding the PG-14 27-kDa protein (Galjart *et al.*, 1987). This sequence, illustrated in Fig. 4, demonstrated that the coding region of the gene, except for a single base change, C to G, at position 310, is identical to the sequence of the gene encoding the 27-kDa protein of *B. thuringiensis* subsp. *israelensis* determined by Waalwijk *et al.* (1985). This single base change results in a single amino acid substitution, proline to alanine, at position 82 (Fig. 4).

In regard to other proteins within the parasporal body of PG-14, as in *B. thuringiensis* subsp. *israelensis*, dissolution and proteolytic cleavage under alkaline conditions generates a series of additional peptides. Most of these occur in minor quantities in the intact parasporal body and fall within the following size ranges; 24/25 kDa, 35–40 kDa, 53 kDa, and 67 kDa. In addition to these, several peptides in the range of 50–60 kDa occur in PG-14 and are thought to be derived from the 144-kDa protein (Ibarra and Federici, 1986b).

C. Toxicity

The purified parasporal body of PG-14 is very similar in toxicity to that of *B. thuringiensis* subsp. *israelensis*. In parallel bioassays against fourth instars of *A. aegypti*, Ibarra and Federici (1986b) obtained LC_{50} values of 2.97 ± 0.63 and 3.36 ± 0.48 ng/ml, respectively, for the parasporal bodies of PG-14 and *B. thuringiensis* subsp. *israelensis*.

The proteins of PG-14 have not yet been isolated or tested for toxicity

a

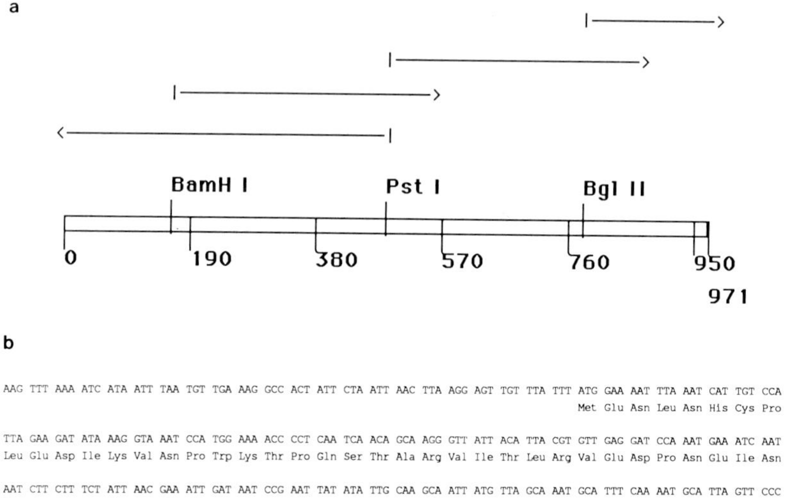

b

```
AAG TTT AAA ATC ATA ATT TAA TGT TGA AAG GCC ACT ATT CTA ATT AAC TTA AGG AGT TGT TTA TTT ATG GAA AAT TTA AAT CAT TGT CCA    90
                                                                                         Met Glu Asn Leu Asn His Cys Pro

TTA GAA GAT ATA AAG GTA AAT CCA TGG AAA ACC CCT CAA TCA ACA GCA AGG GTT ATT ACA TTA CGT GTT GAG GAT CCA AAT GAA ATC AAT   180
Leu Glu Asp Ile Lys Val Asn Pro Trp Lys Thr Pro Gln Ser Thr Ala Arg Val Ile Thr Leu Arg Val Glu Asp Pro Asn Glu Ile Asn

AAT CTT CTT TCT ATT AAC GAA ATT GAT AAT CCG AAT TAT ATA TTG CAA GCA ATT ATG TTA GCA AAT GCA TTT CAA AAT GCA TTA GTT CCC   270
Asn Leu Leu Ser Ile Asn Glu Ile Asp Asn Pro Asn Tyr Ile Leu Gln Ala Ile Met Leu Ala Asn Ala Phe Gln Asn Ala Leu Val Pro

ACT TCT ACA GAT TTT GGT GAT GCC CTA CGC TTT AGT ATG GCA AAA GGT TTA GAA ATC GCA AAC ACA ATT ACA CCG ATG GGT GCT GTA GTG   360
Thr Ser Thr Asp Phe Gly Asp Ala Leu Arg Phe Ser Met [Ala] Lys Gly Leu Glu Ile Ala Asn Thr Ile Thr Pro Met Gly Ala Val Val

AGT TAT GTT GAT CAA AAT GTA ACT CAA ACG AAT AAC CAA GTA AGT GTT ATG ATT AAT AAA GTC TTA GAA GTG TTA AAA ACT GTA TTA GGA   450
Ser Tyr Val Asp Gln Asn Val Thr Gln Thr Asn Asn Gln Val Ser Val Met Ile Asn Lys Val Leu Glu Val Leu Lys Thr Val Leu Gly

GTT GCA TTA AGT GGA TCT GTA ATA GAT CAA TTA ACT GCA GCA GTT ACA AAT ACG TTT ACA AAT TTA AAT ACT CAA AAA AAT GAA GCA TGG   540
Val Ala Leu Ser Gly Ser Val Ile Asp Gln Leu Thr Ala Ala Val Thr Asn Thr Phe Thr Asn Leu Asn Thr Gln Lys Asn Glu Ala Trp

ATT TTC TGG GGC AAG GAA ACT GCT AAT CAA ACA AAT TAC ACA TAC AAT GTC CTG TTT GCA ATC CAA AAT GCC CAA ACT GGT GGC GTT ATG   630
Ile Phe Trp Gly Lys Glu Thr Ala Asn Gln Thr Asn Tyr Thr Tyr Asn Val Leu Phe Ala Ile Gln Asn Ala Gln Thr Gly Gly Val Met

TAT TGT GTA CCA GTT GGT TTT GAA ATT AAA GTA TCA GCA GTA AAG GAA CAA GTT TTA TTT TTC ACA ATT CAA GAT TCT GCG AGC TAC AAT   720
Tyr Cys Val Pro Val Gly Phe Glu Ile Lys Val Ser Ala Val Lys Glu Gln Val Leu Phe Phe Thr Ile Gln Asp Ser Ala Ser Tyr Asn

GTT AAC ATC CAA TCT TTG AAA TTT GCA CAA CCA TTA GTT AGC TCA AGT CAG TAT CCA ATT GCA GAT CTT ACT AGC GCT ATT AAT GGA ACC   810
Val Asn Ile Gln Ser Leu Lys Phe Ala Gln Pro Leu Val Ser Ser Ser Gln Tyr Pro Ile Ala Asp Leu Thr Ser Ala Ile Asn Gly Thr

CTC TAA TCT TAG TAG CTA TAT TTA TTA AAG ATG GTA ATA TCA CAA GTA TAA ATA CTT GTG GTA TTA CCT ACC ATT CTT AAA TTA TAT CCA   900
Leu

AAA TCA TGC GTT AAT CTA CAT TCC CCT TTC TCT AAA ATT TGT TCT TCA CAC ATC CAC ATT TTT CGA TAA AA                            971
```

Fig. 4. The nucleotide sequence of the gene encoding the 27.3-kDa cytolytic protein of *Bacillus thuringiensis* subsp. *morrisoni* (PG-14). (a) The sequencing strategy used to determine the nucleotide sequence of the gene and its immediate flanking regions. (b) The nucleotide sequence of the gene and the predicted amino acid sequence from the coding region of the gene. The amino acid sequence differs from the 27-kDa protein of *B. thuringiensis* subsp. *israelensis* only in that in PG-14 alanine is substituted for proline at position 82 due to a single base change.

individually. However, the immunological relatedness of the 25- and 65-kDa proteins of PG-14 to the respective proteins of *B. thuringiensis* subsp. *israelensis* indicates that their mosquitocidal and cytolytic properties will be very similar to those of the homologous proteins in the latter.

IV. PARASPORAL BODY OF *Bacillus thuringiensis* subsp. *darmstadiensis* (73-E-10-2)

The interest in isolates of *B. thuringiensis* that produce mosquitocidal parasporal bodies makes it appropriate to examine the recent but limited data regarding the parasporal body of *B. thuringiensis* subsp. *darmstadiensis*, serotype 10. The mosquitocidal isolate discussed here, 73-E-10-2, was obtained from silkworm litter collected in Japan and first determined to be toxic to mosquitoes by Padua *et al.* (1980).

A. Structure

The parasporal body of this subspecies is spherical, approximately 1 μm in diameter, and appears to consist of only two major inclusions (Mikkola *et al.*, 1982; Iizuka *et al.*, 1983). These inclusions are about equal in size and hemispherical, but our recent studies have demonstrated a considerable difference in their ultrastructure (Fig. 1e). One of the inclusions is of low electron density, fairly uniform throughout, and often exhibits a repetitive lattice. The other inclusion, though generally more electron dense, consists of fine interconnected dense regions intermixed with electron-translucent lacunae. As in *B. thuringiensis* subsp. *israelensis* and subsp. *morrisoni*, these inclusions are surrounded by a meshlike envelope, but it is not known whether each inclusion is also individually enveloped.

B. Protein Composition

Analysis of *B. thuringiensis* subsp. *darmstadiensis* parasporal bodies by SDS-PAGE reveals around five major proteins with sizes of 25, 49, 70, 125, and 144 kDa (Fig. 2a). Though similar in size to proteins that occur in the parasporal bodies of *B. thuringiensis* subsp. *israelensis* and PG-14, our recent Western blot analysis of the 25-kDa protein using polyclonal antibodies prepared against the 25-kDa protein of *B. thuringiensis* subsp. *israelensis* revealed no immunological cross-activity. In addition to the major proteins, a series of minor proteins are found in parasporal body preparations of *B. thuringiensis* subsp. *darmstadiensis*. The parasporal bodies have proved more difficult to purify than those of *B. thuringiensis* subsp. *israelensis* and PG-14, and these minor proteins may result from contamination or degradation.

C. Toxicity

Kim *et al.* (1984) purified a 67-kDa protein from the spore–parasporal body complex of *B. thuringiensis* subsp. *darmstadiensis*. This protein had an LC_{50} of 16.8 μg/ml for 4-day-old larvae at *A. aegypti* and was not related immunologically to a protein of similar size from the subspecies *israelensis*. Interestingly, these investigators found the *B. thuringiensis* subsp. *israelensis* protein to be about 20 times more toxic than the *B. thuringiensis* subsp. *darmstadiensis* protein.

In our preliminary studies, the partially purified parasporal body had an LC_{50} of about 110 ng/ml when assayed against first instars of *A. aegypti*. Although toxins in particulate form would be expected to be more toxic, this is much more toxic than the 67-kDa protein and, as in *B. thuringiensis* subsp. *israelensis* and PG-14, the toxicity may be due to more than one protein.

V. CLOSING REMARKS

The mosquitocidal parasporal bodies of *B. thuringiensis* have been shown to be much more complex in regard to structure and protein composition than the parasporal bodies of subspecies active against lepidopterous insects. If this complexity is shown to be essential for high mosquitocidal activity, determination of the function of each essential protein becomes of interest because this knowledge may provide insights into enhancing the toxicity of bacterial insecticides in general. For example, if the 25-kDa protein of *B. thuringiensis* subsp. *israelensis* potentiates the toxicity of parasporal body proteins by distorting membrane integrity, it may be capable of enhancing many other insecticidal toxins.

Although the 25-kDa protein of both *B. thuringiensis* subsp. *israelensis* and PG-14 is of major interest because of its broad cytolytic activity, the functions of the 65-, 126-, 135-, and 144-kDa proteins are also of interest because these proteins are similar in size to the protoxins or δ-endotoxins of other subspecies of *B. thuringiensis*, yet differ considerably in host range. Thus, genetic analysis should contribute to defining how specific amino acid changes in these proteins affect toxin specificity. Determination of the toxicity and specificity of the 144-kDa protein in PG-14 will be of particular interest because this protein appears to be capable of forming a true bipyramidal crystal, yet has no known toxicity to lepidopterous larvae.

REFERENCES

Armstrong, J. L., Rohrmann, G. F., and Beaudreau, G. S. (1985). Delta endotoxin of *Bacillus thuringiensis* subsp. *israelensis*. *J. Bacteriol.* **161**, 39–46.

Charles, J.-F., and de Barjac, H. (1982). Sporulation et cristallogénèse de *Bacillus thuringiensis* var. *israelensis* en microscopie électronique. *Ann. Microbiol.* (*Paris*) **133,** 425–442.

Charles, J.-F., and de Barjac, H. (1983). Action des cristaux de *Bacillus thuringiensis* var. *israelensis* sur l'intestin moyen des larves de *Aedes aegypti* L., en microscopie électronique. *Ann. Microbiol.* (*Paris*) **134,** 197–218.

Chestukhina, G. G., Zalunin, I. A., Kostina, L. I., Bormatova, M. E., Klepikova, F. S., Khodova, O. M., and Stepanov, V. M. (1985). Structural features of crystal-forming proteins produced by *Bacillus thuringiensis* subspecies *israelensis*. *FEBS Lett.* **190,** 345–348.

Cheung, P. Y. K., and Hammock, B. D. (1985). Separation of three biologically distinct activities from the parasporal crystal of *Bacillus thuringiensis* var. *israelensis*. *Curr. Microbiol.* **12,** 121–126.

Chilcott, C. N., Kalmakoff, J., and Pillai, J. S. (1983). Characterization of proteolytic activity associated with *Bacillus thuringiensis* var. *israelensis* crystals. *FEMS Microbiol. Lett.* **18,** 37–41.

Davidson, E. W., and Yamamoto, T. (1984). Isolation and assay of the toxic component from the crystal of *Bacillus thuringiensis* var. *israelensis*. *Curr. Microbiol.* **11,** 171–174.

de Barjac, H. (1978). Une nouvelle variété de *Bacillus thuringiensis* tres toxique pour les moustiques: *B. thuringiensis* var. *israelensis* serotype 14. *C. R. Hebd. Seances Acad. Sci., Ser. D* **286,** 797–800.

de Barjac, H. (1981). Identification of H-serotypes of *Bacillus thuringiensis*. *In* "Microbial Control of Pests and Plant Diseases 1970–1980" (H. D. Burges, ed.), pp. 35–43. Academic Press, London.

Faust, R. M., and Bulla, L. A., Jr. (1982). Bacteria and their toxins as insecticides. *In* "Microbial and Viral Insecticides" (E. Kurstak, ed.), pp. 75–208. Dekker, New York.

Galjart, N. J., Sivasubramanian, N., and Federici, B. A. (1987). Plasmid location, cloning, and sequence analysis of a gene encoding the 27.3-kDa cytolytic protein of *Bacillus thuringiensis* subsp. *morrisoni* (PG-14). *Curr. Microbiol.* (in press).

Gill, S. S., Hornung, J. M., Ibarra, J. E., Singh, G. J. P., and Federici, B. A. (1987). Cytolytic activity and immunological similarity of the toxins of *Bacillus thuringiensis* subsp. *israelensis* and *morrisoni* (PG-14). *Appl. Environ. Microbiol.* **53,** 1251–1256.

Goldberg, L. J., and Margalit, J. (1977). A bacterial spore demonstrating rapid larvicidal activity against *Anopheles sergentii, Uranotaenia unguiculata, Culex univitatus, Aedes aegypti*, and *Culex pipiens*. *Mosq. News* **37,** 355–358.

Heimpel, A. M. (1967). A critical review of *Bacillus thuringiensis* var. *thuringiensis* and other crystalliferous bacteria. *Annu. Rev. Entomol.* **12,** 287–322.

Huber, H. E., and Luethy, P. (1981). *Bacillus thuringiensis* delta-endotoxin: Composition and activation. *In* "Pathogenesis of Invertebrate Microbial Diseases" (E. W. Davidson, ed.), pp. 209–233. Allanheld, Osmun, Totowa, New Jersey.

Hurley, J. M., Lee, S. G., Andrews, R. E., Jr., Klowden, M. J., and Bulla, L. A., Jr. (1985). Separation of the cytolytic and mosquitocidal proteins of *Bacillus thuringiensis* subsp. *israelensis*. *Biochem. Biophys. Res. Commun.* **126,** 961–965.

Ibarra, J. E., and Federici, B. A. (1986a). Isolation of a relatively nontoxic 65-kilodalton protein inclusion from the parasporal body of *Bacillus thuringiensis* subsp. *israelensis*. *J. Bacteriol.* **165,** 527–533.

Ibarra, J. E., and Federici, B. A. (1986b). Parasporal bodies of *Bacillus thuringiensis* subsp. *morrisoni* (PG-14) and *Bacillus thuringiensis* subsp. *israelensis* are similar in protein composition and toxicity. *FEMS Microbiol. Lett.* **34,** 79–84.

Iizuka, T., Faust, R. M., and Ohba, M. (1983). Comparative profiles of plasmid DNA and morphology of parasporal crystals in four strains of *Bacillus thuringiensis* subsp. *darmstadiensis*. *Appl. Entomol. Zool.* **18,** 486–494.

Insell, J. P., and Fitz-James, P. C. (1985). Composition and toxicity of the inclusion of *Bacillus thuringiensis* subsp. *israelensis*. *Appl. Environ. Microbiol.* **50,** 56–62.

Kim, K.-H., Ohba, M., and Aizawa, K. (1984). Purification of the toxic protein from *Bacillus thuringiensis* serotype 10 isolate demonstrating a preferential larvicidal activity to the mosquito. *J. Invertebr. Pathol.* **44,** 214–219.

Lacey, L. A. (1985). *Bacillus thuringiensis* serotype H-14. *In* "Biological Control of Mosquitoes" (H. C. Chapman, ed.), Bull. No. 6, pp. 132–158. Am. Mosq. Control Assoc., Fresno, California.

Lacey, L. A., and Undeen, A. H. (1986). Microbial control of black flies and mosquitoes. *Annu. Rev. Entomol.* **31,** 265–296.

Lecadet, M. M., and Dedonder, R. (1967). Enzymatic hydrolysis of the crystals of *Bacillus thuringiensis* by proteases of *Pieris brassicae*. I. Preparation and fractionation of the lysates. *J. Invertebr. Pathol.* **9,** 310–321.

Lee, S. G., Eckblad, W., and Bulla, L. A., Jr. (1985). Diversity of protein inclusion bodies and identification of mosquitocidal protein in *Bacillus thuringiensis* subsp. *israelensis*. *Biochem. Biophys. Res. Commun.* **126,** 953–960.

Margalit, J., and Dean, D. (1985). The story of *Bacillus thuringiensis* var. *israelensis* (*B.t.i.*). *J. Am. Mosq. Control Assoc.* **1,** 1–7.

Mikkola, A. R., Carlberg, G. A., Vaara, T., and Gyllenberg, H. G. (1982). Comparison of inclusions in different *Bacillus thuringiensis* strains. An electron microscope study. *FEMS Microbiol. Lett.* **13,** 401–408.

Padua, L. E., Ohba, M., and Aizawa, K. (1980). The isolates of *Bacillus thuringiensis* serotype 10 with a highly preferential toxicity to mosquito larvae. *J. Invertebr. Pathol.* **36,** 180–186.

Padua, L. E., Gabriel, B. P., Aizawa, K., and Ohba, M. (1981). *Bacillus thuringiensis* isolated in the Philippines. *Philipp. Entomol.* **5,** 199–208.

Padua, L. E., Ohba, M., and Aizawa, K. (1984). Isolation of a *Bacillus thuringiensis* strain (serotype 8a : 8b) highly and selectively toxic against mosquito larvae. *J. Invertebr. Pathol.* **44,** 12–17.

Pfannenstiel, M. A., Ross, E. J., Kramer, V. C., and Nickerson, K. W. (1984). Toxicity and composition of protease-inhibited *Bacillus thuringiensis* var. *israelensis* crystals. *FEMS Microbiol. Lett.* **21,** 39–42.

Sekar, V. (1986). Biochemical and immunological characterization of the cloned crystal toxin of *Bacillus thuringiensis* var. *israelensis*. *Biochem. Biophys. Res. Commun.* **137,** 748–751.

Sriram, R., Kamdar, H., and Jayaraman, K. (1985). Identification of the peptides of the crystals of *Bacillus thuringiensis* involved in the mosquito larvicidal activity. *Biochem. Biophys. Res. Commun.* **132,** 19–27.

Thomas, W. E., and Ellar, D. J. (1983a). *Bacillus thuringiensis* var. *israelensis* crystal δ-endotoxin: Effects on insect and mammalian cells *in vitro* and *in vivo*. *J. Cell Sci.* **60,** 181–197.

Thomas, W. E., and Ellar D. J. (1983b). Mechanisms of action of *Bacillus thuringiensis* var. *israelensis* insecticidal δ-endotoxin. *FEBS Lett.* **154,** 362–368.

Tyrell, D. J., Davidson, L. I., Bulla, L. A., Jr., and Ramoska, W. A. (1979). Toxicity of parasporal crystals of *Bacillus thuringiensis* subsp. *israelensis* to mosquitoes. *Appl. Environ. Microbiol.* **38,** 656–658.

Tyrell, D. J., Bulla, L. A., Jr., Andrews, R. E., Jr., Kramer, K. J., Davidson, L. I., and Nordin,

P. (1981). Comparative biochemistry of entomocidal parasporal crystals of selected *Bacillus thuringiensis* strains. *J. Bacteriol.* **145,** 1052–1062.

Undeen, A. H., and Nagel, W. L. (1978). The effect of *Bacillus thuringiensis* ONR-60A strain (Goldberg) on *Simulium* larvae in the laboratory. *Mosq. News* **38,** 524–527.

Visser, B., van Workman, M., Dullemans, A., and Waalwijk, C. (1986). The mosquitocidal activity of *Bacillus thuringiensis* var. *israelensis* is associated with M_r 230000 and 130000 crystal proteins. *FEMS Microbiol. Lett.* **30,** 211–214.

Waalwijk, C., Dullemans, A. M., van Workum, M. E. S., and Visser, B. (1985). Molecular cloning and the nucleotide sequence of the M_r 28,000 crystal protein gene of *Bacillus thuringiensis* subsp. *israelensis*. *Nucleic Acids Res.* **13,** 8207–8217.

Wu, D., and Chang, F. N. (1985). Synergism in mosquitocidal activity of 26 and 65 kDa proteins from *Bacillus thuringiensis* subsp. *israelensis* crystal. *FEBS Lett.* **192,** 232–235.

Yamamoto, T., Iizuka, T., and Aronson, J. N. (1983). Mosquitocidal protein of *Bacillus thuringiensis* subsp. *israelensis*, identification and partial isolation of the protein. *Curr. Microbiol.* **9,** 279–284.

10

Current Status of the Microbial Larvicide Bacillus sphaericus

SAMUEL SINGER

I. INTRODUCTION

Bacillus sphaericus has been awaiting its day for over a decade. There are signs now that that day may be at hand. Currently there are three *B. sphaericus* candidates of note, strains 1593, 2297, and 2362. Strain 1593 was isolated in 1975 (Singer, 1975a; Singer and Murphy, 1976), strain 2297 was isolated in 1980 (Wickremesinghe and Mendis, 1980), and strain 2362, although available earlier, was reported in 1984 (Weiser, 1984).

An informal consultation on the development of *B. sphaericus* as a microbial larvicide was recently held in Geneva under the auspices of the World Health Organization's (WHO's) special program (TDR) and

BIOTECHNOLOGY IN INVERTEBRATE
PATHOLOGY AND CELL CULTURE

WHO's Division of Vector Biology and Control. The most noteworthy conclusion was that "the most toxic *B. sphaericus* isolates are more effective than *B.ti. Bacillus thuringiensis* subsp. *israelensis* against species of *Culex, Mansonia* and some species of anopheline mosquitoes and on this basis alone warrant further development and evaluation" [World Health Organization (WHO), 1985]. These cultures did not spring forth full grown but were the end result of an international effort starting in academia in the United States, during the past decade. This effort for many years lay in the shadow of the successes of *B. thuringiensis* subsp. *israelensis*. With the isolation of this subspecies, industry's interest was aroused. Here was a situation where technology was in place. Industrial companies had been producing *B. thuringiensis* material for agricultural use for over 30 years. It took only a slight modification of their technology to be able to produce *B. thuringiensis* subsp. *israelensis*. With industry's recognition, interest in microbial insecticides spread. In contrast, *B. sphaericus* in the past required a sizable industrial investment in its development. The future of development of *B. sphaericus* will be on the economic base of the success of *B. thuringiensis* subsp. *israelensis*.

This chapter will relate the more recent events in the development of *B. sphaericus* and will discuss in part the previous decade's efforts that brought us to today's era of biotechnology. Finally, the chapter will conclude with a discussion of where these events are leading us and their future implications.

II. GENERAL CHARACTERISTICS

Unlike other pathogens or parasites such as viruses, protozoa, nematodes, and some fungi, bacilli such as *B. sphaericus* do not need to grow in the living cell since they are not obligate parasites. Rather, they grow on most laboratory media and inexpensive industrial raw materials. In producing them, one can take advantage of all of the existing technology of submerged fermentation.

The members of the genus *Bacillus* in general and *B. sphaericus* in particular are ubiquitous saprophytes, occurring universally in nature (Gibson and Gordon, 1974). *Bacillus sphaericus* is an aerobic rod-shaped endospore-forming bacterium with the endospores in a swollen terminal position. Until recently it was thought that *B. sphaericus* did not form a parasporal crystal like that of *B. thuringiensis* (Singer, 1977; Myers and Yousten, 1978; Davidson, 1979; Myers *et al.*, 1979). However, Davidson and Myers (1981) found that some insecticidal strains, notably those that are highly toxic in the spore stage, produce parasporal inclusions that resemble crystals of *B. thuringiensis. Bacillus sphaericus* does not ferment glucose or other sugars or starch. Rather, it uses amino acids as its carbon and nitrogen sources (Singer *et al.*, 1966). Recently in our laboratory we found that under some combinations of media components, the sugars such as glucose may inhibit growth of strain 1593 and thereby inhibit insecticidal activity (Lee, 1983). On the other hand, *B. thuringiensis* can use these sugars to enhance its growth in the presence of proteins and amino acid sources and, like *B. sphaericus, B. thuringiensis* prefers amino acids as its carbon and nitrogen source.

Using the classical biochemical identification methods, it is impossible to distinguish differences between the insecticidal and noninsecticidal varieties of *B. sphaericus*. There is no simple distinguishing test that one can utilize for the initial examination of populations of these bacteria when they are freshly isolated from dead larvae from the wild. Nor is there any distinguishing characteristic that can be used to judge potential differences in biological activity among these strains. The laborious procedures of cloning (colony picking), media and fermentation evaluation, and bioassaying have had to be used in the development of this group. After initial cloning, the strains isolated from the various accessions are designated as separate cultures by giving them a number corresponding to their accession number.

III. DEVELOPMENT OF INSECTICIDAL POTENCY

A. Isolates

As of this writing, 45 out of 186 strains isolated and examined show some toxicity to mosquito larvae and provide 100% mortality in 48 hr at a concentration of 10^7 cells/ml (WHO, 1985). Most of these are shown in Table I, along with their country of origin.

The first of this array of strains, strain K, was isolated by Kellen (Kellen and Meyers, 1964; Kellen *et al.*, 1965). Strain K and strain Q, which was derived from it, can be said to be the first reported active *B. sphaericus* isolates. Strain K was not generally available initially for proprietary reasons. Because of low larvicidal activity and other commercial considerations, work was discontinued in 1967. From 1970 on, a series of strains of interest were isolated from field material sent to my laboratory by Dr. John Briggs' Collaborating Center for Biological Control. Strain SSII-1 (1321), one of the earlier isolated from India, was isolated (Singer, 1973) and represents the first generally available ac-

TABLE I
Insecticidal Strains of *Bacillus sphaericus* and Their Country of Origin

Strain	Country of origin
2602	Czechoslovakia
1691, 1881	El Salvador
2537-2, 2533-1 (K1), 2533-1 (K2)	Guyana
2601	Hungary
SSII-1, 2173, 2377	India
1593	Indonesia
1883, 1885-1893, 1894-1896	Israel
2362	Nigeria
1404, 2115, 2117-2	Philippines
2013-4, 2013-6	Roumania
2297	Sri Lanka
2500, 2501, 2314-2, 2317-3, 2315	Thailand
K, Q	United States

tive *B. sphaericus* strain, with which much of the early development work was done (Davidson *et al.*, 1975, 1977; Ramoska *et al.*, 1977, 1978; Singer, 1974, 1975b, 1977). Considering the primitive nature of the fermentation effort (when compared to present-day biotechnology) broth culture preparations of strain SSII-1 were often unstable in terms of population and toxin production.

In 1975, strain 1593, the first fermentation and population stable strain, was derived from CCBC material. This strain still remains one of the three key field candidates. Wickremesinghe and Mendis (1980) isolated strain 2297 (which they called MR-4), the second of the present-day key field candidates. Weiser (1984) isolated strain 2362 from *Simulium* adults in Nigeria. Strain 2362 is the third of the three key field candidates and is reported to be as good as or better than strains 1593 and 2297 (Yousten, 1984a; WHO, 1985).

Another series of recent isolates that should be noted are the isolates of Lysenko *et al.* (1985), which are reported to be as active as 1593 and 2362. What makes them (as well as strain 2362) of interest is that they were isolated from nonmosquito sources.

The remaining strains listed in Table I are of more than historical interest. They represent a genetic pool from which material for future genetic manipulation may be derived. Although a majority of research and development effort concentrated on the few strains listed above, many of these "other" strains are as active as the ones just mentioned. They, and similar strains yet to be isolated, represent the potential to satisy the local national need for "endogenously derived" strains.

As mentioned above, it is difficult to distinguish between the insecticidal and noninsecticidal strains of *B. sphaericus*. The *B. sphaericus* group itself is a very heterogeneous collection with little work having been done in this taxonomic area (Gordon *et al.*, 1973; Gordon, 1979). Several approaches have been developed to differentiate between these strains, as well as for use in the verification of the identity of the strains. Yousten developed a combination of bacteriophage for this purpose (Yousten *et al.*, 1980; Yousten, 1984b), allowing grouping of the strains into seven bacteriophage type groups. Using H-antigens, de Barjac developed a serotyping system to divide the strains (de Barjac *et al.*, 1980; Yousten *et al.*, 1980). There is excellent correlation between the two methods (Yousten, 1984a). Differentiation of the insecticidal *B. sphaericus* strains, based on toxicity to mosquito larvae, has been suggested (de Barjac *et al.*, 1985), dividing the strains into low, moderate, and high toxicity, but this suffers from difficulties of standardization of the bioassay and target larvae (WHO, 1985). Similarly, de Barjac has been able to get some separation of strains using an auxanographic approach with a large number of substrates analyzed by numerical taxonomy techniques (de Barjac *et al.*, 1980). Although all of these methods have given good practical results, they do not bear on genetic relationships within this "species." This will be discussed further when we examine the systematics and taxonomy of the group.

B. Laboratory Potency and Activity Spectrum

Bioassays of supernatant and cells of the final whole cultures (FWC) of strain SSII-1 showed that all insecticidal activity was associated with the cells and not with the supernatant. Chloroform was used to treat FWCs (Singer, 1974, 1977; Davidson *et al.*, 1975). One part chloroform to 10 parts FWC lowered the total viable count 10^3 to 10^6-fold down to about the spore count, while the potency dropped less than tenfold. Therefore, viable cell numbers may be irrelevant to the expression of *B. sphaericus* potency since dead cells can still kill insects. Myers and Youston verified the effects of chloroform and have found similar effects with ultraviolet light (Myers and Yousten, 1978).

Replication of *B. sphaericus* in the gut of the larvae is not necessary for biological activity. Myers and Yousten (1978) found that 100 μg/ml of bacitracin and 1 μg/ml of rifampin inhibited bacterial replication but did not affect normal larval growth and development. The antibiotics did not alter mortality due to strain SSII-1 in bioassays, indicating that bacterial replication did not occur in the assay system before death of the larvae. Studies of the gut population (Davidson *et al.*, 1975) indicated that at death the numbers of SSII-1 began to rise and ultimately reached more than 10^5 cells/larva accompanied by a rise in normal flora.

The toxin content of preparation of *B. sphaericus* varies noticeably according to the mode of preparation of the active material. The number of vegetative cells and/or spores per unit of broth or powder varies according to the mode of production. I have reported elsewhere (Singer, 1974, 1979, 1980, 1981) that fermentation conditions and the medium, in particular, have a significant influence on the final bacterial count and culture activity of strain SSII-1 while having less influence on the activity and growth of 1593 (and presumably 2297 and 2362, but this will have to be verified). In addition, living (vegetative) cells are not needed for toxicity (see above chloroform discussion). This variation in biological activity depending on mode of preparation should not be surprising—it is a fact of life of the fermentation art and reflects the variability of final whole cultures even under the best "standardized conditions." It should be noted, however, by those wishing to engage in "local" production in malaria-endogenous areas. Stability is brought to this situation when preparations are compared to standard material and the results are reported in terms of a standard unit of activity (Dulmage, 1975), or when chemical identification of toxin per se allows comparison to be made on a weight basis. The state of the art has not progressed to the point where the latter is possible, although the possibility of producing various biotechnological "probes" and the use of various "blotting" techniques may be available in the future. The former (an international recognized unit based on standard material) have been prepared. De Barjac and the Institut Pasteur have prepared two such materials. These are *B. sphaericus* 1593-RB80, with a defined potency of 1000 toxic units (TU) per milligram; and SPH 84, which is based on strain 2297 and has an established potency of 1500 TU/mg in relation to RB80. Both standards were formulated from dried sporulated cultures of the specific *B. sphaericus* strain and have been shown to be stable particularly in regard to heat (WHO, 1985). Three main units for quantification of the insecticidal activity of *B. sphaericus* preparations have been reported in the past. It has been expressed in terms of (1) how the final whole culture can be diluted and still kill 50% of the test insect population; (2) number of bacterial cells or spores/ml of the preparation capable of killing 50% of the test insect population, dilution of 10^6–10^9 times being approximately equivalent to 10–1000 living or dead cells/ml (Singer, 1981); and (3) milligrams per liter of the active ingredient or experimental formulation under evaluation (Singer, 1980, 1981). In light of the insecticidal activity of chloroform-treated *B. sphaericus*, the first unit for evaluating broth cultures is preferred. For evaluating powders the third unit based on comparison with one of the two international standards mentioned above is preferred.

Standardization of both a bioassay and the preparation of culture material (whether a final whole culture broth or a dried powder of formulated material) is absolutely necessary. The World Health Organization (1985) suggests one such bioassay scheme which is similar to what I have been using in my laboratory for 20 years (Singer, 1980, 1981). We differ in that I prefer second instar larvae for a quick survey of experimental broths because of logistical problems, particularly handling large numbers of larvae in a small laboratory. A carefully standardized bioassay meeting local needs is what is important. Of greater importance is standardization of the preparation of the bacterial material, particularly final whole culture. In my laboratory we use a carefully standardized inoculum buildup preexplored to suit the individual nutritional and fermentation needs of the strain under investigation (Singer, 1980, 1981). This is done whether the purpose of the experiments is fermentation or genetic manipulation. This has been, for example, one of the factors that has allowed us to successfully protoplast many *B. sphaericus* and *B. thuringiensis* cultures that have given other laboratories major problems. It is anathema to a bacteriologist to read of colleagues scraping growth off a petri plate for use in their experiments.

In contrast to *B. thuringiensis* subsp. *israelensis* existing isolates of *B. sphaericus* are virtually nontoxic to blackflies, *Simulium*. They are, however, quite toxic to most of the important mosquito groups with the exception of some of the *Aedes* such as *Aedes aegypti* (Singer, 1980; Yousten, 1984a). With the recent availability of new primary powders and formulations, it would be more germane to focus our attention on the recently reported (WHO, 1985) laboratory results on primary powder of strains 1593, 2297 and 2362. For a discussion of the older materials, I direct your attention to several recent reviews (Yousten, 1984a; Davidson, 1985; Singer, 1985).

In Table II I have summarized the general activity of primary powders of *B. sphaericus* against a variety of mosquito species. Since much of the data [with the exception of that of Lacey and Singer (1982)] is yet to appear in print, I have chosen to take the overview approach. In general, strain 2362 was the most effective against all species and instars tested, followed by 1593, which in turn was more effective than 2297. Strains 1593, 2297, and 2362 all proved toxic to species of *Culex*, *Anopheles*, *Mansonia*, and *Psorophora*. None of the strains was significantly toxic to *A. aegypti*, although *Aedes melanimon*, *Aedes triseriatus*, and *Aedes nigromaculis* were susceptible, indicating that *B.*

TABLE II
Summary of Activity of Primary Powders of *Bacillus sphaericus* against a Variety of Mosquito Species in the Laboratory

Source of primary powder	*B. sphaericus* strain used	Susceptible species[a]	Reference
Singer, Western Illinois University, Macomb, Ill.	2013-4 2013-6	*Cx. quinquefasciatus* *An. albimanus* *An. quadrimaculatus* *Ae. triseriatus* *Ps. columbiae*	Lacey and Singer, 1982
Dulmage, USDA, Brownsville, Texas	1593	*An. gambiae* *Ps. columbiae* *Ae. nigromaculis*	WHO, 1985
RB-80, Institut Pasteur, Paris	1593	*Cx. quinquefasciatus* *Cx. gelidus* *An. albimanus* *An. quadrimaculatus* *Mn. uniformis*	WHO, 1985
Dulmage, USDA, Brownsville, Texas	2297	*Cx. quinquefasciatus* *Cx. modestus* *An. albimanus* *An. quadrimaculatus* *Mn. uniformis* *Ae. nigromaculis*	WHO, 1985
Dulmage, USDA, Brownsville, Texas	2362	*Cx. modestus* *An. gambiae*	WHO, 1985
Abbott Laboratories, North Chicago, Ill.	2362	*Cx. quinquefasciatus* *An. albimanus* *An. quadrimaculatus* *Ps. columbiae* *Ae. nigromaculatus*	WHO, 1985
Solvay and Co., Brussels	2362	*Cx. gelidus* *Mn. uniformis*	WHO, 1985

[a]*Cx., Culex; An., Anopheles; Ae., Aedes; Ps., Psorophora; Mn., Mansonia.*

sphaericus may prove useful as a larvicide against some aedine mosquitoes (WHO, 1985). Of interest is the effectiveness of primary powders of 1593 and 2362 against *Mansonia*, with strain 2297 coming in a relatively poor third.

C. Mammalian Safety Testing

Although safety testing of *B. sphaericus* has not been exhaustive, a significant amount of mammalian safety data exists. The data are reassuring and indicate that several entomopathogenic *B. sphaericus* strains are not hazardous to mammals (WHO, 1985). Several *B. sphaericus* strains (including strains SSII-1, 1593, and 1404) have been tested for safety. None of them appears to affect test mammals adversely via the usual routes of administration (Shadduck *et al.*, 1980). When Shadduck injected these strains intracerebrally, they produced mild lesions in the brains of rats, and injected intraocularly produced more severe lesions in the eye of rabbits, but the presence of lesions in the brain and eye following injection of autoclaved preparations indicates that the lesions may result from the injection of high concentrations of foreign proteins. *Bacillus sphaericus* is capable of surviving in mammalian tissue but is cleared rapidly. Since lesions occurred only at the highest doses in the most vulnerable mammalian target sites and were in large measure the result of injection of foreign material, the evidence indicates that the isolates of *B. sphaericus* studied to date are avirulent for mammals; it seems highly unlikely that they pose any hazard to humans.

D. Nontarget Organisms

It appears that the *B. sphaericus* strains tested are highly specific, almost exclusively affecting mosquito larvae. This may be related to the nature of the toxin, which is normally released only under the special conditions of the mosquito larval gut. In a 3-month study by Davidson *et al.* (1977), it was shown that strain SSII-1 had no adverse effect on the longevity of newly emerged honey bees (*Apis mellifera*), on brood production of colonies, or on production of honey by the bees. Lack of an adverse effect against the honey bee has been confirmed by Cantwell and Lehnert (1979). In addition, nontarget organisms showing no adverse effect in the presence of *B. sphaericus* strains 1593 or SSII-1 were crayfish, larvivorous fish, Nepidae, Libellulidae, Hydrophilidae, fruit fly, blackfly, Indian meal moth, cigarette beetle, greater wax moth, tadpoles, mosquito fish (and their fry), damsel fly naiads, chironomids, and many others listed in a review by Singer (1985). Similar studies have been done using strains 2297 and 2362 (WHO, 1985).

E. Longevity, Persistence, Recycling

It is generally accepted that in considering the economics of vector control it is not the cost of the agent (whether chemical or biological) that is critical, but rather the cost of the application. Prolonged larvicidal activity of *B. sphaericus* may be due to recycling and amplification of spores in larval cadavers and certain aquatic situations or may simply be due to the long-term persistence of sufficient and accessible toxin in the environment or a combination of both of the above. The level of organic material in water and physical factors play a role in the longevity and persistence in the wild. Physical factors with variable effects on *B. sphaericus* larvicidal activity were studied by Mulligan *et al.* (1980). Exposure to sunlight and pH 10 reduced or eliminated larvicidal activity. However, studies by Burke *et al.* (1983) demonstrate that exposure to sterilizing UV irradiation does not reduce larvicidal activity or strain 1593 spores even though the spores are rendered inviable. Persistence of *B. sphaericus* spores and larvicidal activity in larval mosquito habitats has been well documented (Davidson *et al.*, 1984; Des Rochers and Garcia, 1984; Hertlein *et al.*, 1979; Mulla *et al.*, 1984a,b; Mulligan *et al.*, 1980; Silapanuntakul *et al.*, 1983; Singer, 1980). Rapid settling of the toxic entities and the failure of subsequent generations of larvae to come into contact with them (Mulligan *et al.*, 1980) interfere with possible residual activity in many habitats. Hornby *et al.* (1984), however, obtained residual activity and apparent recycling in sewage treatment systems. Silapanuntakul *et al.* (1983) observed prolonged larvicidal activity (9 months) in tap water in shallow jars, as did Singer in tree holes (Singer, 1980). Indirect evidence for recycling in dead larvae is presented by Silapanuntakul *et al.* (1983) and Des Rochers and Garcia (1984). In spite of all of the above, definitive evidence demonstrating recycling (whether in the larval cadaver, the water column, or the substrate at the bottom of the water column) awaits a good ecological series of experiments which will examine the normal physiological role of *B. sphaericus* in the microbial community of the mosquito larval habitat.

IV. FIELD TRIALS

Successful use of isolates of *B. sphaericus* against field populations of mosquitoes is beginning to be documented (Davidson *et al.*, 1981; Mulla *et al.*, 1984a,b; Mulligan *et al.*, 1978, 1980; Ramoska *et al.*, 1978; WHO, 1985).

Final whole culture preparations of strain 1593 were applied to field populations of *Culex nigripalpus* and *Psorphora columbiae* in roadside ditches and flooded field depressions in Ft. Myers, Florida. The larval population was reduced by nearly 90% when the bacterial suspension was applied in 3.1×10^{-4} to 1.9×10^{-3} dilution of the FWCs (Ramoska *et al.*, 1978). Field trials of strain 1593 FWC in California provided excellent control of a natural population of *Culex tarsalis* Coq. at 10^4 cells/ml (Mulligan *et al.*, 1978). High mortality of *Cx. tar-*

salis larvae was also observed in a 1-acre water recharge pond in California when 0.6 lb/acre of dry powder formulation of 1593 resulted in operational control of *Cx. tarsalis* at 0.84 kg/ha (Mulligan *et al.*, 1978).

Dry powder commercial formulations of *B. sphaericus* 1593 were compared to similar dry powder preparations of *B. thuringiensis* subsp. *israelensis* by Davidson *et al.* (1981) in test ponds at dosages equivalent to 0.5, 1, and 2 kg/ha. Natural populations of the Australian encephalitis vectors, *Culex quinquefasciatus* and *Culex annulirostris* Skuse, and a potential malaria vector, *Anopheles annulipes* Walker, were controlled at 1 and 2 kg/ha. Although *B. sphaericus* 1593 has been reported to be larvicidal against *Anopheles* species larvae in the laboratory (Singer, 1980), this report was among the first to demonstrate control of *Anopheles* species by both *B. sphaericus* and *B. thuringiensis* subsp. *israelensis* in the field. These workers also noted that *Anopheles annulipes* larvae were as sensitive to *B. sphaericus* as to *B. thuringiensis* subsp. *israelensis* and died quicker than *Cx. quinquefasciatus* larvae. Davidson *et al.* (1981) observed that the *B. thuringiensis* subsp. *israelensis* and *B. sphaericus* preparations were similar in efficacy and persistence against the vectors when the bacterial preparations were compared side by side.

Mulla *et al.* (1984b) report field tests where *B. sphaericus* strain 2362 was slightly more active than strain 1593 against *Cx. tarsalis* and Cx. quinquefasciatus. Both yielded excellent control (95%+) at rates of 0.18–0.36 kg/ha with no acute adverse effects on prevailing macroinvertebrate fauna.

Field trials conducted in clear water with spray-dried primary powder of strain 2362 at 0.22 kg/ha resulted in nearly complete control of *Cx. tarsalis* for over 4 weeks (Mulla *et al.*, 1984a). In habitats that were heavily enriched with organic material, control was reduced and brief (Mulla *et al.*, 1984a). Control of *Culex* spp. was obtained in sewage effluent by Yu *et al.* (1983) using a rather elevated dosage of 1 kg/ha. This raises an interesting point. Using higher initial doses of bacterial insecticides in field application is usually resisted by field workers. This fear may be related to the need in the past to use reduced levels of toxic chemical insecticides. Since the bacterial insecticides have been shown to be environmentally safe, this concern should not color the judgment of the applicators. Rather, a judgment should be made in terms of what is the necessary dosage. Economics say that the cost of application is far greater than the cost of the agent. Habits (previous good judgment) apparently are hard to break.

Larvicidal activity, application and coverage, persistence in the mosquito habitat, and storage life of *B. sphaericus* could undoubtedly be improved through better formulation (Lacey, 1984). To date, little research on formulation of *B. sphaericus* has been conducted. This picture may be changing if we are to believe the preliminary results reported in Geneva (WHO, 1985). The following is from the report.

Field evaluations were conducted in several habitats, in organically enriched or clear water, and in shaded habitats or in full sun against natural populations of *Culex* spp., *Mansonia* spp., *Anopheles* spp., and *Psorophora* spp. Effective control (95–100%) was achieved by two primary powders of the 2362 isolate (see Table II), the 2297 isolate and a flowable concentrate (Solvay & Co.) of 2362 at 0.25 kg/ha against the *Culex* spp. The flowable concentrate (12% primary power) provided 82% reduction of *An. quadrimaculatus,* in mature rice fields when applied aerially at 1.0 kg/ha. Virtually 100% control of *P. columbiae* was achieved with the flowable concentrate at 0.5 kg/ha in recently flooded rice fields. When preparations of the 1593 and 2362 isolates were applied to sewage effluent at concentrations of 7.5 and 8.2 mg/l, respectively, they resulted in 94 and 96% reduction of *An. gambiae.*

According to the report (WHO, 1985) some of the most impressive results achieved in the field to date with experimental formulations of *B. sphaericus* have been obtained against *Cx. quinquefasciatus* in highly polluted waters in the Ivory Coast and in the United Republic of Tanzania. In the latter country, for example, the formulated 2362 provided effective control of *Cx. quinquefasciatus* in cesspits and latrines for as long as 6–10 weeks when applied at a rate of 10 g/m^2. In addition to efficacy against *Cx. quinquefasciatus,* the formulated preparations of 2362 have been shown to be relatively effective against *Mansonia uniformis,* an important vector of Brugian filariasis in South-east Asia. At a rate of 1 kg/ha, both these formulations provided 80% reduction of larval populations for as long as 14 days after application.

V. MODE OF ACTION

The larvicidal activity of the *B. sphaericus* strains was originally thought to be associated with the cell wall of the bacterial cell. But soon it was discovered that parasporal inclusions are produced in some strains (Davidson and Myers, 1981) and apparently contain larvicidal toxins (Payne and Davidson, 1984). Myers and Yousten (1980) concluded that most of the toxin in strain 1593 sporulating cells is found in the cell wall, although whole spores were more insecticidal than purified cell walls. Payne and Davidson (1984) indicated in their studies on strain 1593 that the parasporal crystalline inclusions are the major source of larvicidal toxin. The pathogenicity does not reflect invasiveness but appears to be a matter of toxicity. Mosquito larvae normally feed on bacteria and other particulate materials (Dadd, 1971). Once the bacilli are ingested, they, along with normal gut flora, remain within the peritrophic membrane, where they are digested [and the crystallike inclusions that may be present are rapidly dissolved (Yousten and Davidson, 1982)]. The broad outline of the pathogenesis of *B. sphaericus* SSII-1 has been revealed by histological studies using light and electron microscopy (Davidson *et al.*, 1975; Davidson, 1977; Kellen *et al.*, 1965). Bacterial invasion of host tissue only occurred long after death—in many larvae after autolysis of some body tissues. The cells of the midgut swell shortly after ingestion of *B. sphaericus*, separate from one another at their bases, and contain large numbers of lytic vacuoles. Death of the larvae may occur in as little as 4 hr following exposure to high concentrations of spores; at lower concentrations full expression of activity is seen within 48 hr. The confinement of the bacteria entirely within the peritrophic membrane, even in some dead larvae, and the lack of insecticidal activity of the supernatant from bacterial broth cultures led to the conclusion that one or more cell-associated toxins are involved. *Bacillus sphaericus* cells are digested or modified within the peritrophic membrane of the gut, and the toxin is released to pass through the peritrophic membrane and kill the larvae. The ability of chloroform-killed SSII-1 bacteria to kill larvae and the drop in bacterial numbers in larvae after feeding (Davidson *et al.*, 1975) support these conclusions. The exact mode of action of the larvicidal toxin(s) of *B. sphaericus* is unknown.

Bacillus sphaericus toxic material can be solubilized by freezing and thawing the spore–crystal complex or by solubilizing it with 50 m*M* NaOH (Bourgouin *et al.*, 1984; Davidson, 1983; Tinelli and Bourgouin, 1982). These same workers found that partial purification of such material by column chromatography indicated that the toxic protein might have a molecular mass of about 35, 55, or 100 kDa. Baumann *et al.* (1985) purified crystals from spore–crystal complexes of *B. sphaericus* 2362 by disruption in a French pressure cell followed by centrifugation through 48% (wt/vol) NaBr. Crystals from such preparations had a 50% lethal concentration of 6 ng protein/ml for the larvae of the mosquito *Culex pipiens*. Polyacrylamide gel electrophoresis (under denaturing conditions) indicated that the principal protein bands had molecular masses of about 43, 63, 98, 110, and 125 kDa. After solubilization of the crystal in alkali, the 43- and 63-kDa proteins were purified by column chromatography. The 43- but not the 63-kDa protein was toxic for 2nd–3rd instar larvae of *C. pipiens* (LC_{50}=35 ng/ml). Considerable difference in the amino acid composition between these two proteins was detected. By electrophoretically separating the crystal proteins and then electroblotting onto nitrocellulose paper and visualizing the bands with antisera to the 43- and 63-kDa proteins in conjunction with an immunoblot assay, it was found that the high-molecular-mass crystal proteins (98–124 kDa) contained antigenic determinants of both proteins. These results suggested that the lower-molecular-weight crystal proteins detected in polylacrylamide gels after electrophoresis under denaturing conditions were derivatives of one or more of the higher-molecular-weight crystal proteins. *In vivo* studies of the products of crystal degradation by larvae of *Cx. pipiens* indicated that the high-molecular-weight proteins and the 63-kDa antigenic determinants were rapidly degraded and that a 40-kDa protein related to the 43-kDa toxin persisted for the duration of the experiment (4 hr). Some of the studies performed with *B. sphaericus* 2362 were extended to strains 1593, 1691, and 2297 of this species with results which indicated a high degree of similarity between the crystal proteins of all these larvicidal strains.

Studies of the pathological changes caused by the *B. sphaericus* toxin in susceptible larvae have suggested that the midgut is the primary target organ of this toxin (Davidson, 1979, 1981; Davidson *et al.*, 1975). However, the mode of action of the *B. sphaericus* toxin is not known. The small size and aquatic nature of the mosquito larva make study of the toxin mode of action in this host quite difficult. No other organism is known to be susceptible to this toxin. Davidson (1986) therefore has examined an *in vitro* assay system utilizing cultured mosquito cells to provide a convenient alternative to the host for the study of the effects of this toxin on mosquito cells. The cytotoxic activity of extracts of *B. sphaericus* strain 1593 was found by Davidson (1986) not necessarily to correlate with insecticidal activity. Cytotoxicity and larvicidal activity were neutralized by immune rabbit serum prepared by her against crude toxin extracts as well as by serum prepared against purified toxin from strain 2362. This purified toxin was also found by her to be cytotoxic. Activation with mosquito larval gut homogenates enhanced cytotoxicity of both 1593 extracts and purified toxin from 2362. The activity of cytotoxic preparations against three mosquito cell lines paralleled the activity of *B. sphaericus* spores against larvae of these mosquito species. The results suggest the presence of a prototoxin and one or more cytotoxic proteins derived from it (Davidson, 1986).

VI. SYSTEMATICS OF *Bacillus sphaericus*

Systematics is concerned with the scientific study of the diversity of organisms and the relationship between them. Systematics includes taxonomy, which is the science of classification, its principles, procedures, and purposes. Systematics includes elements of evolution and as a consequence is concerned with the interaction of genes and their products. Classification of organisms can be either phenetic, based on shared properties, or phylogenetic, according to their degree of common ancestry (Goodfellow and Minnikin, 1985). There have been recent attempts to bring the two together (Schliefer and Stackebrandt, 1983).

At the epigenetic level analysis of homologous proteins via functional studies is an excellent and reliable way to differentiate closely related strains of a species as well as closely related species (Schliefer and Stackebrandt, 1983). The zymogram technique has been used to estimate genetic relationships between subspecies or related species of a very wide spectrum of living organisms (Baptist *et al.*, 1978). The results of the use of the zymogram technique, if a sufficient number of enzymes (five or more) are compared, show that two individuals from different species will differ in the electrophoretic mobility of about 50% or more of these enzymes, whereas two members of the same species will usually differ in 20% or less (Baptist *et al.*, 1978). The zymogram technique has been particularly useful to differentiate between bacterial species and subspecies (Baptist *et al.*, 1969, 1971, 1978; Goullet, 1973, 1978, 1981; Norris and Burges, 1963; Norris, 1964; Watson, 1976).

In addition to using enzymes as taxonomic tools, DNA homology studies have also been considered a major modern epigenetic approach (Schliefer and Stackebrandt, 1983). Schliefer and Stackebrandt (1983) reviewed the use of different DNA/DNA hybridization methods in gram-positive and gram-negative bacteria and in Archaebacteria. Bradley (1980) reviewed these methods for use in bacteria and Priest *et al.* (1979) did the same for the genus *Bacillus*, as did Seki *et al.* (1975). As a specific example of the use of DNA homology studies pertinent to *B. sphaericus*, Krych *et al.* (1980) examined 67 strains of *B. sphaericus* and were able to group these strains into five homology groups, with group II divided into two subgroups. According to these authors, each of the homology groups identified in their study contains strains having a sufficiently high level of genetic relatedness to justify speciation. The inability to distinguish these strains by readily determined phenotypic tests suggested to them that it would be premature to propose the establishment of new species.

More usual has been the phenetic (classical) bacteriological approach to systematics and classification, where phenotypic characterizations of bacteria are based on biochemical properties. This approach may not discriminate between homologous and analogous metabolic avenues or reflect as directly the genetic relationship as DNA homology, zymogram, and the other new techniques. Actually, the ultimate data for studying the systematics of any group would be nucleotide sequencing of 5 S and/or 25 S components of rRNA, or DNA itself, as this has become accepted as a modern systematic tool (Schliefer and Stackebrandt, 1983). Unfortunately, at present, the latter techniques are time-consuming, expensive, and require investigators with a great deal of expertise for routine use in most laboratories.

Examples of more recent phenetic approaches in studying *Bacillus* are use of the microtube/API-style systems (Logan and Berkeley, 1979, 1984), pyrolysis gas–liquid chromatographic studies (O'Donnell and Norris, 1979), studies of thermophilic strains of *Bacillus* (Wolf and Sharp, 1979), insect pathogens in *Bacillus* (de Barjac, 1979), characterization of insecticidal *B. sphaericus* based on toxicity to mosquito larvae (de Barjac *et al.*, 1985), and differentiation of insecticidal *B. sphaericus* based on bacteriophage typing (Yousten, 1984b; Yousten *et al.*, 1980; Yousten and Hedrick, 1982). The systematics of *B. sphaericus* has been examined mainly in a phenetic manner.

I have arranged the strains of *B. sphaericus* (Table III) in light of available information in terms of DNA homology (Krych *et al.*, 1980) and bacteriophage typing of the insecticidal strains (Yousten *et al.*, 1980). It should be mentioned that de Barjac's serotype groupings coincide almost exactly with the phage-type groupings (Yousten *et al.*, 1980).

An example of the use of the zymogram technique for systematic purposes is the work going on in my laboratory (Cole and Singer, 1986; Hessler and Singer, 1986). The genetic relatedness among 20 strains from the five DNA homology groups (by Hessler) and among 21 strains from the seven phage groups of the insecticidal *B. sphaericus* (by Cole) is being examined. Electromorph mobility is being examined with more than 15 enzymes representing almost 25 loci, using horizontal starch gel or vertical polyacrylamide gel electrophoresis systems. The genetic relatedness among the strains is being estimated by making a phenogram from unweighted pair group cluster analysis with arithmetic averages (UPGMA, Sneath and Sokal, 1973) from matrices derived from a comparison of percent similarities of electromorph mobilities (Lessel and Holt, 1970). The percent similarity between each strain is calculated by dividing the number of similar characteristics between two strains by the total number of characteristics studied. From these data I have constructed two phenograms examining genetic relatedness of at least one representative strain from each of the DNA homology groups (Fig. 1) and at least one representative strain from

TABLE III
Arrangement of Strains of *Bacillus sphaericus* in Light of Available Information

DNA Homology Groups[a]					
I	IIA[b]	IIB	III	IV	V
ATCC 14577[c]	1593	ATCC 7055	NRS 592	NRS 400	NRS 1198
ATCC 10208	SSII-1	ATCC 7054	ATCC 12123	NRS 717	NRS 1199
NRS 967	1404	ATCC 12300	P1	NRS 1529	NRS 1184
NCTC 9602	Kellen K	NRS 718	NRS 800	ATCC 13805	
	Kellen Q	NRS 1191	NRS 1692	ATCC 245	
	1881	NRS 1196	NRS 593	NRS 1090	
	1691	ATCC 7063	NRS 1195	NRS 1693	
		NRS 1194	NRS 1193	NRS 1307	
		NRS 1200	NRS 1197		
		NRS 1201	NRS 1187		
		NRS 1192	NRS 1023		
		NRS 156	NRRL B4197		
			NRS 810		
			NRS 719		
			ATCC 4978		
			NRS 1223		

Phage-type Groups[d]						
1	2	3	4	5	6	7
Kellen K	SSII-1	1593	2297	1894	2115	2315
Kellen Q	1404	1691	2173			
	1883	1881	2377			
	1885	2013-4	2314-2			
	1886	2013-6	2317-3			
	1887	2117-2				
	1888	2362				
	1889	2500				
	1890	2501				
	1891	2537-2				
	1892	2533-1 (K1)				
	1893	2533-1 (K2)				
	1895	2601				
	1896	2602				

[a]From Krych *et al.* (1980).
[b]Members of this group all insecticidal.
[c]Underlined strains = type strain DNA homology group.
[d]From Yousten (1984a).

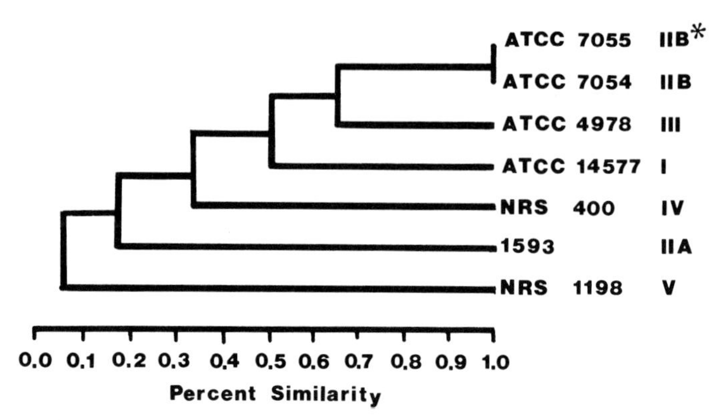

Fig. 1. Phenograms of percent similarities among 7 *Bacillus sphaericus* strains representative of DNA homology groups. *DNA homology groups (according to Krych *et al.*, 1980).

each of the bacteriophage groups (Fig. 2). The data represent electromorphs of the enzymes alanine dehydrogenase, esterase (two substrates), and shikimate dehydrogenase (a total of seven loci) for the DNA homology groups and electromorphs of the enzymes alanine dehydrogenase, esterase (one substrate), shikimate dehydrogenase, and glutamate dehydrogenase (a total of seven loci) for the bacteriophage groups. The small sample used in each case is undoubtedly a biased

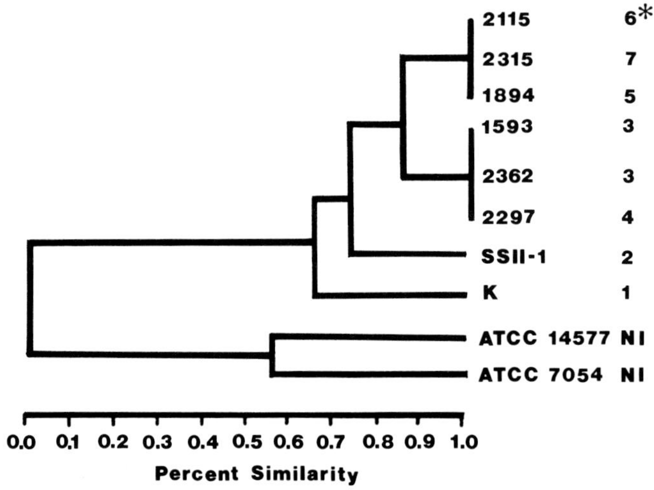

Fig. 2. Phenograms of percent similarities among 10 *Bacillus sphaericus* strains, 8 of which are representative of insecticidal bacteriophage groups. *Bacteriophage groups (according to Yousten, 1984a). NI, Noninsecticidal.

sample but is illustrative of the technique and conforms in general to the pattern obtained with larger samples. Analyzing a larger number of strains and using a large number of enzyme systems, chosen at random, would lead to higher percent similarity values.

In Fig. 1 we can see that the individual DNA homology groups separate out (cluster separately, if one can use the term "cluster" when dealing with one or two strains). The insecticidal group (DNA homology group IIA) is readily separable from the noninsecticidal DNA homology groups. This is likewise true in the second phenogram. This remains true even when large samples (more strains, more enzyme systems) are compared. On the other hand, when the insecticidal group represented by individuals from each of the bacteriophage groups is examined (Fig. 2), the phage groups are relatively more tightly clustered. We see clustering of phage groups 5, 6, and 7, as well as phage groups 3 and 4, while representatives of phage groups 1 and 2 (as well as the representative of noninsecticidal strains) are separated. Considering that this is a small biased sample, note the relatively greater similarity of the insecticidal strains (67.3% similarity) and the low level of similarity of the noninsecticidal strains to the insecticidal strains (0.1% similarity). The similarity between the two noninsecticidal strains remains approximately what it was in the first phenogram. If we are to believe the second phenogram, then there may be up to four subgroups among the insecticidal strains, not seven as seen by the number of bacteriophage groups (or more by the serotype groups). Since enzymes (proteins) are a more direct reflection of the gene than the differences in surface configurations (represented by bacteriophage attachment sites), we should not expect a clear separation by bacteriophage patterns. We need to investigate further the number of "clones" (Hartl and Dykhuizen, 1984; Whittam *et al.*, 1983) emerging from this "subspecies." It is not stretching the imagination to visualize that by obtaining sufficient enzyme systems and/or the addition of other phenetic information we could eventually be able to separate every strain. This brings up a practical point and why I chose some of these enzymes for this analysis. Many of these enzyme systems or combinations of enzyme systems can be used for diagnostic purposes. Two, or at the most three, of these enzymes can separate the strains by DNA homology groups, and a similar number and set can separate many of the insecticidal strains. This makes this procedure interesting not only from a systematics point of view but also from a practical diagnostic point of view. It is a different set of characteristics that most laboratories will be able to utilize for their own purposes.

VII. BIOTECHNOLOGY OF *Bacillus sphaericus*

Biotechnology consists of the isolation or construction of a useful bacterial strain (e.g., bacterial insecticide) on the one hand and the ability to mass-produce and package (formulate) a product (for the field) on the other. The construction of the strain relates to molecular genetics, while the production of the strain with the property of interest relates to fermentation. We will deal with fermentation first.

A. Fermentation

The ability to mass produce strains of *B. sphaericus* existed in 1965 when strain K was first isolated. What was not available then were the potent strains, the better fermentation raw material substrates, some of the industrial instrumentation, and, most important, the need. Need, in this case, is either an available market or an endogenous malarial situation stripped bare of effective chemical insecticide on one hand and effective antimalarials on the other. Only recently [as seen in our discussion of preliminary field trials reported in Geneva (WHO, 1985)] has industry been motivated sufficiently to make the necessary development effort in this direction. Because of the proprietary nature of industrial development, we may never be privy to the progress made or to be made in this particular fermentation. What we can say is that some extremely effective formulated products containing *B. sphaericus* appear to be arriving on the scene.

There have been in the past several explanations of the basic *B. sphaericus* fermentation (Bourgouin *et al.*, 1984; Kalfon *et al.*, 1983; Mulla *et al.*, 1984a; Singer, 1974, 1975a, 1977, 1979, 1980, 1981, 1982; Yousten, 1984a,b), an excellent review on the subject (Yousten, 1984a), as well as attempts at formulation of field product (Lacey, 1984; Lacey and Undeen, 1984; Lacey *et al.*, 1984). To bring us up to date, we need only to discuss briefly the ability of local groups to serve endogenous malarial needs and to formulate the questions that will need to be answered when genetically constructed strains become available.

Attempts at local (cottage) production have been made (Dharmsthiti *et al.*, 1985; Hertlein *et al.*, 1981; Obeta and Okafor, 1983). The Thais are the most pressed to combat antimalarial–chemical–insectide resistance problems being carried across the border from Cambodia by refugees. This need, combined with knowledgeable professional scientists and technologists, will go far toward solving their problems. Local production need not mean sloppy technology. Any local production still demands good science, good technology, comparison of potency with international standard materials, and an eye for the suitable inexpensive local commodity that can be ued in the fermentation. Recent conversations with the Thais assure me that they are aware of these responsibilities.

Just as we cannot assume that the fermentation lessons learned from *B. thuringiensis* can be directly applied to the fermentation of *B. sphaericus*, we also cannot assume that the fermentation lessons learned from *B. sphaericus* (or *B. thuringiensis*) can be directly applied to the new genetically constructed strains. One can predict that these newly constructed strains will be strange beasts indeed. They will not be the *B. sphaericus* (or *B. thuringiensis*) strains whose characterisics we are familiar with, but rather a composite form that is more than *B. sphaericus*, more than *B. thuringiensis* and perhaps a hybrid and chimera that is more than both. One needs to remember that (1) the strain found in nature is the best adapted to that combination of ecological necessities and (2) the added genetically derived features may cost not only in terms of ecology but also in terms of fermentation compatibility. Lessons from the recent past history of *B. sphaericus* tell us that we had better develop a whole series of constructed strains in order to obtain a few that will be ecologically and fermentation compatible.

B. Molecular Biology

There has been some effort to study the molecular biology of *B. sphaericus*. Attempts to define the toxic moiety have been described in Section V. Similarly, the DNA homology studies of Krych *et al.* (1980) have been discussed in Section VI.

δ-Endotoxin of *B. thuringiensis* is generally associated with large plasmids and these large plasmids are reputed to be present in low copy numbers (Gonzalez *et al.*, 1981). At present it is not known whether large plasmids are associated with insecticidal activity in *B. sphaericus*. The search for plasmids present in *B. sphaericus* has been incomplete and the results are somewhat conflicting (Yousten, 1984a). The presence of small plasmids has been reported (Davidson *et al.*, 1982; Abe *et al.*, 1983; Yoshimura *et al.*, 1983). Davidson *et al.* (1982) reported the presence of a single large plasmid in strains 1593 and 1881 but not in 1691 or 2362, while Abe *et al.* (1983) found a large plasmid in 1881 but not in 1593 or 1691. Abe *et al.* (1983) found five plasmids in strain K while Davidson found none. These differences are probably due to differences in techniques, with plasmid isolation techniques undoubtedly resulting in the loss of large plasmids while direct lysis of the whole protoplast in the wells of the agarose gel (Eckhardt, 1978)

TABLE IV
Insecticidal and Noninsecticidal Strains of *Bacillus sphaericus* and Their Plasmids

Strain	Homology or bacteriophage group	Number of plasmids[b]	Plasmid mass (MDa)
Insecticidal bacteriophage group			
K	1	1	29
SSII-1	2	1	75
1593	3[a]	1	75
2362	3	1	75
2297	4	3	75, 3.4, 3.2
1894	5	1	5.7
2115	6	1	4.2
2315	7	7	75, 40, 15, 7.6, 5.7, 3.7, 2.9
Noninsecticidal homology group			
14577	I	—	—
7054	IIB	2	5.2, 3.4
1191	IIB	3	75, 5.2, 3.4
4978	III	—	—
1090	IV	3	35, 9.2, 2.7
1198	V	—	—

[a]Homology group IIA.
[b]None detected (—).

preserves the large plasmids. Recent work in my laboratory (Shelley and Singer, 1986) using the Eckhardt (1978) technique indicates that most of the highly active *B. sphaericus* strains contain one or more large plasmids (Table IV, Fig. 3), but so do several of the noninsecticidal strains. Table IV illustrates the plasmid pattern of at least one insecticidal strain from each of the bacteriophage groups and at least one noninsecticidal strain from each of the DNA homology groups, with strain 1593 (bacteriophage group 3) representing DNA homology group IIA.

Although no plasmids were detected for the strains from DNA homology groups I, III, and V (chosen for illustration), other strains of the same groups (not shown here) do possess plasmids. Note also that no plasmids were detected in strain 1691. This did not surprise us. We could not detect plasmids in strain SSII-1 the first time we examined it. When we examined SSII-1 from Yousten (who originally obtained the strain from us) we found it possessed the 75-MDa plasmid. Obviously, plasmids were lost (or the plasmid copy number was reduced) during

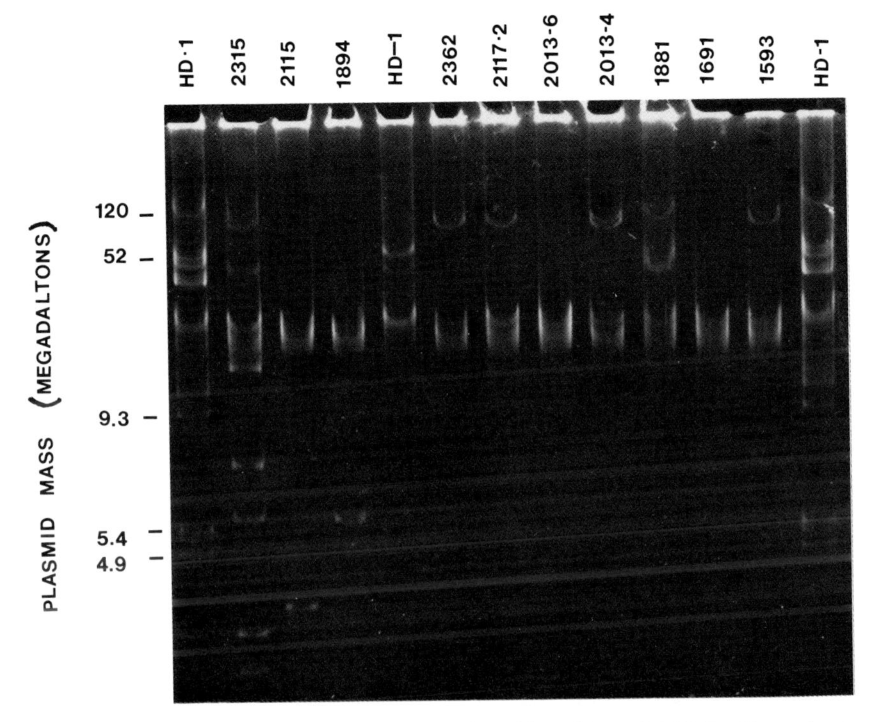

Fig. 3. Agarose gel (0.5%) showing plasmid profiles of insecticidal *Bacillus sphaericus* strains. Marker strain *B. thuringiensis* HD-1 (lanes 1, 5, and 13).

slant-to-slant passage over the long period that we had been working with it. Media variation not only appears to affect the growth and fermentation of these strains but also may have an effect on plasmid replication. In general we have had less plasmid loss with the *B. sphaericus* strains than we have had with the *B. thuringiensis* strains. We used *B. thuringiensis* strain HD-1 as our marker strain in lanes 1, 5, and 13 (Fig. 3). The cultures for all three lanes were derived from the same initial slant in our inoculum buildup (but different subsequent shake flasks), yet HD-1 in lane 5 shows a loss of plasmids. An apparent loss of plasmids in one out of three cultures is not uncommon in our work with *B. thuringiensis*.

The most unusual finding, however, was the distinctive pattern of seven plasmids in strain 2315. This insecticidal strain would make an excellent marker culture for both ecological studies and genetic work. We have also selected antibiotic-resistant variants of this strain in a manner similar to that done with several of the most active *B. sphaericus* (Yurks, 1985).

Kalfon *et al.* (1983) grew strains 1593 and 2297 under three different conditions known to promote plasmid curing in other bacteria. None of the resulting 200 colonies lost the ability to form parasporal bodies and none demonstrated any decrease in insecticidal activity. Ganesan *et al.* (1983) claim to have cloned a 3.7-kilobase fragment of *B. sphaericus* in *Escherichia coli* C600 using plasmid pHV33 as a vector and selected the transformants with resistance to chloramphenicol and ampicillin. One clone out of 100 tested was found to possess toxicity against two species of mosquitoes. The first reported genetic transfer in *B. sphaericus* was that of McDonald and Burke (1984). They introduced plasmids pUB110 and pBC16 (*B. subtilis* plasmids) by plasmid transformation into polyethylene glycol (PEG)-treated protoplast of *B. sphaericus* 1593. The transformants expressed the antibiotic-resistant determinants present in the plasmids and exhibited sporulation frequencies and larvicidal toxicity equivalent to those characteristic of the parent strains.

As noted above, direct genetic manipulation via appropriate vector systems awaits the results of further genetic work with *B. sphaericus*. A more direct approach favored by my laboratory is that of protoplast fusion of appropriate *B. thuringiensis* and *B. sphaericus* parent strains. Protoplast fusion is a versatile general technique to induce genetic recombination in a variety of prokaryotes and eukaryotes. It is particularly useful for prokaryotes which have not been subjected to extensive genetic analysis since it does not require transducing phage, plasmid sex factors, or competency development. The technique has its own problems, however, in terms of generating and fusing stable protoplasts, as well as regenerating viable cells from the fusants (Matsushima and Baltz, 1986).

In my laboratory we had little difficulty in obtaining protoplasts once careful attention was paid to the conditions of lysozyme treatment of cultures grown in a carefully standardized inoculum buildup. There have been and are some recalcitrant strains such as *B. sphaericus* strain 2297 and one or two of the *B. thuringiensis* strains, and even these are succumbing to a reexamination of conditions necessary for protoplasting. We have examined conditions for the regeneration of the protoplasts and the fusants and find that the methods of Akamatsu and Sekiguchi (1984) are quite satisfactory once minor adjustments are made for the specific cultures. We look forward, in the near future, to examining newly derived hybrid strains for what we hope will be new or enhanced activity.

VIII. SUMMARY AND FUTURE EFFORTS

One must understand that the bottom line of research and development of *B. sphaericus* (as well as the other bacterial insecticides) is an agent that will have a direct bearing on reduction of the incidence of malaria and related vector-borne diseases. However, this should not preclude pursuit of the basic research necessary to answer many of the questions that remain, whether they are basic or applied.

The advent of new formulations of old strains by industry should not halt the local search for endogenously isolated strains. The number of sources from which isolates were obtained in the past is small compared to the number that can yet be sampled. In addition, it should be well noted that the most active strain available (strain 2362) was obtained from a nonmosquito source. This is one avenue that perhaps the next generation of isolates should be directed toward. Until a mature field-proven product is obtained, we must continue to pursue the development of new formulations of both old and new isolates.

Along the same lines, we glibly speak of genetic manipulation of these strains. The question of what should be the goals of genetic manipulation needs to be asked. Placing *B. thuringiensis* or *B. sphaericus* toxin genes in *E. coli* is nice basic genetics, but this "product" would not serve us well in the field. It seems to me that the best of both worlds would be to place *B. thuringiensis* toxin genes into a *B. sphaericus* host, thereby gaining the wider spectrum of *B. thuringiensis* in a bacterium capable of surviving, persisting, and perhaps even recycling in the environment. To accomplish this by what has become the classical recombinant DNA approach, we need more vectors capable of doing this as well as more information on the genetics of *B. sphaericus* than is currently available. If we take the more direct approach of protoplast fusion, we face the question of what to call (name) the beast (the fusant). Will we obtain a "*B. sphaericus*," a "*B. thuringiensis*," or something quite different? We would truly be blind men feeling the elephant. The hope in this situation is a clear definition of what *B. sphaericus* and *B. thuringiensis* are, based on a genetic or epigenetic approach. In this situation we would need more than phage or serotyping to guide us. Whether we use neoclassic genetic manipulation or portoplast fusion, the naive hope would be to see synergisms accrue and new larvicidal activities develop.

In terms of mode of action, we are closer to "knowing" what the active toxic moiety produced by *B. sphaericus* is. We have a weight (or range of weights). But we still do not know whether the gene is chromosomal, plasmid [which plasmid(s)], or both. We also still do not know, at the molecular level, the sensitive site of action in the host larva. But then again, after all these years we still do not know for certain what the mode of action (at the molecular level) is for the crystals of *B. thuringiensis.*

In spite of all this, the future looks bright. There are still many basic and applied questions that need to be answered.

REFERENCES

Abe, K., Faust, R. M., and Bulla, L. A., Jr. (1983). Plasmid deoxyribonucleic acid in strains of *Bacillus sphaericus* and in *Bacillus moritai. J. Invertebr. Pathol.* **41,** 328–335.

Akamatsu, T., and Sekiguchi, J. (1984). An improved method of protoplast regeneration for *Bacillus* species and its application to protoplast fusion and transformation. *Agric. Biol. Chem.* **48,** 651–655.

Baptist, J. N., Shaw, C. R., and Mandel, M. (1969). Zone electrophoresis of enzymes in bacterial taxonomy. *J. Bacteriol.* **99,** 180–188.

Baptist, J. N., Shaw, C. R., and Mandel, M. (1971). Comparative zone electrophoresis of enzymes of *Pseudomonas solanacearum* and Pseudomonas cepacia. J. Bacteriol. **108,** 799–803.

Baptist, J. N., Mandel, M., and Gherna, R. L. (1978). Comparative zone electrophoresis of enzymes in the genus *Bacillus. Int. J. Syst. Bacteriol.* **28,** 229–244.

Baumann, P., Unterman, B. M., Baumann, L., Broadwell, A. H., Abbene, S. J., and Bowditch, R. D. (1985). Purification of the larvicidal toxin of *Bacillus sphaericus* and evidence for high molecular weight percursors. *J. Bacteriol.* **163,** 738–747.

Bourgouin, C., Tinelli, R., Bouvet, J. P., and Pires, R. (1984). *Bacillus sphaericus:* 1593-4, purification of fractions toxic for mosquito larvae. *FEMS Symp.* **24,** 387–388.

Bradley, S. G. (1980). DNA reassociation and base composition. *In* "Microbiological Classification and Identification" (J. Goodfellow and R. G. Board, eds.), pp. 11–26. Academic Press, London.

Burke, W. F., McDonald, K. O., and Davidson, E. W. (1983). Effect of UV light on spore viability and mosquito larvicidal activity of *Bacillus sphaericus* 1593. *Appl. Environ. Microbiol.* **46,** 954–956.

Cantwell, G. E., and Lehnert, T. (1979). Lack of effect of certain microbial insecticides on the honeybee. *J. Invertebr. Pathol.* **33,** 381–382.

Cole, C. L., and Singer, S. (1986). A comparison of zymograms among insecticidal phage groups and noninsecticidal *Bacillus sphaericus* strains. *Abstr., 86th Annu. Meet., Am. Soc. Microbiol.,* p. 240.

Dadd, R. H. (1971). Effects of size and concentration of particles on rates of ingestion of latex particulates by mosquito larvae. *Ann. Entomol. Soc. Am.* **64,** 687–692.

Davidson, E. W. (1977). Pathogenesis of bacterial diseases of vectors. *DHEW Publ.* (NIH) (*U.S.*) **NIH 77-1180,** 19–30.

Davidson, E. W. (1979). Ultrastructure of midgut events in the pathogenesis of *Bacillus sphaericus* SSII-1 infections of *Culex pipiens quinquefasciatus* larvae. *Can. J. Microbiol.* **25,** 178–184.

Davidson, E. W. (1981). A review of the pathology of bacilli infecting mosquitoes, including an ultrastructure study of larvae fed *Bacillus sphaericus* 1593 spores. *Dev. Ind. Microbiol.* **22,** 69–81.

Davidson, E. W. (1983). Alkaline extraction of toxin from spores of the mosquito pathogen, *Bacillus sphaericus* strain 1593. *Can. J. Microbiol.* **29,** 271–275.

Davidson, E. W. (1985). *Bacillus sphaericus* as a microbial control agent for mosquito larvae. *Integr. Mosq. Control Methodol.* **2,** 213–226.
Davidson, E. W. (1986). Effects of *Bacillus sphaericus* 1593 and 2362 spore/crystal toxin on cultured mosquito cells. *J. Invertebr. Pathol.* **47,** 21–31.
Davidson, E. W., and Myers, P. (1981). Parasporal inclusions in *Bacillus sphaericus*. *FEMS Microbiol. Lett.* **10,** 261–265.
Davidson, E. W., Singer, S., and Briggs, J. D. (1975). Pathogenesis of *Bacillus sphaericus* strain SSII-1 infecLions in *Culex pipiens quinquefasciatus* larvae. *J. Invertebr. Pathol.* **25,** 179–184.
Davidson, E. W., Morton, H. L., Moffett, J. O., and Singer, S. (1977). Effect of *Bacillus sphaericus* strain SSII-1 on honey bees, *Apis mellifera*. *J. Invertebr. Pathol.* **29,** 344–346.
Davidson, E. W., Sweeney, A. W., and Cooper, R. (1981). Comparative field trials of *Bacillus sphaericus* strain 1593 and *Bacillus thuringiensis* var. *israelensis* commercial powder formulations. *J. Econ. Entomol.* **74,** 350–354.
Davidson, E. W., Spizizen, J., and Yousten, A. A. (1982). Recent advances in the genetics of *Bacillus sphaericus*. *Proc. Int. Colloq. Invertebr. Pathol, 3rd,* pp. 14–17.
Davidson, E. W., Urbina, M., Payne, J., Mulla, M. S., Darwazeh, H., Dulmage, H. T., and Correa, J. A. (1984). Fate of *Bacillus sphaericus* 1593 and 2362 spores used as larvicides in the aquatic environment. *Appl. Environ. Microbiol.* **47,** 125–129.
de Barjac, H. (1979). Insect pathogens in the genus *Bacillus*. *In* "The Aerobic Endospore-Forming Bacteria: Classification and Identification" (R. C. W. Berkeley and M. Goodfellow, eds.), pp. 242–250. Academic Press, London.
de Barjac, H., Vernon, M., and Dumanoir, V. C. (1980). Caractérization biochimique et sérologique de souches de *Bacillus sphaericus* pathogenes ou non pour les moustiques. *Ann. Microbiol.* (*Paris*) **131B,** 191–201.
de Barjac, H., Larget-Thiery, I., Dumanoir, V. C., and Pipouteau, H. (1985). Serological classification of *Bacillus sphaericus* strains on the basis of toxicity to mosquito larvae. *Appl. Microbiol. Biotechnol.* **21,** 85–90.
Des Rochers, B., and Garcia, R. (1984). Evidence for persistence and recycling of *Bacillus sphaericus*. *Mosq. News* **44,** 160–165.
Dharmsthiti, S. C., Pantuwatana, S., and Bhumiratana, A. (1985). Production of *Bacillus thuringiensis* var. *israelensis* and *Bacillus sphaericus* strain 1593 on media using by-product from a monosodium glutamate factory. *J. Invertebr. Pathol.* **46,** 231–238.
Dulmage, H. T. (1975). The standardization of formulations of the endotoxins produced by *Bacillus thuringiensis*. *J. Invertebr. Pathol.* **25,** 279–281.
Eckhardt, T. (1978). A rapid method for the identification of plasmid deoxyribonucleic acid in bacteria. *Plasmid* **1,** 584–588.
Ganesan, S., Kamdar, H., Jayaraman, K., and Sjulmajster, J. (1983). Cloning and expression in *Escherichia coli* of a DNA fragment from *Bacillus sphaericus* coding for biocidal activity against mosquito larvae. *Mol. Gen. Genet.* **189,** 181–183.
Gibson, T., and Gordon, R. E. (1974). Genus I. Bacillus. *In* "Bergey's Manual of Determinative Bacteriology" (R. E. Buchanan and N. E. Gibbons, eds.), pp. 529–555. Williams & Wilkins, Baltimore, Maryland.
González, J. M., Dulmage, H. T., and Carleton, B. C. (1981). Correlation between specific plasmids and delta-endotoxin production in *Bacillus thuringiensis*. *Plasmid* **5,** 351–365.
Goodfellow, M., and Minnikin, D. E. (1985). Introduction to chemosystematics. *In* "Chemical Methods in Bacterial Systematics" (M. Goodfellow and D. E. Minnikin, eds.), pp. 1–15. Academic Press, London.

Gordon, R. E. (1979). One hundred and seven years of the genus *Bacillus*. *In* "The Aerobic Endospore-forming Bacteria: Classification and Identification" (R. C. W. Berkeley and M. Goodfellow, eds.), pp. 1–15. Academic Press, London.

Gordon, R. E., Haynes, W. C., and Hov-Nay Pang, C. (1973). The genus *Bacillus*. *U.S., Dep. Agric., Agric. Handb.* **427.**

Goullet, P. H. (1973). An esterase zymogram of *Escherichia coli*. *J. Gen Microbiol.* **77,** 27–35.

Goullet, P. H. (1978). Characterization of *Serratia marcescens, S. liquefaciens, S. plymuthica* and *S. marinoruba* by the electrophoretic patterns of their esterases. *J. Gen. Microbiol.* **108,** 275–281.

Goullet, P. H. (1981). Characterization of *Serratia odorifera, S. fonticola,* and *S. ficaria* by the electrophoretic patterns of their esterases. *J. Gen. Microbiol.* **127,** 161–167.

Hartl, D. L., and Dykhuizen, D. E. (1984). The population genetics of *Escherichia coli*. *Annu. Rev. Genet.* **18,** 31–68.

Hertlein, B. C., Levy, R., and Miller, T. W., Jr. (1979). Recycling potential and selective retrieval of *Bacillus sphaericus* from soil in a mosquito habitat. *J. Invertebr. Pathol.* **33,** 217–221.

Hertlein, B. C., Hornby, J., Levy, R., and Miller, T. W., Jr. (1981). Prospects of spore forming bacteria for vector control with special emphasis on their local production potential. *Dev. Ind. Microbiol.* **22,** 53–60.

Hessler, P., and Singer, S. (1986). Zymogram differences in strains from homology groups of *Bacillus sphaericus*. *Abstr., 86th Annu. Meet., Am. Soc. Microbiol.*, p. 240.

Hornby, J. A., Hertlein, B. C., and Miller, T. W., Jr. (1984). Persistent spores and mosquito larvicidal activity of *Bacillus sphaericus* 1593 in well water and sewage. *J. Ga. Entomol. Soc.* **19,** 165–167.

Kalfon, A., Larget-Thiery, I., Charles, J. F., and de Barjac, H. (1983). Growth, sporulation and larvicidal activity of *Bacillus sphaericus*. *Eur. J. Appl. Microbiol. Biotechnol.* **18,** 168–173.

Kellen, W. R., and Meyers, C. M. (1964). *Bacillus sphaericus* Neide as a pathogen of mosquitoes. *Proc. Annu. Conf. Calif. Mosq. Control. Assoc.* **32,** 37.

Kellen, W. R., Clark, T. B., Lindegren, J. E., Ho, B. C., Rogoff, M. H., and Singer, S. (1965). *Bacillus sphaericus* Neide as a pathogen of mosquitoes. *J. Invertebr. Pathol.* **7,** 442–448.

Krych, V. K., Johnson, J. C., and Yousten, A. A. (1980). Deoxyribonucleic acid homologies among strains of *Bacillus sphaericus*. *Int. J. Syst. Bacteriol.* **30,** 476–484.

Lacey, L. A. (1984). Production and formulation of *Bacillus sphaericus*. *Mosq. News* **44,** 153–159.

Lacey, L. A., and Singer, S. (1982). Larvicidal activity of new isolates of *Bacillus sphaericus* and *Bacillus thuringiensis* (H-14) against anopheline and culicine mosquitoes. *Mosq. News* **42,** 537–543.

Lacey, L. A., and Undeen, A. H. (1984). Effect of formulation, concentration, and application time on the efficacy of *Bacillus thuringiensis* (H-14) against black fly (Diptera: Simuliidae) larvae under natural conditions. *J. Econ. Entomol.* **77,** 412–418.

Lacey, L. A., Urbina, M. J., and Heitzman, C. M. (1984). Sustained release formulations of *Bacillus sphaericus* and *Bacillus thuringiensis* (H-14) for control of container breeding *Culex quinquefasciatus*. *Mosq. News* **44,** 26–32.

Lee, E. J. (1983). Analysis of nutritional interrelationships of media components needed for the growth, sporulation, and larvicidal activity of *Bacillus sphaericus* strains 1593 and 2013-4. Master's Thesis, Western Illinois University, Macomb (unpublished).

Lessel, E. F., and Holt, J. G. (1970). Presenting and interpreting the results. *In* "Methods for Numerical Taxonomy" (W. R. Lockhart and J. Liston, eds.), pp. 50–58. Am. Soc. Microbiol., Bethesda, Maryland.

Logan, N. A., and Berkeley, R. C. W. (1979). Classification and identification of members of the genus *Bacillus* using API tests. *In* "The Aerobic Endospore-Forming Bacteria: Classification and Identification" (R. C. W. Berkeley and N. Goldfellow, eds.), pp. 105–140. Academic Press, London.

Logan, N. A., and Berkeley, R. C. W. (1984). Identification of *Bacillus* strains using API system. *J. Gen Microbiol.* **130,** 1871–1882.

Lysenko, O., Davidson, E. W., Lacey, L. A., and Yousten, A. A. (1985). Five new mosquito larvicidal strains of *Bacillus sphaericus* from non-mosquito origins. *J. Am. Mosq. Control Assoc.* **1,** 369–371.

McDonald, L. O., and Burke, W. F., Jr. (1984). Plasmid transformation of *Bacillus sphaericus* 1593. *J. Gen. Microbiol.* **130,** 203–208.

Matsushima, O., and Baltz, R. H. (1986). Protoplast fusion. *In* "Manual of Industrial Microbiology and Biotechnology" (A. L. Demain and N. A. Solomon, eds.), pp. 170–183. Am. Soc. Microbiol., Washington, D.C.

Mulla, M. S., Darwazeh, H. A., Davidson, E. W., and Dulmage, H. T. (1984a). Efficacy and persistence of the microbial agent *Bacillus sphaericus* for the control of mosquito larvae in organically enriched habitats. *Mosq. News* **44,** 166–173.

Mulla, M. S., Darwazeh, H. A., Davidson, E. W., Dulmage, H. T., and Singer, S. (1984b). Larvicidal activity and field efficacy of *Bacillus sphaericus* strains against mosquito larvae and their safety to non-target organisms. *Mosq. News* **44,** 336–342.

Mulligan, F. S., III, Schaefer, C. H., and Miura, T. (1978). Laboratory and field evaluation of *Bacillus sphaericus* as a mosquito control agent. *J. Econ. Entomol.* **71,** 774–777.

Mulligan, F. S., III, Schaefer, C. H., and Wilder, W. H. (1980). Efficacy and persistence of *Bacillus sphaericus* and *B. thuringiensis* H-14 against mosquitoes under laboratory and field conditions. *J. Econ. Entomol.* **73,** 684–688.

Myers, P. S., and Yousten, A. A. (1978). Toxic activity of *Bacillus sphaericus* SSII-1 for mosquito larvae. *Infect. Immun.* **19,** 1047–1053.

Myers, P. S., and Yousten, A. A. (1980). Localization of a mosquito-larval toxin of *Bacillus sphaericus* 1593. *Appl. Environ. Microbiol.* **39,** 1205–1211.

Myers, P. S., Yousten, A. A., and Davidson, E. W. (1979). Comparative studies of the mosquito-larval toxin of *Bacillus sphaericus* SSII-1 and 1593. *Can. J. Microbiol.* **25,** 1227–1231.

Norris, J. R. (1964). The classification of *Bacillus thuringiensis*. *J. Appl. Bacteriol.* **27,** 439–447.

Norris, J. R., and Burges, H. D. (1963). Esterases of crystalliferous bacteria pathogenic for insects: Epizootiological applications. *J. Insect Pathol.* **5,** 460–472.

Obeta, J. A. N., and Okafor, N. (1983). Production of *Bacillus sphaericus* strain 1593 primary powder on media made from locally obtainable Nigerian agricultural products. *Can. J. Microbiol.* **29,** 704–709.

O'Donnell, A. G., and Norris, J. R. (1979). Pyrolysis gas–liquid chromatographic studies. *In* "The Aerobic Endospore-Forming Bacteria: Classification and Identification" (R. C. W. Berkeley and M. Goodfellow, eds.), pp. 141–179. Academic Press, London.

Payne, J. M., and Davidson, E. W. (1984). Insecticidal activity of the crystalline parasporal inclusions and other components of the *Bacillus sphaericus* 1593 spore complex. *J. Invertebr. Pathol.* **43,** 383–388.

Priest, F. G., Goodfellow, M., and Todd, C. (1979). The genus *Bacillus:* A numerical analysis. *In* "The Aerovic Endospore-Forming Bacteria: Classification and Identification" (R. C. W. Berkeley and M. Goodfellow, eds.), pp. 92–103. Academic Press, London.

Ramoska, W. A., Singer, S., and Levy, R. (1977). Bioassay of three strains of *Bacillus sphaericus* on field-collected mosquito larvae. *J. Invertebr. Pathol.* **30,** 151–154.

Ramoska, W. A., Burges, J., and Singer, S. (1978). Field applications of a bacterial insecticide. *Mosq. News* **38,** 57–60.

Schliefer, K. H., and Stackebrandt, E. (1983). Molecular systematics of prokaryotes. *Annu. Rev. Microbiol.* **37,** 143–187.

Seki, T., Oshima, T., and Oshima, Y. (1975). Taxonomic study of *Bacillus* by deoxyribonucleic acid hybridization and interspecific transformation. *Int. J. Syst. Bacteriol.* **25,** 258–270.

Shadduck, J. A., Singer, S., and Lause, S. (1880). Lack of mammalian pathogenicity of entomocidal isolates of *Bacillus sphaericus. Environ. Entomol.* **9,** 403–407.

Shelley, T., and Singer, S. (1986). Plasmid profiles of insecticidal, noninsecticidal and antibiotic resistant strains of *Bacillus sphaericus. Abstr., 86th Annu. Meet., Am. Soc. Microbiol.*, p. 155.

Silapanuntakul, S., Pantuwanta, S., Bhumiratana, A., and Charoensiri, K. (1983). The comparative persistence of toxicity of *Bacillus sphaericus* strain 1593 and *Bacillus thuringiensis* serotype H-14 against mosquito larvae in different kinds of environments. *J. Invertebr. Pathol.* **42,** 387–392.

Singer, S. (1973). Insecticidal activity of recent bacterial isolates and their toxins against mosquito larvae. *Nature (London)* **244,** 110–111.

Singer, S. (1974). Entomogenous bacilli against mosquito larvae. *Dev. Ind. Microbiol.* **15,** 187–194.

Singer, S. (1975a). Isolation and development of bacterial pathogen vectors. NIAID/NIH Workshop on "The Biological Regulation of Vectors." Tidewater Inn, Easton, Maryland (see Singer, 1977).

Singer, S. (1975b). Use of bacteria for control of aquatic insect pests. *In* "Impact of the Use of Microorganisms on the Aquatic Environment," Natl. EPA Ecol. Res. Ser., EPA 660-3-75-001. pp. 5–22. U.S. Environ. Prot. Agency, Washington, D.C.

Singer, S. (1977). Isolation and development of bacterial pathogens of vectors. *DHEW Publ. (NIH) (U.S.)* **NIH 77–1180,** 3–17.

Singer, S. (1979). Use of entomogenous bacteria against insects of public health importance. *Dev. Ind. Microbiol.* **20,** 117–122.

Singer, S. (1980). *Bacillus sphaericus* for the control of mosquitoes. *Biotechnol. Bioeng.* **22,** 1335–1355.

Singer, S. (1981). Potential of spore-formers without crystals for pest control. *In* "Microbial Control of Insects, Mites, and Plant Diseases 1970–1980" (H. D. Burges, ed.), pp. 283–298. Academic Press, London.

Singer, S. (1982). The biotechnology for strains of *Bacillus sphaericus* with vector control potential. *Proc. Int. Colloq. Invertebr. Pathol., 3rd,* pp. 485–489.

Singer, S. (1985). *Bacillus sphaericus* (Bacteria). *In* "Biological Control of Mosquitoes" (H. C. Chapman, ed.) Bull. No. 6, pp. 123–131. Am. Mosq. Control Assoc., Fresno, California.

Singer, S., and Murphy, D. J. (1976). New insecticidal strains of *Bacillus sphaericus* useful against *Anopheles albimanus* larvae. *Abstr., 76th Annu. Meet., Am. Soc. Microbiol.*, p. 181.

Singer, S., Goodman, N. S., and Rogoff, M. H. (1966). Defined media for the study of bacilli pathogenic for insects. *Ann. N. Y. Acad. Sci.* **139,** 16–23.

Sneath, P. H. A., and Sokal, R. R. (1973). "Numerical Taxonomy." Freeman, San Francisco, California.

Tinelli, R., and Bourgouin, C. (1982). Larvicidal toxin from *Bacillus sphaericus* spores—isolation of toxic components. *FEBS Lett.* **142,** 155–158.

Watson, R. R. (1976). Substrate specifications of aminopeptidases: A specific method for microbial differentiation. *In* "Methods in Microbiology" (J. R. Norris, ed.), Vol. 9, pp. 1–14. Academic Press, New York.

Weiser, J. (1984). A mosquito-virulent *Bacillus sphaericus* in adult *Simulium damnosum* from northern Nigeria. *Zentralbl. Mikrobiol.* **139,** 57–60.

Whittam, T. S., Ochman, H., and Selander, R. K. (1983). Multilocus genetic structure in natural populations of *Escherichia coli*. *Proc. Natl. Acad. Sci. U.S.A.* **80,** 1751–1755.

Wickremesinge, R. S. B., and Mendis, C. L. (1980). *Bacillus sphaericus* spore from Sri Lanka demonstrating rapid larvicidal activity on *Culex quinquefasciatus*. *Mosq. News* **40,** 387–389.

Wolf, J., and Sharp, R. J. (1979). Taxonomic and related aspects of thermophiles within the genus *Bacillus*. *In* "The Aerobic Endospore-Forming Bacteria: Classification and Identification" (R. C. W. Berkeley and M. Goodfellow, eds.), pp. 252–296. Academic Press, London.

World Health Organization (WHO). (1985). "Information Consultation on the Development of *Bacillus sphaericus* as a Microbial Larvicide," TDR/BCV/ SPHAERICUS/ 85.3 (mimeo, doc). WHO, Rome.

Yoshimura, K., Yamamoto, O., Seki, T., and Oshima, Y. (1983). Distribution of heterogeneous and homologous plasmids in *Bacillus* spp. *Appl. Environ. Microbiol.* **45,** 1733–1740.

Yousten, A. A. (1984a). *Bacillus sphaericus:* Microbiological factors related to its potential as a mosquito larvicide. *Adv. Biotechnol. Processes* **3,** 315–343.

Yousten, A. A. (1984b). Bacteriophage typing of mosquito pathogenic strains of *Bacillus sphaericus*. *J. Invertebr. Pathol.* **43,** 124–125.

Yousten, A. A., and Davidson, E. W. (1982). Ultrastructural analysis of spores and parasporal crystals formed in *Bacillus sphaericus* 2297. *Appl. Environ. Microbiol.* **44,** 1449–1455.

Yousten, A. A., and Hedrick, J. (1982). Bacteriophage typing of mosquito pathogenic strains of *Bacillus sphaericus*. *Proc. Int. Colloq. Invertebr. Pathol., 3rd*, pp. 476–482.

Yousten, A. A., de Barjac, H., Dumanoir, V. C., and Myers, P. (1980). Comparison between bacteriophage typing and serotyping for the differentiation of *Bacillus sphaericus* strains. *Ann. Microbiol. (Paris)* **131B,** 297–308.

Yu, H. S., Lee, D. K., Na, J. O., and Ban, S. J. (1983). Integrated control of mosquitoes by combined use of *Bacillus thuringiensis* var. *israelensis* and larvivorous fish *Aplocheilus latipes* in simulated rice paddies in South Korea. *Korean J. Entomol.* **13,** 75–84.

Yurks, M. A. S. (1985). A study of antibiotic resistant, insecticidal and noninsecticidal strains of *Bacillus sphaericus*. Master's Thesis, Western Illinois University, Macomb (unpublished).

II

Mass Production of Microbial and Viral Insecticides

11

Production of Viral Agents in Invertebrate Cell Cultures

ROBERT R. GRANADOS, KATHLEEN G. DWYER,
AND ANJA C. G. DERKSEN

I. INTRODUCTION

More than 200 cell lines have been established from approximately 70 species of insects (Hink, 1980). The majority of these cell lines have been described from Lepidoptera, Diptera, Orthoptera, Homoptera, Hemiptera, Coleoptera, and Hymenoptera. Many established cell lines from lepidopteran species have proved to be invaluable tools for the *in vitro* propagation of insect-pathogenic viruses. In particular, during the past 10 years significant progress has been made in understanding the replication and molecular biology of baculoviruses in cell culture, and these studies are providing the basis for understanding the nature of virus–host interactions including pathogenicity, host range, virulence, and latency.

In this chapter the current status of the infection and replication of insect pathogenic viruses in cell culture will be reviewed. Only the cytoplasmic polyhedrosis viruses, entomopoxviruses, and baculoviruses will be considered here. For additional information on these and other viruses, the reader is referred to the reviews by Granados (1976), Knudson and Buckley (1977), Paschke and Webb (1976), Longworth (1978), and Volkman and Knudson (1986).

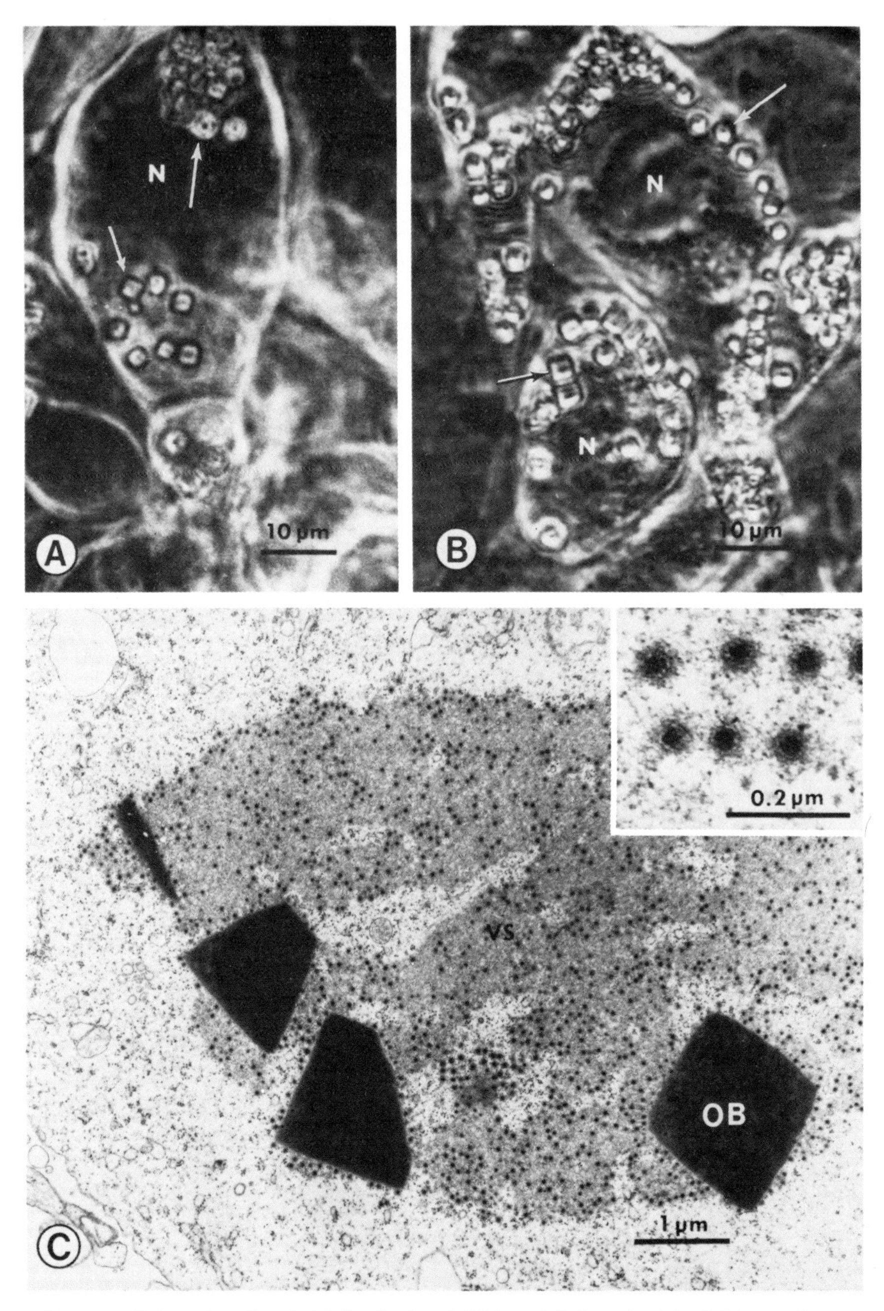

Fig. 1. Cultured cells of *Trichoplusia ni* (TN-368) infected with CPV. (A) Phase-contrast photograph of cell 5 days postinoculation. Note occlusion bodies (arrows) in cell cytoplasm. N, Nucleus. (B) Infected cells at 7 days postinoculation. Numerous occlusion bodies can be seen in cytoplasm (arrows). N, Nucleus. (C) Electron micrograph of infected cell showing a virogenic stroma (VS) containing numerous virus particles. Occlusion bodies (OB) are formed at the periphery of the virogenic stroma. (Insert) Higher magnification of virions in the virogenic stroma (from Granados *et al.*, 1974).

II. VIRUS REPLICATION IN CELL CULTURE

A. Cytoplasmic Polyhedrosis Viruses

At least four different cytoplasmic polyhedrosis viruses (CPVs) have been grown in established cell lines (Granados *et al.*, 1974; Longworth, 1981; Belloncik and Chagon, 1980; Inoue and Mitsuhashi, 1985). Two of these CPVs have an extended host range *in vitro*. For example, *Trichoplusia ni* CPV will replicate in *Spodoptera frugiperda* (IPLB-Sf21), *Lymantria dispar* (IPLB-65Z), and *Estigmene acrea* (BTI-EAA) cell lines in addition to TN-368 cells (Fig. 1) (Granados *et al.*, 1974; Granados, 1975). *Euxoa scandens* CPV (EuCPV) has been grown in 10 lepidopteran cell lines with *L. dispar* cells being the most susceptible (Arella *et al.*, 1984; Belloncik *et al.*, 1985). Belloncik and co-workers have published several papers on EuCPV in cell culture and have clearly demonstrated that the available cell culture systems are suitable for studies on the molecular biology of CPV replication *in vitro* (Grancher-Barray *et al.*, 1981; Arella *et al.*, 1984). CPVs can be titrated in culture by end point dilution methods, infected focus assays, and ELISA (Longworth, 1981; Belloncik and Changon, 1980; Payment *et al.*, 1982). A plaque assay for CPVs has not been developed to date, but this development should not be a major problem since lateral infection can occur (Longworth, 1981). Replication of CPVs in cell cultures does not lead to cell lysis (Granados *et al.*, 1974; Belloncik and Chagon, 1980; Longworth, 1981) and only low levels of virus particles appear to bud through the cell membrane and into the medium (Longworth, 1981). Since many infectious particles are not released into the medium it is necessary to disrupt and sonicate infected cells in order to passage the virus *in vitro*. Several CPV/cell culture systems are now available for study of the molecular aspects of virus replication and gene expression.

B. Entomopoxviruses

Only two entomopoxviruses (EPVs) have been grown in established cell lines: the *Amsacta moorei* EPV (Granados and Naughton, 1976a,b; Granados, 1981; Quiot *et al.*, 1975) and *Adoxophyes* sp. EPV (Sato, 1985). The *A. moorei* EPV (AmEPV) replicates very efficiently in a cell line, BTI-EAA, which was established from hemocytes of *E. acrea* larvae (Fig. 2). In addition, Zhiyu and Granados (1986) reported that AmEPV could replicate in *L. dispar* (IPLB 65Z), *S. frugiperda* (IPLB SF-21), and *Manduca sexta* cell lines. In standard infection studies using a multiplicity of infection of 5, the percentage of cells infected at

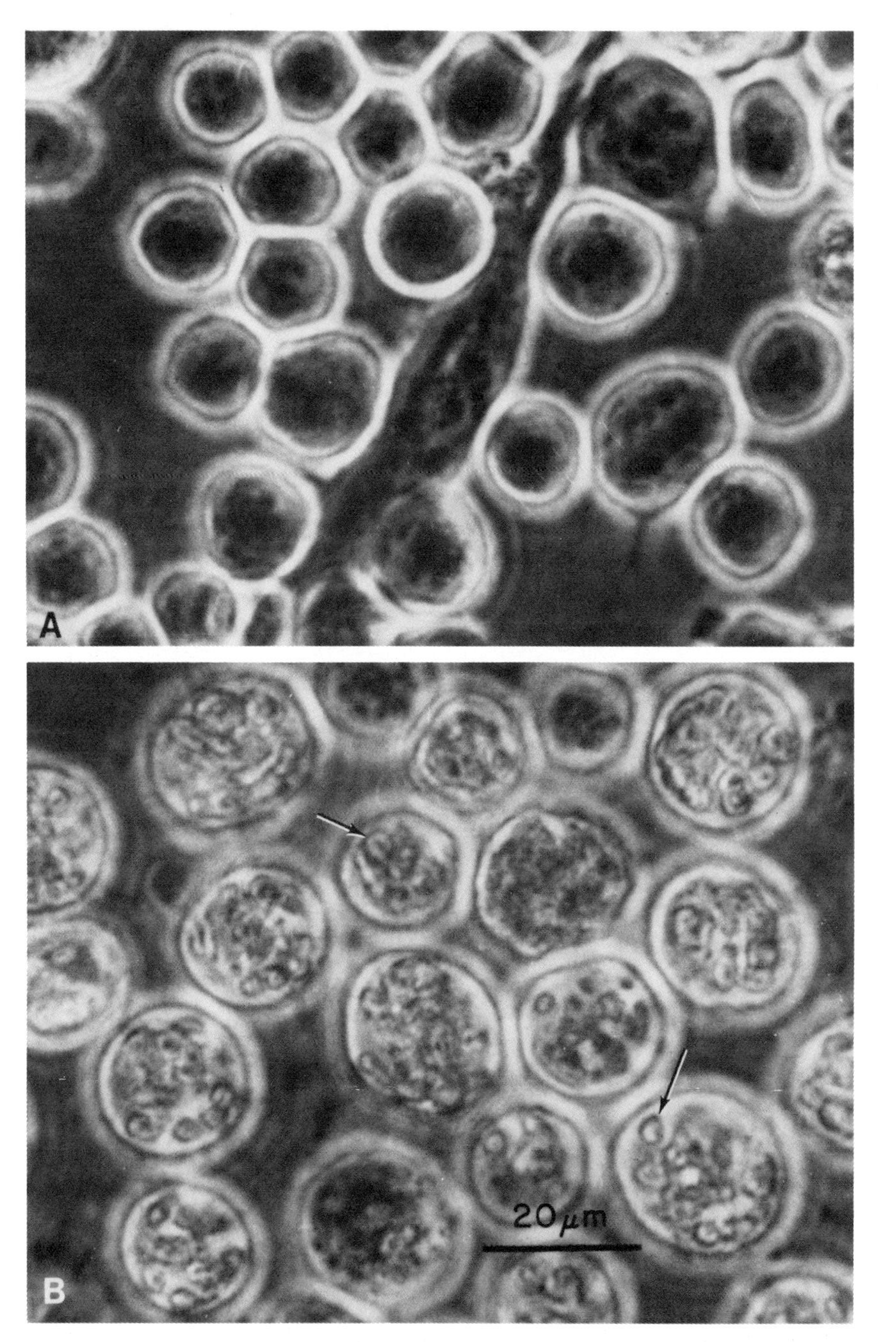

Fig. 2. Healthy and entomopoxvirus-infected cultured cells of *Estigmene acrea* (BTI-EAA). (A) Healthy cultured cells established from hemocytes of 5th instar *E. acrea* larvae. Cells grow in suspension. (B) BTI-EAA infected with *Amsacta moorei* entomopoxvirus 7 days postinoculation. At this stage more than 95% of the cells are swollen and infected. Note occlusion bodies (arrows) in cytoplasm of cells.

7 days postinoculation were 94% (BTI-EAA), 65% (IPLB-65Z), 2.9% (IPLB-SF21), and 0.7% (*M. sexta*). Lysis of infected cells does not occur, but infectious virus is released by budding at the cell membrane. The AmEPV can easily be titrated by the dilution end point method or by a plaque assay using an agarose overlay (Granados and Naughton, 1976a). This EPV/cell culture system represents an excellent model for future studies on the molecular biology and genetics of this interesting group of insect viruses.

C. Baculoviruses

1. Subgroup A (Nuclear Polyhedrosis Viruses)

Viruses in subgroup A are usually subdivided into two morphotypes: the single-nucleocapsid NPV (SNPV), in which only one nucleocapsid is found per envelope, and the multinucleocapsid NPV (MNPV), in which several nucleocapsids are packaged per envelope. The most extensively studied M-type baculovirus is the *Autographa californica* nuclear polyhedrosis virus (AcMNPV) (Fig. 3). This virus has a relatively broad *in vivo* host range and infects over 30 insect species (van der Beek, 1980; Capinera and Kanost, 1979). Similarly, AcMNPV can be replicated in several lepidopteran cell lines (van der Beek, 1980; Volkman and Knudson, 1986; Lynn and Hink, 1980), and many eloquent studies on the molecular biology of this virus have been published (Miller, 1984, 1986; Cochran *et al.*, 1986). In particular, the use of this virus as a vector for propagating and expressing foreign genes in eukaryotic cells (Smith *et al.*, 1983; Pennock *et al.*, 1984) has focused additional attention on this and other similar NPVs. More than 14 different MNPVs have been grown in different cell lines. In addition to AcMNPV, NPVs from *Bombyx mori*, *L. dispar*, and *S. frugiperda* grow readily in cell cultures, are easily plaqued, and should be amenable to genetic and molecular biological analysis (see the chapter by L. K. Miller in this volume).

Until recently, the *Heliothis zea* SNPV (HzSNPV) was the only SNPV to have been grown in an established cell line (Goodwin *et al.*, 1973; 1982; Granados *et al.*, 1981; Yamada and Maramorosch, 1981; Yamada *et al.*, 1982). This status had led many virologists to believe that SNPVs may be more difficult to grow in cell cultures than MNPVs. However, since 1984 at least three new SNPVs (*Heliothis armigera* SNPV, *Orgyia leucostigma* SNPV, and *T. ni* SNPV) have been cultivated *in vitro* (Zhu and Zhang, 1985; Sohi *et al.*, 1984; Granados *et al.*, 1986). Granados *et al.* (1986) established 36 new *T. ni* cell lines from embryonic tissue and 29 lines supported replication of *T. ni* SNPV (TnSNPV) (Fig. 4). Infec-

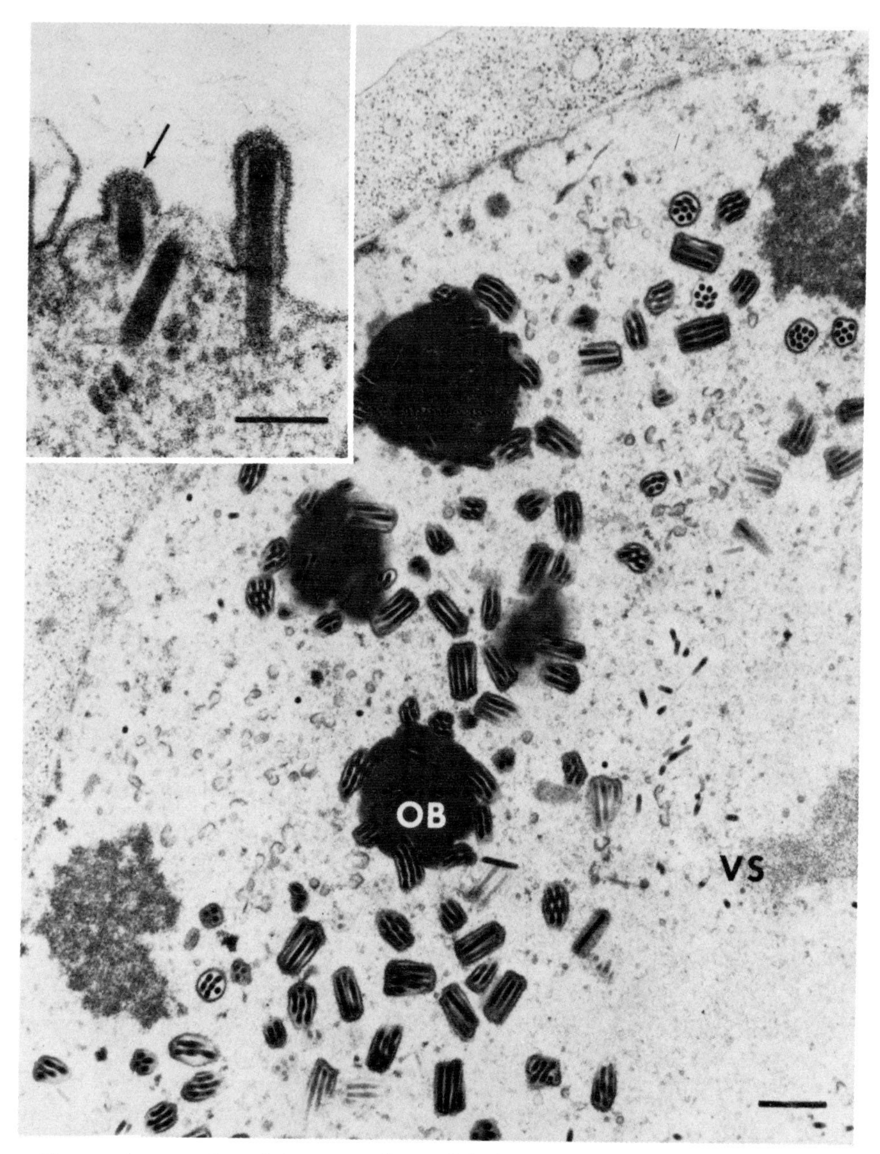

Fig. 3. *Autographa californica* nuclear polyhedrosis virus in TN-368 cell culture at 24 hr postinoculation. The virogenic stroma (VS) is located in the central region of the nucleus and the developing occlusion bodies (OB) are present in a ring zone between the stroma and the nuclear envelope. Insert: nucleocapsids budding through the plasma membrane at 16 hr postinoculation. A peplomer structure (arrow) can be observed at one end of the viral envelope. Bar, 0.1 μm. From Granados (1980).

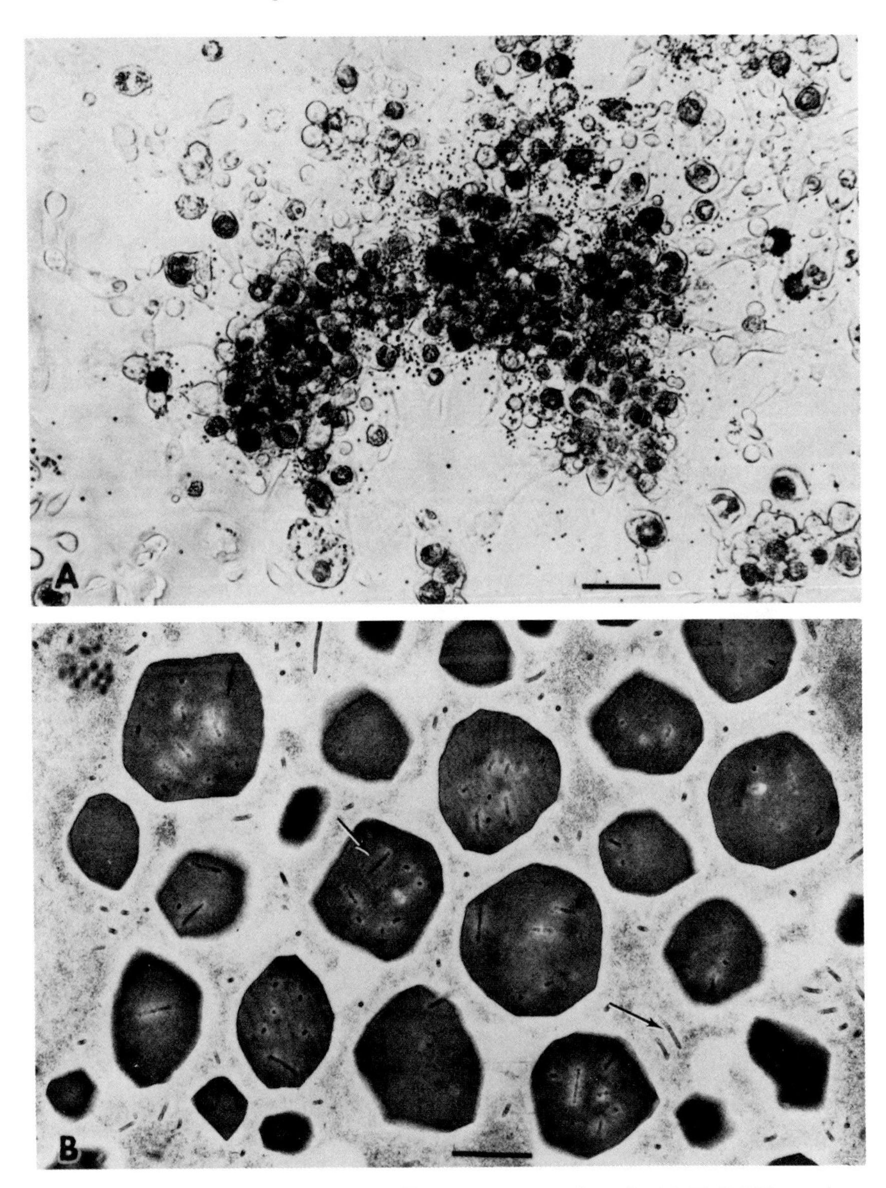

Fig. 4. *Trichoplusia ni* cultured cells (BTI-TN5F2) infected with TnSNPV at 5 days postinoculation. (A) Phase-contrast photograph of infected cells showing dark nuclei that are filled with viral occlusion bodies (OBs). OBs from lysed cells can be seen in the surrounding medium. Bar, 50 μm. (B) Electron micrograph of the nucleus containing numerous OBs. Single-nucleocapsid virions (arrows) can be observed within the OBs or nucleoplasm. Bar, 0.5 μm. From Granados *et al.* (1986).

tion rates in the various cell lines ranged from <1 to 60% infection at 5 days postinoculation. Infection of selected cell lines with TnSNPV was recorded at cell passages 12, 18, 30, and 43; therefore, it appeared that susceptibility of these lines to TnSNPV was stable. All of the new cell lines were highly susceptible (>95% of cells infected) to AcMNPV infection and several were susceptible to *T. ni* granulosis virus (TnGV) (see below). The ability of many of these new cell lines to support growth of different baculoviruses may be related to the tissues used to initiate the cultures. Embryonic egg tissue yields many different cell types, which would enhance the probability of establishment of cell lines with varying virus susceptibilities. It is clear that several new cell culture/SNPV systems are now available for in-depth biological and molecular studies.

2. Subgroup B (Granulosis Viruses)

Prior to 1984 attempts to replicate GVs in primary organ cultures or established cell lines had met with minimal or no success. In larvae the virus replicates mainly in the midgut and fat body tissues. Most established insect cell lines are derived from hemocytes and ovarian cells and thus may not be appropriate cells for GV replication, i.e., lacking viral receptors or missing host enzymes required for replication. Attempts to establish lepidopteran cell lines from midgut and fat body organs have failed. In 1984, Miltenburger and co-workers in Germany reported the first successful *in vitro* replication of *Cydia pomonella* (codling moth) GV (CpGV) in primary cell lines derived from *C. pomonella* eggs (Naser *et al.*, 1984; Miltenburger *et al.*, 1984). From a total of 200 new established primary cell lines of *C. pomonella*, 81 were screened for CpGV replication and 9 were susceptible. Cell line IZD-Cp-3300 showed 20% infection of cells at 8 days postinoculation (Naser *et al.*, 1984). The replication was confirmed by light and electron microscopy, and dot immuno assays with monoclonal antibodies.

A recent development in our laboratory was the successful establishment of several new *T. ni* cell lines which were susceptible to TnGV (Granados *et al.*, 1986). From a total of 36 new *T. ni* embryonic cells lines, 15 different cell lines and 3 sublines were susceptible to TnGV as determined by the peroxidase–antiperoxidase (PAP) assay. Of the original 15 susceptible cell lines (approximate cell line passage = 12), only 2 lines and 2 sublines (approximate cell line passage = 25) are now susceptible to TnGV. The data strongly indicated that on passage there is a selection for nonsusceptible cells. The 4 susceptible lines, BTI-TN4B1, BTI-TN5F2, BTI-TN5F2P, and BTI-TN5F2A, were slow-grow-

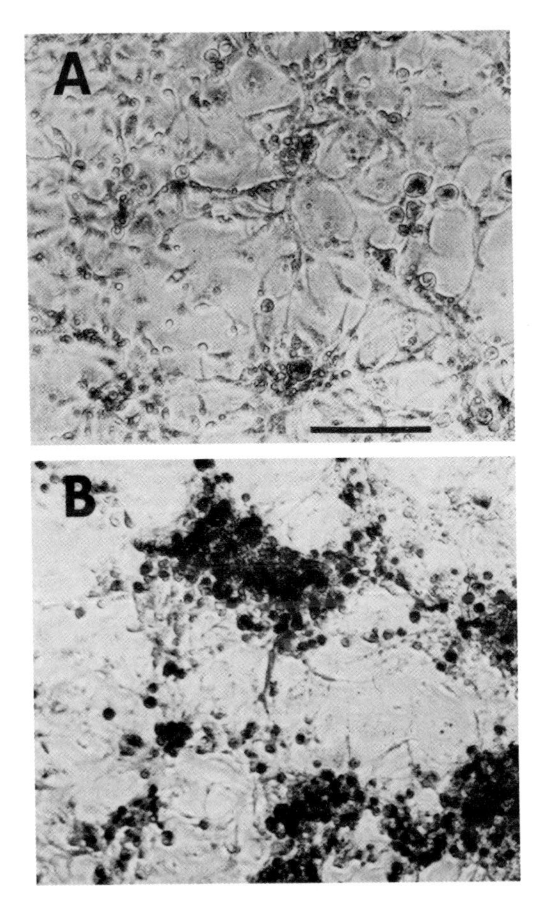

Fig. 5. *Trichoplusia ni* cell line (BTI-TN5G2A1) uninfected (A) and infected (B) with TnGV. The peroxidase–antiperoxidase-stained TnGV-infected cells (B) are surrounded by unstained, uninfected cells. Bar, 100 μm.

ing (weekly split ratio of 1 : 5) in comparison with the lines that lost susceptibility (weekly split ratio of 1 : 10). All susceptible cells were attached, but not all attached cells were susceptible. Aside from minor granulation and clumping of infected cells, no strong cytopathogenic effects were observed by phase-contrast microscopy. PAP assays of inoculated BTI-TN5F2P cells showed that newly synthesized viral proteins could be detected at 7, 9, and 12 days postinoculation (Fig. 5) but not at 5 days postinoculation.

An electron microscopy examination of cell cultures (lines TN5G2A1 and TN5F2P) infected for 11 or 12 days showed all stages of virus replication (Fig. 6). Infected cells contained numerous virus-containing

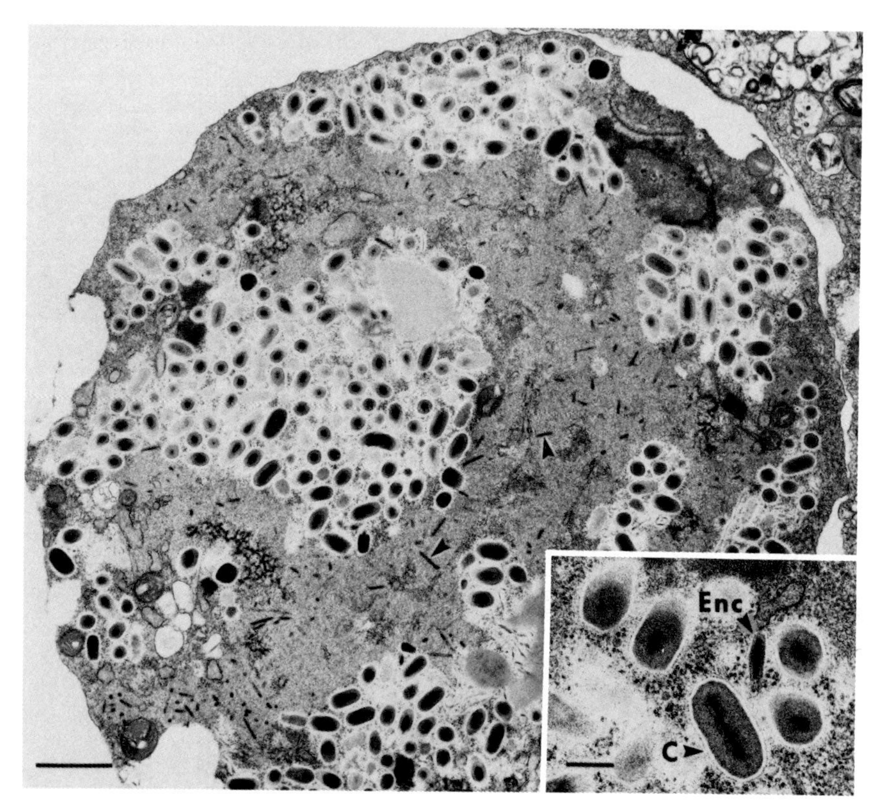

Fig. 6. Electron micrograph of a *Trichoplusia ni* cultured cell (BTI-TN5G2A1 cell line) at 12 days postinfection with TnGV. Note the absence of a nuclear envelope, leading to mixing of nuclear and cytoplasmic components. Numerous virus-containing capsules and virus nucleocapsids (arrows) are present. Bar, 1.0 μm. (Insert) High magnification of virus-containing capsules (C) and an enveloped nucleocapsid (Enc) found in infected cells. Bar, 0.2 μm. From Granados *et al.* (1986).

GV capsules, enveloped nucleocapsids, and nucleocapsids. The cytopathology was typical of a GV infection in that the nuclear envelope was disrupted, resulting in mixing of nuclear and cytoplasmic components. The cell-free virus was passaged three times in the Tn5G2A1 cell line without loss of infectivity.

Virus-specific DNA replication in TN5F2P and TN5F2A sublines was followed by DNA slot-blot hybridization. Viral DNA present above the basal level (2 hr postinoculation) of the viral inoculum was first detected 5 days postinoculation and continued to increase with each successive time point. These results showed that viral DNA synthesis had commenced by 5 days and proceeded at approximately the same rate from 5 to 9 days postinoculation.

Two important observations from these studies were that (1) none of the floating cells was susceptible to TNGV, which confirms a similar finding reported by Miltenburger *et al.* (1984), and (2) these new cell lines, which represent many different embryonic cell types, may change their viral susceptibility properties on passage. Loss of TnGV susceptibility appeared to correlate with faster cell growth. Indeed, the four susceptible cell lines had a lower rate of growth. The experimental approach developed by Miltenburger *et al.* (1984) and Naser *et al.* (1984) and confirmed in this study suggests that other new cell lines from different insect species may be developed for the growth of new GVs. Such cell systems would allow studies of GV replication at the molecular level, genetic manipulation of GV, and use of GV as a recombinant DNA vector for the expression of foreign genes (see chapter by L. K. Miller in this treatise).

3. Subgroup C (Nonoccluded Viruses)

Subgroup C baculoviruses have been reported from a variety of arthropod hosts including insects, mites, and crustaceans (Crawford and Granados, 1982). To date, only two nonoccluded baculoviruses, *Oryctes* and HZ-1, have been shown to replicate in cell culture. The first report of *Oryctes* baculovirus in cell culture was in a primary cell culture derived from *O. rhinocerus* (Quiot *et al.*, 1973). The *Oryctes* baculovirus replicates efficiently in an established cell line from *Heteronychus arator* (DSIR-HA-1179) (Crawford, 1982), and *in vitro* studies on viral morphogenesis, the infectivity of cell-associated and budded virus, and the synthesis of viral proteins have been reported (Crawford and Sheehan, 1985).

The HZ-1 virus, which persistently infects the *H. zea* IMC-HZ-1 cell line, can be replicated in lepidopteran cell lines (Granados *et al.*, 1978; Ralston *et al.*, 1981). Several published papers on molecular events of HZ-1 replication in cell cultures suggest that this virus–cell culture system may be ideal for the study of the molecular mechanisms of baculovirus persistence (Burand *et al.*, 1983; 1986; Huang *et al.*, 1982; Burand and Wood, 1986).

III. SUMMARY AND CONCLUSIONS

In the past decade, considerable advances have been made in developing cell lines for the growth of baculoviruses, entomopoxviruses, and cytoplasmic polyhedrosis viruses. There are no longer any major obstacles to progress in understanding the genetics and molecular biology of these viruses. Many cell lines are established for most major lepidopteran and dipteran pests; however, other important insect pest groups, such as beetles, grasshoppers, and sawflies, are not well represented. There is also a need to develop the *in vitro* technology for the culture of specific differentiated tissue, e.g., fat body, muscle, intestinal, and epidermal cells. Finally, the prospects for eventual commercialization of viral pesticides and the recent developments in the use of baculoviruses as vectors for propagating and expressing exogenous DNAs in insect cells suggest an important application for large-scale production of insect cells. Research on the growth of viruses in large-scale systems has not received adequate attention in recent years; however, the technology is available (Tramper *et al.*, 1986; Weiss *et al.*, 1981) and major advances in this area of cell culture research will be made in the near future.

ACKNOWLEDGMENTS

We are grateful to Ms. Lisa Austin and Mr. G. Li for their skilled technical assistance and Ms. Sue Noti for excellent secretarial help. This study was supported by Grant 83-CRCR-1230 from the U.S. Department of Agriculture.

REFERENCES

Arella, M., Belloncik, K. S., and Devauchelle, G. (1984). Protein synthesis in a *Lymantria dispar* cell line infected by cytoplasmic polyhedrosis virus. *J. Virol.* **52,** 1024–1027.

Belloncik, K. S., and Chagon, A. (1980). Titration of a cytoplasmic polyhedrosis virus by a tissue microculture assay: Some applications. *Intervirology* **13,** 28–32.

Belloncik, K. S., Rocheleau, H., Su, D.-M, and Arella, M. (1985). Replication of a cytoplasmic polyhedrosis virus (CPV) in cultured insect cells. *Abstr., Int. Cell Cult. Congr., 3rd,* p. 65.

Burand, J. P., and Wood, H. A. (1986). Intracellular protein synthesis during standard and defective HZ-1 virus replication. *J. Gen. Virol.* **67,** 167–173.

Burand, J. P., Stiles, B., and Wood, H. A. (1983). Structural and intracellular proteins of the nonoccluded baculovirus HZ-1. *J. Virol.* **46,** 137–142.

Burand, J. P., Kawanishi, C. Y., and Huang, Y.-S. (1986). Persistent baculovirus infections. *In* "The Biology of Baculoviruses" (R. R. Granados and B. A. Federici, eds.), Vol. 1, pp. 159–175. CRC Press, Boca Raton, Florida.

Capinera, J. L., and Kanost, M. R. (1979). Susceptibility of the zebra caterpillar to *Autographa californica* nuclear polyhedrosis virus. *J. Econ. Entomol.* **72,** 570–572.

Cochran, M. A., Brown, S. E., and Knudson, D. L. (1986). Organization and expression of the baculovirus genome. *In* "The Biology of Baculoviruses" (R. R. Granados and B. A. Federici, eds.), Vol. 1, pp. 239–258. CRC Press, Boca Raton, Florida.

Crawford, A. M. (1982). A coleopteran cell line derived from *Heteronychus arator* (Coleoptera: Scarabaeidae). *In Vitro* **18,** 813–816.

Crawford, A. M., and Granados, R. R. (1982). Non-occluded baculoviruses. *Proc. Int. Colloq. Invertebr. Pathol., 3rd,* pp. 154–159.

Crawford, A. M., and Sheehan, C. (1985). Replication of *Oryctes* baculovirus in cell culture: Viral morphogenesis, infectivity and protein synthesis. *J. Gen. Virol.* **66,** 529–539.

Goodwin, R. H., Vaughn, J. L., Adams, J. R., and Louloudes, S. J. (1973). The influence of insect cell lines and tissue culture media on *Baculovirus* polyhedra production. *Misc. Pub. Entomol. Soc. Am.* **9,** 66–72.

Goodwin, R. H., Tompkins, G. J., Gettig, R. R., and Adams, J. R. (1982). Characterization and culture of virus replicating continuous insect cell lines from the bollworm, *Heliothis zea* (Boddie). *in Vitro* **18,** 843–850.

Granados, R. R. (1975). Multiplication of a cytoplasmic polyhedrosis virus (CPV) in insect tissue cultures. *Abstr. Int. Congr. Virol., 3rd,* 1975, pp. 1–98.

Granados, R. R. (1976). Infection and replication of insect pathogenic viruses in tissue culture. *Adv. Virus Res.* **20,** 189–236.

Granados, R. R. (1980). Infectivity and mode of action of baculoviruses. *Biotechnol. Bioeng.* **22,** 1377–1405.

Granados, R. R. (1981). Entomopoxvirus infections in insects. *In* "Pathogenesis of Invertebrate Microbial Diseases" (E. Davidson, ed.), pp. 101–126. Allanheld, Osmun, Totowa, New Jersey.

Granados, R. R., and Naughton, M. (1976a). Replication of *Amsacta moorei* entomopoxvirus in a continuous cell culture from *Estigmene acrea*. *Proc. Int. Colloq. Invertebr. Pathol., 1st,* pp. 113–117.

Granados, R. R., and Naughton, M. (1976b). Replication of *Amsacta moorei* entomopoxvirus and *Autographa californica* nuclear polyhedrosis virus in hemocyte cell lines from *Estigmene acrea*. *In* "Invertebrate Tissue Culture: Research Applications" (K. Maramorosch, ed.), pp. 379–389. Academic Press, New York.

Granados, R. R., McCarthy, W. J., and Naughton, M. (1974). Replication of a cytoplasmic polyhedrosis virus in an established cell line of *Trichoplusia ni* cells. *Virology* **59,** 584–586.

Granados, R. R., Nguyen, T., and Cato, B. (1978). An insect cell line persistently infected with a baculovirus-like particle. *Intervirology* **10,** 309–317.

Granados, R. R., Lawler, K. A., and Burand, J. P. (1981). Replication of *Heliothis zea* baculovirus in an insect cell line. *Intervirology* **16,** 71–79.

Granados, R. R., Derksen, A. C. G., and Dwyer, K. G. (1986). Replication of the *Trichoplusia ni* granulosis and nuclear polyhedrosis viruses in cell cultures. *Virology* **152,** 472–476.

Grancher-Barray, S., Boisvert, J., and Belloncik, K. S. (1981). Electrophoretic characterization of proteins and RNA of cytoplasmic polyhedrosis virus (CPV) from *Euxoa scandens*. *Arch. Virol.* **70,** 55–61.

Hink, W. F. (1980). The 1979 compilation of invertebrate cell lines and culture media. *In* "Invertebrate Systems in Vitro" (E. Kurstak, K. Maramorosch, and A. Dübendorfer, eds.), pp. 553–578. Elsevier/North-Holland Biomedical Press, Amsterdam.

Huang, Y.-S., Hedberg, M., and Kawanishi, C. Y. (1982). Characterization of the DNA of a nonoccluded baculovirus, HZ-1V. *J. Virol.* **43,** 174–181.

Inoue, H., and Mitsuhashi, J. (1985). A new continuous cell line from embryos of *Bombyx mori*. *Proc. Symp. Appl. Invertebr. Cells in Vitro, Int. Cell Cult. Congr., 3rd,* p. 26.

Knudson, D. L., and Buckley, S. M. (1977). Invertebrate cell culture methods for the study of invertebrate-associated animal viruses. *Methods Virol.* **6,** 323–391.

Longworth, J. F. (1978). Small isometric viruses of invertebrates. *Adv. Virus Res.* **23,** 103–157.

Longworth, J. F. (1981). The replication of a cytoplasmic polyhedrosis virus from

Chrysodeixis eriosoma (Lepidoptera: Noctuidae) in *Spodoptera frugiperda* cells. *J. Invertebr. Pathol.* **37,** 54–61.

Lynn, D. E., and Hink, W. F. (1980). Comparison of nuclear polyhedrosis virus replication in five lepidopteran cell lines. *J. Invertebr. Pathol.* **35,** 234–240.

Miller, L. K. (1984). Exploring the gene organization of baculoviruses. *Methods Virol.* **7,** 227–258.

Miller, L. K. (1986). The genetics of baculovirus. *In* "The Biology of Baculoviruses" (R. R. Granados and B. A. Federici, eds.), Vol. 1, pp. 217–238. CRC Press, Boca Raton, Florida.

Miltenburger, H. G., Naser, W. L., and Harvey, J. P. (1984). The cellular substrate: A very important requirement for baculovirus *in vitro* replication. *Z. Naturforsch., C: Biosci.* **39C,** 993–1002.

Naser, W. L., Miltenburger, H. G., Harvey, J. P., Huber, J., and Huger, A. M. (1984). In vitro replication of the *Cydia pomonella* (codling moth) granulosis virus. *FEMS Microbiol. Lett.* **24,** 117–121.

Paschke, J. D., and Webb, S. R. (1976). Fundamental studies on insect icosahedral cytoplasmic deoxyribovirus in continually propagated *Aedes aegypti* cells. *In* "Invertebrate Tissue Culture: Research Applications" (K. Maramorosch, ed.), pp. 269–293. Academic Press, New York.

Payment, P., Arora, D. J. S., and Belloncik, K. S. (1982). An enzyme-linked immunosorbent assay for the detection of cytoplasmic polyhedrosis virus. *J. Invertebr. Pathol.* **40,** 55–60.

Pennock, G. D., Shoemaker, C., and Miller, L. K. (1984). Strong and regulated expression of *Escherichia coli* β-galactosidase in insect cells with a baculovirus vector. *Mol. Cell. Biol.* **4,** 399–406.

Quiot, J.-M., Monsarrat, P., Meynadier, G., Croizier, G., and Vago, C. (1973). Infection des cultures cellulaires de coléoptères par le virus *Oryctes*. *C. R. Hebd. Seances Acad. Sci.* **276,** 3229–3231.

Quiot, J.-M., Bergoin, M., and Vago, C. (1975). Développement et pathogénèse d'un entomopoxvirus en culture cellulaire de Lépidoptère. *C.R. Hebd. Seances Acad. Sci.* **280,** 2273–2275.

Ralston, A. L., Huang, Y.-S., and Kawanishi, C. Y. (1981). Cell culture studies with the IMC-Hz-1 nonoccluded virus. *Virology* **115,** 33–44.

Sato, T. (1985). Establishment of several cell lines from neonate larvae of tortricids (Lepidoptera) and their susceptibility to insect viruses. *Proc. Symp. Appl. Invertebr. Cells in Vitro, Int. Cell Cult. Congr., 3rd,* pp. 6–7.

Smith, G. E., Summers, M. D., and Fraser, M. J. (1983). Production of human beta interferon in insect cells infected with a baculovirus expression vector. *Mol. Cell. Biol.* **3,** 3156–2165.

Sohi, S. S., Percy, J., Arif, B. M., and Cunningham, J. C. (1984). Replication and serial passage of a singly enveloped baculovirus of *Orgyia leucostigma* in homologous cell lines. *Intervirology* **21,** 50–60.

Tramper, J., Williams, J. B., Joustra, D., and Vlak, J. M. (1986). Shear sensitivity of insect cells in suspension. *Enzyme Microb. Technol.* **8,** 33–36.

van der Beek, C. P. (1980). On the origin of the polyhedral protein of the nuclear polyhedrosis virus of *Autographa californica*. Ph.D. Thesis, pp. 1–74. H. Veenman and Zonen B. V., Wageningen, The Netherlands.

Volkman, L. E., and Knudson, D. L. (1986). In vitro replication of baculoviruses. *In* "The Biology of Baculoviruses" (R. R. Granados and B. A. Federici, eds.), Vol 1, pp. 109–127. CRC Press, Boca Raton, Florida.

Weiss, S. A., Smith, G. C., Kalter, S. S., and Vaughn, J. L. (1981). Improved method for the production of insect cell cultures in large volume. *In Vitro* **17,** 495–592.

Yamada, K., and Maramorosch, K. (1981). Plaque assay of *Heliothis zea* baculovirus employing a mixed agarose overlay. *Arch. Virol.* **67,** 187–189.

Yamada, K., Sherman, K. E., and Maramorosch, K. (1982). Serial passage of *Heliothis zea* singly embedded nuclear polyhedrosis virus in a homologous cell line. *J. Invertebr. Pathol.* **39,** 185–191.

Zhiyu, L., and Granados, R. R. (1986). Replication of *Amsacta moorei* entomopoxvirus in insect cell lines. *Proc. Int. Colloq. Invertebr. Pathol., 4th. In* "Fundamental and Applied Aspects of Invertebrate Pathology" (R. A. Samson, I. M. Vlak, and D. Peters, eds.) Aug. 18–22, Veldhoven, The Netherlands, p. 407.

Zhu, G., and Zhang, H. (1985). The multiplication characteristics of *Heliothis armigera* baculovirus in the established cell lines. *Abstr., Int. Cell Cult. Congr., 3rd,* p. 66.

12

Morphogenesis of Germinating Conidia and Protoplast-Associated Structures in Entomophthoralean Fungi

JOJI AOKI

I. INTRODUCTION

Fungal protoplasts are very useful for studies of fungal morphology, biochemistry, and genetics. The entomopathogenic Entomophthorales fungi have a highly specialized process of protoplast release. Sorokin (1883) reported that a naked mass of protoplasm with an ameboid movement commonly occurred in the hemolymph of a cricket infected with *Entomophthora colorata*. Thaxter (1888) observed that the conjugating hyphal bodies of *Empusa fresenii* from which zygospores form have no proper wall. Tyrrell and MacLeod (1972) more recently observed the protoplast stage of *Entomophthora egressa* in tissue culture. Later, Tyrrell (1977) found this stage in larval hemolymph of the spruce budworm, *Choristoneura fumiferana*. Dunphy and Nolan (1977a,b) studied the subsequent morphogenesis of *E. egressa* protoplasts under a variety of *in vitro* conditions with light microscopy. Ultrastructural

BIOTECHNOLOGY IN INVERTEBRATE
PATHOLOGY AND CELL CULTURE

observations have been conducted by Butt *et al.* (1981) on the life cycle of *Erynia neoaphidis* protoplasts formed naturally within the body of their aphid host, *Acrythosiphon pisum*. Kobayashi *et al.* (1984) also investigated the ultrastructure of protoplast regeneration of *E. neoaphidis* in the green peach aphid, *Myzus persicae*, with an electron microscope.

The Entomophthorales fungi have one more peculiar property: they show different patterns of morphological development in their conidial germinations. Some of the *Erynia* species show three different germinating forms, namely germ tube extension, secondary conidium budding, and capillary tube formation. These phenomena have physiological importance in the fungal growth.

In the present chapter, morphogenesis of protoplast-associated structures of several *Erynia* species is compared with that of germinating conidia studied with electron microscopy and fluorescent markers. Protoplast-associated structures of *Conidiobolus obscurus* are also investigated ultrastructurally. These comparative studies may provide interesting material related to the significance of wall composition and ultrastructure and the involvement of the plasma membrane as determinants of cell morphology.

On the grounds of biochemical evidence, cell walls of zygomycetes were classified as chitin–chitosan type, and it was assumed that the chitosan replaces β-1,3-glucan (Sengbusch *et al.*, 1983). Therefore, fluorescein isothiocyanate (FITC)-labeled wheat germ agglutinin (WGA), which binds to *N*-acetyl-β-1,4-D-glucosamine, and FITC-labeled concanavalin A (Con A), which binds to α-D-glucose and α-D-mannose (Taso, 1970), were used to investigate carbohydrate constituents of the fungal cell wall. Moreover, 4′,6-diamidino-2-phenylindole (DAPI), which has a high affinity for AT-rich sections in DNA (Lemke *et al.*, 1978), was used to stain nuclear material in the fungal cells.

II. CONIDIA IN DORMANT STAGE

FITC-labeled WGA and Con A intensively bind the periphery of dormant conidia (Fig. 1),* and DAPI receptors are detected mainly at the center position (Fig. 2). A nucleus is observed ultrastructurally in the central zone of the conidia and is surrounded by many lipid globules

*The following abbreviations are used in the figures: N, nucleus; L, lipid globule; G, glycogen deposit; D, dictyosomal vesicle; Mi, mitochondrion; FC, fluffy coat; CL, compact layer; S, septum; R, ribosome; MB, multivesicular body; ER, endoplasmic reticulum; PG, protein granule; C, coat.

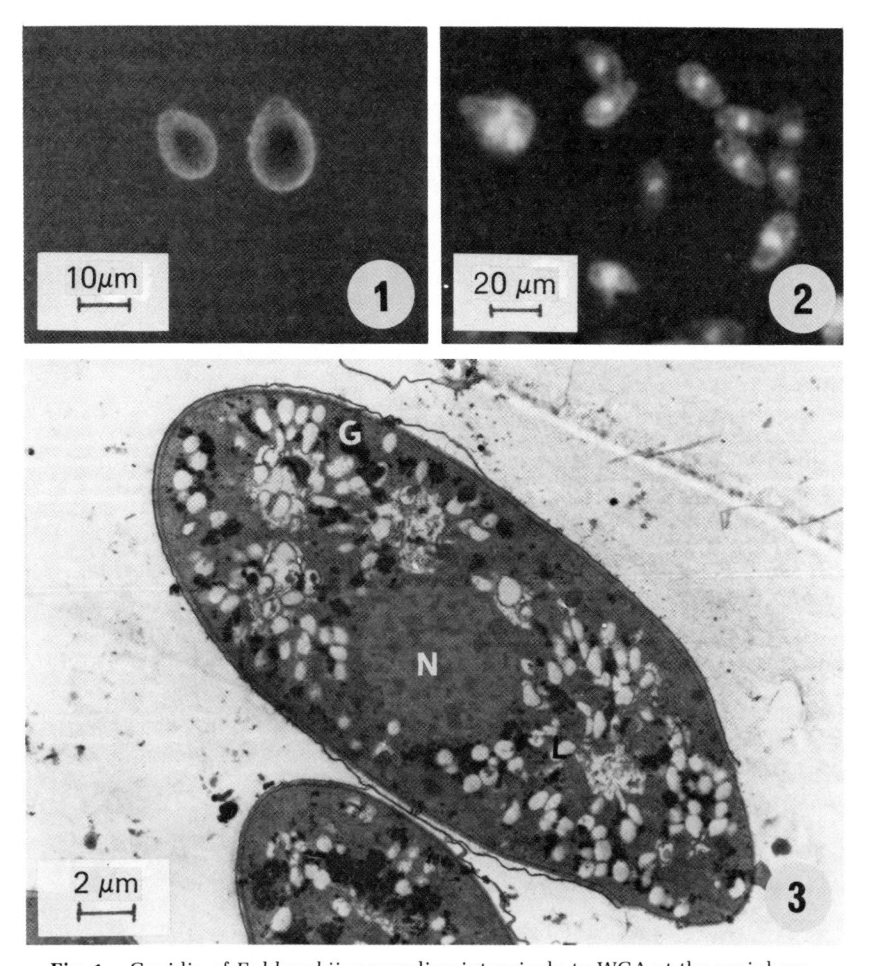

Fig. 1. Conidia of *E. blunckii* responding intensively to WGA at the periphery.

Fig. 2. Conidia of *E. blunckii* responding intensively to DAPI in the central position.

Fig. 3. Conidium of *E. radicans*. A centrally positioned nucleus is surrounded by many lipid globules and glycogen deposits.

and glycogen deposits. Plasma membrane inner and outer layers are distinguishable in the wall (Fig. 3).

III. GERM TUBE EXTENSION

Most of the conidia of *Erynia radicans* and *Erynia blunckii* extend their germ tubes on Sabouraud dextrose agar supplemented with 0.2% yeast extract (SDAY) under either light or dark conditions at 20°C.

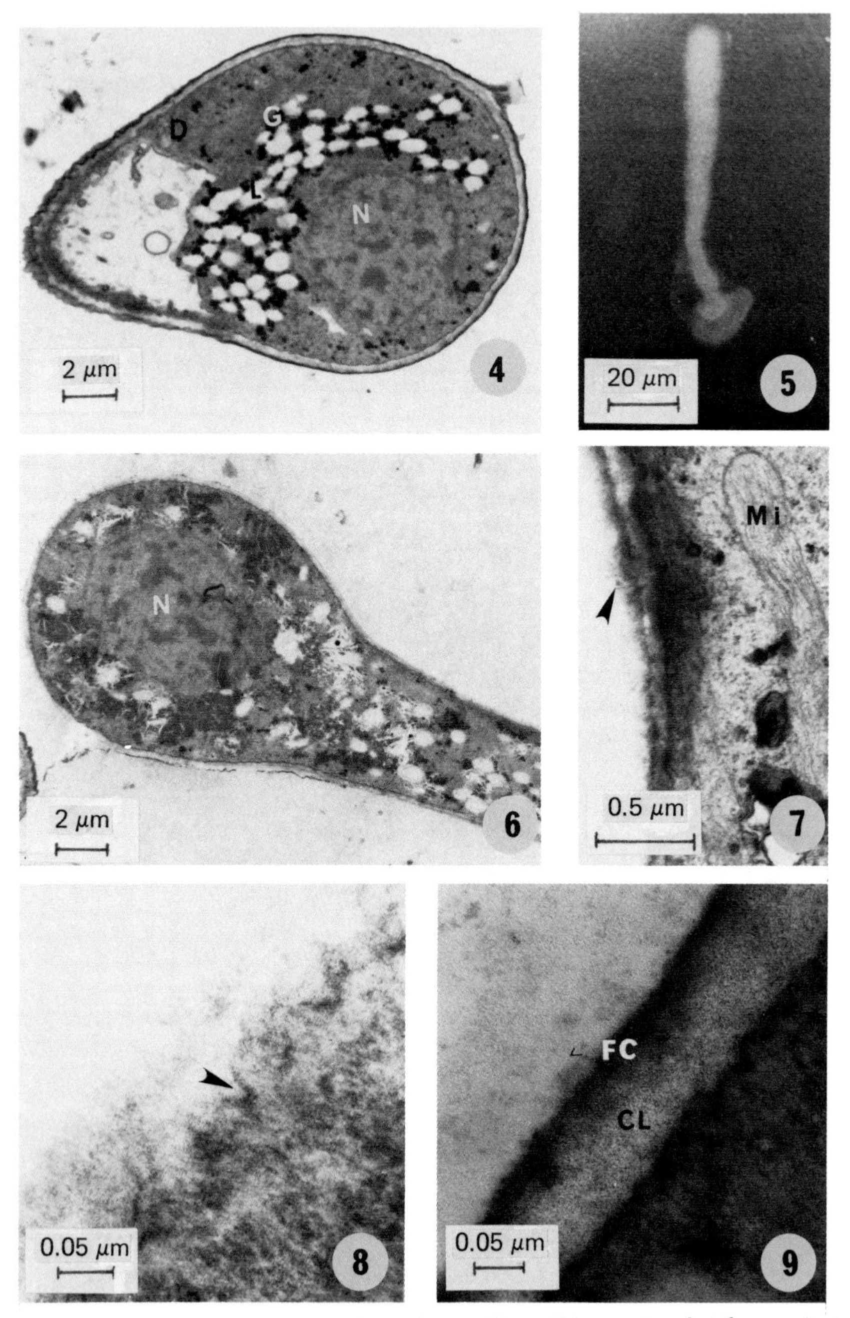

Fig. 4. Germinating conidium of *E. radicans*. The wall has ruptured at the germinating pole and dictyosomal vesicles are moving toward the pole.

Fig. 5. Germ tube derived from a conidium of *E. blunckii* and the positive WGA reaction of the whole tube.

Fig. 6. Germinating conidium of *E. radicans*. One nucleus remains in the conidial cell.

When germination occurs, the wall layers are ruptured at the germinating pole, and many dictyosomal vesicles concentrate near the pole (Fig. 4). The conidial cells and germ tubes are stained equally by DAPI; however, WGA and Con A stain the germ tubes intensively (Fig. 5). The nucleus remains in the conidial cell after germination even though most of the lipid globules pass into the germ tube and concentrate at the tip (Fig. 6). Mitochondria increase in number during germination, and many of them are observed near the plasma membrane with glycogen deposits. Electron-transparent amorphous substances border the plasma membrane of the extending germ tubes (Fig. 7). Invagination structures are observed in the membrane, and the amorphous substances seem to be secreted from there. However, the structures are indistinct compared with those formed in the protoplast regeneration of the fungus (Fig. 8). Finally, the amorphous substances develop into two layers, a fluffy coat and an electron-transparent compact layer (Fig. 9).

IV. SECONDARY CONIDIUM BUDDING

Most of the conidia (primary) of *E. radicans* and *E. blunckii* bud secondary conidia morphologically similar to themselves when immersed in deionized water in darkness at 20°C. The budding secondary conidia are strongly stained by DAPI, WGA, and Con A (Figs. 10 and 11). First of all, a nucleus migrates into the budding secondary conidium when the budding occurs. The wall ruptures at the tip of the bud and a small quantity of amorphous substances border the tip (Fig. 12). When the nucleus migrates, many lipid globules, glycogen deposits, and small mitochondria follow it. The cell wall at the tip of the bud reforms quickly, and a septum is formed between the secondary conidium and the emptied primary one after all of the cytoplasm has migrated. Many lipid globules and glycogen deposits remain intact and surround the nucleus in the secondary conidium (Fig. 13). This suggests that the secondary conidia control their metabolic activity and retain an energy source for the coming germination. Accordingly, the secondary conidia are considered as a kind of resting form.

Fig. 7. Part of an extending hypha of *E. radicans* bordered with amorphous substances (arrow). Large mitochondria with distinct cristae appear near the plasma membrane.

Fig. 8. Invaginations (arrow) of plasma membrane in a germinating tube of *E. radicans*. Amorphous substances appear near the membrane.

Fig. 9. Part of an extending germ tube of *E. radicans*. Amorphous substances have developed into two distinct wall layers. The wall consists of an electron-dense fluffy coat and an electron-transparent compact layer.

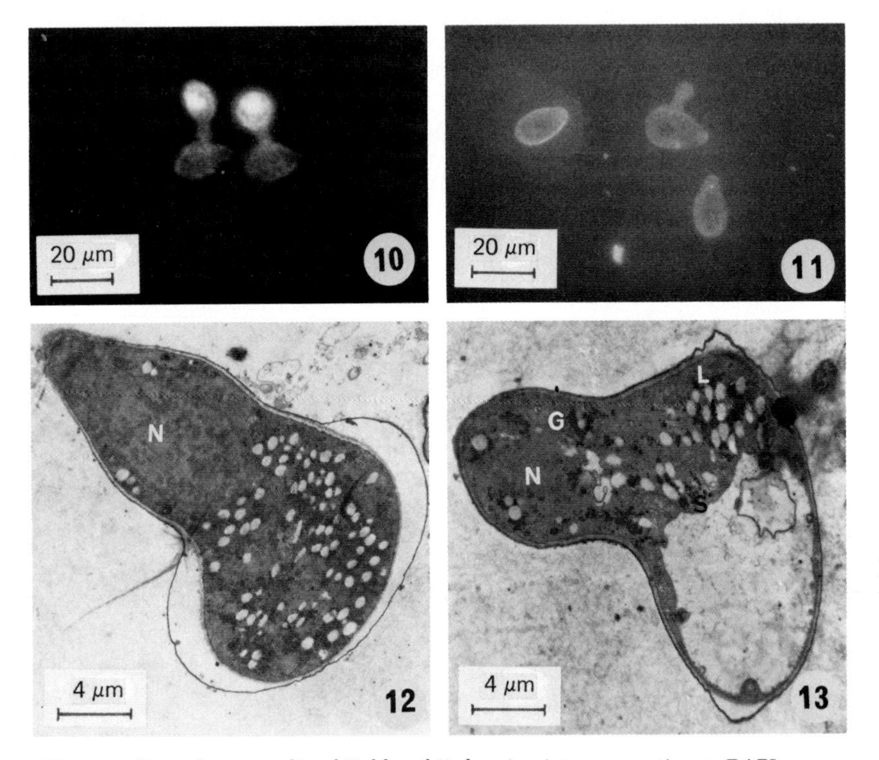

Fig. 10. Secondary conidia of *E. blunckii* showing intense reaction to DAPI.

Fig. 11. Budding of secondary conidium of *E. blunckii* giving a weak response to Con A.

Fig. 12. Budding of secondary conidium of *E. radicans*. One nucleus has migrated quickly into the budded portion and the cell wall is slightly ruptured at the tip of the bud.

Fig. 13. Budding of secondary conidium of *E. radicans* containing many lipid globules and glycogen deposits. A septum has formed between the secondary conidium and the emptied primary conidium.

V. CAPILLARY TUBE FORMATION

Long thin capillary tubes develop largely from conidia of *E. radicans* immersed in deionized water under white light irradiation at 20°C. When the capillary tubes emerge from the conidia, lipid globules, multivesicular bodies, and dark vesicles from dictyosomes are observed near the emerged portion. The wall of the emerged portion disintegrates and amorphous substances are overlaid at the tip (Fig. 14). This suggests that the wall of the capillary tubes ruptures and then reforms in the same way as germ tubes and secondary conidia. The capillary tubes are weakly stained by WGA or Con A; however, they respond

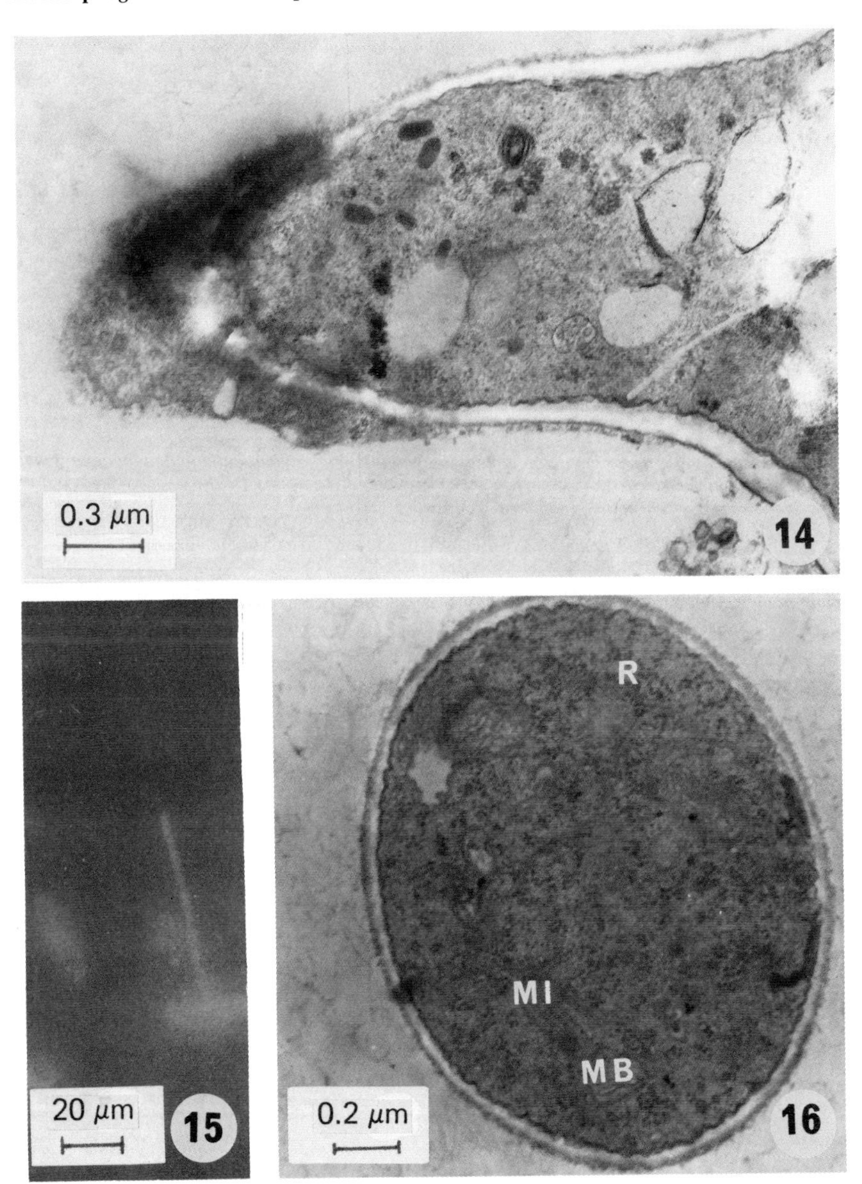

Fig. 14. Tip of a newly emerged capillary tube of *E. radicans*. The wall has disintegrated and amorphous substances are overlaid at the tip.

Fig. 15. Capillary tube of *E. radicans* stained intensively by DAPI.

Fig. 16. Cross section of a capillary tube of *E. radicans* filled with a large number of ribosomes, mitochondria, and multivesicular bodies.

intensively to DAPI (Fig. 15). This shows that the wall of the capillary tubes is weak compared with that of the germ tubes, but a large quantity of DNA is contained in the capillary tubes. Timing of the movement of the nucleus from the primary conidia to the capilliconidia formed at the tip of the capillary tubes is variable. In some cases the nuclei remain, even though most of the cytosomes transfer into the capilliconidia and remarkable vacuolations occur in the primary conidia. In other cases the nuclei quickly pass through to the capilliconidia. The wall of the capillary tubes is divided into three distinct layers, but the layers are very thin compared with those of the germ tubes and the secondary conidia. The capillary tubes are filled with a large number of ribosomes, mitochondria, and multivesicular bodies (Fig. 16) and therefore seem to serve as transportation pipes for cytosomes that stimulate the metabolic activity for subsequent germination of the capilliconidia. Glare *et al.* (1985) concluded that the capilliconidia of *Zoophthora phalloides* are the infective propagules to the aphid *M. persicae*. They also observed that many aphids, when exposed to primary conidia of the fungus, bore accumulations of capilliconidia on their lower legs as little as 48 hr later.

VI. PROTOPLAST REGENERATION

Protoplasts of *E. blunckii* and of *C. obscurus* were obtained by immersion of conidia produced on SDAY in a mixture of digestive enzymes (cellulase Onozuka R-10, 10 mg/ml; Macerozyme R-10, 10 mg/ml; Driselase X-10, 10 mg/ml) included in a stabilizing medium (0.2 *M* phosphate buffer solution, pH 6.2, containing 0.6 *M* KCl and 2 m*M* $MgCl_2$). The protoplasts were washed with the stabilizing medium by gentle centrifugation and then transferred into a regeneration medium (Bacto-peptone, 1%; yeast extract, 0.2%; 0.6 *M* mannitol) in a glass well. Protoplasts of *E. neoaphidis* formed within the bodies of *M. persicae* were also investigated ultrastructurally.

Newly formed protoplasts respond intensively to DAPI, especially in the central zone (Fig. 17). However, the response to DAPI occurs equally in the reversional hyphae and in the regenerated cells (Fig. 18). This implies that the protoplast-associated structure is at a high level of metabolic activity during the regeneration. Intensive reactions to DAPI are observed sporadically in the extended reversional hyphae (Fig. 19). The regenerated cells give a slight autofluorescence, but newly formed protoplasts give no autofluorescence in the absence of WGA and Con A under filter combinations of a fluorescent microscope. The newly

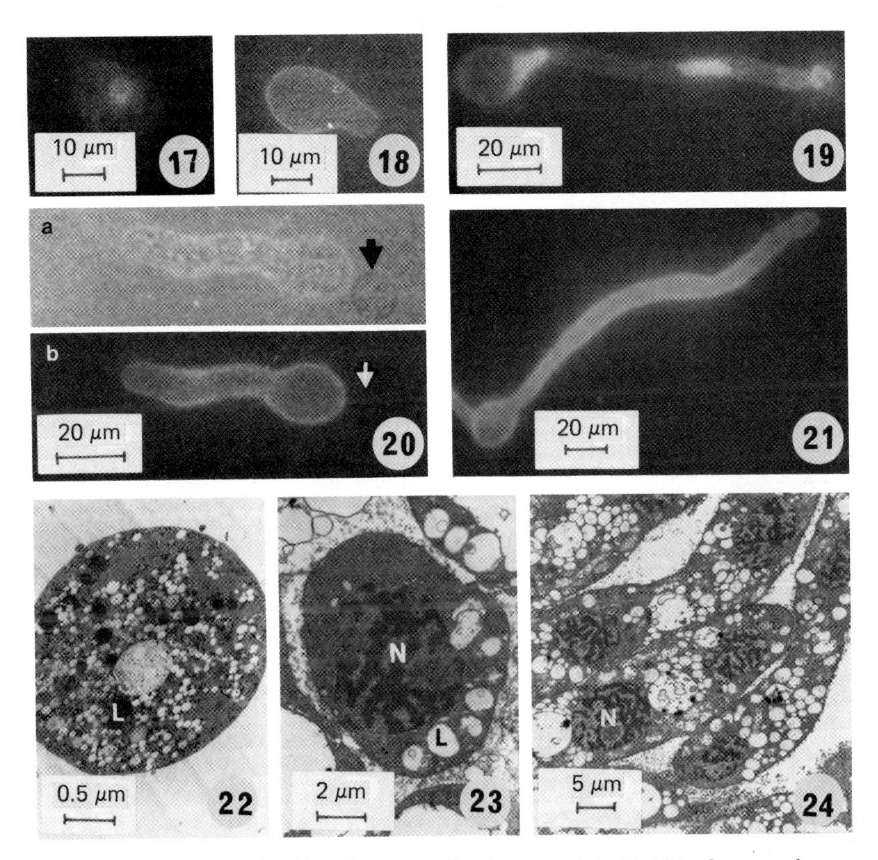

Fig. 17. Protoplast of *E. blunckii* responding intensively to DAPI in the central zone.

Fig. 18. Protoplast of *E. blunckii* producing a reversional hypha shows intense response to DAPI.

Fig. 19. Reversional hypha derived from a regenerated cell of *E. blunckii* reacting sporadically to DAPI.

Fig. 20. (a) Newly formed protoplast (arrow) and reversional hypha arising from a regenerated cell of *E. blunckii* under a fluorescent microscope without any fluorescent markers. (b) The protoplast (arrow) shows no response but the reversional hypha responds intensively to WGA.

Fig. 21. Reversional hypha of *E. blunckii* extending from a regenerated cell. Response to WGA becomes weak toward the tip.

Fig. 22. Protoplast of *C. obscurus* containing a large number of lipid globules.

Fig. 23. Protoplast of *E. neoaphidis* containing a centrally positioned nucleus with dense chromatin substances and many lipid globules.

Fig. 24. Divided nuclei in reversional hyphae of *E. neoaphidis* having no wall layer.

formed protoplasts are not stained by WGA and Con A, whereas the regenerated cells show intense responses to WGA and Con A along the lateral cell wall (Fig. 20). This suggests that the cell wall formation occurs rapidly when regeneration is initiated. The reactions to WGA and Con A become weak toward the tip of the reversional hyphae (Fig. 21).

Protoplasts of the entomophthoralean fungi are filled with lipid globules (Fig. 22). In the protoplasts of *E. neoaphidis*, a nucleus with dense chromatin substances is centrally positioned (Fig. 23). Nuclear division occurs in the reversional hyphae that do not yet have a complete wall layer (Fig. 24). Endoplasmic reticulum underlies the plasma membrane and a fibrillar material projects on the membrane (Fig. 25). The fibrillar material increases in quantity and becomes clavate (Fig. 26). Kreger and Kopecká (1975) demonstrated that the fibrillar net produced by protoplasts of *Saccharomyces cerevisiae* is composed of chitin and a highly crystalline β-1,3-linked glucan, with subsequent incorporation of an α-1,3-glucan as regeneration proceeds. Van der Valk and Wessels (1976) also reported that fibrillar material was produced on a plasma membrane of *Schizophyllum commune* as a net of chitin, which later was covered by a fluffy α-glucan layer.

The plasma membrane frequently exhibits invaginations at this stage, and amorphous substances appear near the invagination structures along the membrane surface (Fig. 27). As mentioned above, these structures are observed on the plasma membrane of germ tubes. Butt *et al.* (1981) observed invagination on the plasma membrane of protoplasts of *E. neoaphidis* and considered it a pinocytotic phenomenon of the protoplasts. However, the invagination structure observed in the present investigations differs in shape from that reported by Butt *et al.* A similar invaginated structure has been observed by Dorward and Powell

---

Fig. 25. Projection of fibrillar materials (arrows) on the plasma membrane of a reversional hypha of *E. neoaphidis*. Endoplasmic reticulum underlies the membrane.

Fig. 26. Projection of clavate fibrillar materials (arrow) on the plasma membrane of a reversional hypha of *C. obscurus*.

Fig. 27. Invaginations (small arrows) of plasma membrane of a reversional hypha of *E. neoaphidis*. Amorphous substances (large arrow) appear near the membrane.

Fig. 28. Invagination structure on the plasma membrane of a reversional hypha of *C. obscurus*. One dictyosomal vesicle approachs the membrane.

Fig. 29. Amorphous substances (arrow) appearing in an invagination structure on the plasma membrane of a reversional hypha of *C. obscurus*.

Fig. 30. Connection of protein granules to the plasma membrane (arrows) of a coated reversional hypha of *E. neoaphidis*.

Fig. 31. Part of a reversional hypha of *E. neoaphidis* with a completely formed cell wall.

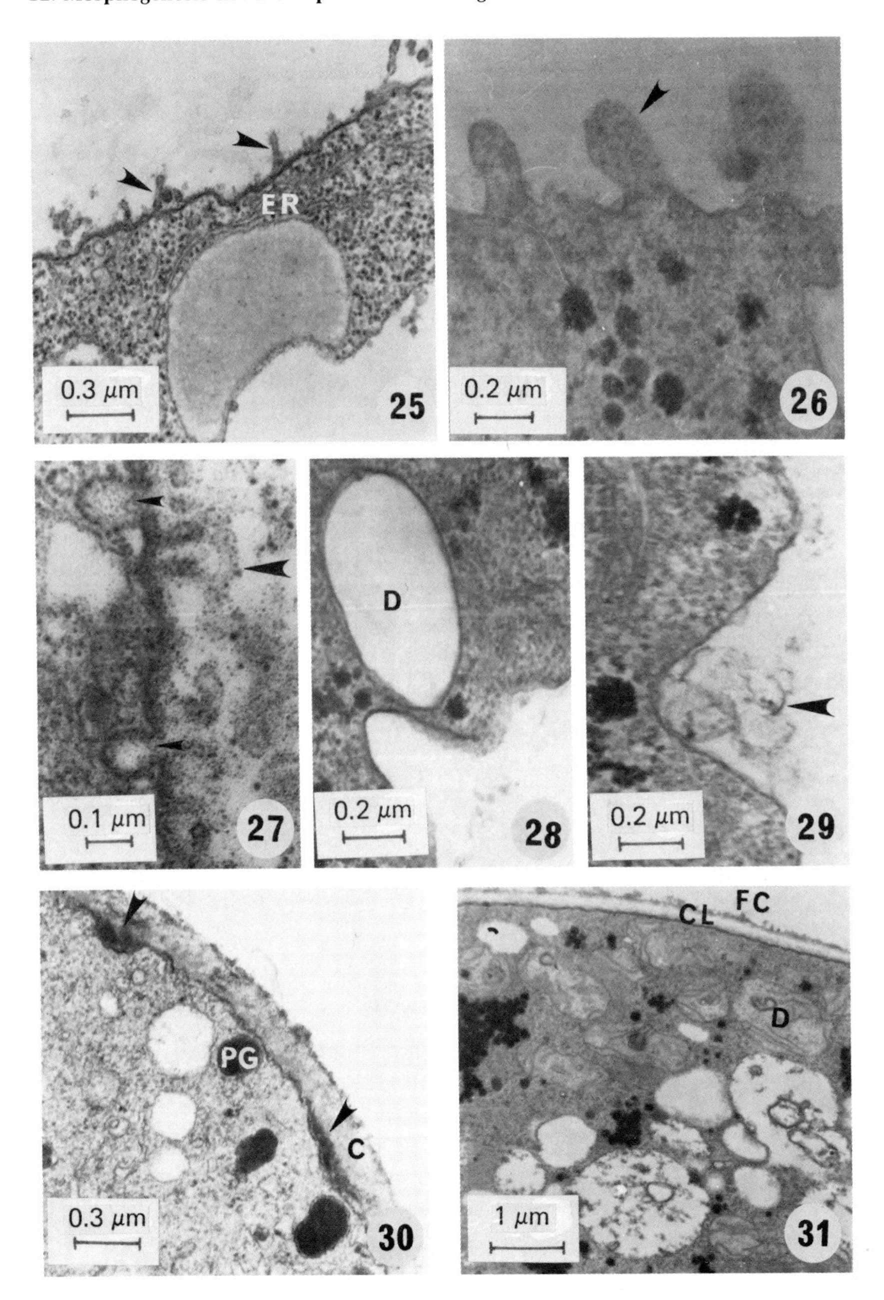
ER
0.3 μm
25
0.2 μm
26
0.1 μm
27
D
0.2 μm
28
0.2 μm
29
PG
C
0.3 μm
30
CL
FC
D
1 μm
31

(1983) in cell coat formation of zoospores of *Chytriomyces aureus*. They considered that the structure was a fusion profile between plasma membrane and dictyosomal vesicle, and it appeared to deposit material directly onto the cell surface. In the present investigations, such phenomena are observed in the reversional hyphae of *C. obscurus* (Figs. 28 and 29).

The amorphous substances are electron-transparent and homogeneous and coat the membrane as a layer. Larger mitochondria with distinct cristae and many electron-dense bodies underlie the membrane at this time; some of the electron-dense bodies are connected to the membrane (Fig. 30). A similar structure has been found by Lambiase and Yendol (1977) in *Entomophthora apiculata* as a formation of the conidial germ tube; they considered it to contain enzymes needed for the penetration of the insect cuticle. Powell (1976) also reported such an electron-dense body in the thallus of *Coelomomyces punctatus* to be a protein granule. According to Nolan (1985), greater levels of L-histidine, L-alanine, L-proline, and DL-serine are required for protoplast growth of *E. egressa*.

The amorphous substances finally divide into two compact layers, namely an electron-dense outer layer and an electron-transparent inner layer of uniform thickness (Fig. 31). De Vries and Wessels (1975) showed that the regenerating protoplasts of *S. commune* manufactured three main wall polymers; S-glucan (α-1,3-glucan) and chitin were the first wall components to be synthesized, while synthesis of *R*-glucan (β-1,3- or β-1,6-glucan) was yet to be done. Van der Valk and Wessels (1976) reported that the S-glucan layer on the protoplasts of the fungus has a fluffy appearance in the primary cell, but the wall of the hyphal tube arising from the primary cell has a compact layer of S-glucan.

VII. CONCLUSION

The level of secretory or metabolic activities seemed remarkably higher in the protoplasts than in the walled germinating conidia in *Erynia* species. Therefore, the physiological function of the reversional hyphae from protoplasts is probably very different from that of the walled cells. On the other hand, no difference was observed in the morphogenesis of cell wall formation between protoplast-associated structures and walled cells such as germ tubes, secondary conidia, and capillary tubes. Wall disintegration was, however, restricted to a limited area in those walled cells. Moreover, the disintegrated wall reformed quickly, so the morphogenesis could be traced clearly only in the protoplasts. The developmental aspects of protoplast reversion in entomophthoralean fungi thus provide significant understanding of the fungal group.

REFERENCES

Butt, T. M., Beckett, A., and Wilding, N. (1981). Protoplasts in the *in vivo* life cycle of *Erynia neoaphidis*. *J. Gen. Microbiol.* **127,** 417–421.

de Vries, O. M. H., and Wessels, J. G. H. (1975). Chemical analysis of cell wall regeneration and reversion of protoplasts from *Schizophyllum commune*. *Arch. Microbiol.* **102,** 209–218.

Dorward, D. W., and Powell, M. J. (1983). Cytochemical detection of polysaccharides and the ultrastructure of the cell coat of zoospores of *Chytriomyces aureus* and *Chytriomyces hyalinus*. *Mycologia* **75,** 209–220.

Dunphy, G. B., and Nolan, R. A. (1977a). Regeneration of protoplasts of *Entomophthora egressa*, a fungal pathogen of the eastern hemlock looper. *Can. J. Bot.* **55,** 107–113.

Dunphy, G. B., and Nolan, R. A. (1977b). Morphogenesis of protoplasts of *Entomophthora egressa* in simplified culture media. *Can. J. Bot.* **55,** 3046–3053.

Glare, T. R., Chilvers, G. A., and Milner, R. J. (1985). Capilliconidia as infective spores in *Zoophthora phalloides* (Entomophthorales). *Trans. Br. Mycol. Soc.* **85,** 463–470.

Kobayashi, Y., Mogami, K., and Aoki, J. (1984). Ultrastructural studies on the hyphal growth of *Erynia neoaphidis* in the green peach aphid, *Myzus persicae*. *Trans. Mycol. Soc. Jpn.* **25,** 425–434.

Kreger, D. R., and Kopecká, M. (1975). On the nature and formation of the fibrillar nets produced by protoplasts of *Saccharomyces cerevisiae* in liquid media: An electron-microscopic, X-ray diffraction and chemical study. *J. Gen. Microbiol.* **92,** 207–220.

Lambiase, J. T., and Yendol, W. G. (1977). The fine structure of *Entomophthora apiculata* and its penetration of *Trichoplusia ni*. *Can J. Microbiol.* **23,** 452–464.

Lemke, P. A., Kugelman, B., Morimoto, H., Jacobs, E. C., and Ellison, J. R. (1978). Fluorescent staining of fungal nuclei with a benzimidazole derivative. *J. Cell Sci.* **29,** 77–84.

Nolan, R. A. (1985). Protoplasts from Entomophthorales. *In* "Fungal Protoplasts" (J. F. Peberdy and L. Ferenczy, eds.), pp. 87–112. Dekker, New York.

Powell, M. J. (1976). Ultrastructural changes in the cell surface of *Coelomomyces punctatus* infecting mosquito larvae. *Can. J. Bot.* **54,** 1419–1437.

Sengbusch, P. V., Hechler, J., and Müller, U. (1983). Molecular architecture of fungal cell walls. An approach by use of fluorescent markers. *Eur. J. Cell Biol.* **30,** 305–312.

Sorokin, N. (1883). *Entomophthora*. *In* "Plant Parasites of Men and Animals as a Cause of Infectious Diseases," Vol. 2, pp. 191–240. St. Petersburg (in Russian).

Taso, P. H. (1970). Application of the vital fluorescent labelling technique with brighteners to studies of the saprophytic behaviour of *Phytophthora* in soil. *Soil Biol. Biochem.* **2,** 247–256.

Thaxter, R. (1888). The Entomophthoreae of the United States. *Mem. Boston Soc. Nat. Hist.* **4,** 133–201.

Tyrrell, D. (1977). Occurrence of protoplasts in the natural life cycle of *Entomophthora egressa*. *Exp. Mycol.* **1,** 259–263.

Tyrrell, D., and MacLeod, D. M. (1972). Spontaneous formation of protoplasts by a species of *Entomophthora*. *J. Invertebr. Pathol.* **19,** 354–360.

van der Valk, P., and Wessels, J. G. H. (1976). Ultrastructure and localization of wall polymers during regeneration and reversion of protoplasts of *Schizophyllum commune*. *Protoplasma* **90,** 65–87.

13

Prospects for Development of Molecular Technology for Fungal Insect Pathogens

O. C. YODER, K. WELTRING, B. G. TURGEON,
R. C. GARBER, AND H. D. VANETTEN

I. INTRODUCTION

Little is known about the genetic and molecular bases of fungal pathogenicity to insects. One of the major reasons for this is the scarcity of developed genetic systems for fungal insect pathogens. However, with the recent advances in molecular technology for filamentous fungi, it seems reasonable to assume that fungal insect pathogens can now be analyzed and manipulated by recombinant DNA techniques with or without a conventional genetic system.

This chapter summarizes recent literature on molecular technology

BIOTECHNOLOGY IN INVERTEBRATE
PATHOLOGY AND CELL CULTURE

for filamentous fungi, including some of those which have economic significance such as plant pathogens and industrial fungi. The emphasis is on genes that have been isolated from filamentous fungi, including those which have been used as markers for transformation, and on various types of transformation systems which have been developed for filamentous fungi. The literature survey is not exhaustive, but rather focuses on some of the best known and most recently studied fungal genes and transformation systems. The description of work done in our own laboratories is derived from a previous summary (Yoder *et al.*, 1986) and is included to illustrate how molecular technology can be applied to relatively undeveloped organisms such as fungal plant or insect pathogens.

II. CLONED GENES FROM FILAMENTOUS FUNGI

A. *Neurospora*

Among the first filamentous fungal genes clones was the *qa-2* gene of *Neurospora crassa*, which encodes inducible catabolic 5-dehydroquinate hydroxylase (EC 4.2.1.10), an enzyme involved in the metabolism of quinic acid (Vapnek *et al.*, 1977). More recently the *aro-9* gene, which encodes constitutive catabolic enzyme, was cloned (Catcheside *et al.*, 1985). Both genes were isolated by complementing an *Escherichia coli aroD6* mutant, which lacked the enzyme activity, with a library of *N. crassa* DNA constructed in either a plasmid or a phage. The inducible *qa-2* gene is one of three structural genes located in a cluster on chromosome VII of *N. crassa*, along with the regulatory gene *qa-1*, which has been found actually to be two interacting genes, *qa-1S* and *qa-1F* (Huiet, 1984). The other inducible structural genes, *qa-3* and *qa-4*, encode quinate: NAD^+ oxidoreductase (EC 1.1.1.24) and dehydroshikimate dehydratase, respectively. All four genes were isolated on a 36.5 kilobase (kb) fragment of DNA cloned in a cosmid vector (Schweizer *et al.*, 1981). The entire inducible *qa* gene cluster has been sequenced (Giles *et al.*, *1985; Rutledge, 1984), revealing that there are seven genes in the cluster. In addition to those already mentioned, two presumed structural genes, qa*-x and *qa-y*, were found to correspond to mRNAs of unknown function that are regulated by quinic acid. The entire set of genes in the inducible cluster occupies 17.3 kb of DNA. Interestingly, the cloned *N. crassa qa-2* gene has been used as a probe to isolate a corresponding cluster of quinic acid utilization genes from *Aspergillus nidulans* (Hawkins *et al.*, 1985).

The *N. crassa trp-1* gene was cloned independently in two different laboratories by complementing an *E. coli trpF* mutant with a library of *N. crassa* genomic DNA carried in a plasmid vector (Keesey and DeMoss, 1982; Schechtman and Yanofsky, 1983). The *trp-1* gene is trifunctional in *N. crassa* (and perhaps in filamentous Ascomycetes generally), encoding the enzymes glutamine amidotransferase (GAT), indoleglycerol-phosphate synthase (IGPS), and phosphoribosylanthranilate isomerase (PRAI). These activities correspond to the *E. coli* functions *trpG, trpC,* and *trpF*, respectively. All three of the enzyme activities were shown to be encoded on the *N. crassa trp-1* gene by complementing individual *N. crassa* mutants, each lacking one of three enzymes, with the cloned fragment (Schechtman and Yanofsky, 1983). Although all three enzymes are present in the *N. crassa trp-1* clone, the clone did not complement an *E. coli trpC* mutation. However, if a bacterial ribosome-binding site was fused to the 5′ end of the *trp-1* gene, both *trpC* and *trpF* mutants of *E. coli* could be complemented (Schechtman and Yanofsky, 1983). The gene corresponding to *N. crassa trp-1* has also been cloned from *A. nidulans* (Yelton *et al.*, 1983) and sequenced (Mullaney *et al.*, 1985) and from *Cochliobolus heterostrophus* (Turgeon *et al.*, 1986). The native *C. heterostrophus* gene functions in *E. coli* and *A. nidulans*, and in yeast if the 5′ end of the gene is deleted.

The *am* gene of *N. crassa*, which determines NADP-specific glutamate dehydrogenase (GDH), was isolated using a synthetic DNA probe (Kinnaird *et al.*, 1982). To construct the probe, the first 17 nucleotides of the coding sequence of *am* were deduced and a synthetic oligonucleotide corresponding to these nucleotides was prepared. The 17-mer probe hybridized to a 9-kb *Hin*dIII fragment from wild-type *N. crassa* but not to genomic DNA from a mutant in which most of *am* was deleted. The *am* gene was localized to a 2.7 kb *Bam*HI fragment and was found to have at least one intron of 67 bp.

Another *N. crassa* gene cloned by complementation of *E. coli* is *pyr-4*, which specifies orotidine-5′-phosphate carboxylase (Buxton and Radford, 1983). A plasmid library of *N. crassa* genomic DNA was transformed into each of four *E. coli* mutants: *pyrC, pyrD, pyrE,* and *pyrF*. Cells were selected for pyrimidine prototrophy. Colonies were obtained only in the case of transformed *pyrF* cells. From these, two plasmids were recovered and cloned. Each plasmid carried a *N. crassa* insert which had on it the structural *pyr-4* gene.

Recently, an unusual gene of very small size was isolated from *N. crassa*. The copper MT gene encodes a polypeptide of 26 amino acid residues which is a copper metallothionine, i.e., it specifically binds to copper ions (Munger *et al.*, 1985). Interestingly, the MT gene has an intron of 94 nucleotides. The gene was cloned by probing a genomic plasmid library of *N. crassa* DNA with a synthetic 21-mer which corresponded to a region of the mRNA.

Two genes have been cloned from *N. crassa* using the sib selection procedure (Akins and Lambowitz, 1985). The genes, *nic-1* (nicotinic acid auxotrophy) and *inl* (inositol auxotrophy) were isolated, not by the usual complementation or probing procedures, but rather by subdividing an *N. crassa* genomic library, transforming *N. crassa* $ni\text{-}1^{-}$ inl^{-} cells with various pools of clones, identifying the pool that contained the gene of interest, and then continuing to subdivide that pool until a single cloned fragment was found which carried the gene being sought. This approach to gene isolation avoids the necessity to recover the gene from genomic DNA and takes advantage of the fact that the *N. crassa* transformation frequency can be very high (10,000–50,000 transformants/μg DNA).

A group of conidiation-specific genes were isolated from *N. crassa* by screening a genomic library with a cDNA probe enriched for sequences expressed during conidiation (Berlin and Yanofsky, 1985). Of 12 clones recovered that represented 22 transcripts, 11 transcripts were found only in conidiating cultures, 8 transcripts were found in both mycelia and conidia (but at higher levels in conidia), and 3 were nonspecific. The genes mapped to six of the seven chromosomes; those on the same chromosome were so tightly linked that they appeared to be clustered. None of the cloned genes seemed to represent any of the known mutations that affect conidium production.

B. *Aspergillus*

Conidiation-specific genes were first isolated from a filamentous fungus (*A. nidulans*) using a cDNA probe (Zimmermann *et al.*, 1980). The probe was prepared by a "cascade hybridization" technique in which poly(A)$^{+}$ RNAs were enriched for sequences specific to conidia prior to synthesis of the cDNA. At least 350 clones were recovered, many of which encoded more than one RNA. Genes specific for conidia were found to be clustered in the genome (Orr and Timberlake, 1982; Timberlake and Barnard, 1981). The functions of the conidiation-specific genes are unknown. However, a 53 kb region on one clone, designated the spoC1 gene cluster, has been transcriptionally analyzed (Gwynne *et al.*, 1984). At least 19 transcripts are encoded in this region, most of which accumulate specifically in conidia. Transcripts that are found in both conidia and hyphae are encoded toward the borders of the cluster.

The *amdS* gene of *A. nidulans* encodes acetamidase, which converts acetamide to acetate and ammonia. The gene was isolated by virtue of the fact that clones carrying it hybridized to cDNA prepared from wild-type RNA and not to cDNA prepared from a mutant with a large *amdS* deletion (Hynes *et al.*, 1983). Also on the clone is the 5′ regulatory element called *amdI;* the transcript is about 1700 bp long.

Another *A. nidulans* gene, *argB*, was isolated by complementation of a yeast *arg3* mutant (lacking ornithine carbamoyltransferase, OCT) with a library of *A. nidulans* wild-type DNA constructed in a plasmid vector (Berse *et al.*, 1983). The gene was subcloned on a 2 kb fragment and its polypeptide product was found to react with antibody prepared against OCT. The gene did not function in *E. coli* unless rearranged.

Glucoamylase (1,4-α-D-glucan glucohydrolase) genes, which code for enzymes that catalyze the release of glucose from starch, have been cloned from both *A. niger* (Boel *et al.*, 1984) and *A. awamori* (Nunberg *et al.*, 1984). Both genes were isolated from genomic libraries by probing with cDNA prepared from glucoamylase-specific mRNA. The enzyme is inducibly synthesized in the presence of starch but not xylose, and regulation is at the level of transcription. The *A. niger* gene has five introns (four of 55 to 75 bp and one of 169 bp), while the *A. awamori* gene has four introns, apparently missing the one of 169 bp. The *A. awamori* gene was modified so that it could be expressed by yeast (Innis *et al.*, 1985). To achieve this, the introns were removed from the coding sequence, which was then fused to yeast promoter and terminator regions. Yeast cells containing this construction not only synthesized the glucoamylase polypeptide but also processed its leader sequence and glycosylated and secreted it into the medium.

The *A. niger* gene encoding β-glucosidase is one of the few filamentous fungal genes that has been isolated by expression in yeast (Penttila *et al.*, 1984). A library of *A. niger* genomic DNA was prepared in a yeast cosmid shuttle vector and then used to transform yeast cells deficient in a gene for β-glucosidase synthesis. The correct clones were identified in colonies that produced β-glucosidase. The usefulness of this strategy for gene isolation may be quite restricted, since attempts to clone four other *A. niger* genes by this method all failed.

The 3-phosphoglycerate kinase (*PGK*) gene is highly expressed in yeast and encodes a highly conserved polypeptide. The *PGK* gene of *A. nidulans* was cloned from a phage genomic library using a probe corresponding to the yeast *PGK* gene (Clements and Roberts, 1985). The *A. nidulans PGK* gene has one intron of 57 bp whereas the yeast *PGK* gene has none.

The *alcA* gene of *A. nidulans* encodes alcohol dehydrogenase and is positively regulated by *alcR*, to which it is tightly linked. Both *alcA* and *alcR* have been cloned on a single fragment of DNA (Doy *et al.*, 1985; Lockington *et al.*, 1985). Both groups identified the fragment carrying *alcA-alcR* by screening genomic libraries with differential cDNA probes made to RNA of induced and uninduced cultures.

Recently, technology has been developed to isolate *A. nidulans* genes by complementation of *A. nidulans* mutants with libraries of *A. nidulans* DNA constructed in plasmid or cosmid vectors. Yelton *et al.* (1985) prepared a library of wild-type DNA in a cosmid that was selectable in *E. coli* and in *A. nidulans*. Transformants, which appeared at about 10/μg library DNA, were screened for expression of the *yA* gene, which determines spore color. Cosmids from two of the transformants had inserts carrying the *yA* gene. The advantages of cosmid libraries for gene isolation by complementation include small library size (with 35–40-kb inserts only about 2000 transformants are needed for 0.98 probability of full genome coverage) and easy recovery of transforming DNA from fungal chromosomes by rescue in *E. coli*. However, plasmid libraries can also be used. Johnstone *et al.* (1985) recovered transformants after treatment of *A. nidulans* protoplasts with a plasmid library and visually screened them for expression of the *brlA1* gene, which controls conidiation. Three transformants (of 2×10^4 screened) harbored the *brlA1*$^+$ allele, which was recovered by rescue in *E. coli*.

C. *Nectria*

Only two fungal virulence genes have been cloned to date and they are both from the pea pathogen *Nectria haematococca* mating population VI. One of these encodes cutinase, an inducible enzyme required by *N. haematococca* to penetrate the plant cuticle. First a cDNA clone was obtained by differential hybridization with RNA from induced and uninduced cultures (Soliday *et al.*, 1984). The cDNA was used as a probe to isolate the genomic clone from a lambda library (Kolattukudy *et al.*, 1985). The gene has a 516 bp intron and codes for a polypeptide of 230 amino acids. All *N. haematococca* strains tested have at least one copy of the cutinase gene, but highly virulent, high-cutinase producers have two copies. The other virulence gene cloned from *N. haematococca* is *PDA*, which encodes a cytochrome *P*-450 monooxygenase; this work is described in detail in Section IV,B.

III. TRANSFORMATION SYSTEMS FOR FILAMENTOUS FUNGI

A. *Neurospora*

The first report of successful transformation of a filamentous fungus was from Case *et al.* (1979), who transformed *N. crassa* using the *qa-2* gene (Section II,A) as a selectable marker. In all transformants the transforming plasmid integrated into chromosomal DNA, either at the *qa-2* locus or elsewhere. There was no indication of autonomous replication; in each transformant analyzed the introduced *qa-2*$^+$ allele segregated as a single gene in meiosis. Most *N. crassa* transformations have been done with *qa-2* although the *am* gene has also been used as a selectable marker for transformation (Kinsey and Rambosek, 1984; Grant *et al.*, 1984). It is most common to prepare protoplasts to facilitate uptake of DNA, but a method of transformation based on uptake of DNA by germinated conidia after treatment with lithium acetate has been reported (Dhawale *et al.*, 1984).

Apparent autonomous replication of *qa-2*-based vectors has been observed (Stohl and Lambowitz, 1983), although plasmid DNA recovered from transformants was often rearranged (Stohl *et al.*, 1984; Kuiper and De Vries, 1985). There have been attempts to isolate chromosomal replicators from *N. crassa* to promote autonomous replication, but although the frequency of transformation can be improved by inclusion of certain chromosomal fragments in the vector, the tendency of such vectors is to integrate rather than replicate autonomously (Buxton and Radford, 1984; Paietta and Marzluf, 1985a).

The majority of transformation events in *N. crassa* appear to be by integration of transforming DNA into nonhomologous chromosomal sites. However, approximately 10% of events are homologous, which suggests that precise gene replacement is possible in this fungus (Dhawale and Marzluf, 1985; Paietta and Marzluf, 1985b).

B. *Aspergillus*

Transformation of *A. nidulans* was first achieved by complementing a mutant deficient in orotidine-5′-phosphate decarboxylase with a plasmid carrying the *N. crassa pyr-4* gene (see above), which encodes that enzyme (Ballance *et al.*, 1983). At about the same time Tilburn *et al.* (1983) used the *A. nidulans amdS* gene (see above) to transform an *amdS*$^-$ mutant to *amdS*$^+$. The *amdS* gene has also been used as a selectable marker to transform *A. niger* (Kelly and Hynes, 1985) and *C. heterostrophus* (see below). Yelton *et al.* (1984) transformed an *A. nidulans trpC*$^-$ strain with the cloned *trpC*$^+$ gene (see above). In all three systems, the transforming DNA integrates into nuclear chromosomes, where it is stably maintained in both mitosis and meiosis. The vector can integrate at a single locus as a single copy or as a tandem repeat; occasionally there are multiple locus integrations (Wernars *et al.*, 1985).

Most of the systems that can be used successfully to transform *As-*

pergillus do so rather inefficiently. Several factors have contributed to increasing the transformation frequency. First it was observed that the *A. nidulans argB* gene (see above), which can be selected in either *A. nidulans* (John and Peberdy, 1984) or *A. niger* (Buxton *et al.*, 1985), transforms at a higher frequency than genes such as *trpC* and *amdS*. The vector itself seems to be important, since the *argB* gene in pUC8 transforms better (500 transformants/μg DNA) than it does in pBR322 (Johnstone *et al.*, 1985). An additional influence on transformation efficiency is inclusion in the vector of certain *A. nidulans* chromosomal sequences. Ballance and Turner (1985) found that a particular *A. nidulans* fragment, which had been isolated by virtue of its ability to act as an autonomously replicating sequence (ARS) in yeast, increased transformation frequency some 50-fold; the mechanism by which this occurs is unknown. Other *A. nidulans* ARSs had no beneficial effect.

If there is substantial homology between the vector and the *A. nidulans* chromosome, the majority of integration events occur by homologous recombination. If a single crossover occurs, the result is a nontandem duplication, whereas certain double crossover events result in precise gene replacement (Miller *et al.*, 1985). Duplications can be resolved in some cases to yield a chromosomal structure that is indistinguishable from a gene replacement. This technique provides a powerful tool for analyzing the biological functions of cloned genes.

Aspergillus nidulans can be cotransformed simply by incubating competent protoplasts with a mixture of two plasmids, each bearing a different marker gene (Timberlake *et al.*, 1985). Thus, cloned genes can be quickly inserted into the *A. nidulans* genome without first recombining them into a vector that is selectable in *A. nidulans*.

C. *Mucor*

Transformation of *Mucor circinelloides* was accomplished by first cloning a selectable gene from *M. circinelloides* by complementation of *M. circinelloides* protoplasts with a library of its own DNA constructed in an *E. coli*/yeast shuttle vector (VanHeeswijck and Roncero, 1984). The gene selected, *leu*, was recovered from *M. circinelloides* transformants, cloned in *E. coli*, and found to transform a *leu*$^-$ mutant of *M. circinelloides* to *leu*$^+$ at a frequency of 600 transformants/μg DNA.

D. *Podospora*

The *ura5* gene of *Podospora anserina*, encoding orotidylic acid pyrophosphorylase (orotate phosphoribosyltransferase), was isolated by complementing the corresponding mutant of *E. coli* (Begueret *et al.*, 1984). This gene was then used to transform *P. anserina*, which occurred at low frequency. Some transformation events occurred by homologous recombination but the majority were heterologous. Brygoo and Debuchy (1985) also transformed *P. anserina* but by a different approach. Instead of transforming with a structural gene, they inserted a plasmid carrying a *leu1* tRNA suppressor into chromosomal DNA of a *leu1* mutant and selected for *leu1*$^+$. Again, in this case most of the transformation events involved heterologous recombination.

E. *Cephalosporium*

There are at least two reports of *Cephalosporium acremonium* transformation. In one case, the prokaryotic gene encoding aminoglycoside 3′-phosphotransferase from Tn903 was used to transform protoplasts to resistance to the drug G418 (Penalva *et al.*, 1985). In the other case, the *E. coli* gene for hygromycin B phosphotransferase, which codes for an enzyme that inactivates the drug hygromycin B (Kaster *et al.*, 1984), was fused to the promoter from the *C. acremonium* isopenicillin N synthetase gene (Samson *et al.*, 1985) and used to transform protoplasts of *C. acremonium* (S. W. Queener and T. D. Ingolia, personal communication). Stable transformants resistant to hygromycin B were obtained.

IV. FUNGAL MOLECULAR TECHNOLOGY

In our laboratory work has focused on two fungal plant pathogens, *Cochliobolus heterostrophus* and *Nectria haematococca*. In the sections below we first describe our progress in developing transformation systems for *Cochliobolus* and then deal with the cloning of a virulence gene from *Nectria*.

A. *Cochliobolus* Transformation

To pursue the study of the molecular bases of fungal pathogenicity to plants we have used the maize pathogen *C. heterostrophus* (anamorph: *Helminthosporium maydis* = *Bipolaris maydis*) as a model. This fungus is amenable to molecular investigations (Garber and Yoder, 1983, 1984; Garber *et al.*, 1984; Turgeon *et al.*, 1985) because it is easy to culture (Leach *et al.*, 1982a) and can be genetically manipulated both sexually (Leach *et al.*, 1982a; Yoder and Gracen, 1975) and asexually (Leach and Yoder, 1982, 1983; Leach *et al.*, 1982b). We describe transformation of *C. heterostrophus* in some detail since the work is very recent (Turgeon *et al.*, 1985; Yoder *et al.*, 1986). Our strategy may have relevance for development of molecular tools for insect pathogens since it is based on use of a genetically undomesticated fungus.

1. *Cochliobolus* Transformation with the *E. coli hygB* Gene

The antibiotic hygromycin B binds to both 70 S and 80 S ribosomes, inhibiting growth of both prokaryotes and eukaryotes. It is produced by the prokaryote *Streptomyces hygroscopicus*, which protects itself by synthesizing the enzyme hygromycin B phosphotransferase. This enzyme phosphorylates the drug, making it inactive in the host (Pardo *et al.*, 1985). Genes encoding hygromycin B phosphotransferase have been cloned from both *S. hygroscopicus* (Malpartida *et al.*, 1983) and *E. coli* (Rao *et al.*, 1983; Kaster *et al.*, 1983). The gene (*hygB*) from *E. coli* confers hygromycin B resistance to *E. coli* when under control of a prokaryotic promoter and to yeast when fused to a yeast promoter (Gritz and Davies, 1983; Kaster *et al.*, 1984). Moreover, *hygB* has been coupled to appropriate eukaryotic promoters and shown to function as a selectable marker for transformation of mammalian cells (Blochinger and Diggelmann, 1984; Santerre *et al.*, 1984) and plant cells (Van Den Elzen *et al.*, 1985; Waldron *et al.*, 1985). These observations suggested to us that *hygB* might function in cells of filamentous fungi, such as *C. heterostrophus*, if placed under control of a fungal promoter.

Cochliobolus heterostrophus is completely inhibited by hygromycin B at 100 μg/ml. Since there was no readily available strong *Cochliobolus* promoter to fuse to *hygB*, we constructed a "promoter library" of *Cochliobolus* genomic DNA to serve as a source of promoters. Random fragments from an *Mbo*I partial digest of genomic DNA were fractionated on a sucrose density gradient and 0.5–1.5-kb fragments ligated into a unique *Bam*HI site located at the 5′ end of a *hygB* gene that lacked its own promoter on a vector modified from one constructed by Kaster *et al.* (1984). Protoplasts of *Cochliobolus* were transformed with the library and plated in nonselective regeneration agar that was overlaid several hours later with agar containing hygromycin B. After 6–8 days at 32°C colonies appeared on plates containing protoplasts exposed to library DNA but not on plates containing control protoplasts. When these colonies were transferred to fresh hygromycin B-containing medium, they continued to grow at near-normal rates whereas wild-type cells did not grow at all.

DNA was isolated (Garber and Yoder, 1983) from transformed and untransformed colonies, digested, and probed with pBR322. There was no hybridization of the probe to DNA of untransformed cells but distinct hybridization to DNA of transformed cells. In all cases the transforming DNA appeared to be integrated into a chromosome, either as a single copy or as multiple copies, perhaps in a tandem array.

Genetic analysis of transformants was performed by crossing transformants with wild type, using standard procedures (Leach *et al.*, 1982a). Random ascospores were isolated, and colonies derived from them were tested for sensitivity to hygromycin B. The progeny of hygTX1 segregated 28 resistant : 29 sensitive and the progeny of hygTX2 segregated 16 resistant : 19 sensitive. Neither ratio is significantly different from 1 : 1, indicating that in each transformant the transforming DNA integrated at a single chromosomal locus.

Our predominant interest in transformants was to identify the *Cochliobolus* promoters isolated by this procedure and compare their abilities to drive transcription of the hygromycin B-phosphotransferase gene in *C. heterostrophus*. We recovered the transforming DNA by preparing lambda libraries of genomic DNA from several transformants. Plaques were probed with pBR322 to identify clones that contained transforming sequences. DNA was prepared from lambda clones that hybridized with the probe and was mapped with restriction enzymes. A fragment carrying the *hygB* gene fused to the *Cochliobolus* promoter insert, the *E. coli amp* gene, and the *E. coli* origin of replication was eluted from a gel, circularized by ligation, cloned in *E. coli*, and used to retransform *Cochliobolus* protoplasts. Colonies appeared at a frequency of 100–1000/μg DNA within 3–4 days on plates containing protoplasts treated with the cloned *Sal*I fragment, whereas no colonies were found on plates containing protoplasts treated with the vector without the *Cochliobolus* promoter fragment. This evidence indicated that we had a cloned *Cochliobolus* promoter which will drive the *hygB* gene in *Cochliobolus* so that the recombinant plasmid can be used as the basis of an efficient gene isolation vector.

2. *Cochliobolus* Transformation with the *Aspergillus nidulans amdS* Gene

We have found that *Cochliobolus* grows poorly on acetamide medium and that its genomic DNA shows no significant homology with the cloned *amdS* gene, p3SR2 (see above). Thus, it was a good candidate for transformation using *amdS* as a selectable marker. *Cochliobolus* protoplasts were transformed with p3SR2 and plated in a minimal medium containing only acetamide as a nitrogen source. Colonies appeared on plates with DNA-treated protoplasts approximately 1 week later, whereas plates with control protoplasts had no colonies (Turgeon *et al.*, 1985). When transferred to complete medium, the colonies grew at about the same rate as wild-type colonies, whereas on acetamide medium growth of transformants was much faster than that of wild type. To test for mitotic stability, transformant colonies were grown to maturity on complete (nonselective) medium and then returned to acetamide medium, where they grew at the normal transformant rate, indicating that they were stable under nonselecfive conditions.

The organization of the transforming DNA was determined in five transformants grown in either acetamide or complete medium; similar results were obtained with both media. DNAs were prepared (Garber and Yoder, 1983), digested with either *Bam*HI (one site in p3SR2) or *Xho*I (no sites in p3SR2), separated on an agarose gel, blotted to nitrocellulose paper, and probed with ^{32}P-labeled plasmid. Identical results were obtained using either pBR322 or p3SR2 as the probe.

Under conditions of moderate stringency, there was no detectable homology between either probe and wild-type *Cochliobolus* genomic DNA. If genomic DNA from transformed mycelium was undigested, only the high-molecular-weight DNA hybridized to the probe; there was no evidence that p3SR2 existed as a free plasmid. There was clear hybridization between either probe and genomic DNA from each transformant. Two patterns of hybridization were observed. In the first case, if genomic DNA was digested with *Bam*HI, two hybridizing bands were seen, neither of which was 8.8 kb (the size of p3SR2). If digested with *Xho*I, one band was found which was larger than 8.8 kb but smaller than 23 kb. Our interpretation of these two observations is that in this class of transformant a single copy of p3SR2 integrated at one chromosomal locus. In the second class, *Bam*HI digestion of genomic DNA yielded a very strongly hybridizing band of 8.8 kb and two weaker bands of various sizes. Digestion with *Xho*I resulted in one band of high molecular weight. These results indicate that this second class of transformant has multiple copies of p3SR2, arranged in a head-to-tail orientation, integrated at a single locus in a chromosome. So far we have not found multiple-locus integrations of p3SR2 in *Cochliobolus*. Such events have been observed in *A. nidulans* (Wernars *et al.*, 1985) and *A. niger* (Kelly and Hynes, 1985), along with the single-copy integrations and tandem repeats we see in *Cochliobolus*.

To determine whether *Cochliobolus* transformants were heterokaryons, single conidia, which resolve heterokaryons in *Cochliobolus* (Leach and Yoder, 1982), were isolated from transformant and wild-type colonies and plated on acetamide medium. Transformant conidia were 70% $amdS^{+}$, 20% $amdS^{-}$, and 10% nonviable. Conidia from wild type were 100% $amdS^{-}$. Additional single conidia were isolated from the purified transformant and wild-type colonies and plated on acetamide medium. Of those from the transformant, 100% were now *amdS*$^+$, whereas those from wild type remained 100% *amdS*$^-$. Thus, it appears that only a portion of the nuclei in each cell of a transformant actually carry foreign DNA. The remainder may be maintained by activity of the transformed nuclei. It is easy, however, to purify transformants by isolation of single conidia. *Aspergillus nidulans* transformants can also be heterokaryotic, with the frequency of transformed nuclei ranging from 0.1 to 100% (Wernars *et al.*, 1985).

Meiotic segregation of the *amdS* gene was analyzed by crossing *Cochliobolus amdS*$^+$ transformants carrying either single or multiple copies of p3SR2 with wild type. Ascospores were isolated randomly or as complete tetrads (four sets of twins). Progeny of a cross between a single-copy transformant and wild type segregated 35 *amdS*$^+$: 43 *amdS*$^-$ and progeny of a cross between a multiple-copy transformant and wild type segregated 31 *amdS*$^+$: 29 *amdS*$^-$. Both of these ratios approximate 1 : 1; indicating that *amdS* segregates as a single gene in genomes carrying either a single copy or multiple copies of p3SR2. All complete tetrads that were scored segregated 4 : 4 (*amdS*$^+$: *amdS*$^-$), as expected when a single gene is involved.

Two independent *amdS*$^+$ transformants were crossed with each other to determine whether integration of p3SR2 was at the same chromosomal locus in each transformant. Ten tetrads were dissected and scored for *amdS* and for *Alb1*, a gene-controlling pigment production that was heterozygous in the cross. When *amdS*$^+$: *amdS*$^-$ ratios among the ten tetrads were scored, three were found to be 8 : 0, four were 6 : 2, and three were 4 : 4. These frequencies indicate that two genes were segregating and permit the conclusion that p3SR2 integrated at different, unlinked loci in these two transformants. Segregation at *Alb1* was 4 : 4 for all ten of the tetrads.

3. Construction of a *Cochliobolus* Gene-Isolation Vector

Now that we have two genes, *amdS* and *hygB*, that can be used as selectable markers for *Cochliobolus* transformation, we will construct a cosmid vector that transforms *Cochliobolus* at high frequency and will be suitable for construction of *Cochliobolus* genomic libraries from which virulence genes can be isolated by complementation of *Cochliobolus* itself. In the cosmid, the selectable marker will be *hygB* fused to a *Cochliobolus* promoter. There will be a unique *Bam*HI site that will accept DNA fragments after partial digestion with *Sau*3A or *Mbo*I. The *E. coli amp* gene and origin of DNA replication will permit cloning in *E. coli*, and the lambda *cos* site (Yelton *et al.*, 1985) will modify the vector to accept only large (35–45 kb) DNA fragments so that the number of transformants needed to represent the entire *Cochliobolus* genome will be relatively small (approximately 2000). A small library is important when transformants must be screened for the presence of a particular virulence gene since virulence assays are often inefficient, especially if they involve inoculation of plants.

To isolate a virulence gene using such a vector, two strains of the pathogen would be chosen—one which had a dominant allele of the virulence gene and one which had a recessive allele. It would be desirable to have the two strains nearly isogenic with each other, a condition that could be achieved by backcrossing. A library of DNA fragments from a partial digest prepared from genomic DNA of the strain with the dominant allele would be inserted into the *Bam*HI site. The library would be used to transform protoplasts of the strain with the recessive allele and transformants would be selected by their resistance to hygromycin conferred by *hygB*. Approximately 2000–3000 transformants would then be screened for a change in phenotype that reflected expression of the virulence gene. When such a transformant is found, the transforming cosmid would be recovered from the genome by lambda packaging and transduction of *E. coli* as described for *Aspergillus* (Yelton *et al.*, 1985). The fragment bearing the virulence gene itself would be subcloned by the cotransformation technique (Timberlake *et al.*, 1985).

B. Cloning a *Nectria* Virulence Gene

The *N. haematococca PDA* gene controls production of pisatin demethylase, a cytochrome *P*-450 monooxygenase that is responsible for detoxification of the phytoalexin pisatin by *O*-demethylation (VanEtten *et al.*, 1975; Matthews and VanEtten, 1983). Phytoalexins are antimicrobial compounds produced by plants in response to infection and are considered to be responsible for an active mechanism of disease resistance in plants (Albersheim and Valent, 1978; Bailey and Mansfield, 1982; Darvill and Albersheim, 1984; Dixon *et al.*, 1983; Kuc and Rush, 1985; Paxton, 1980; Smith and Ingham, 1981). The ability of a plant pathogen to detoxify its host's phytoalexin is thought to be one factor required for pathogenicity in some host–parasite interactions (Smith *et al.*, 1982, 1984; VanEtten, 1982; VanEtten *et al.*, 1982).

A survey of field isolates of *N. haematococca* revealed several different phenotypes with regard to pisatin-demethylating ability (pda) (Tegtmeier and VanEtten, 1983; VanEtten and Matthews, 1984; VanEtten *et al.*, 1980). The phenotypes are referred to as Pda$^-$ (no ability to demethylate), Pda^i (inducible Pda), and Pda^n (noninducible Pda), distinguished by their rates of demethylation after preinduction with pisatin of 0, 260, and <15 pmol/min·mg fresh weight, respectively. These Pda phenotypes are correlated with virulence of *N. haematococca* on peas in that Pda^i isolates are highly to moderately virulent while almost all Pda^n isolates are of low virulence. Pda^- isolates are virtually nonpathogenic. Genetic analysis has verified this correlation and led to the identification of several unlinked genes that independently confer a specific Pda phenotype in *N. haematococca* (Kistler and VanEtten, 1984a,b).

We have begun to isolate the *Pda* genes to examine whether the different *Pda* loci code for different structural genes or one structural gene and several regulatory genes. A genomic library of *N. haematococca* isolate T-9 (which is Pda^i) was constructed by ligating 30–40 kb genomic DNA fragments into the cosmid vector pKBY2. This cosmid contains the *trpC* gene of *A. nidulans* and was constructed for transformation of *trpC*$^-$ mutants of *A. nidulans* to allow direct isolation of fungal genes (Yelton *et al.*, 1985). The availability of this vector and the fact that *A. nidulans* does not demethylate pisatin made it possible to look for expression of a *PDA* gene in *Aspergillus*. Moreover, we knew from reconstruction experiments that *A. nidulans* has a nonspecific NADPH–cytochrome *P*-450 reductase (required for pda) that will complement the *N. haematococca* cytochrome *P*-450 *in vitro* such that pisatin-demethylating activity is produced.

Transformation of *A. nidulans* strain UCD1 (*trpC*$^-$) with the *Nectria* library yielded transformants at a rate of about 3/μg DNA. Individual transformants were transferred to medium containing [3-*O-methyl*-^{14}C]pisatin and incubated for 1 week. One transformant (III-202) among 1250 transformants tested was able to demethylate pisatin, as indicated by a decrease in radioactivity. The rate of demethylation after preinduction of the *A. nidulans* transformant was less than that of *Nectria* strain T-9, the source of the cloned gene. Transformant III-202 was less sensitive than *A. nidulans* Pda^- transformants to growth inhibition by pisatin.

The cosmid containing the *PDA* gene, designated pNhT9 (47 kb), was recovered by lambda-packaging of total DNA from transformant III-202 as described by Yelton *et al.*, (1985). After retransformation of *A. nidulans* strain UCD1 with pNhT9, 99% of the *trpC*$^+$ transformants had the Pda^+ phenotype, indicating that the insert carried a gene that determined pda. The presence of vector DNA in the transformant was verified by Southern hybridization of a restriction enzyme digest of genomic DNA from transformant III-202, with pKBY2 as a probe. Furthermore, in Southern hybridization experiments only pNhT9, but not pKBY2, hybridized to *Nectria* strain T-9 genomic DNA.

After subcloning of the 37 kb insert of pNhT9 to obtain the smallest fragment containing the functional *PDA* gene it will be possible to isolate and compare other *PDA* genes from strains with different Pda phenotypes. Furthermore, the isolation of a *PDA* gene is the first step toward the most stringent test for evaluating the importance of pisatin demethylation in virulence: the introduction of isolated *PDA* genes into *N. haematococca* isolates with different Pda phenotypes, i.e., construction of pairs of isolates that are isogenic except for a particular *PDA* gene.

In addition, the molecular study of the *Pda* genes will contribute to the general understanding of cytochrome *P*-450 systems and the genes that control their function. Knowledge of the structure and regulation of fungal cytochrome *P*-450s should also facilitate comparative studies on the evolution of these genes in prokaryotic and eukaryotic organisms.

V. CONCLUSION

A perusal of the reference list for this chapter reveals that there has been an explosion in development of molecular technology for filamentous fungi in the past 2 years. This technology is not restricted to the well-developed genetic models such as *A. nidulans* and *N. crassa* but extends to a number of fungi for which there is little or no genetic information and which are economically important. Many of the protocols needed for molecular manipulation of fungi (e.g., isolation of viable protoplasts, preparation of high-quality DNA, transformation) are now simple and routine. It is not uncommon to find that a protocol developed for one filamentous fungus will work with little or no modification for a different fungus. Thus, it appears that the groundwork has been laid for the application of molecular technology to problems in fungal insect pathology. One of the most immediate potential applications is the genetic engineering of fungi for use as biological control agents of insect pests.

ACKNOWLEDGMENTS

The experimental investigations described in this article were supported by grants to O. C. Yoder from the U.S. National Science Foundation, U.S. Department of Agriculture

Competitive Research Grants Office, and Pioneer Hibred International and to H. D. Van-Etten from the U.S. Department of Energy and the U.S. Department of Agriculture Competitive Research Grants Office. K. Weltring was supported by the German Academic Exchange Service (DAAD). We wish to thank Thomas Ingolia for providing pIT213 with a promoterless *hyg*B gene, Michael Hynes for p3SR2 with *amdS*, and William Timberlake for the cosmid pKBY2 and *A. nidulans* strain UCD1.

REFERENCES

Akins, R. A., and Lambowitz, A. (1985). General method for cloning *Neurospora crassa* nuclear genes by complementation of mutants. *Mol. Cell. Biol.* **5**, 2272–2278.

Albersheim, P., and Valent, B. S. (1978). Host–pathogen interactions in plants. Plants, when exposed to oligosaccharides of fungal origin, defend themselves by accumulating antibiotics. *J. Cell Biol.* **78**, 627–643.

Bailey, J. A., and Mansfield, J. W., eds. (1982). "Phytoalexins." Blackie and Son, Glasgow and London.

Ballance, D. J., and Turner, G. (1985). Development of a high-frequency transforming vector for *Aspergillus nidulans*. *Gene* **36**, 321–331.

Ballance, D. J., Buxtron, F. P., and Turner, G. (1983). Transformation of *Aspergillus nidulans* by the orotidine-5′-phosphate decarboxylase gene of *Neurospora crassa*. *Biochem. Biophys. Res. Commun.* **112**, 284–289.

Begueret, J., Razanamparanu, V., Perrot, M., and Barreau, C. (1984). Cloning gene *ura*5 for the orotidylic acid pyrophosphorylase of the filamentous fungus *Podospora anserina:* Transformation of protoplasts. *Gene* **32**, 487–492.

Berlin, V., and Yanofsky, C. (1985). Isolation and characterization of genes differentially expressed during conidiation of *Neurospora crassa Mol. Cell. Biol.* **5**, 839–848.

Berse, B., Dmochowska, A., Skrzypek, M., Weglenski, P., Bates, M. A., and Weiss, R. L. (1983). Cloning and characterization of the ornithine carbamoyltransferase gene from *Aspergillus nidulans*. *Gene* **25**, 109–117.

Blochinger, K., and Diggelmann, H. (1984). Hygromycin B phosphotransferase as a selectable marker for DNA transfer experiments with higher eukaryotic cells. *Mol. Cell. Biol.* **4**, 2929–2931.

Boel, E., Hansen, M. T., Hjort, I., Høegh, I., and Fiil, N. P. (1984). Two different types of intervening sequences in the glucoamylase gene from *Aspergillus niger*. *EMBO J.* **3**, 1581–1585.

Brygoo, Y., and Debuchy, R. (1985). Transformation by integration in *Podospora anserina* I. Methodology and phenomenology. *Mol. Gen. Genet.* **200**, 128–131.

Buxton, F. P., and Radford, A. (1983). Cloning of the structural gene for orotidine 5′-phosphate carboxylase of *Neurospora crassa* by expression in *Escherichia coli*. *Mol. Gen. Genet.* **190**, 403–405.

Buxton, F. P., and Radford, A. (1984). The transformation of mycelial spheroplasts of *Neurospora crassa* and the attempted isolation of an autonomous replicator. *Mol. Gen. Genet.* **196**, 339–344.

Buxton, F. P., Gwynne, D. I., and Davies, R. W. (1985). Transformation of *Aspergillus niger* using the *argB* gene of *Aspergillus nidulans*. *Gene* **37**, 207–214.

Case, M. E., Schweizer, M., Kushner, S. R., and Giles, N. H. (1979). Efficient transformation of *Neurospora crassa* by utilizing hybrid plasmid DNA. *Proc. Natl. Acad. Sci. U.S.A.* **76**, 5259–5263.

Catcheside, D. E. A., Storer, P. J., and Klein, B. (1985). Cloning of the ARO cluster gene of *Neurospora crassa* and its expression in *Escherichia coli*. *Mol. Gen. Genet.* **199**, 446–451.

Clements, J. M., and Roberts, C. F. (1985). Molecular cloning of the 3-phosphoglycerate kinase (PGK) gene from *Aspergillus nidulans*. *Curr. Genet.* **9**, 293–298.

Darvill, A. G., and Albersheim, P. (1984). Phytoalexins and their elicitors—A defense against microbial infection in plants. *Annu. Rev. Plant Physiol.* **35**, 243–275.

Dhawale, S. S., and Marzluf, G.A. (1985). Transformation of *Neurospora crassa* with circular and linear DNA and analysis of the fate of the transforming DNA. *Curr. Genet.* **10**, 205–212.

Dhawale, S. S., Paietta, J. V., and Marzluf, G. A. (1984). A new, rapid and efficient transformation procedure for *Neurospora*. *Curr. Genet.* **8**, 77–79.

Dixon, R. A., Dey, P. M., and Lamb, C. J. (1983). Phytoalexins: Enzymology and molecular biology. *Adv. Enzymol.* **55**, 1–136.

Doy, C. H., Pateman, J. A., Olsen, J. E., Kane, H. J., and Creaser, E. H. (1985). Genomic clones of *Aspergillus nidulans* containing alcA, the structural gene for alcohol dehydrogenase, and alcR, a regulatory gene for ethanol metabolism. *DNA* **4**, 105–114.

Garber, R. C., and Yoder, O. C. (1983). Isolation of DNA from filamentous fungi and separation into nuclear, mitochondrial, ribosomal, and plasmid components. *Anal. Biochem.* **135**, 416–422.

Garber, R. C., and Yoder, O. C. (1984). Mitochondrial DNA of the filamentous ascomycete *Cochliobolus heterostrophus*. Characterization of the mitochondrial chromosome and population genetics of a restriction enzyme polymorphism. *Curr. Genet.* **8**, 621–628.

Garber, R. C., Turgeon, B. G., and Yoder, O. C. (1984). A mitochondrial plasmid from the plant pathogenic fungus *Cochliobolus heterostrophus*. *Mol. Gen. Genet.* **196**, 301–310.

Giles, N. H., Case, M. E., Baum, J., Geever, R., Huiet, L., Patel, V., and Tyler, B. (1985). Gene organization and regulation in the qa (quinic acid) gene cluster of *Neurospora crassa*. *Microbiol. Rev.* **49**, 338–358.

Grant, D. M., Lambowitz, A. M., Rambosek, J. A., and Kinsey, J. A. (1984). Transformation of *Neurospora crassa* with recombinant plasmids containing the cloned glutamate dehydrogenase (*am*) gene: Evidence for autonomous replication of the transforming plasmid. *Mol. Cell. Biol.* **4**, 2041–2051.

Gritz, L., and Davies, J. (1983). Plasmid-encoded hygromycin B resistance: the sequence of hygromycin B phosphotransferase gene and its expression in *Escherichia coli* and *Saccharomyces cerevisiae*. *Gene* **25**, 179–188.

Gwynne, D. I., Miller, B. L., Miller, K. Y., and Timberlake, W. E. (1984). Structure and regulated expression of the SpoCl gene cluster from *Aspergillus nidulans*. *J. Mol. Biol.* **180**, 91–109.

Hawkins, A. R., Francisco Da Silva, A. J., and Roberts, C. F. (1985). Cloning and characterization of the three enzyme structural genes *QUTB*, *QUTC*, and *QUTE* from the quinic acid utilization gene cluster in *Aspergillus nidulans*. *Curr. Genet.* **9**, 305–311.

Huiet, L. (1984). Molecular analysis of the *Neurospora qa*-I regulatory region indicates that two interacting genes control *qa* gene expression. *Proc. Natl. Acad. Sci. U.S.A.* **81**, 1174–1178.

Hynes, M. J., Corrick, C. M., and King, J. A. (1983). Isolation of genomic clones containing the *amd*S gene of *Aspergillus nidulans* and their use in the analysis of structural and regulatory mutations. *Mol. Cell. Biol.* **3**, 1430–1439.

Innis, M. A., Holland, M. J., McCabe, P. C., Cole, G. E., Wittman, V. P., Tal, R., Watt, K. W., Gelfand, D. H., Holland, J. P., and Meade, J. H. (1985). Expression, glycosylation, and secretion of an *Aspergillus* glucoamylase by *Saccharomyces cerevisiae*. *Science* **228**, 21–26.

John, M. A., and Peberdy, J. F. (1984). Transformation of *Aspergillus nidulans* using the *argB* gene. *Enzyme Microb. Technol.* **6**, 386–389.

Johnstone, I. L., Hughes, S. G., and Clutterbuck, A. J. (1985). Cloning an *Aspergillus nidulans* developmental gene by transformation. *EMBO J.* **4**, 1307–1311.

Kaster, K. R., Burgett, S. G., Rao, R. N., and Ingolia, T. D. (1983). Analysis of a bacterial hygromycin B resistance gene by transcriptional and translational fusions and by DNA sequencing. *Nucleic Acids Res.* **11**, 6895–6911.

Kaster, K. R., Burgett, S. G., and Ingolia, T. D. (1984). Hygromycin B resistance as dominant selectable marker in yeast. *Curr. Genet.* **8**, 353–358.

Keesey, J. K., Jr., and DeMoss, J. A. (1982). Cloning of the *trp-1* gene from *Neurospora crassa* by complementation of a *trpC* mutation in *Escherichia coli*. *J. Bacteriol.* **152**, 954–958.

Kelly, J. M., and Hynes, M. J. (1985). Transformation of *Aspergillus niger* by the *amdS* gene of *Aspergillus nidulans*. *EMBO J.* **4**, 475–479.

Kinnaird, J. H., Keighren, M. A., Kinsey, J. A., Eaton, M., and Fincham, J. R. S. (1982). Cloning of the *am* (glutamate dehydrogenase) gene of *Neurospora crassa* through the use of a synthetic DNA probe. *Gene* **20**, 387–396.

Kinsey, J. A., and Rambosek, J. A. (1984). Transformation of *Neurospora crassa* with the cloned *am* (glutamate dehydrogenase) gene. *Mol. Cell. Biol.* **4**, 117–122.

Kistler, H. C., and VanEtten, H. D. (1984a). Three nonallelic genes for pisatin demethylation in the fungus *Nectria haematococca*. *J. Gen. Microbiol.* **130**, 2595–2603.

Kistler, H. C., and VanEtten, H. D. (1984b). Regulation of pisatin demethylase in *Nectria haematococca* and its influence on pisatin tolerance and virulence. *J. Gen. Microbiol.* **130**, 2605–2613.

Kolattukudy, P. E., Soliday, C. L., Woloshuk, C. P., and Crawford, M. (1985). Molecular biology of the early events in the fungal penetration into plants. *In* "Molecular Genetics of Filamentous Fungi" (W. E. Timberlake, ed.), pp. 421–438. Alan R. Liss, Inc., New York.

Kuc, J., and Rush, J. S. (1985). Phytoalexins. *Arch. Biochem. Biophys.* **236**, 455–472.

Kuiper, M. T. R., and De Vries, H. (1985). A recombinant plasmid carrying the mitochondrial plasmid sequence of *Neurospora intermedia* LaBelle yields new plasmid derivatives in *Neurospora crassa* transformants. *Curr. Genet.* **9**, 471–477.

Leach, J., and Yoder, O. C. (1982). Heterokaryosis in *Cochliobolus heterostrophus*. *Exp. Mycol.* **6**, 364–374.

Leach, J., and Yoder, O. C. (1983). Heterokaryon incompatibility in the plant pathogenic fungus, *Cochliobolus heterostrophus*. *J. Hered.* **74**, 149–152.

Leach, J., Lang, B. R., and Yoder, O. C. (1982a). Methods for selection of mutants and *in vitro* culture of *Cochliobolus heterostrophus*. *J. Gen. Microbiol.* **128**, 1719–1729.

Leach, J., Tegtmeier, K. J., Daly, J. M., and Yoder, O. C. (1982b). Dominance at the *Tox1* locus controlling T-toxin production by *Cochliobolus heterostrophus*. *Physiol. Plant Pathol.* **21**, 327–333.

Lockington, R. A., Sealy-Lewis, H. M., Scazzocchio, C., and Davies, R. W. (1985). Cloning and characterization of the ethanol utilization regulon in *Aspergillus nidulans*. *Gene* **33**, 137–149.

Malpartida, F., Zalacain, M., Jimenez, A., and Davies, J. (1983). Molecular cloning and expression in *Streptomyces lividans* of a hygromycin B phosphotransferase gene from *Streptomyces hygroscopicus*. *Biochem. Biophys. Res. Commun.* **117**, 6–11.

Matthews, D. E., and VanEtten, H. D. (1983). Detoxification of the phytoalexin pisatin by a fungal cytochrome P-450. *Arch. Biochem. Biophys.* **224**, 494–505.

Miller, B. L., Miller, K. Y., and Timberlake, W. E. (1985). Direct and indirect gene replacements in *Aspergillus nidulans*. *Mol. Cell. Biol.* **5**, 1714–1721.

Mullaney, E. J., Hamer, J. E., Roberti, K. A., Yelton, M. M., and Timberlake, W. E. (1985). Primary structure of the *trpC* gene from *Aspergillus nidulans*. *Mol. Gen. Genet.* **199**, 37–45.

Munger, K., Germann, U. A., and Lerch, K. (1985). Isolation and structural organization of the *Neurospora crassa* copper metallothionein gene. *EMBO J.* **4**, 2655–2668.

Nunberg, J. H., Meade, J. H., Cole, G., Lawyer, F. C., McCabe, P., Schweickart, V., Tal, R., Wittman, V. P., Flatgaard, J. E., and Innis, M. A. (1984). Molecular cloning and characterization of the glucoamylase gene of *Aspergillus awamori*. *Mol. Cell. Biol.* **4**, 2306–2315.

Orr, W. C., and Timberlake, W. E. (1982). Clustering of spore-specific genes in *Aspergillus nidulans*. *Proc. Natl. Acad. Sci. U.S.A* **79**, 5976–5980.

Paietta, J., and Marzluf, G. A. (1985a). Plasmid recovery from transformants and the isolation of chromosomal DNA segments improving plasmid replication in *Neurospora crassa*. *Curr. Genet.* **9**, 383–388.

Paietta, J., and Marzluf, G. A. (1985b). Gene disruption by transformation in *Neurospora crassa*. *Mol. Cell. Biol.* **5**, 1554–1559.

Pardo, J. M., Malpartida, F., Rico, M., and Jimenez, A. (1985). Biochemical basis of resistance to hygromycin B in *Streptomyces hygroscopicus*—the producing organism. *J. Gen. Microbiol.* **131**, 1289–1298.

Paxton, J. (1980). A new working definition of the term "Phytoalexin." *Plant Dis.* **64**, 734.

Penalva, M. A., Tourino, A., Patino, C., Sanchez, F., Fernandez Sousa, J. M., and Rubio, V. (1985). Studies on transformation of *Cephalosporium acremonium*. *In* "Molecular Genetics of Filamentous Fungi" (W. E. Timberlake, ed.) pp. 59–68. Alan R. Liss, Inc., New York.

Penttila, M. E., Nevalainen, K. M. H., Raynal, A., and Knowles, J. K. C. (1984). Cloning of *Aspergillus niger* genes in yeast—expression of the gene coding *Aspergillus* beta-glucosidase. *Mol. Gen. Genet.* **194**, 494–499.

Rao, R. N., Allen, N.E., Hobbs, J. N., Jr., Alborn, W. E., Jr., Kirst, H. A., and Paschal, J. W. (1983). Genetic and enzymatic basis of hygromycin B resistance in *Escherichia coli*. *Antimicrob. Agents Chemother.* **24**, 689–695.

Rutledge, B. J. (1984). Molecular characterization of the *qa-4* gene of *Neurospora crassa*. *Gene* **32**, 275–287.

Samson, S. M., Belagaje, R., Blankenship, D. T., Chapman, J. L., Perry, D., Skatrud, P. L., VanFrank, R. M., Abraham, E. P., Baldwin, J. E., Queener, S. W., and Ingolia, T. D. (1985). Isolation, sequence determination and expression in *Escherichia coli* of the isopenicillin N synthetase gene from *Cephalosporium acremonium*. *Nature (London)* **318**, 191–194.

Santerre, R. F., Allen, N. E., Hobbs, J. N., Jr., Rao, R. N., and Schmidt, R. J. (1984). Expression of prokaryotic genes for hygromycin B and G418 resistance as dominant-selection markers in mouse L cells. *Gene* **30**, 147–156.

Schechtman, M. G., and Yanofsky, C. (1983). Structure of the trifunctional *trp-1* gene from *Neurospora crassa* and its aberrant expression in *Escherichia coli*. *J. Mol. Appl. Genet.* **2**, 83–99.

Schweizer, M., Case, M. E., Dykstra, C. C., Giles, N. H., and Kushner, S. R. (1981). Identification and characterization of recombinant plasmids carrying the complete *qa* gene cluster from *Neurospora crassa* including the *qa-1+* regulatory gene. *Proc. Natl. Acad. Sci. U.S.A.* **78**, 5086–5090.

Smith, D. A., and Ingham, J. L. (1981). Phytoalexins and plant disease resistance. *Biologist* **28**, 69–74.

Smith, D. A., Harrer, J. M., and Cleveland, T. E. (1982). Relation between production of extracellular kievitone hydratase by isolates of *Fusarium* and their pathogenicity on *Phaseolus vulgaris*. *Phytopathology* **72**, 1319–1323.

Smith, D. A., Wheeler, H. E., Banks, S. W., and Cleveland, T. E. (1984). Association between lowered kievitone hydratase activity and reduced virulence to bean in variants of *Fusarium solani* f. sp. *phaseoli*. *Physiol. Plant Pathol.* **25**, 135–147.

Soliday, C. L., Flurkey, W. H., Okita, T. W., and Kolattukudy, P. E. (1984). Cloning and structure determination of cDNA for cutinase, an enzyme involved in fungal penetration of plants. *Proc. Natl. Acad. Sci. U.S.A.* **81**, 3939–3943.

Stohl, L. L., and Lambowitz, A. M. (1983). Construction of a shuttle vector for the filamentous fungus, *Neurospora crassa*. *Proc. Natl. Acad. Sci. U.S.A.* **80**, 1058–1062.

Stohl, L. L., Akins, R. A., and Lambowitz, A. M. (1984). Characterization of deletion derivatives of an autonomously replicating *Neurospora* plasmid. *Nucleic Acids Res.* **12**, 6169–6178.

Tegtmeier, K. J., and VanEtten, H. D. (1983). The role of pisatin tolerance and degradation in the virulence of *Nectria haematococca* on peas: A genetic analysis. *Phytopathology* **72**, 608–612.

Tilburn, J., Scazzocchio, C., Taylor, G. G., Zabicky-Zissman, J. H., Lockington, R. A., and Davies, R. W. (1983). Transformation by integration in *Aspergillus nidulans*. *Gene* **26**, 205–221.

Timberlake, W. E., and Barnard, E. C. (1981). Organization of a gene cluster expressed specifically in the asexual spores of *A. nidulans*. *Cell (Cambridge, Mass.)* **26**, 29–37.

Timberlake, W. E., Boylan, M. T., Cooley, M. B., Mirabito, P. M., O'Hara, E. B., and Willett, C. E. (1985). Rapid identification of mutation-complementing restriction fragments from *Aspergillus nidulans* cosmids. *Exp. Mycol.* **9**, 351–355.

Turgeon, B. G., Garber, R. C., and Yoder, O. C. (1985). Transformation of the fungal maize pathogen *Cochliobolus heterostrophus* using the *Aspergillus nidulans amdS* gene. *Mol. Gen. Genet.* **201**, 450–453.

Turgeon, B. G., MacRae, W. D., Garber, R. C., Fink, G. R., and Yoder, O. C. (1986). A cloned tryptophan synthesis gene from the ascomycete *Cochliobolus heterostrophus* functions in *Escherichia coli*, yeast, and *Aspergillus nidulans*. *Gene* **42**, 79–88.

Van Den Elzen, P. J. M., Townsend, J., Lee, K. Y., and Bedbrook, J. R. (1985). A chimaeric hygromycin resistance gene as a selectable marker in plant cells. *Plant Mol. Biol.* **5**, 299–302.

VanEtten, H. D. (1982). Phytoalexin detoxification by monooxygenases and its importance for pathogenicity. *In* "The Physiological and Biochemical Basis of Plant Infection" (Y. Asada, W. R. Bushnell, S. Ouchi, and C. P. Vance, eds.), pp. 315–327. Japan Scientific Societies Press and Springer-Verlag.

VanEtten, H. D., and Matthews, P. S. (1984). Naturally-occurring variation in the induction of pisatin demethylating ability in *Nectria haematococca* mating population VI. *Physiol. Plant Pathol.* **25**, 149–160.

VanEtten, H. D., Pueppke, S. G., and Kelsey, T. C. (1975). 3,6a-Dihydroxy-8,9-methylenedioxypterocarpan as a metabolite of pisatin produced by *Fusarium solani* f. sp. *pisi*. *Phytochemistry* **14**, 1103–1105.

VanEtten, H. D., Matthews, P. S., Tegtmeier, K. J., Dietert, M. F., and Stein, J. I. (1980). The association of pisatin tolerance and demethylation with virulence on pea in *Nectria haematococca*. *Physiol. Plant Pathol.* **19**, 257–271.

VanEtten, H. D., Matthews, D. E., and Smith, D. A. (1982). Metabolism of phytoalexins. *In*

"Phytoalexins" (J. A. Bailey and J. W. Mansfield, eds.), pp. 181–217. Blackie and Son Ltd., Glasgow and London.

Van Heeswijk, R. and Roncero, M. I. G. (1984). High frequency transformation of *Mucor* with recombinant plasmid DNA. *Carlsberg Res. Commun.* **49**, 691–702.

Vapnek, D., Hautala, J. A., Jacobson, J. W., Giles, N. H., and Kushner, S. R. (1977). Expression in *Escherichia coli* K-12 of the structural gene for catabolic dehydroquinase of *Neurospora crassa*. *Proc. Natl. Acad. Sci. U.S.A.* **74**, 3508–3512.

Waldron, C., Murphy, E. B., Roberts, J. L., Gustafson, G. D., Armour, S. L., and Malcolm, S. K. (1985). Resistance to hygromycin B: A new marker for plant transformation studies. *Plant Mol. Biol.* **5**, 103–108.

Wernars, K., Goosen, T., Wennekes, L. M. J., Visser, J., Bos, C. J., van den Broek, H. W. J., van Gorcom, R. F. M., van den Hondel, C. A. M. J. J., and Pouwels, P. H. (1985). Gene amplification in *Aspergillus nidulans* by transformation with vectors containing the *amdS* gene. *Curr. Genet.* **9**, 361–368.

Yelton, M. M., Hamer, J. E., de Souza, E. R., Mullaney, E. J., and Timberlake, W. E. (1983). Developmental regulation of the *Aspergillus nidulans trpC* gene. *Proc. Natl. Acad. Sci. U.S.A.* **80**, 7576–7580.

Yelton, M. M., Hamer, J. E., and Timberlake, W. E. (1984). Transformation of *Aspergillus nidulans* by using a *trpC* plasmid. *Proc. Natl. Acad. Sci. U.S.A.* **81**, 1470–1474.

Yelton, M. M., Timberlake, W. E., and van den Hondel, C. A. M. J. J. (1985). A cosmid for selecting genes by complementation in *Aspergillus nidulans*: Selection of the developmentally regulated *yA* locus. *Proc. Natl. Acad. Sci. U.S.A.* **82**, 834–838.

Yoder, O. C., and Gracen, V. E. (1975). Segregation of pathogenicity types and host-specific toxin production in progenies of crosses between races T and O of *Helminthosporium maydis (Cochliobolus heterostrophus)*. *Phytopathology* **65**, 273–276.

Yoder, O. C., Weltring, K., Turgeon, B. G., Garber, R. C., and VanEtten, H. D. (1986). Technology for molecular cloning of fungal virulence genes. *In* "Biology and Molecular Biology of Plant–Pathogen Interactions" (J. A. Bailey, ed.), pp. 371–384. Plenum, New York.

Zimmermann, C. R., Orr, W. C., Leclerc, R. F., Barnard, E. C., and Timberlake, W. E. (1980). Molecular cloning and selection of genes regulated in *Aspergillus* development. *Cell (Cambridge, Mass.)* **21**, 709–715.

III

Gene Manipulation and Cell Culture

14

Expression of Human Interferon α in Silkworms with a Baculovirus Vector

S. MAEDA

Recent progress in biotechnology has enabled the expression of foreign genes in different hosts. Although microorganisms are generally used for this purpose, it is difficult to produce some foreign gene products in prokaryotic systems. To overcome this problem in prokaryotes, many eukaryotic *in vitro* systems have been established for the expression of eukaryotic genes. A wide variety of expression vectors have been constructed, mainly from infectious viruses. For the mass production of foreign gene products at low costs, *in vivo* systems, i.e., higher organisms such as plants, are now being considered as future potential hosts. Limited bacterial genes have been successfully expressed in plants using a Ti plasmid (1) or a cauliflower mosaic virus (2) as the vectors. Insects have not been considered as hosts for genetic engineering. However, the silkworm *Bombyx mori* has several advantageous characteristics which differ from those of the other insects. We have

BIOTECHNOLOGY IN INVERTEBRATE PATHOLOGY AND CELL CULTURE

developed a host–vector system using the silkworm and a baculovirus vector.

I. *Bombyx mori* NUCLEAR POLYHEDROSIS VIRUS AS A VECTOR

Insect viruses that are pathogenic to insects are classified into seven virus families (3). Nuclear polyhedrosis viruses (NPVs) are members of Baculoviridae and are typical insect pathogenic viruses. NPV has a large [>100 kilobase (kb)] double-stranded circular DNA genome within its rod-shaped capsid. It was previously stated by Miller *et al.* (4) that NPV has the following advantages as an expression vector: (1) a closed circular double-stranded DNA genome, (2) a rod-shaped capsid, (3) a nonessential protein, polyhedrin, coded on the viral genome, (4) a strong promoter of the polyhedrin which is turned on at a late stage of infection, and (5) a genetic marker, polyhedral production, which is easy to select for recombinant virus. As expected, several mammalian and bacterial genes have been expressed.

Autographa californica NPV (AcNPV), which has a wide host range and replicates in various established insect cell lines, has been studied as a model of baculovirus replication at the molecular level. *Bombyx mori* NPV (BmNPV), in contrast, has a narrow host range and will not grow in other insects found in the field. This feature is advantageous in a vector from the aspect of pollution and environmental safety. Furthermore, it is also speculated that the foreign gene products can be produced in an *in vivo* system, silkworms, as well as in established cell lines. Since there were few reports of successful replication of BmNPV *in vitro* (5), first we had to find a cell line which would support efficient growth of BmNPV. After screening many insect and mammalian established cell lines, we found that the BmN cell line provided by Dr. Loy E. Volkman (6) was susceptible to BmNPV and produced large amounts of polyhedra at a late stage of infection. Recently, BoMo (7) and e21 cells (Ninagi, personal communication) susceptible to BmNPV were also established from the silkworm.

To isolate viral clones, especially recombinant viruses, it is necessary to obtain a plaque assay system using established cells. By methods similar to those described for AcNPV (8), clear plaques were formed on BmN cells using SeaPlaque agarose as an overlay (9). BmNPV T3 isolate was cloned from a stock used in our laboratory. Restriction digests of DNA of the T3 isolate were analyzed on a 0.7% agarose gel. None of the submolar bands were revealed, indicating that the virus is a pure isolate and its DNA is very stable. The whole T3 DNA was estimated as 130 kb by summing the molecular weights of the fragments for each digest. For insertion of foreign genes into the BmNPV genome, we first had to determine the coding region of the polyhedrin gene on the viral DNA.

II. PROTEIN SYNTHESIS IN THE FAT BODY OF THE SILKWORM

Larvae of lepidopteran insects play a role in the synthesis and storage of proteins, lipids, and sugars in preparation for metamorphosis. In the silkworm several predominant proteins, namely storage proteins 1 and 2, vitellogenin, and 30K proteins, have been found (10). These proteins are produced mainly by fat bodies, stored in the hemolymph, transferred to target organs, and finally destroyed and used as a source of amino acids. In the silkworm several milligrams of a specific protein were synthesized within a day. The synthesis of these proteins is under the control of specific hormones.

During metamorphosis the protein synthesis of fat bodies in the silkworm was examined very carefully on sodium dodecyl sulfate (SDS)-polyacrylamide gels by incorporation of [^{35}S]methionine. The patterns of synthesis were found to be divided into three phases: (1) various polypeptides were synthesized at the early stage of the last instar; (2) two or three specific polypeptides were predominantly synthesized from the late stage of the last instar to the spinning stage; and (3) a few polypeptides were synthesized at an extremely low rate from the spinning to the pupal stage (protein synthesis almost stopped after pupation).

If a larva at the spinning stage is ligatured between the head and thorax, the larva can shed its cuticle. However, the abdomen of such a pupa remains in a steady state; i.e., further differentiation of the ovary and other organs is not detected. After subcutaneous infection of the isolated pupal abdomen with BmNPV, changes in polypeptide composition of the fat bodies were examined by SDS-polyacrylamide gel electrophoresis (PAGE). Figure 1 shows the polypeptide patterns separated on an SDS gel stained with Coomassie Brilliant Blue. The major protein in the fat bodies, storage protein 2, was almost degraded 4 days postinfection and the newly synthesized polyhedral protein replaced it as the major polypeptide. This finding indicates that viral protein synthesis is controlled by a different mechanism involving virus-specific

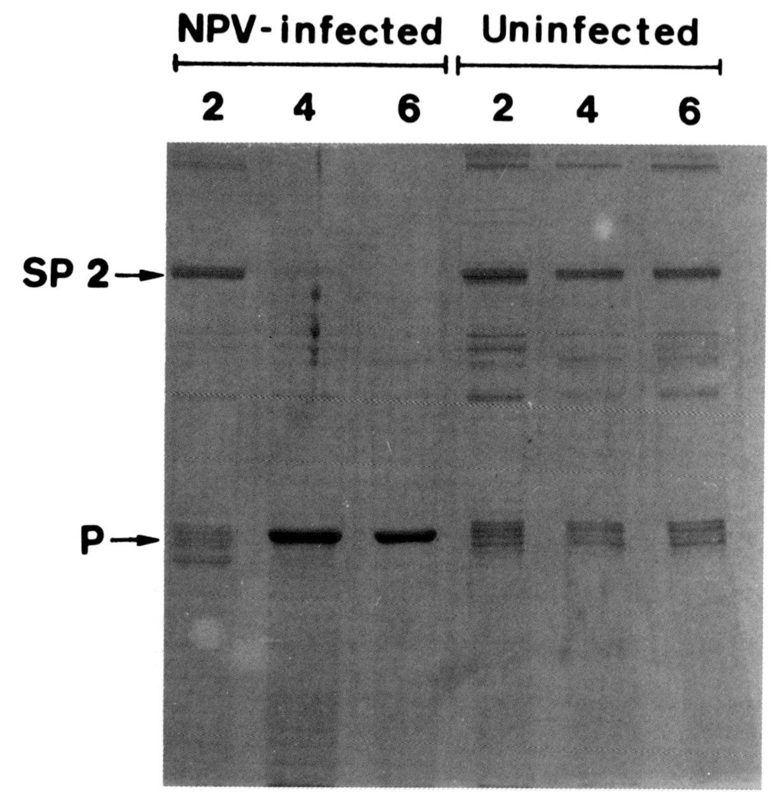

Fig. 1. Changes in polypeptide compositions in isolated pupal abdomens infected with BmNPV. Fat bodies were disrupted with SDS, separated on a 10% gel, stained with Coomassie Brilliant Blue, and destained. Samples were collected 2, 4, and 6 days after mock infection (Uninfected) or infection (NPV-infected). SP 2 and P indicate storage protein 2 and polyhedral protein, respectively.

enzyme catalysis and that polyhedrin is highly expressed, differing from other viral polypeptides perhaps by a strong promoter. The fat body at this stage was used for the isolation of mRNA to identify the location of the polyhedrin gene of BmNPV.

Studies on baculovirus replication at the molecular level have been done in *in vitro* systems, using established cell lines, because of simplicity. *In vivo* systems such as the pupal abdomen will be useful in the investigation of viral multiplication, which might be covered in the *in vitro* system.

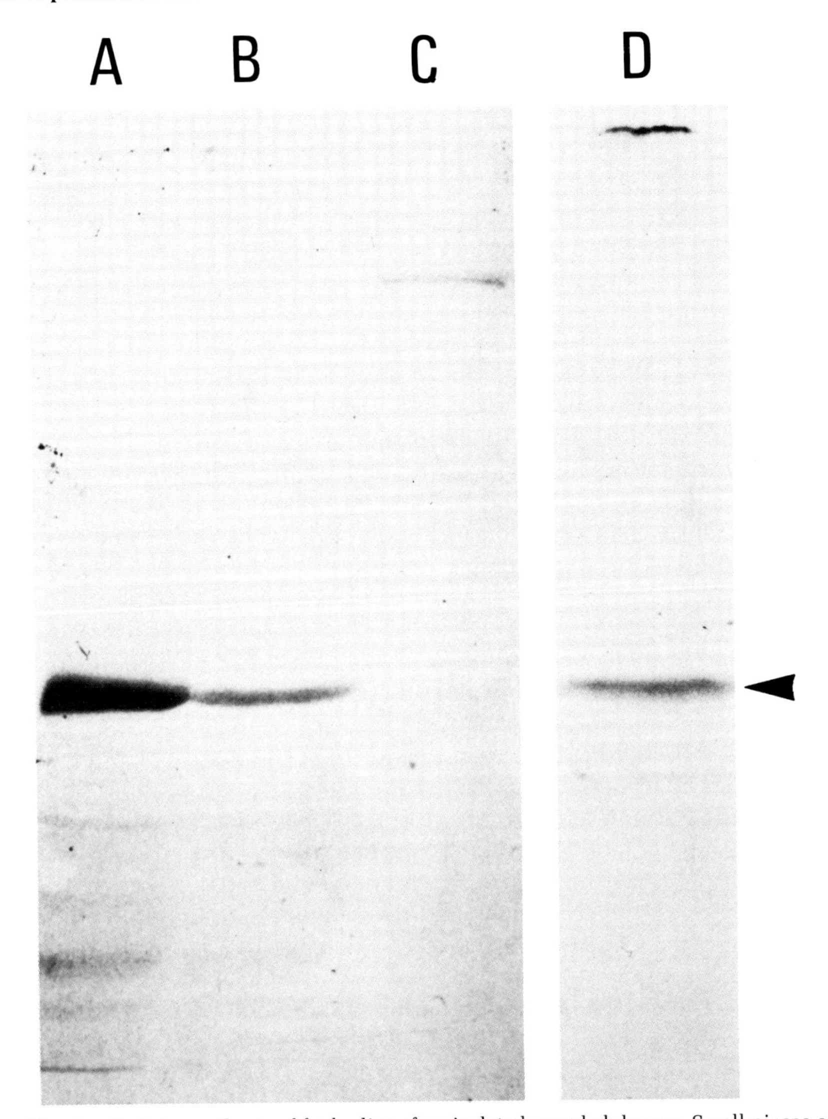

Fig. 2. Protein synthesis of fat bodies of an isolated pupal abdomen. Small pieces of the fat bodies were incubated in 20 μl of Grace's medium with [^{35}S]methionine for 20 min and analyzed by electrophoresis on a 10% SDS gel. Lanes A and B, infected fat body; lane C, uninfected fat body. *In vitro* translation of mRNA extracted from the infected isolated pupal abdomen in a reticulocyte lysate (lane D). Arrow indicates the polyhedral protein of BmNPV.

III. IDENTIFICATION AND CLONING OF THE POLYHEDRIN GENE OF BmNPV

To identify the polyhedrin gene, a cDNA probe made from mRNA isolated from infected fat bodies of the isolated pupal abdomen at a late stage of infection was used. The polyhedral protein was a major protein synthesized as shown in Fig. 2. *In vitro* translation experiments with the extracted RNA in reticulocyte lysates indicated that an mRNA for polyhedral protein was also predominant (Fig. 2). Poly(A)-containing mRNA was isolated by oligo(dT) column chromatography.

Viral DNA was cleaved by several restriction endonucleases, separated on a 0.7% agarose gel, and transferred onto nitrocellulose filters. The cDNA made from the above mRNA using oligo(dT) as a probe was then hybridized to the nitrocellulose filters. Exposure of labeled DNA to X-ray film after washing showed that a *Hin*dIII 3.9 kb and an *Eco*RI 10.5 kb fragment hybridized best to the probe (Fig. 3). The *Eco*RI 10.5 kb fragment was cloned into pBR322 at the *Eco*RI site (pBmE36). Further hybridization experiments indicated that the *Hpa*I–*Hin*dIII fragment in the *Hin*dIII 3.9 kb fragment of this plasmid contained polyhedrin gene (Fig. 4).

The entire nucleotide sequence of 1763 bases from the *Hpa*I–*Hin*dIII fragment was determined by the dideoxy chain-termination procedure (11,12). By comparing the sequencing data with the amino acid sequences reported by Serebryani *et al.* (13), the polyhedrin gene was found in this fragment, starting at 580 bases from the *Hpa*I site and terminating at 448 bases from the *Hin*dIII site (14).

When the nucleotide sequence was compared to the polyhedrin gene of AcNPV (15,16), there were several nucleotide changes in the coding region of the polyhedrin gene and many changes in the predicted promoter and its upstream region (14). Interestingly, 71 bases upstream from the translational start ATG was completely conserved. This conserved sequence seems to be important for a high level of expression of the polyhedrin gene. Actually, recombinant virus having the conserved sequence in which 29 bases were deleted from the translational start ATG produced about one-fifth less of an inserted foreign gene product (human interferon α) than was produced by recombinant viruses in which 19 or 5 bases were deleted from the ATG of the conserved sequence (16a).

IV. CONSTRUCTION OF A TRANSFER PLASMID VECTOR

Because of the large size of the BmNPV genome, marker rescue, i.e., allelic replacement, is the only available technique for insertion of

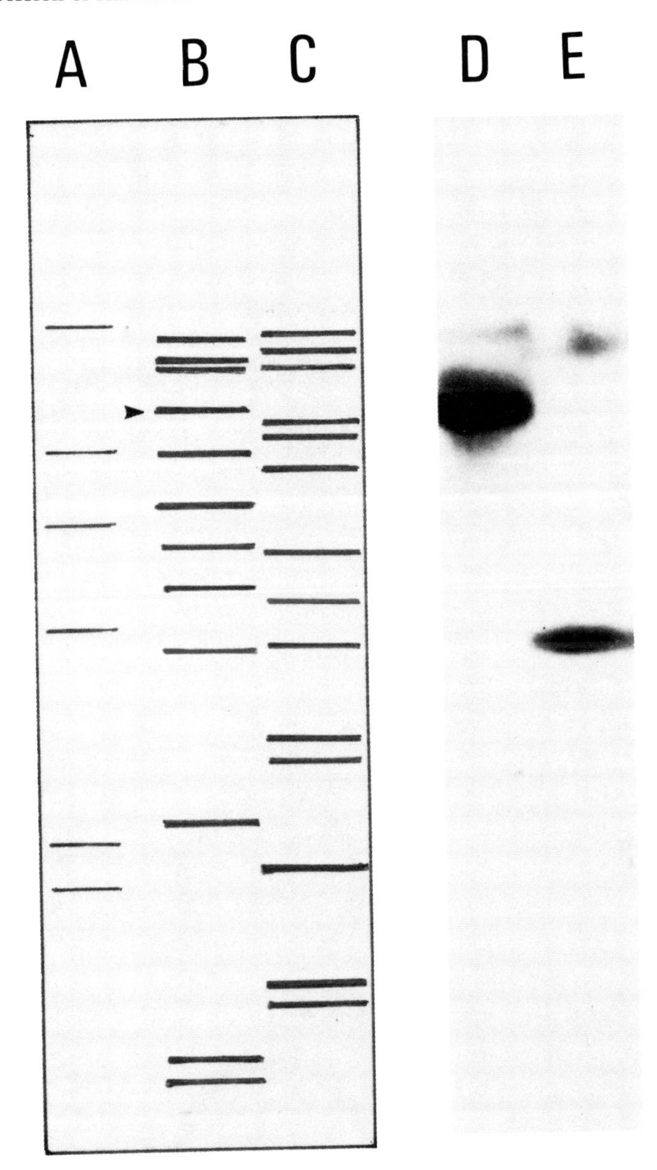

Fig. 3. Southern blot hybridization of viral DNA to cDNA made from mRNA from isolated pupal abdomen. Lane A, diagrams of molecular weight marker, λ DNA digested with *Hin*dIII; lanes B and C, diagrams of viral DNA digested with *Eco*RI and *Hin*dIII, respectively; lanes D and E, hybridization of viral DNA digested with *Eco*RI and *Hin*dIII, respectively.

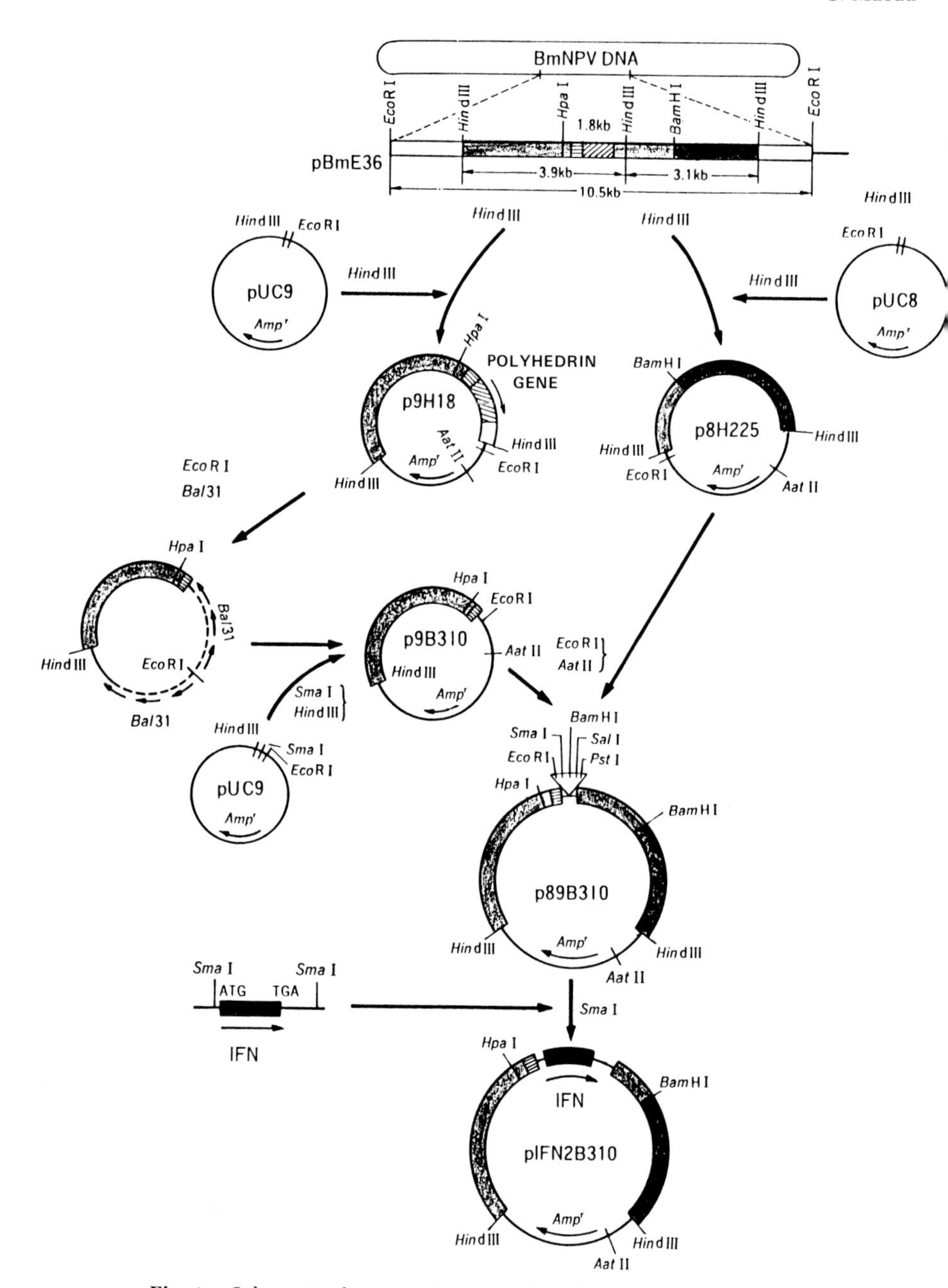

Fig. 4. Schematic diagram of construction of a recombinant plasmid.

foreign genes into NPV genome. A transfer plasmid vector which contained about 3 kb each of the upstream and downstream regions of the polyhedrin gene and a polylinker for the insertion of foreign genes was constructed as shown in Fig. 4. Since pUC8 and pUC9 plasmids are identical except for the order of the polylinker, constructed plasmids which contain the upstream and downstream regions of the polyhedrin gene in the correct order can generate the polylinker derived from pUC8. A new transfer plasmid vector, p89B310, constructed by this strategy was cleaved with *Sma*I, and a foreign gene, human interferon α, was inserted at this site. The resulting plasmid, pIFN2B310, contained the interferon gene in the same direction as the polyhedrin gene (Fig. 4).

V. CONSTRUCTION OF A RECOMBINANT VIRUS WITH INSERTION OF THE INTERFERON GENE

Recombinant plasmid pIFN2B310 and viral DNA isolated from viral particles were cotransfected into BmN cells in the presence of calcium ion. The molar ratio of plasmid to viral DNA was about 50 : 1. The transfected cells were cultured for several days in complete medium TC-10 containing 10% fetal calf serum (FCS).

Four or five days after cotransfection, polyhedral production was detected in the nuclei of several cells. The cell fluid was collected 6 days posttransfection and subjected to plaque assay for cloning of recombinant viruses. The titer in the fluid usually reached almost 10^7 plaque-forming units (PFU)/ml. About 0.5% of the plaques produced by a nonoccluded virus in the fluid lacked polyhedral production. These plaques were picked with a Pasteur pipette and suspended in 1 ml of TC-10 medium containing 1% FCS. Plaque purification was repeated twice to ensure homogeneity of the clones. The recombinant viruses thus obtained were propagated on BmN cells for the following experiments.

VI. CHARACTERISTICS OF RECOMBINANT VIRUSES

In the recombinant virus with insertion of a foreign gene, characteristics such as growth, infectivity, heat stability, and genome stability are important for the production of foreign gene products. It is anticipated that the recombinant viruses have properties similar to the wild-type virus except for polyhedral production. DNA of recombinant virus BmIFN2B310 was extracted and analyzed by restriction enzyme digestion. The patterns on an agarose gel after cleavage with *Eco*RI, *Hind*III, or *Bam*HI indicated that the interferon gene was inserted at the expected site. The result was confirmed by hybridization experiments using ^{32}P-labeled DNA of the interferon gene. Several serial *in vitro* passages of the recombinant virus did not affect the pattern of the viral DNA, showing that the genome of the recombinant virus is considerably stable.

When viral growth *in vitro* and *in vivo* was compared, the growth curve of the recombinant virus appeared to be almost the same as that of the BmNPV T3 isolate. Heat stability was examined at 30°, 40°, 50°, and 60°C for 5, 15, and 60 min. Neither wild-type T3 nor the recombinant IFNBmB310 was inactivated at 30° and 40°C for 60 min. However at 50° and 60°C both viruses were immediately inactivated to less than 10^{-2} (50°C) or less than 10^{-3} (60°C) even after a 5-min incubation.

Cytopathic effects of the recombinant virus were compared with those of the T3 isolate. Except for polyhedral production, all infected BmN cells showed cytopathic effects typical of the wild-type viral infection and were destroyed within several days. If the silkworm was infected subcutaneously with 10^5 PFU of the recombinant virus, all the infected larvae died within 7 days, like those with the wild-type infection. Symptoms of recombinant viral infection were almost the same except that the color of the body was darker. The hemolymph in recombinant viral infection did not show the milky turbidity due to polyhedral production which appeared in wild-type viral infection. However, the hemolymph became slightly trubid, probably because of the degradation of infected cells.

VII. PURIFICATION OF THE INTERFERON FROM THE SILKWORM

Even when large amounts of foreign gene products were synthesized in *Escherichia coli*, it was very difficult and sometimes impossible to purify the product. Generally, eukaryotic cells with an appropriate expression vector produce relatively small quantities of products. However, the proteins so produced were easily purified. When the BmN cell line with BmNPV vector was used for the production of interferon, the product was purified by the methods commonly used for biochemical experiments, such as Sephadex or ion-exchange column chromatography.

It is important that the products so synthesized can be easily purified in unaltered form from the foreign host, such as silkworm, used in the *in vivo* system. One of the reasons for the few reports on expression of foreign genes in other organisms, especially plants, might be the degradation of the products by host enzymes.

We demonstrated experimentally purification of the interferon from silkworm larvae. Almost all activities of the interferon produced in the larvae were found in the hemolymph. The hemolymph, which could be considered as a storage tissue, constitutes 30% of the total volume of the body (17). Abdominal legs of the infected body were pierced with a pin and the oozing hemolymph was collected in a test tube kept in an ice bath. The hemolymph was diluted with an appropriate buffer and processed for purification or stored at −80°C until used for purification. The interferon was partially purified by either Sephadex G-100 or ion-exchange FPLC-mono Q column chromatography. Scanning of these purified samples for UV absorption showed one major peak of interferon activity separated from other hemolymph proteins (18). When the infected hemolymph was applied to an affinity column with a monoclonal antibody against mature interferon and the eluted sample was collected and analyzed by SDS-PAGE, the gel stained with Coomassie Brilliant Blue showed a major band with mobility at the expected molecular weight. About 50 μg of the pure interferon was obtained from one larva without apparent loss of activity (14).

Analysis of 11 amino acid sequences from the N terminus indicated that the interferon produced in the silkworm was identical to natural interferon (14). This finding indicated that the human signal peptide for secretion was recognized and properly removed in the silkworm as well as in BmN cells. Similar results with the insect cell line SF21AE and an AcNPV vector with insertion of a human gene, interleukin 2, were recently reported (19).

To examine this phenomenon more closely, a recombinant virus without the signal sequence for secretion was constructed and used to infect BmN cells. Throughout the infection process about 90% of the total activity was found predominantly inside the cells. Furthermore, the total activity was less than 10% of that obtained with recombinant virus with the signal sequence (18), probably because the interferon produced in the cytoplasm was degraded by endogenous proteases.

VIII. CONCLUSION

Nuclear polyhedrosis viruses are now considered most attractive vectors for expression of foreign genes. One interesting example is the expression of the E2 gene of bovine papillomavirus-1. Although E2 protein has been predicted by DNA sequencing and cDNA cloning, no one has been able to detect the protein or antibody against it even when recombinant DNA technology was used. By using BmNPV as a vector, E2 fused to the polyhedrin gene was expressed at an extremely high level and easily purified using the antibody against the polyhedral protein (19a). Many other bacterial, mammalian, and viral genes have been expressed by using AcNPV or BmNPV vectors: β-galactosidase (16), human interferon β (20), mouse interleukin 3 (20a), c-*myc* (21), hemagglutinin of influenza virus (W. Doerfler, personal communication) (see following chapter by Kuroda *et al.*), and structural proteins of rotavirus (M. Estes, personal communication).

For the mass production of foreign gene products, the silkworm has many advantageous characteristics, as discussed earlier. The silkworm can be cultured under sterile conditions using an artificial diet throughout the larval stage. Furthermore, culture of the silkworm can be controlled automatically and the process can be scaled up easily for mass production. Because the silkworm cannot survive in the wild, it is preferred as a host for bioengineering from a safety point of view. Silkworms with the BmNPV vector will be suitable hosts for mass production of foreign gene products in the near future.

REFERENCES

1. Herrera-Estrella, L., Depicker, A., Van Montagu, M., and Schell, J. (1983). Expression of chimaeric genes transferred into plant cells using a Ti-plasmid-derived vector. *Nature (London)* **303**, 209–213.
2. Brisson, N., Paszkowski, J., Penswick, J. R., Gronenborn, B., Potrykus, I., and Hohn, T. (1984). Expression of a bacterial gene in plants by using a viral vector. *Nature (London)* **310**, 511–514.
3. Matthews, R. E. F. (1982). Classification and nomenclature of viruses. *Intervirology* **17**, 1–199.
4. Miller, L. K., Miller, D. W., and Adang, M. J. (1983). An insect virus for genetic engineering in eukaryotes. *In* "Genetic Engineering in Eukaryotes" (P. F. Lurquin and A. Kleinhofs, eds.), pp. 89–97. Plenum, New York.
5. Raghow, R., and Grace, T. D. C. (1974). Studies on a nuclear polyhedrosis virus in *Bombyx mori* cells in vitro. I. Multiplication kinetics and ultrastructural studies. *J. Ultrastruct. Res.* **47**, 384–399.
6. Volkman, L. E., and Goldsmith, P. A. (1982). Generalized immunoassay for *Autographa californica* nuclear polyhedrosis virus infectivity *in vitro*. *Appl. Environ. Microbiol.* **44**, 227–233.
7. Inoue, H., and Mitsuhashi, J. (1984). A *Bombyx mori* cell line susceptible to a nuclear polyhedrosis virus. *J. Seric. Sci. Jpn.* **53**, 108–113.
8. Brown, M., and Faulkner, P. (1977). A plaque assay for nuclear polyhedrosis viruses using a solid overlay. *J. Gen. Virol.* **36**, 361–364.
9. Maeda, S. (1984). A plaque assay and cloning of *Bombyx mori* nuclear polyhedrosis virus. *J. Seric. Sci. Jpn.* **53**, 547–548.
10. Tomino, S. (1985). Major plasma proteins of *Bombyx mori*. *Zool. Sci.* **2**, 293–303.

11. Sanger, F., Nicklen, S., and Coulson, A. R. (1977). DNA sequencing with chain-terminating inhibitors. *Proc. Natl. Acad. Sci. U.S.A.* **74**, 5463–5467.
12. Guo, L.-H., and Wu, R. (1983). Exonuclease III: Use for DNA sequence analysis and in specific deletions of nucleotides. *In* "Methods in Enzymology" (R. Wu, L. Grossman, and K. Moldave, eds.), Vol. 100, pp. 60–96. Academic Press, New York.
13. Serebryani, S. B., Levitina, T. L., Kautsman, M. L., Radavski, Y. L., Gusak, N. M., Ovander, M. N., Sucharenko, N. V., and Kozlov, E. A. (1977). The primary structure of the polyhedral protein of nuclear polyhedrosis virus (NPV) of *Bombyx mori*. *J. Invertebr. Pathol.* **30**, 442–443.
14. Maeda, S., Kawai, T., Obinata, M., Fujiwara, H., Horiuchi, T., Saeki, Y., Sato, Y., and Furusawa, M. (1985). Production of human α-interferon in silkworm using a baculovirus vector. *Nature (London)* **315**, 592–594.
15. Hooft van Iddekinge, B. J., Smith, G. E., and Summers, M. D. (1983). Nucleotide sequence of the polyhedrin gene of *Autographa californica* nuclear polyhedrosis virus. *Virology* **131**, 561–565.
16. Pennock, G. D., Shoemaker, C., and Miller, L. K. (1984). Strong and regulated expression of *Escherichia coli* galactosidase in insect cells with a baculovirus vector. *Mol. Cell. Biol.* **4**, 399–406.
16a. Horiuchi, T., Marumoto, Y., Saeki, Y., Sato, Y., Furusawa, M., Kondo, A., and Maeda, S. (1987). High-level expression of the Raman-α-gene through the use of an improved baculovirus vector in the silkworm, *Bombyx mori*. *Agric. Biol. Chem.* **51**, 1573–1580.
17. Nagata, M., Seong, S., and Yoshitake, N. (1980). Variation of the haemolymph volume with larval development of the silkworm, *Bombyx mori*. *J. Seric. Sci. Jpn.* **49**, 453–454.
18. Maeda, S., Kawai, T., Obinata, M., Chika, T., Horiuchi, T., Maekawa, K., Nakasuji, K., Saeki, Y., Sato, Y., Yamada, K., and Furusawa, M. (1984). Characteristics of human interferon α produced by a gene transferred by a baculovirus vector in the silkworm, *Bombyx mori*. *Proc. Jpn. Acad. Ser. B* **60**, 423–426.
19. Smith, G. E., Grace, J., Ericson, B. L., Moschera, J., Lahm, H.-W., Chizzonite, R., and Summers, M. D. (1985). Modification and secretion of human interleukin 2 produced in insect cells by a baculovirus expression vector. *Proc. Natl. Acad. Sci. U.S.A.* **82**, 8404–8408.
19a. Fuse *et al.* Submitted for publication.
20. Smith, G. E., Summers, M. D., and Fraser, M. J. (1983). Production of human β-interferon in insect cells infected with a baculovirus expression vector. *Mol. Cell. Biol.* **3**, 2156–2165.
20a. Miyajima, A., Schreurs, J., Otsu, K., Kondo, A., Ken-ichi, A., and Maeda, S. (1987). Use of the silkworm, *Bombyx mori*, and an insect baculovirus vector for high-level expression of biologically active mouse interleukin-3. *Gene*, in press.
21. Miyamoto, C., Smith, G. E., Farrell-Towt, J., Chizzonite, R., Summers, M. D., and Ju, G. (1985). Production of human c-myc protein in insect cells infected with a baculovirus expression vector. *Mol. Cell. Biol.* **5**, 2860–2865.

15

Biologically Active Influenza Virus Hemagglutinin Expressed in Insect Cells by a Baculovirus Vector

KAZUMICHI KURODA, CHARLOTTE HAUSER,
RUDOLF ROTT, HANS-DIETER KLENK, AND
WALTER DOERFLER

The insect baculovirus *Autographa californica* nuclear polyhedrosis virus (AcNPV) has played a major role in studies on the molecular biology of insect DNA viruses. Recently, this virus system has been effectively adapted as a highly efficient eukaryotic vector using insect cells for the expression of several mammalian genes (Smith *et al.*, 1983b, 1985; Miyamoto *et al.*, 1985). A cDNA sequence of the influenza (fowl plague) virus hemagglutinin gene has been inserted into the *Bam*HI site of the pAc373 polyhedrin vector. *Spodoptera frugiperda* cells were cotransfected with this construct, pAc-HA651, and authentic

BIOTECHNOLOGY IN INVERTEBRATE
PATHOLOGY AND CELL CULTURE

AcNPV DNA. Recombinant virus was selected by adsorption of transfected cells to erythrocytes followed by serial plaque passages on *S. frugiperda* cells. We have determined the site of insertion of the hemagglutinin gene into the AcNPV genome by restriction enzyme cleavage and Southern blot–hybridization analyses using hemagglutinin cDNA as probe. The influenza hemagglutinin gene is located in the polyhedrin gene of AcNPV DNA. Immunofluorescent labeling, immunoprecipitation, and immunoblot analyses employing specific antisera revealed that *S. frugiperda* cells produce immune reactive hemagglutinin after infection with the recombinant virus. The hemagglutinin is expressed at the cell surface and has hemolytic capacity that has been activated by posttranslational proteolytic cleavage. When chickens were immunized with *S. frugiperda* cells expressing hemagglutinin, they developed hemagglutination-inhibiting and neutralizing antibodies and were protected from infection with fowl plague virus. These observations demonstrate that hemagglutinin processing in insect cells is similar to that in fowl plague virus-infected vertebrate cells and that the hemagglutinin has full biological activity.

I. INTRODUCTION

The major constituent of the envelope of the influenza virus is hemagglutinin. It induces the formation of neutralizing antibodies and, because of its antigenic variability, it is responsible for the characteristic epidemiology of influenza in humans (for review, see Palese and Young, 1983). Hemagglutinin is also of interest because it initiates virus infection by binding to neuraminic acid-containing receptors of host cells and by promoting penetration of the viral genome into the cytoplasm through fusion of the envelope with cellular membranes (Klenk and Rott, 1980; Klenk *et al.*, 1984). The amino acid sequences of many hemagglutinin subtypes have been elucidated (Ward, 1981), and their conformations have been studied by X-ray crystallography (Wilson *et al.*, 1981). As an integral membrane protein, hemagglutinin is translated at membrane-bound polysomes, translocated by means of an amino-terminal signal sequence into the lumen of the endoplasmic reticulum, and transported from there through the Golgi apparatus to the plasma membrane. In the course of transport, hemagglutinin undergoes posttranslational modifications, including removal of the signal peptide, attachment of N-glycosidic oligosaccharide side chains (Keil *et al.*, 1985), and proteolytic cleavage of the precursor HA into the fragments HA_1 and HA_2. The latter modification, which is essential for realization of the fusion activity, involves the sequential action of two host-dependent enzymes, a trypsinlike endoprotease and carboxypeptidase N (arginine carboxypeptidase) that attack on arginine- or lysine-containing cleavage sites (Klenk *et al.*, 1975; Lazarowitz and Choppin, 1975; Garten and Klenk, 1983).

Among the expression vector systems, the insect baculovirus *A. californica* nuclear polyhedrosis virus (AcNPV) has been successfully developed; human interferon β (Smith *et al.*, 1983b), β-galactosidase from *Escherichia coli* (Pennock *et al.*, 1984), human c-myc protein (Miyamoto *et al.*, 1985), and human interleukin 2 (Smith *et al.*, 1985) have been expressed in insect cell lines infected with recombinant virus.* This expression system has proved to be very valuable because of the efficient polyhedrin promoter of AcNPV. Because of the ability of insect cells to process newly synthesized proteins, biologically active gene products have been obtained. There is evidence that interferon β has been glycosylated, that the signal peptide of interleukin 2 has been cleaved off, and that the c-myc protein has been phosphorylated.

In this chapter we describe the construction of a plasmid containing the polyhedrin promoter, part of the polyhedrin coding sequence (Smith *et al.*, 1983a), and the gene for influenza virus hemagglutinin. On cotransfection of *S. frugiperda* insect cells with this plasmid and intact AcNPV DNA, recombinant virus was isolated which contained in the polyhedrin sequence of its genome the gene for influenza virus hemagglutinin. When *S. frugiperda* cells were infected with recombinant virus, the infected cells produced biologically active influenza hemagglutinin that induced immunity antibodies in chickens.

Experimental details of this research have been published elsewhere (Kuroda *et al., 1986).*

II. HEMAGGLUTININ: RESULTS OF STUDIES

A. Construction of the Recombinant Vector pAc373-HA and the Recombinant Virus AcNPV-HA

The strategy for construction of the insect vector containing the hemagglutinin gene is shown in Fig. 1. The hemagglutinin gene of fowl plague virus mutant *ts651* was cloned into the *Pst*I site of the plasmid vector pUC8. This plasmid (pUC-HA651) contains the whole sequence of the hemagglutinin gene. The construct also carries dG-dC tails of 15 nucleotide pairs at the 5′ end and 21 nucleotide pairs at the 3′ end of the hemagglutinin gene. As the tails were expected to interfere with expression of the hemagglutinin gene (Riedel *et al.*, 1984), they were removed by *Bal*31 exonuclease digestion. The tailored hemagglutinin

*Note: Since this paper was written, a large number of additional proteins have been expressed in this vector-host system.

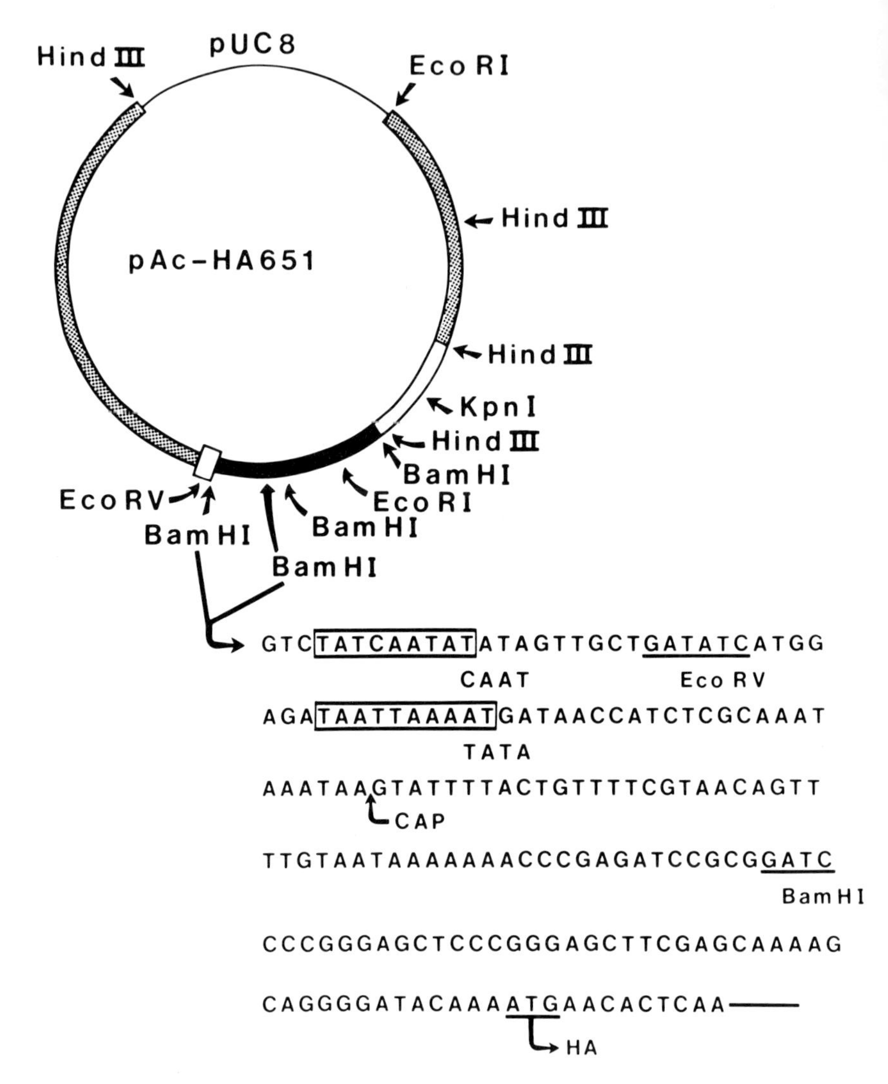

Fig. 1. Construction of the pAc-HA651. Fowl plague virus hemagglutinin and AcNPV sequences are indicated by solid and by dotted bars, respectively. For details see text.

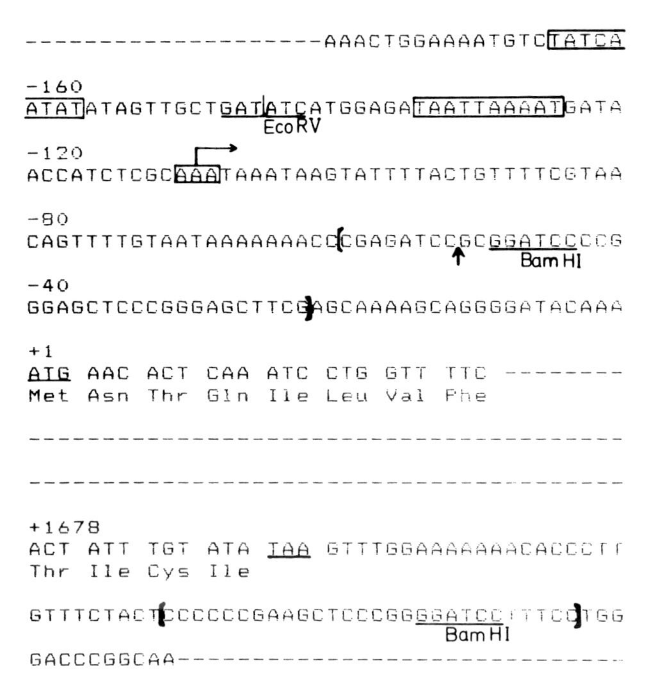

Fig. 2. Nucleotide sequences flanking the hemagglutinin gene in the construct pAc-HA651. The boxes at −165 and −134 designate the nucleotides that resemble the canonical CAAT and TATA box sequences, respectively. The box at −110 marks the bases that code for the 5′ end of polyhedrin mRNA. The brackets at −59 and +1721 indicate the linker and adaptor regions. Sequences between −21 and −1 and between +1694 and +1720 represent the noncoding regions of the hemagglutinin gene. Nucleotide sequences were determined by the method of Sanger *et al.* (1977), as described elsewhere (Kuroda *et al.*, 1986).

gene fragments were cloned into the *Hin*dIII site of pUC13 using *Hin*dIII linkers. The clones that contained the hemagglutinin gene were examined by nucleotide sequence analysis of the flanking regions of the hemagglutinin gene. For further work one clone, pUC-HA651/34, was chosen which had remaining dG-dC residues of one nucleotide pair at the 5′ end and of five nucleotide pairs at the 3′ end. The full-length hemagglutinin gene could be excised from this plasmid by *Hin*dIII cleavage and was recloned into the *Bam*HI site of pAc373 using *Bam*HI/*Sma*I and *Hin*dIII/*Sma*I adaptors. For expression studies (see below) one clone, pAc-HA651, was chosen which showed the correct orientation. In this construct the flanking regions of the hemagglutinin gene were sequenced (Fig. 2). The nucleotide sequence data proved that the hemagglutinin gene had been correctly cloned into the *Bam*HI site of pAc373. The distance between the start site of polyhedrin mRNA transcription and the initiation codon of the hemagglutinin gene is longer than the respective distance in the original polyhedrin promoter-gene region in AcNPV DNA because of the inserted linker, the adaptors, and the noncoding region of the hemagglutinin gene. In the authentic polyhedrin gene that distance amounts to 58 nucleotide pairs and in the pAc373-HA construct it measures 108 nucleotide pairs.

In order to transfer the hemagglutinin gene into the AcNPV genome, *S. frugiperda* cells were cotransfected with pAc-HA651 DNA and the genomic DNA of authentic AcNPV. Because the progeny virus consisted predominantly of wild-type AcNPV, it was difficult to screen for infected cells lacking polyhedra. The transfected cells were therefore adsorbed to chicken erythrocytes fixed on plastic cell culture dishes in order to preselect cells that expressed influenza virus hemagglutinin (Kuroda *et al.*, 1986). After release by neuraminidase treatment, these cells were then seeded on fresh culture dishes, and virus production was allowed to proceed. By this procedure, we were able to enrich the recombinant virus by a factor of 10 to 100. The progeny virus was subjected to two plaque purifications, and *S. frugiperda* cells with polyhedrin-free plaques were selected to obtain pure recombinant virus.

B. Analysis of the Recombinant AcNPV DNA by Restriction Enzyme Cleavage and Southern Blot Hybridization

Recombinant AcNPV was propagated, purified, and the recombinant AcNPV DNA was extracted as described (Carstens *et al.*, 1979; Tjia *et al.*, 1979). The recombinant DNA was cleaved with the restriction endonuclease *Hin*dIII or *Eco*RI, and the fragments were separated by electrophoresis on a 0.7% agarose gel in order to identify the site of insertion of the hemagglutinin gene in the AcNPV genome (Fig. 3A). The DNA fragments were then transferred to a nylon membrane (Southern, 1975), and the fragments containing the hemagglutinin gene were identified by hybridization (Wahl *et al.*, 1979) to a hemagglutinin gene fragment derived from pUC-HA651/34 which was ^{32}P-labeled by nick translation (Rigby *et al.*, 1977). The results demonstrate that the polyhedrin gene-containing fragment *Eco*RI-H (Fig. 3A, lane b, arrow) is absent from the cleavage pattern of the recombinant virus DNA and that two new fragments are generated (Fig. 3A, lane c, arrows) which are absent from the cleavage pattern of AcNPV DNA. It is apparent that the two newly generated DNA fragments hybridize with the hemagglutinin DNA (Fig. 3B, lane c). AcNPV DNA obviously does not hybridize to this probe (Fig. 3B, lane b), nor does control λ DNA (Fig. 3B, lane a). Simi-

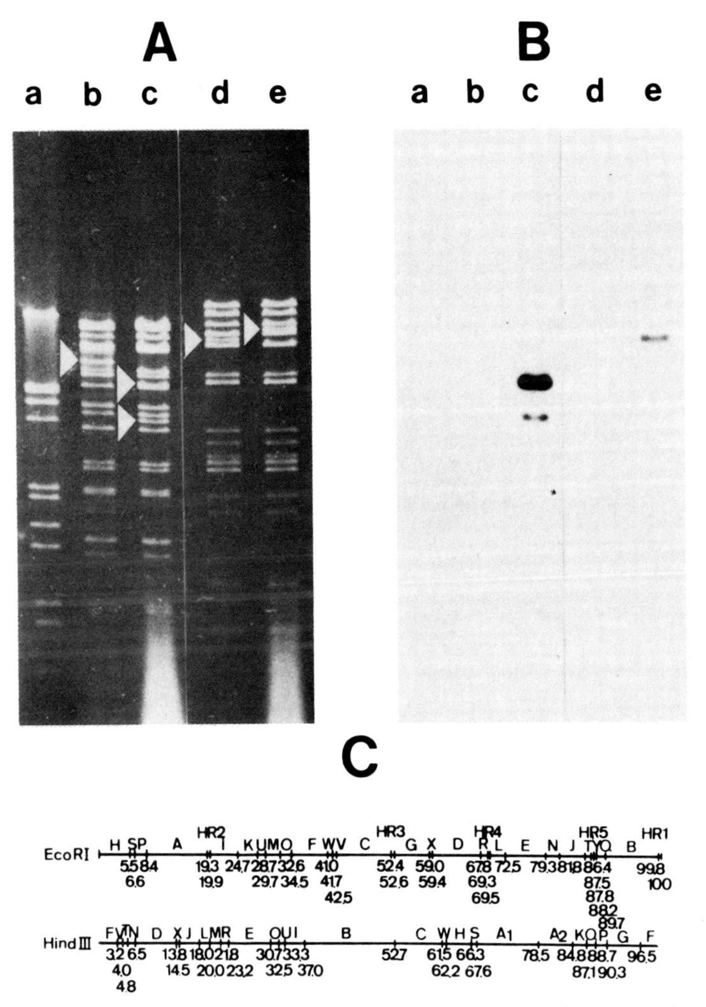

Fig. 3. Analysis of recombinant AcNPV by restriction enzyme cleavage and Southern blot hybridization. Authentic AcNPV DNA (b, d) or recombinant AcNPV DNA (c, e) was digested with *Eco*RI (b, c) or *Hin*dIII (d, e), and the digests were separated on 0.7% agarose gel, followed by staining with ethidium bromide (A). Subsequently the DNA fragments were blotted to a nylon membrane sheet. The blot was hybridized with ^{32}P-labeled hemagglutinin cDNA (B). *Eco*RI/*Hin*dIII-digested λ DNA was used as a size marker (a). The *Eco*RI and *Hin*dIII restriction maps of authentic AcNPV DNA are also shown (C) (Lübbert and Doerfler, 1984).

larly, when the recombinant AcNPV DNA is cut with *Hin*dIII, fragment *Hin*dIII-F is not generated. A new fragment arises (Fig. 3A, lanes d and e, arrow) which hybridizes to the cloned hemagglutinin gene (Fig. 3B, lane e). This result is consistent with the restriction map of the hemagglutinin gene that has an *Eco*RI site but no *Hin*dIII site. The results of this analysis demonstrate that the hemagglutinin gene has been inserted in the polyhedrin gene of AcNPV DNA. So far, the site of insertion has not been mapped more precisely, as determining the exact location of the hemagglutinin gene in the AcNPV vector does not appear relevant for the expression of this gene.

C. Expression of Influenza Hemagglutinin in *Spodoptera frugiperda* Cells

After infection with recombinant virus, *S. frugiperda* cells, unlike cells infected with AcNPV wild-type, did not show nuclear inclusions. Hemagglutinating activity could be detected in homogenates of such cells starting at about 24 hr after inoculation. There was a continuous rise in the hemagglutination titer up to about 96 hr, when most of the cells deteriorated (data not shown). Hemagglutinating activity was not detectable in the medium. Recombinant virus-infected cells also acquired the ability to hemadsorb. At about 48 hr after infection, each cell was heavily loaded with erythrocytes (Fig. 4). The decrease in the amount of adsorbed erythrocytes observed after 48 hr appeared to be due to a substantial number of infected cells being detached from the plastic support at this time. Neither cells infected with wild-type AcNPV nor mock-infected cells displayed hemagglutination or hemadsorption activities (Fig. 4). These observations indicate that after infection with the recombinant virus hemagglutinin is expressed in *S. frugiperda* cells and is transported to the cell surface.

The hemagglutinin synthesized in *S. frugiperda* cells can also be directly recognized by antibodies raised against hemagglutinin derived from fowl plague virions, as can be demonstrated by immunofluorescent labeling (Fig. 5). Each cell expresses hemagglutinin.

The recombinant hemagglutinin was then characterized by sodium dodecyl sulfate (SDS)-polyacrylamide gel electrophoresis. Figure 6 shows the results of immunoprecipitation experiments in which hemagglutinin synthesized in fowl plague virus-infected MDCK cells and in *S. frugiperda* cells infected with recombinant virus was labeled by a 2-hr pulse with [^{35}S]methionine. Hemagglutinin is present in MDCK cells in the cleaved and in the uncleaved forms (Fig. 6A, lane b), whereas mainly uncleaved hemagglutinin can be detected in stained

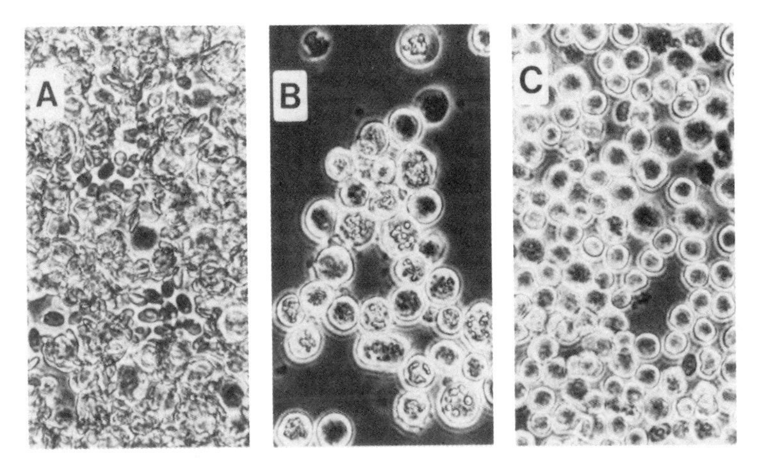

Fig. 4. Hemadsorption of *S. frugiperda* cells after infection with recombinant AcNPV virus. At 2 days after infection with recombinant AcNPV (A), after infection with authentic AcNPV (B), or after mock infection (C), hemadsorption was assayed and cell cultures were photographed under the microscope. Cells infected with recombinant virus are loaded with erythrocytes. Cells infected with authentic virus show polyhedra in their nuclei.

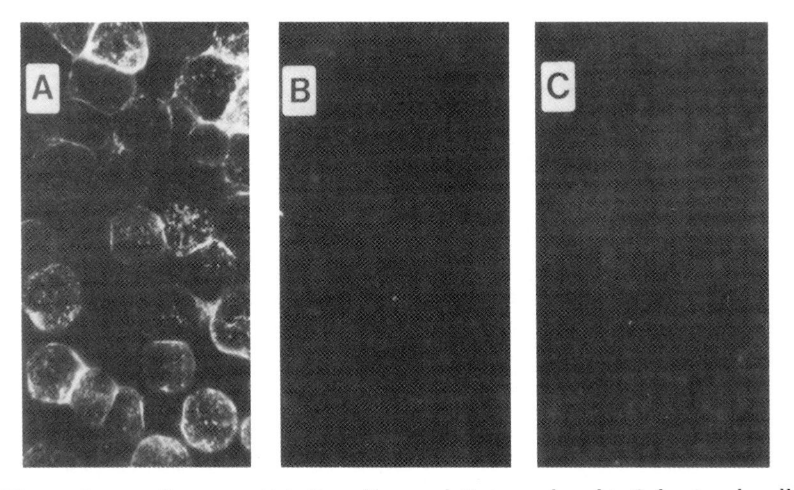

Fig. 5. Immunofluorescent labeling of hemagglutinin produced in *S. frugiperda* cells. Cells were infected with recombinant AcNPV (A) or authentic AcNPV (B) or were mock-infected with PBS (C). After 2 days of incubation, the cells were labeled by indirect immunofluorescence as described (Kuroda *et al.*, 1986). The exposure time for UV photography was 15 sec.

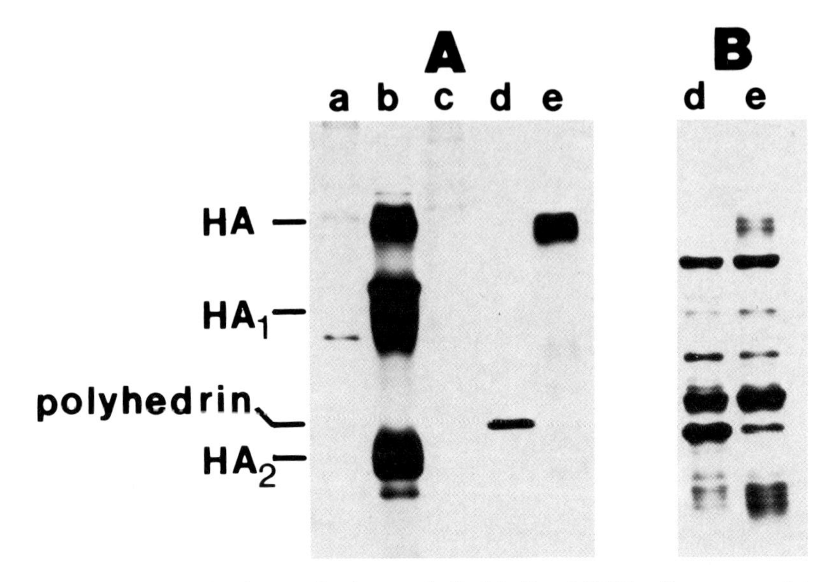

Fig. 6. Analysis of polypeptides by metabolic labeling. MDCK cells were mock-infected (a) or infected with fowl plague virus for 5 hr (b). *Spodoptera frugiperda* cells were infected with authentic AcNPV (d) or recombinant AcNPV (e) or were mock-infected (c) for 2 days. The infected cells were labeled by a 2-hr pulse with [^{35}S]methionine, and the hemagglutinin was immunoprecipitated and analyzed by SDS-polyacrylamide gel electrophoresis as described (Kuroda *et al.*, 1986). Samples derived from 10^6 cells were loaded onto the gel (A). Lysates of insect cells infected with authentic and recombinant AcNPV were also analyzed by gel electrophoresis prior to immunoprecipitation. Samples derived from 10^5 cells were loaded onto the gel (B), and the distribution of polypeptides was determined by direct autoradiography after drying the gel.

gels in recombinant virus-infected *S. frugiperda* cells (Fig. 6A, lane e). The polyhedrin band in the immunoprecipitate obtained from AcNPV-infected control cells (Fig. 6A, lane d) resulted from incomplete solubilization of the occlusions in the immunoprecipitation buffer (Miyamoto *et al.*, 1985).

Gel electrophoretic analyses of lysates not subjected to immunoprecipitation allow comparison of the relative amounts of hemagglutinin synthesized in cells infected with recombinant virus (Fig. 6B, lane e) and polyhedrin synthesized in cells infected with authentic AcNPV (Fig. 6B, lane d). The data indicate that hemagglutinin production does not reach the level of polyhedrin synthesis. As pointed out already, polyhedrin is not made in cells infected with recombinant virus. The minor band comigrating with polyhedrin (Fig. 6A, lane d, and Fig. 6B, lane e) has been identified as a different protein (Pennock *et al.*, 1984).

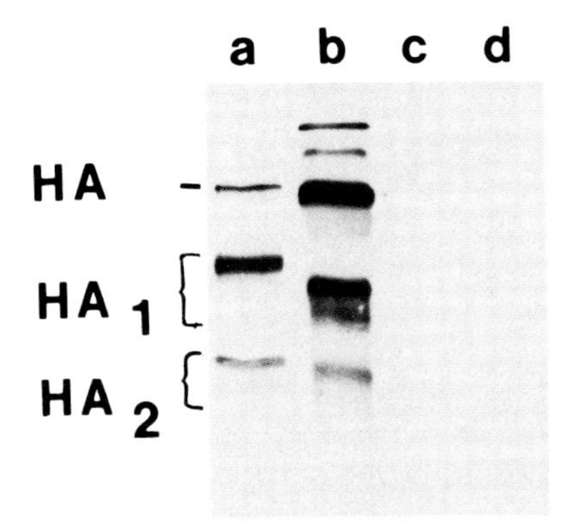

Fig. 7. Analysis of hemagglutinin by immunoblotting. MDCK cells were infected with fowl plague virus (a) for 5 hr. *Spodoptera frugiperda* cells were infected with recombinant AcNPV (b) or authentic AcNPV (c) or were mock-infected (d) for 3 days. The polypeptides were separated by SDS-polyacrylamide gel electrophoresis and then blotted to a nitrocellulose membrane. The blot was analyzed with antihemagglutinin antibodies as described (Kuroda *et al.*, 1986). Samples derived from 10^5 cells were loaded onto the gels.

When the recombinant virus-produced hemagglutinin was analyzed by the immunoblotting technique, it was apparent that the protein was cleaved in *S. frugiperda* cells, though to a lesser extent than in MDCK cells (Fig. 7, lanes a and b). It is also evident that the precursor HA and the cleavage fragments HA_1 and HA_2 are more heterogeneous in size and have a lower molecular weight when derived from *S. frugiperda* cells. These differences are most likely due to smaller and incomplete oligosaccharides. It is clear, however, that the hemagglutinin is glycosylated in *S. frugiperda* cells, because unglycosylated hemagglutinin would form sharp bands considerably smaller than HA, HA_1, and HA_2 shown in lane b, Fig. 7. The lysates analyzed in lanes a and b of Fig. 7 were obtained from the same number (10^5) of MDCK or *S. frugiperda* cells, respectively. The relative intensities of the bands indicate, therefore, that the amount of hemagglutinin produced in insect cells is comparable to that obtained in fowl plague virus-infected vertebrate cells.

D. Immunization of Chickens with Recombinant Hemagglutinin

To test the immune response to recombinant hemagglutinin, chickens were immunized as described previously. The immunized animals produced hemagglutination-inhibiting antibodies at titers of 1:64 to 1:128. The sera neutralized infectivity of fowl plague virus with indices of >500. The immunized chickens survived a challenge infection with 10^4 plaque-forming units of fowl plague virus 3 weeks after immunization without showing signs of fowl plague, whereas unprotected control animals died 2 days after infection, as expected.

III. DISCUSSION

We have analyzed recombinants between AcNPV DNA and the influenza virus hemagglutinin gene which express hemagglutinin in insect cell cultures. The amount of hemagglutinin produced in these cells is comparable to that obtained in vertebrate cells infected with fowl plague virus, but it does not reach the levels of polyhedrin synthesis in *S. frugiperda* cells infected with AcNPV. Several eukaryotic genes have already been successfully expressed in insect cells with the aid of this expression vector system, e.g., human interferon β (Smith *et al.*, 1983b), human c-myc protein (Miyamoto *et al.*, 1985), and human interleukin 2 (Smith *et al.*, 1985). A similar vector has been used to produce human interferon α in silkworms (Maeda *et al.*, 1985). In all instances, the insect system-derived proteins have proved biologically active. Similarly, the influenza virus hemagglutinin synthesized under the conditions described here is able to induce hemadsorption, cause hemolysis, and react with hemagglutinin-specific antibodies. Moreover, the recombinant virus can protect animals against challenge by fowl plague virus. Thus, in its biological activities, the recombinant virus-produced hemagglutinin strikingly resembles authentic fowl plague virus hemagglutinin.

The fowl plague virus hemagglutinin, to our knowledge, represents the first membrane protein among the recombinant AcNPV-encoded vertebrate proteins. Our data show that the recombinant hemagglutinin is membrane-bound in *S. frugiperda* cells and that it is transported to the cell surface in a fashion similar to that in vertebrate cells. This observation suggests that the amino-terminal signal peptide of hemagglutinin has been removed, although we have not shown this directly by biochemical techniques. However, at least partial cleavage of a signal peptide could be demonstrated when interleukin 2 was produced in the same expression vector system (Smith *et al.*, 1985).

Moreover, influenza virus hemagglutinin is the first vertebrate polypeptide that has been shown to undergo posttranslational proteolytic cleavage in *S. frugiperda* cells. The observation that the recombinant hemagglutinin has hemolytic activity indicates that the endoprotease involved in cleavage attacks the correct peptide bond between the carboxy-terminal arginine of the sequence Lys-Lys-Arg-Lys-Lys-Arg and the adjacent glycine (Garten *et al.*, 1982), since fusion capacity can be expressed only if cleavage occurs exactly in this position (Garten *et al.*, 1981). It is of general interest that the fowl plague virus hemagglutinin is activated in *S. frugiperda* cells, because numerous other vertebrate glycoproteins, notably that of human immunodeficiency virus I (HIVI) (Muesing *et al.*, 1985), and a large number of prohormones and proenzymes (Steiner *et al.*, 1980) have similar cleavage sites and should therefore be susceptible to proteolytic activation if expressed in insect cells.

Although the presence of carbohydrates has not been directly demonstrated here, e.g., by methods such as metabolic radiolabeling or glycosidase treatment, the gel electrophoretic analyses indicate that the hemagglutinin is glycosylated. The higher and more heterogeneous electrophoretic mobilities of HA as well as of HA_1 and HA_2 derived from insect cells indicate that the oligosaccharide side chains are not identical to those attached in vertebrate cells. It is known that the biological functions of the hemagglutinin tolerate a relatively high degree of variability in oligosaccharide content (Romero *et al.*, 1983). On the other hand, the presence of glycosyl groups is required to protect the fowl plague virus hemagglutinin from nonspecific proteolytic degradation (Klenk *et al.*, 1974). The glycosylating capacity of *S. frugiperda* cells appears to be sufficient to meet this requirement.

Insect cells do not contain neuraminic acid; i.e., they lack the receptor for influenza viruses. Vector-mediated expression is therefore the only way to produce the hemagglutinin and other myxovirus proteins in these cells. On the other hand, the absence of an influenza virus receptor on insect cells may facilitate release of the hemagglutinin. From a biotechnological point of view this may be a considerable advantage over expression systems in vertebrate cells, which contain neuraminic acid and thus would tend to retain the hemagglutinin.

We have shown that in recombinant AcNPV the hemagglutinin gene has been inserted into the *Eco*RI-H fragment that encodes the polyhedrin gene (Fig. 3). As proposed earlier (Smith *et al.*, 1983b), recombination between the AcNPV genome and the pAc373 construct is mediated by homologous recombination within the polyhedrin gene. We have not yet determined the exact site of recombination in the recombinant analyzed here. It will be interesting to investigate whether homologous recombination is the only means of inserting foreign genes into AcNPV DNA, or whether other sites in the AcNPV genome can also be utilized, e.g., by heterologous recombination. The homologous recombination in this system appears to be efficient. Hence this system may offer possibilities for the detailed study of mechanisms of homologous recombination.

ACKNOWLEDGMENTS

We are indebted to Gale E. Smith and Max D. Summers, Texas A & M University, for providing the pAc373 construct. We thank Gerti Meyer zu Altenschildesche for media preparation and Elfriede Otto and Werner Berk for technical assistance. Petra Böhm and Anneliese Muth rendered invaluable editorial help. This work was supported by grants PTB 03 8498 to W. D. and PBE 8723 to H.-D. K. from the Bundesministerium für Forschungsgemeinschaft through SFB47-B3 (H.-D. K). K. K. was a recipient of a research fellowship of the Alexander von Humboldt Foundation. The figures reproduced in this chapter were taken from Kuroda *et al.*, 1986, with permission.

REFERENCES

Carstens, E. B., Tjia, S. T., and Doerfler, W. (1979). Infection of *Spodoptera frugiperda* cells with *Autographa californica* nuclear polyhedrosis virus. I. Synthesis of intracellular proteins after virus infection. *Virology* **99**, 386–398.

Garten, W., and Klenk, H.-D. (1983). Characterization of the carboxypeptidase involved in the proteolytic cleavage of the influenza haemagglutinin. *J. Gen. Virol.* **64**, 2127–2137.

Garten, W., Bosch, F. X., Linder, D., Rott, R., and Klenk, H.-D. (1981). Proteolytic activation of the influenza virus hemagglutinin: The structure of the cleavage site and the enzymes involved in cleavage. *Virology* **115**, 361–374.

Garten, W., Linder, D., Rott, R., and Klenk, H.-D. (1982). The cleavage site of the hemagglutinin of fowl plague virus. *Virology* **122**, 186–190.

Keil, W., Geyer, R., Dabrowski, J., Dabrowski, U., Niemann, H., Stirm, S., and Klenk, H.-D. (1985). Carbohydrates of influenza virus. Structural elucidation of the individual glycans of the FPV hemagglutinin by two-dimensional ^{1}H n.m.r. and methylation analysis. *EMBO J.* **4**, 2711–2720.

Klenk, H.-D., and Rott, R. (1980). Cotranslational and posttranslational processing of viral glycoproteins. *Curr. Top. Microbiol. Immunol.* **90**, 19–48.

Klenk, H.-D., Wöllert, W., Rott, R., and Scholtissek, C. (1974). Association of influenza virus proteins with cytoplasmic fractions. *Virology* **57**, 28–41.

Klenk, H.-D., Rott, R., Orlich, M., and Blödorn, J. (1975). Activation of influenza A viruses by trypsin treatment. *Virology* **68**, 426–439.

Klenk, H.-D., Garten, W., and Rott, R. (1984). Structure and function of a viral membrane protein. *Z. Anal. Chem.* **317**, 614–615.

Kuroda, K., Hauser, C., Rott, R., Klenk, H.-D., and Doerfler, W. (1986). Expression of the influenza virus haemagglutinin in insect cells by a baculovirus vector. *EMBO J.* **5**, 1359–1365.

Lazarowitz, S. G., and Choppin, P. W. (1975). Enhancement of the infectivity of influenza A and B viruses by proteolytic cleavage of the hemagglutinin polypeptide. *Virology* **68**, 440–455.

Lübbert, H., and Doerfler, W. (1984). Mapping of early and late transcripts encoded by the *Autographa californica* nuclear polyhedrosis virus genome: Is viral RNA spliced? *J. Virol.* **50**, 497–506.

Maeda, S., Kawai, T., Obinata, M., Fujiwara, H., Horiuchi, T., Saeki, Y., Sato, Y., and Furusawa, M. (1985). Production of human α-interferon in silkworm using a baculovirus vector. *Nature (London)* **315**, 592–594.

Miyamoto, C., Smith, G. E., Farrell-Towt, J., Chizzonite, R., Summers, M. D., and Ju, G. (1985). Production of human c-myc protein in insect cells infected with a baculovirus expression vector. *Mol. Cell. Biol.* **5**, 2860–2865.

Muesing, M. A., Smith, D. H., Cabradilla, C. D., Benton, C. V., Lasky, L. A., and Capon, D. J. (1985). Nucleic acid structure and expression of the human AIDS/lymphadenopathy retrovirus. *Nature (London)* **313**, 450–458.

Palese, P., and Young, J. F. (1983). Molecular epidemiology of influenza virus. *In* "Genetics of Influenza Viruses" (P. Palese and D. W. Kingsbury, eds.), pp. 321–336. Springer-Verlag, Berlin and New York.

Pennock, G. D., Shoemaker, C., and Miller, L. K. (1984). Strong and regulated expression of *Escherichia coli* β-galactosidase in insect cells with a baculovirus vector. *Mol. Cell. Biol.* **4**, 399–406.

Riedel, H., Kondor-Koch, C., and Garoff, H. (1984). Cell surface expression of fusogenic vesicular stomatitis virus G protein from cloned cDNA. *EMBO J.* **3**, 1477–1483.

Rigby, P. W. J., Dieckmann, M., Rhodes, C., and Berg, P. (1977). Labeling deoxyribonucleic acid to high specific activity in vitro by nick translation with DNA polymerase I. *J. Mol. Biol.* **113**, 237–251.

Romero, P. A., Datema, R., and Schwarz, R. T. (1983). *N*-methyl-1-deoxynojirimycin, a novel inhibitor of glycoprotein processing and its effect on fowl plague virus maturation. *Virology* **130**, 238–242.

Sanger, F., Nicklen, S., and Coulsen, A. R. (1977). DNA sequencing with chain-terminating inhibitors. *Proc. Natl. Acad. Sci. U.S.A.* **74**, 5463–5467.

Smith, G. E., Fraser, M. J., and Summers, M. D. (1983a). Molecular engineering of the *Autographa californica* nuclear polyhedrosis virus genome: Deletion mutations within the polyhedrin. *J. Virol.* **46**, 584–593.

Smith, G. E., Summers, M. D., and Fraser, M. J. (1983b). Production of human beta interferon in insect cells infected with a baculovirus expression vector. *Mol. Cell. Biol.* **3**, 2156–2165.

Smith, G. E., Ju, G., Ericson, B. L., Moschera, J., Lahm, H.-W., Chizzonite, R., and Summers, M. D. (1985). Modification and secretion of human interleukin 2 produced in insect cells by a baculovirus expression vector. *Proc. Natl. Acad. Sci. U.S.A.* **82**, 8404–8408.

Southern, E. M. (1975). Detection of specific sequences among DNA fragments separated by gel electrophoresis. *J. Mol. Biol.* **98**, 503–517.

Steiner, D. F., Quinn, P. S., Chan, S. J., Marsuk, J., and Tager, H. S. (1980). Processing mechanism in the biosynthesis of proteins. *Ann. N.Y. Acad. Sci.* **343**, 1–39.

Tjia, S. T., Carstens, E. B., and Doerfler, W. (1979). Infection of *Spodoptera frugiperda* cells with *Autographa californica* nuclear polyhedrosis virus. II. The viral DNA and the kinetics of its replication. *Virology* **99**, 399–409.

Wahl, G. M., Stern, M., and Stark, G. R. (1979). Efficient transfer of large DNA fragments from agarose gels to diazobenzyloxymethyl-paper and rapid hybridization by using dextran sulfate. *Proc. Natl. Acad. Sci. U.S.A.* **76**, 3683–3687.

Ward, C. W. (1981). Structure of the influenza virus hemagglutinin. *Curr. Top. Microbiol. Immunol.* **94**, 1–74.

Wilson, I. A., Skehel, J. J., and Wiley, D. C. (1981). Structure of the haemagglutinin membrane glycoprotein of influenza virus at 4 Å resolution. *Nature (London)* **289**, 366–373.

16

Transfection of Drosophila melanogaster Transposable Elements into the Drosophila hydei Cell Line

TADASHI MIYAKE, NAOMI MAE,
TADAYOSHI SHIBA, SHUNZO KONDO,
MANABU TAKAHISA, AND RYU UEDA

I. INTRODUCTION

To introduce a gene into cultured cells is one of the most efficient ways to elucidate its function and structure at the molecular level. Success in such attempts in *Drosophila* was first reported in 1983 by Bourouis and Jarry (dihydrofolate reductase gene, *dhfr*), Di Nocera and Dawid (chloramphenicol acetyltransferase gene, *cat*), and Sinclair *et al.* (xanthine guanine phosphoribosyltransferase gene, *gpt*). Since there were no selectable markers of *Drosophila* origin, they utilized the prokaryotic genes combined to either the *Drosophila copia* promoter or the heat shock protein (HSP) promoter. Their method was based on "calcium phosphate transfection," and the expression of introduced genes

BIOTECHNOLOGY IN INVERTEBRATE
PATHOLOGY AND CELL CULTURE

was proved not only at the phenotypic level but also at the mRNA and enzyme levels. The introduced genes were present in the transfected cells as extrachromosomal units in the case of the *gpt* gene or as integrated large oligomers in the case of the *dhfr* gene.

On the other hand, we have been analyzing for several years the viruslike particle (VLP) in cultured cells of *Drosophila melanogaster* and found that the VLP is a retroviruslike particle which contains RNA hybridizable to a *Drosophila* transposable element, *copia* (Saigo *et al.*, 1980; Shiba *et al.*, 1980; Shiba and Saigo, 1983). This finding is of importance because it suggests the relationship between the retrovirus and the transposable element and also the mechanism of transposition and amplification. To clarify the mechanism in more detail, however, it is helpful to develop a system to introduce a transposable element into cultured cells which do not contain the transposable element and related sequences.

Based on such a consideration, we will show the procedure for introducing *Drosophila* transposable elements *copia* or P into a cell line of *Drosophila hydei* and present the results of analysis of the transfectants with regard to the state, fate, and function of the introduced transposable elements.

II. PLASMIDS

The plasmids used in this work were pSVC^{neo}-1, cDm2055, and pπ25.1. Plasmid pSVC^{neo}-1 was constructed by T. Todo from pSV2-*neo* (Southern and Berg, 1982) and a *Hin*dIII fragment containing one of the *copia* long terminal repeat (LTR) sequences recovered from plasmid cDm2055 (T. Todo, personal communication). Due to the expression of the prokaryotic neomycin-resistant gene (*neo*) and the promoter in the *copia* LTR, this plasmid successfully endows the *Drosophila* cell with resistance to the antibiotic G-418 (Geneticin, GIBCO). Therefore it was used as the selectable marker for cotransfection, which will be mentioned later. The entire length of the plasmid is 6.5 kb, and it is cleaved by *Eco*RI digestion into two fragments, of 2.8 and 3.7 kb (Fig. 1). The *neo* gene resides in the 3.7 kb fragment.

In the *copia*-carrying plasmid cDm2055, which was constructed by G. M. Rubin, the *Drosophila* genomic fragment containing the DNA sequence hybridizable to cDNA of *copia* RNA was inserted into the plasmid pBR322 (Rubin *et al.*, 1981). The entire length is 13.6 kb and it is cleaved into five *Eco*RI fragments, of 4.6, 2.8, 2.4, 0.5, and 3.2 kb. The 5 kb *copia* sequence extends over the latter three fragments (Saigo *et al.*, 1981, and Fig. 1).

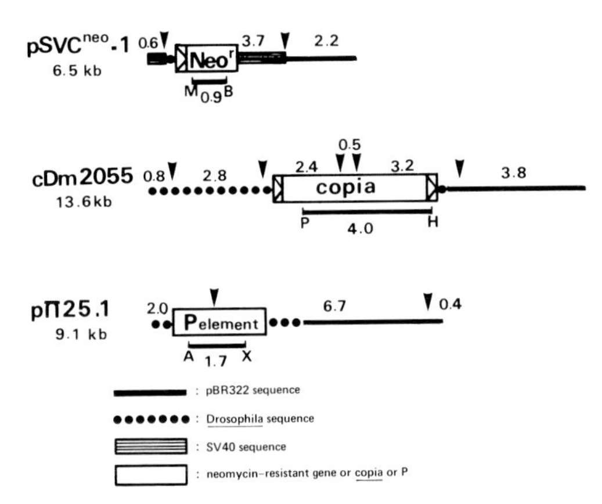

Fig. 1. Structure of the plasmids carrying neomycin resistance gene (pSVC^{neo}-1), *copia* (cDm2055), or P (pπ25.1). Restriction sites: ▼, *Eco*RI site; M, *Sma*I site; B, *Bgl*II site; P, *Pvu*II site; H, *Hinc*II site; A, *Sal*I site; X, *Xho*I site. The upper numerals indicate the length of *Eco*RI fragments and the lower lines indicate the probe fragments for Southern blot analysis.

Plasmid pπ25.1 was constructed by O'Hare and Rubin (1983) from the genomic DNA of the P element-carrying fly strain π_2 and the plasmid pBR322. Its entire length is 9.1 kb and it produces two *Eco*RI fragments, of 6.7 and 2.4 kb. The complete P sequence (2.9 kb) extends over both fragments (Fig. 1).

III. RECIPIENT AND MEDIUM

To transfect the transposable elements *copia* and P, the recipient cells first must be free of these elements. Cell lines from *D. melanogaster* could not be used, because all *D. melanogaster* strains so far tested carried 30–50 *copia* sequences in the genome. In addition, P and Q strains of *D. melanogaster* contain 30–50 copies of P-hybridizable DNA sequences. Although the fly species are not closely related to *D. melanogaster* phylogenetically, the cell line from *Drosophila virilis* [WR75-Dv-1, established and kindly supplied by Schneider and Blumenthal (1978)] and that from *Drosophilia hydei* [KUN-DH-33, established by Sondermeijer *et al.* (1980) and kindly supplied by J. H. Sang] are suited for this purpose. They do not contain any DNA sequence hybridizable to *copia* or P elements. Second, the recipient cells must have a characteristic usable for selection of the transfectants *in vitro*. In *Drosophila* cultured cell systems, not many such characteristics are known. We tested: (1) a HAT method (Sinclair *et al.*, 1983; J. H. Sang, personal communication) developed for the selection of transfectants that had acquired the gene encoding xanthine guanine phosphoribosyl-transferase, (2) a methotrexate selection system for dihydrofolate reductase-positive transfectants (Bourouis and Jarry, 1983) and (3) a G-418 (antibiotic) selection system for neomycin-resistant transfectants (Rio and Rubin, 1985). Among them, the G-418 selection system was the best in terms of strictness, rapidity, and ease of selection. However, of the two cell lines, WR75-Dv-1 and KUN-DH-33, which are free of both *copia* and P, the former was relatively resistant to the drug and the frequency of pseudoresistant colonies in the G-418-containing medium was fairly high. Therefore, KUN-DH-33 was used throughout the work.

Medium M3(BF), bicarbonate-free M3 medium supplemented with heat-inactivated fetal bovine serum (10%), was used (Cross and Sang, 1978). This medium contained the antibiotics streptomycin (streptomycin sulfate, Meiji, 100 μg/ml) and penicillin (penicillin G potassium salt, Meiji, 30 μg/ml). The cells were cultured at 25°C in a humid atmosphere of 5% CO_2–95% air.

IV. TRANSFECTION AND COTRANSFECTION

To introduce foreign DNA into cultured cells, we tested three procedures: the polybrene procedure (Kawai and Nishizawa, 1984), the DEAE-dextran procedure (Lopata *et al.*, 1984), and the calcium phosphate procedure (Wigler *et al.*, 1977). The polybrene procedure was reported to be efficient and reproducible for introducing exogenous DNA (chloramphenicol acetyltransferase gene) into cultured cells of the mosquito species *Aedes albopictus* (Durbin and Fallon, 1985). However, compared with the calcium phosphate procedure, the frequency of transfection was much lower in our *Drosophila* transposable element-DH-33 system even in combination with dimethyl sulfoxide (DMSO) shock. Use of the DEAE-dextran procedure to introduce viruses or their RNA into *Drosophila* cultured cells was reported by Gallagher *et al.* (1983). They succeeded in infecting Schneider line 1 cells sensitized by DEAE-dextran treatment with black beetle viruses or their RNA. KUN-DH-33 cells were sensitized by similar procedures, mixed with pSVC^{neo}-1 plasmid DNA, and, after a 48-hr expression time, transferred into the G-418 medium. The G-418-resistant transfectants, however, were not observed even after 4–6 weeks of culture.

The calcium phosphate procedure is most commonly used in both mammal and insect cultured cell systems. KUN-DH-33 cells were diluted to 0.4×10^6 cells/ml with medium M3 (BF) supplemented with heat-inactivated fetal bovine serum (10%), and each 5 ml was seeded in a culture dish with a diameter of 60 mm (Nunc). After 24 hr at 25°C in a humid atmosphere (5% CO_2–95% air), 500 μl of DNA–calcium phosphate coprecipitate was added to each dish. For DNA–calcium phosphate coprecipitation, 10–30 μg DNA in 125 μl of TE buffer (1 m*M* Tris-HCl, 0.1 mM EDTA, pH 8.0) was first mixed with 125 μl of 500 m*M* $CaCl_2$ in TE buffer. Then the entire amount (250 μl) was gently mixed with 250 μl of 2× HBS (280 m*M* NaCl, 50 m*M* HEPES, 1.5 m*M* Na_2HPO_4, pH 7.12) and incubated at 25°C for 60 min. The sample (500 μl) containing DNA coprecipitated with calcium phosphate was added to each dish. After 16–20 hr incubation at 25°C for DNA uptake, the cells were washed and refed with 5 ml of the medium. They were incubated in the medium for another 48 hr to express the newly introduced gene. The cells were harvested and seeded in 2 ml of the medium containing G-418 (1.5 mg/ml) in a 30-mm dish (Nunc) at densities of 1, 0.5, and 0.25×10^6 per dish. After 3–4 weeks of incubation at 25°C, the G-418-resistant transfectants formed visible colonies.

Since the transposable elements *copia* and P are not the selectable markers, we adopted the cotransfection system to introduce these elements into KUN-DH-33 cells utilizing the neomycin-resistant gene as the selection marker. Usually, 10 μg of $pSVC^{neo}$-1 (selectable gene) and 20 μg of *copia*- or P-containing plasmids (cDm2055 or pπ25.1) per dish were used for cotransfection.

An example of the results of the cotransfection experiment is shown in Table I. In experiment 1, where the recipient KUN-DH-33 cells were transfected with 10 μg of $pSVC^{neo}$-1 DNA, 39 G-418-resistant colonies appeared on the dish seeded with 1×10^6 cells. When DNAs of $pSVC^{neo}$-1 (10 μg) and cDm2055 (20 μg) in experiment 2 or $pSVC^{neo}$-1 and pπ25.1 (20 μg) in experiment 3 were transfected to the recipient, 28 or 75 colonies appeared, respectively. On the dish seeded with 0.5×10^6 cells, the number of transfectants decreased to about 1/10, suggesting a dependence on cell density for efficiency of transfectant colony formation. In control 1, TE buffer used for dissolving and diluting DNA, and in control 2, 10 μg of KUN-DH-33 DNA (self-DNA) was transfected to the recipient. In both cases, no G-418-resistant colonies appeared.

The concentration of G-418 (Geneticin, 1.5 mg/ml) is a rather high dose of antibiotic, but it seems to be in the minimum concentration range for effective selection. If the concentration was decreased to 1 mg/ml some pseudoresistant colonies appeared which could not proliferate further after being transferred in G-418 medium. If the G-418 medium was kept under improper conditions before use, similar pseudoresistant colonies often appeared.

TABLE I

Transfection and Cotransfection of Neomycin-Resistant Gene and Transposable Elements *copia* and P into *Drosophila hydei* Cultured Cells

DNA	Weight (μg)	Recipient cells (KUN-DH-33) per dish	
		1×10^6	0.5×10^6
Experiment			
1. pSVC^{neo}-1	10	39[a]	2
2. pSVC^{neo}-1 cDm2055	10 20	28	4
3. pSVC^{neo}-1 pπ25.1	10 20	75	6
Control			
1. TE buffer		0	0
2. Self-DNA	10	0	0

[a]Average of two dishes.

The transfectants from experiment 1, 2 and 3 have been designated N-1 . . . , NC-1 . . . , and NP-1 . . . , respectively.

V. ANALYSIS OF TRANSFECTANTS

A. Southern Analysis

The transfectants were propagated in the G-418 medium. Their DNAs were extracted, digested with *Eco*RI, electrophoresed, transferred to the filter, and hybridized with appropriate fragments of plasmids ^{32}P-labeled by nick translation. In Fig. 2a, the probe was a 0.9 kb *Bgl*II–*Sma*I fragment of pSVC^{neo}-1, the internal fragment of *neo* gene (see Fig. 1). Since the plasmid pSVC^{neo}-1 was used as the common selection marker, all transfectant DNAs showed hybridization signals to this probe. As expected, this probe did not hybridize to DNA of the recipient, KUN-DH-33. Figure 2 shows a strong signal at the 3.7 kb band, which is expected from *Eco*RI digestion of the plasmid itself; in addition, many signals are seen at bands with various lenghts. The former suggests intactness of the plasmids in the transfectant, although it has not yet been conclusively determined whether the plasmids are integrated as tandem oligomers in the host chromosomes or exist as extrachromo-

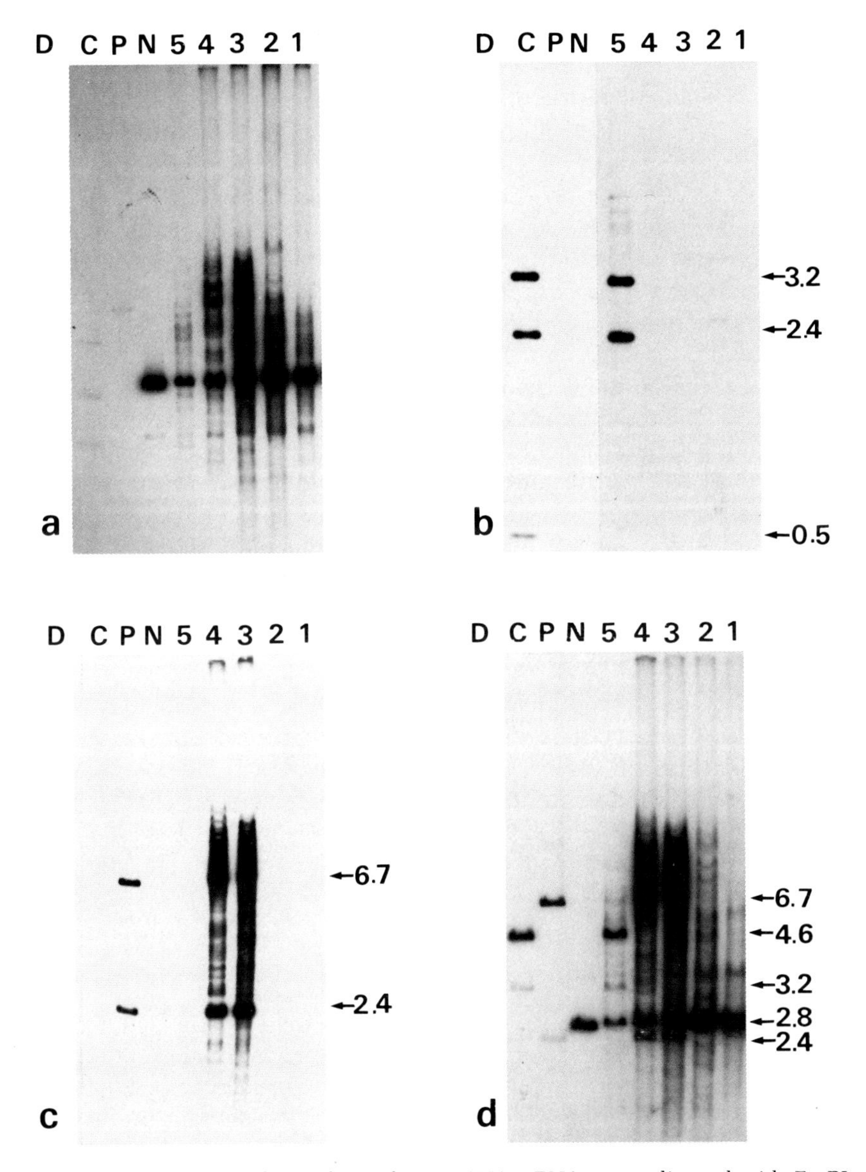

Fig. 2. Southern analysis of transfection DNAs. DNAs were digested with *Eco*RI, electrophoresed, transferred to the filter, and hybridized with ^{32}P-labeled probes. DNAs: lane 1, N-1; lane 2, N-2; lane 3, NP-1; lane 4, NP-2; lane 5, NC-1; lane N, pSVC^{neo}-1; lane P, pπ25.1; lane C, cDm2055; lane D, KUN-DH-33. Probes: (a) 0.9 kb *Sma*I–*Bgl*II fragment of pSVC^{neo}-1; (b) 4.0 kb *Pvu*II–*Hin*cII fragment of cDm2055; (c) 1.7 kb *Sal*I–*Xho*I fragment of pπ25.1; (d) pBR322.

somal entities. On the other hand, the latter indicates that some plasmids were fragmented and integrated in the host chromosomes at various sites. When the *copia* internal fragment (4kb *Pvu*II–*Hinc*II fragment) was used as the probe (Fig. 2b), as expected, hybridization occurred only with the DNA of transfectant NC-1, which received the plasmids pSVC^{neo}-1 and cDm2055 (experiment 2 in Table I). In later experiments (data not shown) this was confirmed, and so far all of the five randomly chosen NC transfectants contained the cotransfected *copia* sequence. These results indicate that, under the conditions we used, the frequency of cotransfection is high. The banding pattern again suggested that a large part of the plasmid cDm2055 retained its structure intact, while part fragmented and integrated in the host chromosome. The P element hybridizable sequence was probed with the 1.7 kb *Sal*I–*Xho*I internal fragment (Fig. 2c). The sequence was detected only in DNAs of NP transfectants. A similar banding pattern, with strong signals at the 6.7 and 2.4 kb bands expected from the intact plasmids and many weaker signals at bands with various lengths, was observed. Since the three plasmids used in this work contained the pBR322 sequence, the transfectant DNAs were also probed with pBR322 whole DNA. As shown in Fig. 2d, lanes N, C, and P, strong hybridization signals were observed at 2.8 kb (pSVC^{neo}-1), 4.6 kb (cDm2055), and 6.7 kb (pπ25.1). In addition, weaker signals were seen at 3.2 kb (cDm2055) and 2.4 kb (pπ25.1). It is clear from Fig. 1 that the weak signals were due to the short pBR322 region in these fragments. The results in Fig. 2d can be explained similarly to those in Figs. 2a–2c. This indicates that the prokaryotic pBR322 sequence behaves in the same manner as the *copia* or P sequence of *Drosophila* origin, even under conditions without any selection advantage.

B. Stability of Introduced Transposable Elements

To determine the stability of these foreign DNAs, transfectants NP-1 and NP-2 were cultured for various periods in medium containing or not containing G-418, and their DNAs were digested with *Eco*RI and analyzed using Southern blots (Fig. 3). In NP-1 the banding patterns were virtually the same after 2, 3, and 6 months of culture in the G-418 medium. Furthermore, a similar pattern was observed even after 1.5 month of culture in the absence of G-418, which followed a 4.5-month culture in the G-418 medium. In NP-2 the patterns were again the same after 2.5, 3, and 6 months of culture in the G-418 medium. However, after culture for 3 months in the G-418 medium and then 3 months in the absence of G-418, only weak signals were observed. Not only the

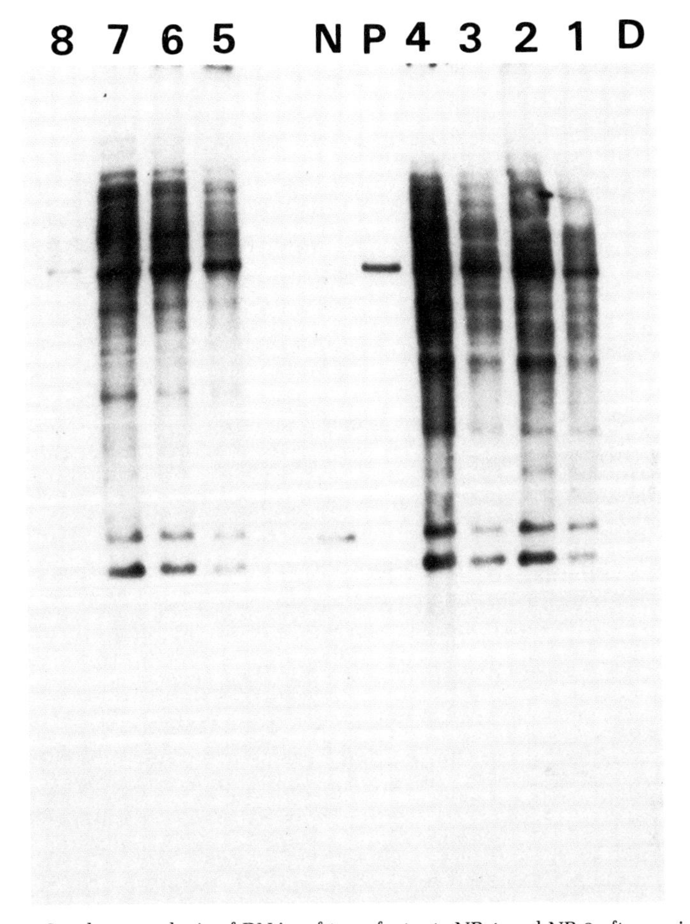

Fig. 3. Southern analysis of DNAs of transfectants NP-1 and NP-2 after various periods in culture. DNAs were digested with *Eco*RI, electrophoresed, transferred to the filter, and hybridized with ^{32}P-labeled pπ25.1. Lanes 1–3, NP-1 DNA after 2, 3, and 6 months of culture in G-418 medium, respectively; lane 4, NP-1 DNA after 6 months of culture (4.5 months in G-418 medium and 1.5 months in G-418-free medium); lanes 5–7, NP-2 DNA after 2.5, 3, and 6 months of culture in G-418 medium, respectively; lane 8, NP-2 DNA after 6 months of culture (3 months in G-418 medium and 3 months in G-418-free medium); lane D, KUN-DH-33 DNA; lane P, pπ25.1; lane N, pSVC^{neo}-1.

signals expected from the neomycin plasmid (pSVC^{neo}-1) but also those from the P-carrying plasmid (pπ25.1), which should have no causal relation to the G-418 selection, were weak. This might suggest that (1) pSVC^{neo}-1 and pπ25.1 behaved as single unit in the transfectant cells or (2) the NP-2 transfectant colony was originally a mixture of G-418-resistant, P-carrying cells and recipient cells which escaped from the cytocidal effect of G-418 by some unknown mechanism. Although analysis at the single-cell level is needed to elucidate this phenomenon, the latter possibility was supported by the fact that most of the NP-2 population cultured in the absence of G-418 was no longer resistant to the drug. Transfectants N-1 and NC-1 were similarly analyzed. In both cases, 6-month culture in the G-418 medium did not affect the banding patterns. NC-1 cultured in the G-418 medium for 3 months, then for an additional 3 months in the G-418-free medium, showed the same pattern. Thus, the foreign DNAs introduced into KUN-DH-33 cells were, in most cases, stably maintained for at least 6 months irrespective of their *Drosophila* or prokaryotic origin.

C. Expression of Introduced Transposable Elements

One of the aims of introducing transposable elements in cultured cells is to elucidate the mechanisms of transposition and amplification and their effects on host functions. For this, the expression of these transposable elements must be studied first. For the expression of the *copia* element, we found from electrophoresis and electron microscopy that the viruslike particles were newly produced in the transfectant, NC-1 (Fig. 4). The particles were detected only in the transfectants in which the *copia*-carrying plasmid cDm2055 was introduced. No such particles were found in any N- or NP- transfectants or KUN-DH-33 recipient cells, although the recipient carried the morphologically distinguishable viruses (Table II) (Sondermeijer *et al.*, 1980). The particles are physically, biochemically, and biologically very similar to the *copia* retroviruslike particles (RVLPs) found in cultured cells of *D. melanogaster* (Saigo *et al.*, 1980; Shiba *et al.*, 1980; Shiba and Saigo, 1983). The particles are polyhedral and 50 nm in diameter. They contain RNAs of low molecular weight (4, 4.5, 5, and 6 S) and high-molecular-weight RNA which is hybridizable to *copia* DNA. Another important property of *D. melanogaster* RVLP, the reverse transcriptase activity, was signficantly low. Since the *copia* sequence in cDm2055 was cloned by Rubin *et al.* (1981) from *D. melanogaster* chromosome using cDNA of *copia* RNA, this finding proves directly that the *copia* viruslike particles in *D. melanogaster* cultured cells were indeed produced by the transposable element *copia* resident in *D. melanogaster* chromosome. The fact that the reverse transcriptase activity was very low could be explained by postulating that the *copia* sequence in plasmid cDm2055 was not complete but that it carried enough of the DNA sequence to produce the viruslike particles but was inadequate for reverse transcriptase produc-

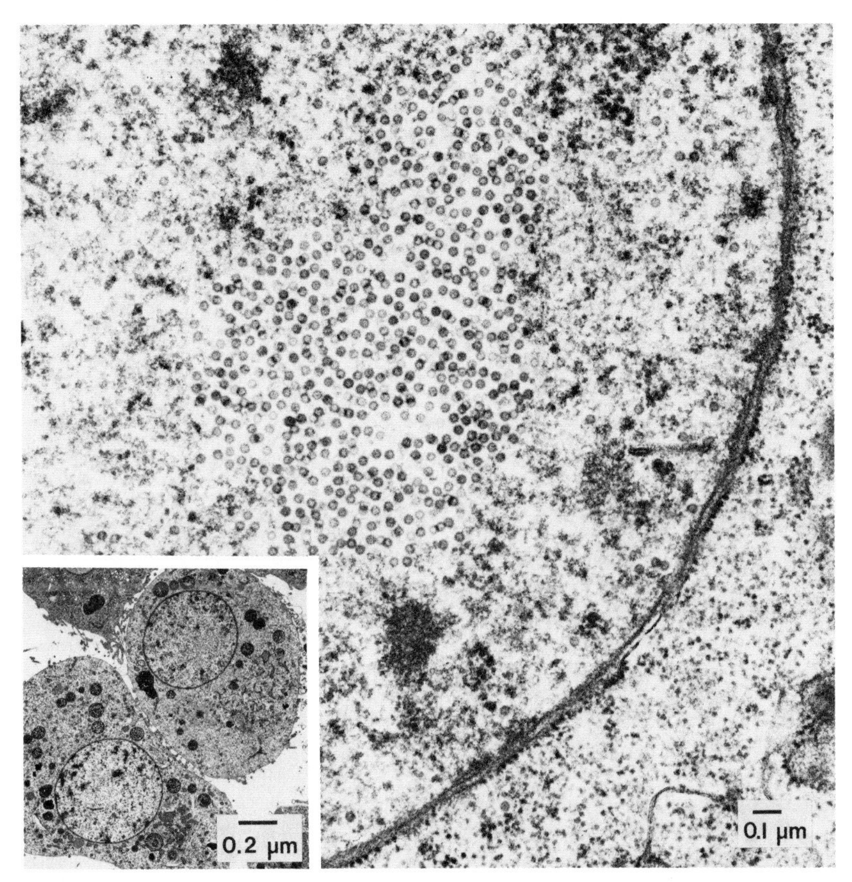

Fig. 4. Electron micrograph of transfectant NC-1 cell. A 200 × g pellet of old NC-1 cells (day 4–5 of stationary phase) was fixed with 2% glutaraldehyde in 0.1 *M* sodium cacodylate buffer, pH 7.4 (4°C, 20 min) and then with 1% OsO_4 in the same buffer (4°C, 1 hr). It was dehydrated with graded ethanol and embedded in Epok 812. Specimens were stained with uranyl acetate and lead compounds and observed with a JEM 100B at an acceleration voltage of 80 kV.

tion and/or activity. It seems very probable that, among 30–50 *copia* sequences in *D. melanogaster* chromosomes, various types of defective *copia* are included, and Rubin picked up the type mentioned above. If the transposable element *copia* was of retroviral origin, the original infectious virus might become defective through the parasitic life cycle in *D. melanogaster*.

Attempts to detect the transcript and protein synthesized by the P transposable element are now in progress.

TABLE II
Viruslike Structures in KUN-DH-33 and Transfectants

Cells	Number examined	*copia* RVLP type		Other type
		Nucleus	Cytoplasm	Cytoplasm
KUN-DH-33	1000	0	0	190
N-1	1000	0	0	108
NP-2	500	0	0	22
NC-1	500	440	2	58

VI. SUMMARY

Transposable elements, *copia* and P, isolated from *D. melanogaster* genome could be introduced into the *D. hydei* cell line KUN-DH-33 by cotransfection with the prokaryotic neomycin-resistant gene. At least part of the transposable element sequence is integrated in the *D. hydei* genome. Most of the transposable element sequence is stably maintained in the transfectants. The *copia* sequence is expressed in the *D. hydei* transfectant cells and produces viruslike particles (VLPs). These VLPs are in many ways similar to the VLPs that spontaneously appear in cultured cells of *D. melanogaster*.

ACKNOWLEDGMENTS

We are deeply grateful to Drs. G. M. Rubin, I. Schneider, P. J. A. Sondermeijer and J. H. Sang for supplying the plasmids and cell lines. We are particularly indebted to Dr. T. Todo of Osaka University for his generosity in allowing us the use of his plasmid pSVC^{neo}-1 before publication and to Drs. J. H. Sang and A. M. Fallon for their advice on testing the HAT selection and polybrene procedures, respectively. We also thank Miss Mami Kawamoto for typing the manuscript.

REFERENCES

Bourouis, M., and Jarry, B. (1983). Vectors containing a prokaryotic dihydrofolate reductase gene transform *Drosophila* cells to methotrexate-resistance. *EMBO J.* **2**, 1099–1104.

Cross, D. P., and Sang, J. H. (1978). Cell culture of individual *Drosophila* embryos. I. Development of wild-type cultures. *J. Embroyl. Exp. Morphol.* **45**, 161–172.

Di Nocera, P. P., and Dawid, I. B. (1983). Transient expression of genes introduced into cultured cells of *Drosophila*. *Proc. Natl. Acad. Sci. U.S.A.* **80**, 7095–7098.

Durbin, J. E., and Fallon, A. M. (1985). Transient expression of the chloramphenicol acetyltransferase gene in cultured mosquito cells. *Gene* **36**, 173–178.

Gallagher, T. M., Friesen, P. D., and Rueckert, R. R. (1983). Autonomous replication and expression of RNA 1 from black beetle virus. *J. Virol.* **46**, 481–489.

Kawai, S., and Nishizawa, M. (1984). New procedure for DNA transfection with polycation and dimethyl sulfoxide. *Mol. Cell. Biol.* **6**, 1172–1174.

Lopata, M. A., Cleveland, D. W., and Sollner-Webb, B. (1984). High level transient expression of a chloramphenicol acetyl transferase gene by DEAE-dextran mediated DNA transfection coupled with a dimethyl sulfoxide or glycerol shock treatment. *Nucleic Acids Res.* **12**, 5707–5717.

O'Hare, K., and Rubin, G. M. (1983). Structures of P transposable elements and their sites of insertion and excision in the *Drosophila melanogaster* genome. *Cell (Cambridge, Mass.)* **34**, 25–35.

Rio, D. C., and Rubin, G. M. (1985). Transformation of cultured *Drosophila melanogaster* cells with a dominant selectable marker. *Mol. Cell. Biol.* **5**, 1833–1838.

Rubin, G. M., Brorein, W. J., Jr., Dunsmuir, P., Flavell, A. J., Levis, R., Strobel, E., Toole, J. J., and Young, E. (1981). *Copia*-like transposable elements in the *Drosophila* genome. *Cold Spring Harbor Symp. Quant. Biol.* **45**, 619–628.

Saigo, K., Shiba, T., and Miyake, T. (1980). Virus-like particles of *Drosophila melanogaster* containing t-RNA and 5S ribosomal RNA I. Isolation and purification from cultured cells and detection of low molecular weight RNAs in the particles. *In* "Invertebrate Systems *in vitro*" (E. Kurstak, K. Maramorosch, and A. Dübendorfer, eds.), pp. 411–424. Elsevier/North-Holland Biomedical Press, Amsterdam.

Saigo, K., Millstein, L., and Thomas, C.A., Jr. (1981). The organization of *Drosophila melanogaster* histone genes. *Cold Spring Harbor Symp. Quant. Biol.* **45**, 815–827.

Schneider, I., and Blumenthal, A. B. (1978). *Drosophila* cell and tissue culture. *In* "The Genetics and Biology of *Drosophila* (M. Ashburner and T. R. F. Wright, eds.), Vol. 2A, pp. 265–315. Academic Press, New York.

Shiba, T., and Saigo, K. (1983). Retrovirus-like particles containing RNA homologous to the transposable element *copia* in *Drosophila melanogaster*. *Nature (London)* **302**, 119–124.

Shiba, T., Saigo, K., and Miyake, T. (1980). Virus-like particles of *Drosophila melanogaster* containing t-RNA and 5S ribosomal RNA II. Isolation and characterization of low molecular weight RNAs. *In* "Invertebrate Systems *in Vitro*" (E. Kurstak, K. Maramorosch, and A. Dübendorfer, eds.), pp. 425–433. Elsevier/North-Holland Biomedical Press, Amsterdam.

Sinclair, J. H., Sang, J. H., Burke, J. F., and Ish-Horowicz, D. (1983). Extrachromosomal replication of *copia*-based vectors in cultured *Drosophila* cells. *Nature (London)* **306**, 198–200.

Sondermeijer, P. J. A., Derksen, J. W., and Lubsen, N. H. (1980). Established cell lines of *Drosophila hydei*. *In Vitro* **16**, 913–914.

Southern, P. J., and Berg, P. (1982). Transfection of mammalian cells to antibiotic resistance with a bacterial gene under control of the SV40 early region promoter. *J. Mol. Appl. Genet.* **1**, 327–341.

Wigler, M., Silverstein, S., Lee, L.,-S., Pellicer, A., Cheng, Y., and Axel, R. (1977). Transfer of purified herpes virus thymidine kinase gene to cultured mouse cells. *Cell (Cambridge, Mass.)* **11**, 223–232.

17

FP Mutation of Nuclear Polyhedrosis Viruses: A Novel System for the Study of Transposon-Mediated Mutagenesis

MALCOLM J. FRASER

I. INTRODUCTION

The insect pathogenic nuclear polyhedrosis viruses (NPVs) represent a unique system for studies of the molecular genetics of DNA viruses. They are among the few eukaryotic DNA viruses which do not produce spliced messenger RNAs (Lubbert and Doerfler, 1984), and they regulate viral-specific gene products as four distinct temporal classes: immediate early, early, late, and very late (Kelly and Lescott, 1981; Friesen and Miller, 1985; Guarino and Summers, 1986).

The very late, delta class of gene products are superfluous to overall replication and extracellular virus production. They provide specialized functions which evolved to ensure environmental stability of these viruses. Stability is conferred through large paracrystalline proteinaceous encapsulating structures called occlusion bodies (OBs). The process of OB formation and encapsidation can be likened to the sporulation functions of bacteria or yeast.

Mutations at several regions of the viral genome related to the production of OBs do not preclude release of infectious extracellular virus (ECV) (Potter and Miller, 1980; Miller and Miller, 1982; Fraser *et al.*, 1983; Smith *et al.*, 1983b,c; Pennock *et al.*, 1984). In addition, these viruses are capable of accommodating large sequences of foreign DNA without affecting their ability to replicate or propagate between cells (Potter and Miller, 1980; Fraser, 1981; Miller and Miller, 1982; Pennock, *et al.*, 1984; Fraser *et al.*, 1983). The ability of NPVs to accommodate foreign DNA has led to their development as experssion vectors for eukaryotic and prokaryotic genes in infected insect cell cultures or larvae (Pennock *et al.*, 1984; Smith *et al.*, 1983c; Maeda *et al.*, 1985).

Evidence is now accumulating which suggests that these two features (nondefective mutation at several loci and accommodation of larger genomes in virion capsids) could permit these viruses to act as vehicles for extracellular transport of host cell fragments, including transposable elements. Many isolates of a spontaneous class of NPV plaque morphology mutants, known as FP mutants, carry insertions of repetitive host cell DNAs (Fraser *et al.*, 1983; Miller and Miller, 1982). In some cases these host DNA insertions are transposons (Miller and Miller, 1982; Fraser *et al.*, 1985). Analysis of this system of transposon-mediated mutagenesis should provide new insights into mechanisms of eukaryotic transposon movement both within and between individuals or species.

II. SPONTANEOUS PLAQUE MORPHOLOGY MUTANTS OF NUCLEAR POLYHEDROSIS VIRUSES

Mutations which affect OB production can be identified on the basis of plaque morphology (Hink and Vail, 1973; Fraser and Hink, 1982a; Fraser, 1981; Fraser and McCarthy, 1984; Smith *et al.*, 1983b). The ease with which this technique can be employed is frequently dependent on both the cell lines and the virus isolates used (Fraser and Hink, 1982a). The first application of the plaque assay technique for an NPV in insect cell lines resulted in the identification of one such mutant plaque morphology (Hink and Vail, 1973). Mutant plaques were distinguished from the wild-type plaques on the basis of their reduced refractivity to oblique lighting. The decreased refractivity resulted from reduced numbers of OBs in mutant infected cells. These plaques were designated FPP (few polyhedra plaques) as opposed to the MPP (many polyhedra plaques) of the wild-type virus. These designations have since been modified to FP and MP, respectively.

Until recently (Wang and Fraser, 1987), FP mutants were the only spontaneous plaque morphology mutants observed in preparations of NPV. Those of *Autographa californica* and *Galleria mellonella* MNPVs (AcMNPV and GmMNPV, respectively) can be detected by any of several plaque assay techniques employing the *Trichoplusia ni* (TN-368) cell line, but are often less easily differentiated using the *Spodoptera frugiperda* (IPLB-SF21AE) cell line as substrate (Fraser and Hink, 1982a). Only one overlay formulation, the 0.9% unbuffered methylcellulose overlay (Volkman and Summers, 1975), does not allow detection of the FP mutant plaques with either cell line (Fraser and Hink, 1982a).

The differential ability of cell lines to detect the FP mutant plaque morphology is related to the cells' relative capacity to form OBs on infection with these mutants. This ability is often dependent on the virus strain used. For example, GmMNPV FP plaques are quite distinct on both cell lines (Fraser and Hink, 1982a), whereas AcMNPV FP are virtually indistinguishable on *S. frugiperda* cells (M. J. Fraser, unpublished). The inability to distinguish AcMNPV FP mutants when cultivated in *S. frugiperda* cells may explain the reported prolonged propagation of AcMNPV in *S. frugiperda* cells without apparent derivation of FP mutants (Burand and Summers, 1982). More recent observations (M. J. Fraser, unpublished) establish that serial propagation of GmMNPV or AcMNPV in the IPLB-SF21AE cell line also results in the derivation of FP mutants. These mutants can be detected when supernatants from infected IPLB-SF21AE cell cultures are assayed on TN-368 cells.

The FP mutants are amplified in serial cell culture passage and were originally thought to arise during *in vitro* propagation. Several studies have also shown that these mutants can arise during normal *in vivo* passage (Fraser *et al.*, 1983), and are amplified if infectious ECV is used as inoculum (Potter *et al.*, 1979; Fraser and Hink, 1982b). In contrast, FP mutants are not favored during serial *in vivo* propagations utilizing OBs because of their reduced ability to produce this type of infectious particle (Fraser and Hink, 1982b; Potter *et al.*, 1979).

The genetic origin of FP mutants was first conclusively demonstrated by Potter *et al.* (1976). Through plaque purification using an agarose overlay they demonstrated that virus generating the FP plaque phenotype could be purified to homogeneity while the MP plaque-forming virus would always generate FP plaques during subsequent passages. Other studies indicated that the type of mutation involved might be reversible. Reversion to the MP phenotype was frequently observed for FP mutants isolated by Hink and Strauss (1976). This might be attributed to the use of the viscous fluid methylcellulose overlay rather than a solid agarose overlay during plaque purifications in these studies. However, the plaque pickings were performed in such a way that cross-contaminations would be minimized if not totally eliminated. Similar plaque pickings were performed on FP mutants from GmMNPV by Fraser and Hink (1982b) with the result that only one FP mutant could not be purified to homogeneity. The latter findings substantiate and corraborate the earlier studies of Hink and Strauss (1976). Later studies (Fraser *et al.*, 1983) clearly demonstrated that some FP mutants are capable of spontaneous reversion at a high frequency while others are relatively stable. The accumulated observations indicate that some FP mutants are revertible with a reversion frequency approaching that of the forward mutation frequency, while others are apparently quite stable and do not revert. A postulated mutation mechanism would have to account for both of these features.

III. BIOLOGICAL PROPERTIES OF FP MUTANTS

Most early studies of FP mutants sought to clarify the general characteristics of the mutant phenotype. In general terms, all FP mutants isolated to date have a common overall phenotype. They all produce fewer OBs than the wild-type virus, and those OBs which are made are less virulent for larvae than wild-type OBs. They generate more ECV than the wild-type virus, which confers a selective advantage during cell culture passages. Finally, they produce the distinctive, less refractive plaque morphology.

Cells infected with FP mutant virus exhibit a normal cytopathology for the first 20 hr of infection, including ring zone formation and nuclear hypertrophy. The cytopathology in certain cell lines (i.e., TN-368) deviates substantially beginning at 20–24 hr postinfection. Cells infected with wild-type virus normally begin OB formation during this time period, while many cells infected with FP mutant begin to lyse. The cell lysis produces distinctive ghost cells having well-defined nuclear envelopes and a central remnant of virogenic stroma. This cytopathology is quite distinct from that observed with engineered or spontaneous OB− mutants, which do not lyse cells (Smith *et al.*, 1983b; Wang and Fraser, 1987).

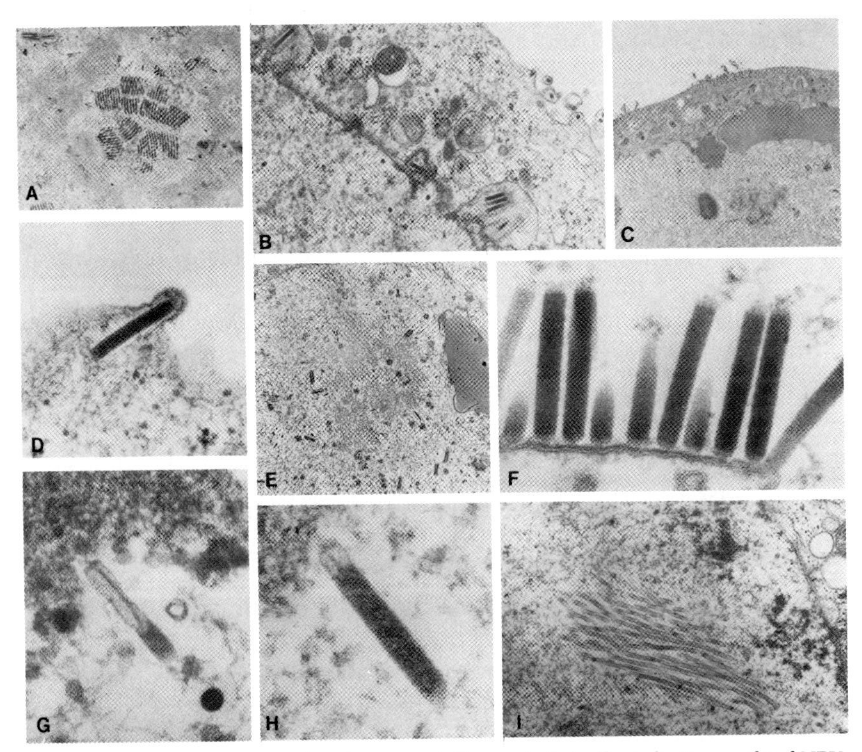

Fig. 1. Electron micrographs of representative periods in the infection cycle of NPV. (A) Arrays of assembled nucleocapsids in pockets of the virogenic stroma at 12 hr postinfection. (B) Budding of nucleocapsids through the nuclear membrane at 12 hr postinfection. (C) Arrays of nucleocapsids budding at the cell membrane in FP-infected cells. (D) Budding of extracellular virus involves attachment of the apical cap structure to the inner surface of the cell membrane. (E) Focus of *de novo* envelope assembly in the nucleus of infected cells. (F) Attachment of nucleocapsids to *de novo* envelopes is mediated through the apical cap structure. (G and H) Apparent mechanism of genome packaging for AcMNPV. The assembled capsid sheath orients toward the virogenic stroma and the nucleoprotein core appears to enter from the apical cap region. A caplike structure appears to be present and seems to restrict lateral movement of the nucleoprotein core as it enters. (I) Arrays of capsid sheathlike material toward the periphery of the nucleus at late times (36 hr) postinfection.

Electron microscopy observations of infected cells confirm this lytic response and indicate that lysis results from massive release of ECV from the cell surface (Fraser, 1981; Fraser and Hink, 1982b). Many nonenveloped nucleocapsids can be seen lining the membranes of FP-infected cells at 16 or 18 hr postinfection (Fig. 1). At 20–24 hr postinfection the budding of ECV proceeds with massive release of progeny virions from the cell surface (Fraser, 1981). Such massive release of

ECV was not seen in MP-infected cells and probably contributes to the rupture of FP-infected cell membranes (Fraser, 1981; Fraser and Hink, 1982b). This correlates well with titrations of infectious plaque forming unit (pfu) at various times postinfection (Fraser and Hink, 1982b) and is apparently related to the inability of nucleocapsids to be occluded in OBs. This is demonstrated more clearly by examination of FP-infected nuclei, in which very few nucleocapsids remain to be occluded in the few OBs being formed (Fraser and Hink, 1982b). The excessive release of ECV from FP-infected cells confers the selective advantage observed in serial cell culture propagations or serial *in vivo* propagations which utilize ECV as inoculum.

The retention of nucleocapsids in the nucleus of NPV-infected cells is apparently associated with the production of *de novo* envelopes, a process that is severely reduced in FP mutant-infected cells (Fraser and Hink, 1982b). By extrapolation it appears that envelopment in *de novo* synthesized envelopes is prerequisite to occlusion. However, at later times postinfection several OBs containing nonenveloped nucleocapsids were also found in FP-infected cells (Fraser, 1981), suggesting that envelopment of nucleocapsids optimizes their occlusion but is not absolutely essential. The process of OB formation is likely enhanced by the presence of these *de novo* envelopes, since it is reduced if *de novo* envelope formation is lacking.

Recent electron microscopy observations (Fig. 1) have established that the process of *de novo* envelope synthesis in AcMNPV-infected *S. frugiperda* cell cultures takes place in localized foci of membranelike vesicles within the nucleus, similar to those observed by Stoltz *et al.* (1973) in *Rhynchosciara angelae* NPV-infected cells. Nucleocapsids associate with these vesicles through the apical cap structure (Fig. 1). This structure is also involved in the attachment of nucleocapsids to the inner surface of the cell membrane during budding of ECV but does not seem to be involved during budding of nucleocapsids through the nuclear membrane (Fig. 1).

If the FP-infected cell remains intact, all other processes such as formation of fibrous material (FM) and OB envelopment will occur at times comparable to those observed in MP-infected cells (Fraser, 1981). This is especially evident in FP-infected *S. frugiperda* cells, which do not undergo the extensive lysis seen in FP-infected *T. ni* cells (M. J. Fraser, unpublished). These observations were among the first indications that FM is not a structural precursor of OBs (Fraser, 1981). Later mutation analysis confirmed that the major OB protein gene, polyhedrin, is not a component of FM (Smith *et al.*, 1983b).

The FP mutant OBs may contain enveloped single or multiple nucleocapsids as do wild-type OBs, but the reduced ability to occlude virions results in reduced virulence of the FP OBs (Hink and Strauss, 1976; Fraser and Hink, 1982b). The degree of reduction was about 3 log ID_{50} in one case (Fraser and Hink, 1982b), which correlated well with the observation that only 3 in 1000 FP OB cross sections examined had any nucleocapsids.

Several investigators attempted to identify structural protein differences between FP and MP viruses. Wood (1980) observed several differences in infected cell-specific proteins between FP- and MP-infected cells but was not able to correlate these proteins with structural components of the virions. Fraser and Hink (1982b) described several apparent differences in envelope proteins between MP and FP virions. They felt that these differences reflected the relative abundance of *de novo* enveloped versus budded enveloped virus in the ECV. Since these ECV were isolated at very late times postinfection, it is possible that extensive cellular lysis (which normally occurs with infected TN-368 cells by 48–72 hr) contributed many *de novo* enveloped virions to the MP ECV population.

Later studies demonstrated a consistent correlation between several FP mutants and the lack of a middle- to late-period 25K protein in viral infected cells (Fraser *et al.*, 1983). The function of this protein in the development of OBs or the structure of virions is still unknown. While the 25K protein is not a major component of ECV, it is conceivable that it is a component of wild-type occluded virion (OV) envelopes. However, this has yet to be defined.

IV. GENETICS OF FP MUTANTS

The studies by Potter *et al.* (1976) and Fraser and Hink (1982b) demonstrated that virus mutants which produce the FP plaque morphology can be purified from those which produce the wild-type, MP plaque morphology. Furthermore, the wild-type isolates always give rise to FP mutants, whereas FP isolates only occasionally revert to the MP or wild-type plaque phenotype. The obvious conclusion from these studies is that the FP virus represents a mutation of the MP virus. Furthermore, the mutation frequency, although not easily determined, is rather high.

There is no reason to believe that only one type of mutation contributes to the FP phenotype. Although it is a rather well-defined phenotype, several genes could easily be involved in reducing OB formation, or several different types of mutations could be involved at single or multiple loci. However, in many cases it appears that the mutation mechanism must fulfill the qualities of high forward mutation frequency and potential reversion. Although point mutations can revert, they do not usually occur with a frequency as high as that seen for FP mutants. Deletions might occur with a high frequency in certain regions of a genome but would revert rarely if at all.

It was conceivable that the genetic aberration producing the FP phenotype might be detected by characterizing mutant viral DNAs with restriction enzymes. Point mutations or very small deletions or insertions would probably go undetected but large deletions or insertions could be detected and mapped. It was hoped that all FP mutants would exhibit similar genetic aberrations or at least have similar regions of the genome affected. This was the original rationale behind two independent examinations of the molecular genetics of FP mutants.

There are several possible approaches to isolation of FP mutants. One way is to take advantage of the fact that FP mutants soon overgrow on serial *in vitro* passages. Thus, isolation of virus after several passages would increase the likelihood of finding FP mutants. In this case the selection criterion would be survival during *in vitro* passages. Such an approach was taken by Potter and Miller (1980). Five mutant plaques were picked after 25 serial *in vitro* passages in the TN-368 cell line. These mutant virus genomes were then characterized with restriction enzymes. Three of the five FP genomes exhibited patterns consistent with the insertion of additional DNA in disparate regions of the viral genome. The other two mutant genomes did not differ significantly from the parental AcMNPV-L1 genotype (Potter and Miller, 1980).

The study of Miller and Miller (1982) established that the inserted DNA in one of these mutants (FP-D) occurs between map positions 86.4 and 86.6 and that it is derived from the host cell genome. As will be discussed in more detail later, this study also established that transposon mutagenesis could occur in baculoviruses.

Because of the high probability of accumulation of mutations during cell culture passages, Fraser (1981) and Fraser *et al.* (1983) employed an alternative strategy in selecting FP mutants. The initial study (Fraser, 1981) focused on the isolation and characterization of both wild-type and FP mutant viruses of GmMNPV. Three plaque-purified wild-type (MP) isolates were independently passaged in the TN-368 cell line. After the third passage, plaque assays contained about 1% FP mutant plaques. In this scheme the criterion for selection was the distinctive plaque morphology. The FP plaques were picked from each lineage and their genotypes characterized with restriction enzymes against their respective MP isolates (Fraser, 1981).

All of the MP isolates had identical genotypes reflected by their restriction enzyme patterns. This parental genotype was designated GmMNPV-C3. Each of the FP mutants had insertions of varying lengths relative to the GmC3 genotype. The exact location of these insertions was impossible to determine without hybridization analysis, but based on restriction digests alone the preliminary results indicated that these insertions might not be viral DNA (Fraser, 1981).

These studies were expanded by Fraser *et al.* (1983). In this case we analyzed several FP mutants isolated from AcMNPV-E2 as well as the original ones isolated from GmMNPV-C3. All but one of the nine FP mutants selected from individual passages of either AcMNPV-E2 or the closely related GmMNPV-C3 were missing a similar 4.95 kbp *Hin*dIII fragment (Fraser *et al.*, 1983). The restriction enzyme maps of this fragment from AcMNPV-E2 and GmMNPV-C3 were identical. Furthermore, hybridization analysis with the cloned 4.95 kbp wild-type fragment identified one or more new bands in eight of the nine FP mutant genomes which were the result of insertions in the 4.95 kbp *Hin*dIII fragment, between 36.0 and 37.0 map units (Fraser *et al.*, 1983).

Hybridization analysis of these inserted sequences revealed that they originated from repetitive TN-368 cell sequences. Several inserts exhibited a dispersed, repetitive pattern when hybridized to genomic digests of TN-368 cell DNAs (Fraser *et al.*, 1983). One insertion sequence, now designated IFP2, was homologous to inserts in two other FP mutants (Fraser *et al.*, 1983). The restriction site maps of this insertion in the mutants designated GmFP2 and AcFP4 are identical (Fig. 2), but the insertion occurs at two different points and in inverse orientations in these two mutant genomes (Fraser *et al.*, 1983).

Through marker rescue with the cloned wild-type *Hin*dIII fragment we established that the inserted DNAs were related to the FP mutation. We also demonstrated that rescue of the FP mutants to the MP phe-

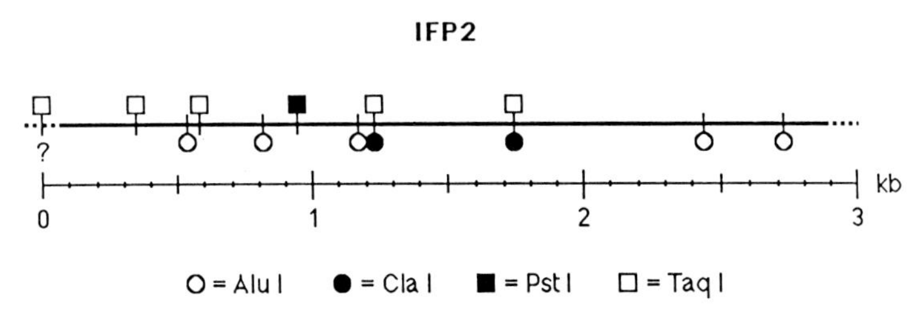

Fig. 2. Physical map of several restriction enzyme sites in the putative transposable insertion element IFP2. Maps were compared for this insertion in both AcFP4 and GmFP2.

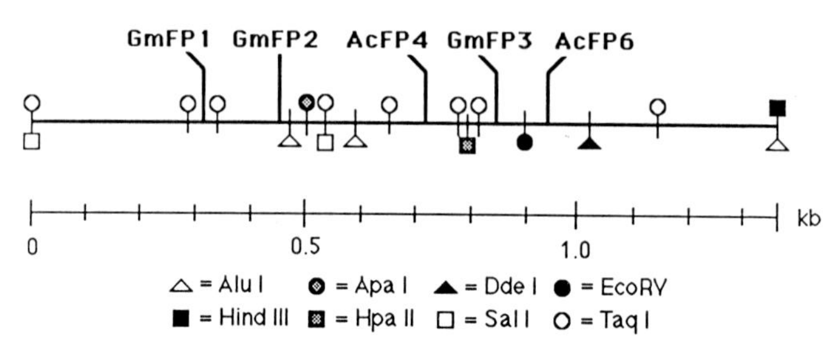

Fig. 3. Detailed restriction enzyme site map of the FP locus in GmMNPV-C3. The relative positions of inserted host sequences from several FP mutants are indicated.

notype was correlated with the reappearance of a 25K infected cell-specific protein that is absent in cells infected with all these FP mutants (Fraser *et al.*, 1983). We could not distinguish whether the inserted DNAs actually caused the mutation through insertion mutagenesis within a gene or through some regulatory mechanism. Given the diversity of insertions at this locus and their localization to a limited region of the *Hind*III fragment (Fig. 3) it is probable that the cause of the FP mutation is insertional inactivation of a gene or genes.

The exact reason for the absence of the 25K protein from FP-infected cells remains a mystery. It seems that the mechanism of mutagenesis through the 25K protein is quite complex. Preliminary findings have demonstrated that some FP mutant viruses produce transcripts virtually identical to those of the wild-type virus in this region of the genome. In fact, no differences in quality or quantity of transcripts could be discerned between wild-type and AcFP6-infected cells (M. J. Fraser, G. E. Smith, and M. D. Summers, unpublished). Analysis of the viral sequences from this region of the *Hind*III-I fragment did not reveal any open reading frames which span the regions of insertion (Fraser *et al.*, 1985). Numerous stop codons are encountered in all reading frames from either direction (Fig. 4).

In a more recent experiment, total cytoplasmic RNAs from GmFP2- and GmC3-infected cells as well as uninfected *T. ni* (TN-368) or *S. frugiperda* (IPLB-SF21AE) cells were probed with labeled IFP2 or wild-type viral sequences from the FP locus. The hybridization profiles (Fig. 5) demonstrate that there are transcripts homologous to IFP2 in both GmFP2-infected cells and uninfected TN-368 cells. They also show again that the FP locus is transcriptionally active, particularly in the rightmost *Sal*/*Hind* fragment.

V. TRANSPOSONS AND TRANSPOSON-MEDIATED MUTAGENESIS OF BACULOVIRUSES

Transposons are highly mobile genetic elements which are capable of movement within the genome of an organism and even between chromosomes of eukaryotic organisms. There are basically three structural types of transposons. The retrotransposons have the basic structure of retrovirus proviruses and move through a similar mechanism using reverse transcriptase. They can exist as extrachromosomal copies and their extrachromosomal form is often linked to viruslike particles in the nucleus of the cell. Examples of retrotransposons include the *copia* and *copia*-like elements of *Drosophila* (e.g. 412, 297, B104, mdg1), the Ty elements of *Saccharomyces cerevisciae*, the TED element of *T. ni*, the Mys elements of *Peromyscus leucopus*, the THE1 family of humans, and the mammalian retroviruses (Table I).

The FB elements are structurally related to foldback elements of *Drosophila*. These structurally related elements are composed of long indirect repeats flanking unique regions of variable length. In many cases the long indirect repeats are themselves composed of direct repeats of a shorter sequence. Examples of such elements are the FB elements of *Drosophila*, the TU family of repeats from *Strongylocentrorus purpuratus*, and the Mu1 element of maize (Table II).

The P-like elements have no internal restriction site symmetry. They consist of internal unique domains bounded by relatively short terminal inverted repetitions. Many similar elements are being described from diverse organisms. Examples include the P elements of *Drosophila*, the Ac/Ds and Cin1 elements of maize, the Tc1 element of *Caenorhabditis elegans*, and the sigma element of *S. cerevisiae* (Table III). *Drosophila* P elements have recently been characterized as carrying a transposase gene which is encoded by the internal domain (Karess and Rubin, 1984). The transposase of the *Drosophila* P elements is translated from a message that is produced only during embryonic stages of development (Laski *et al.*, 1986).

Transposons have structural features which are considered distinctive. These include direct duplications of genomic sequences which they produce at their termini on movement from one site to another in the genome. Nearly all known transposons generate duplications of a target sequence at the site of insertion into new regions of DNA (Tables I–III). These target sites are relatively short (4–30 bp) sequences that are part of the resident sequences at the insertion site and are duplicated in a direct fashion at the ends of the inserted trans-

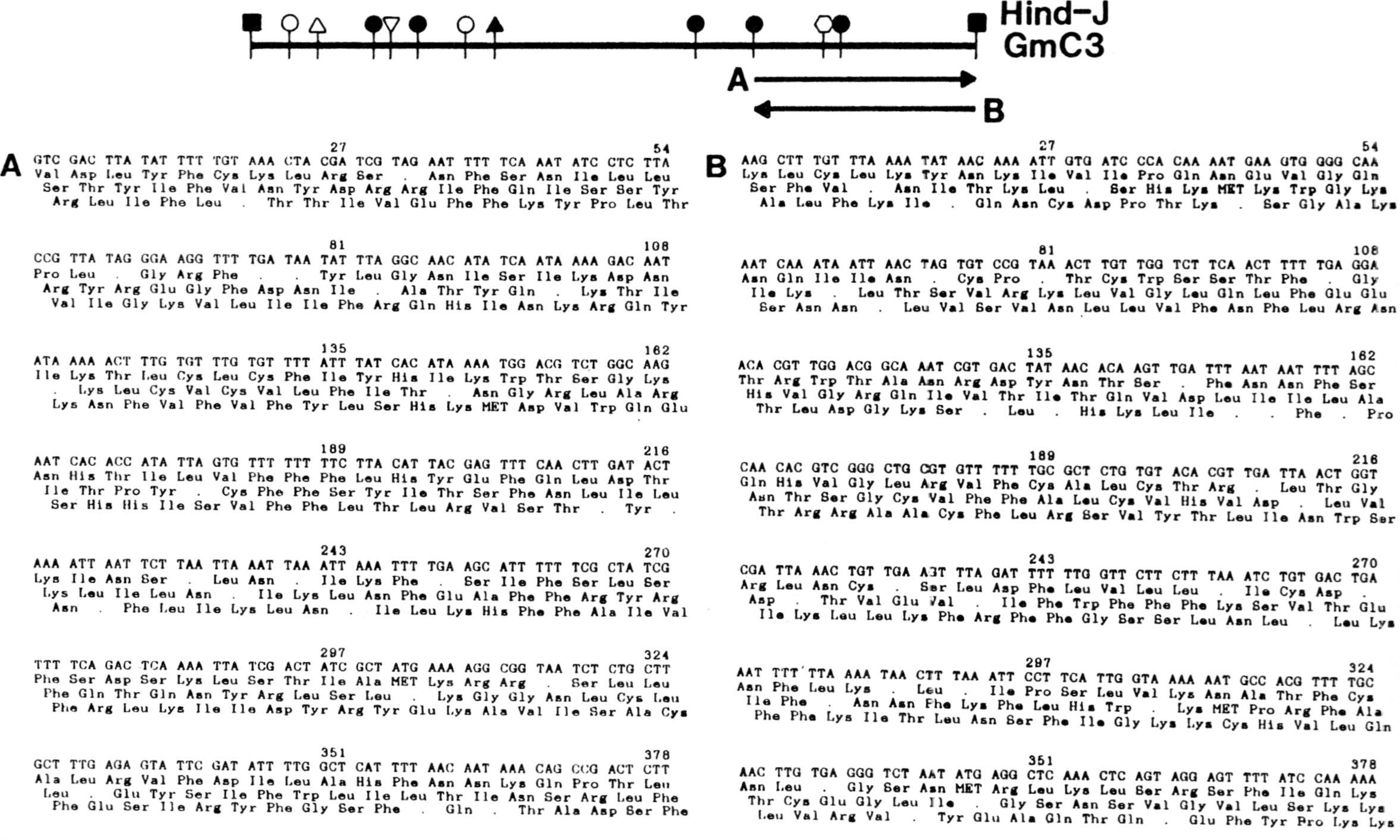

Hind–J
GmC3
A
B
A
27 54
GTC GAC TTA TAT TTT TGT AAA CTA CGA TCG TAG AAT TTT TCA AAT ATC CTC TTA
Val Asp Leu Tyr Phe Cys Lys Leu Arg Ser . Asn Phe Ser Asn Ile Leu Leu
Ser Thr Tyr Ile Phe Val Asn Tyr Asp Arg Arg Ile Phe Gln Ile Ser Ser Tyr
Arg Leu Ile Phe Leu . Thr Thr Ile Val Glu Phe Phe Lys Tyr Pro Leu Thr
81 108
CCG TTA TAG GGA AGG TTT TGA TAA TAT TTA GGC AAC ATA TCA ATA AAA GAC AAT
Pro Leu . Gly Arg Phe . . Tyr Leu Gly Asn Ile Ser Ile Lys Asp Asn
Arg Tyr Arg Glu Gly Phe Asp Asn Ile . Ala Thr Tyr Gln . Lys Thr Ile
Val Ile Gly Lys Val Leu Ile Ile Phe Arg Gln His Ile Asn Lys Arg Gln Tyr
135 162
ATA AAA ACT TTG TGT TTG TGT TTT ATT TAT CAC ATA AAA TGG ACG TCT GGC AAG
Ile Lys Thr Leu Cys Leu Cys Phe Ile Tyr His Ile Lys Trp Thr Ser Gly Lys
. Lys Leu Cys Val Cys Val Leu Phe Ile Thr . Asn Gly Arg Leu Ala Arg
Lys Asn Phe Val Phe Val Phe Tyr Leu Ser His Lys MET Asp Val Trp Gln Glu
189 216
AAT CAC ACC ATA TTA GTG TTT TTT TTC TTA CAT TAC GAG TTT CAA CTT GAT ACT
Asn His Thr Ile Leu Val Phe Phe Phe Leu His Tyr Glu Phe Gln Leu Asp Thr
Ile Thr Pro Tyr . Cys Phe Phe Ser Tyr Ile Thr Ser Phe Asn Leu Ile Leu
Ser His His Ile Ser Val Phe Phe Leu Thr Leu Arg Val Ser Thr . Tyr .
243 270
AAA ATT AAT TCT TAA TTA AAT TAA ATT AAA TTT TGA AGC ATT TTT TCG CTA TCG
Lys Ile Asn Ser . Leu Asn . Ile Lys Phe . Ser Ile Phe Ser Leu Ser
Lys Leu Ile Leu Asn . Ile Lys Leu Asn Phe Glu Ala Phe Phe Arg Tyr Arg
Asn . Phe Leu Ile Lys Leu Asn . Ile Leu Lys His Phe Phe Ala Ile Val
297 324
TTT TCA GAC TCA AAA TTA TCG ACT ATC GCT ATG AAA AGG CGG TAA TCT CTG CTT
Phe Ser Asp Ser Lys Leu Ser Thr Ile Ala MET Lys Arg Arg . Ser Leu Leu
Phe Gln Thr Gln Asn Tyr Arg Leu Ser Leu . Lys Gly Gly Asn Leu Cys Leu
Phe Arg Leu Lys Ile Ile Asp Tyr Arg Tyr Glu Lys Ala Val Ile Ser Ala Cys
351 378
GCT TTG AGA GTA TTC GAT ATT TTG GCT CAT TTT AAC AAT AAA CAG CCG ACT CTT
Ala Leu Arg Val Phe Asp Ile Leu Ala His Phe Asn Asn Lys Gln Pro Thr Leu
Leu . Glu Tyr Ser Ile Phe Trp Leu Ile Leu Thr Ile Asn Ser Arg Leu Phe
Phe Glu Ser Ile Arg Tyr Phe Gly Ser Phe . Gln . Thr Ala Asp Ser Phe
B
27 54
AAG CTT TGT TTA AAA TAT AAC AAA ATT GTG ATC CCA CAA AAT GAA GTG GGG CAA
Lys Leu Cys Leu Lys Tyr Asn Lys Ile Val Ile Pro Gln Asn Glu Val Gly Gln
Ser Phe Val . Asn Ile Thr Lys Leu . Ser His Lys MET Lys Trp Gly Lys
Ala Leu Phe Lys Ile . Gln Asn Cys Asp Pro Thr Lys . Ser Gly Ala Lys
81 108
AAT CAA ATA ATT AAC TAG TGT CCG TAA ACT TGT TGG TCT TCA ACT TTT TGA GGA
Asn Gln Ile Ile Asn . Cys Pro . Thr Cys Trp Ser Ser Thr Phe . Gly
Ile Lys . Leu Thr Ser Val Arg Lys Leu Val Gly Leu Gln Leu Phe Glu Glu
Ser Asn Asn . Leu Val Ser Val Asn Leu Leu Val Phe Asn Phe Leu Arg Asn
135 162
ACA CGT TGG ACG GCA AAT CGT GAC TAT AAC ACA AGT TGA TTT AAT AAT TTT AGC
Thr Arg Trp Thr Ala Asn Arg Asp Tyr Asn Thr Ser . Phe Asn Asn Phe Ser
His Val Gly Arg Gln Ile Val Thr Ile Thr Gln Val Asp Leu Ile Ile Leu Ala
Thr Leu Asp Gly Lys Ser . Leu . His Lys Leu Ile . . Phe . Pro
189 216
CAA CAC GTC GGG CTG CGT GTT TTT TGC GCT CTG TGT ACA CGT TGA TTA ACT GGT
Gln His Val Gly Leu Arg Val Phe Cys Ala Leu Cys Thr Arg . Leu Thr Gly
Asn Thr Ser Gly Cys Val Phe Phe Ala Leu Cys Val His Val Asp . Leu Val
Thr Arg Arg Ala Ala Cys Phe Leu Arg Ser Val Tyr Thr Leu Ile Asn Trp Ser
243 270
CGA TTA AAC TGT TGA AGT TTA GAT TTT TTG GTT CTT CTT TAA ATC TGT GAC TGA
Arg Leu Asn Cys . Ser Leu Asp Phe Leu Val Leu Leu . Ile Cys Asp .
Asp . Thr Val Glu Val . Ile Phe Trp Phe Phe Phe Lys Ser Val Thr Glu
Ile Lys Leu Leu Lys Phe Arg Phe Phe Gly Ser Ser Leu Asn Leu . Leu Lys
297 324
AAT TTT TTA AAA TAA CTT TAA ATT CCT TCA TTG GTA AAA AAT GCC ACG TTT TGC
Asn Phe Leu Lys . Leu . Ile Pro Ser Leu Val Lys Asn Ala Thr Phe Cys
Ile Phe . Asn Asn Phe Lys Phe Leu His Trp . Lys MET Pro Arg Phe Ala
Phe Phe Lys Ile Thr Leu Asn Ser Phe Ile Gly Lys Lys Cys His Val Leu Gln
351 378
AAC TTG TGA GGG TCT AAT ATG AGG CTC AAA CTC AGT AGG AGT TTT ATC CAA AAA
Asn Leu . Gly Ser Asn MET Arg Leu Lys Leu Ser Arg Ser Phe Ile Gln Lys
Thr Cys Glu Gly Leu Ile . Gly Ser Asn Ser Val Gly Val Leu Ser Lys Lys
Leu Val Arg Val . Tyr Glu Ala Gln Thr Gln . Glu Phe Tyr Pro Lys Lys

```
                                405                                 432
TTC GTC GCG TCA CCA TAA CAG TTT TTA CAA AGT AGA AAA TAT ATT TGT AAA ACG
Phe Val Ala Ser Pro  .  Gln Phe Leu Gln Ser Arg Lys Tyr Ile Cys Lys Thr
 Ser Ser Arg His His Asn Ser Phe Tyr Lys Val Glu Asn Ile Phe Val Lys Arg
  Arg Arg Val Thr Ile Thr Val Phe Thr Lys  .  Lys Ile Tyr Leu  .  Asn Ala

                                459                                 486
CAA CAG AGC GTC GCG AGT TTT TTT AAG TAA GCA GCA AAA GCT CCT GTG GCG GCC
Gln Gln Ser Val Ala Ser Phe Phe Lys  .  Ala Ala Lys Ala Pro Val Ala Ala
 Asn Arg Ala Ser Arg Val Phe Leu Ser Lys Gln Gln Lys Leu Leu Trp Arg Pro
  Thr Glu Arg Arg Glu Phe Phe  .  Val Ser Ser Lys Ser Ser Cys Gly Gly His

                                513                                 540
ACA AAG TAT TTT TAC GGG CCC GTC GTA ATT AAT GTT TAA ATT AAA ATT TTT AAG
Thr Lys Tyr Phe Tyr Gly Pro Val Val Ile Asn Val  .  Ile Lys Ile Phe Lys
 Gln Ser Ile Phe Thr Gly Pro Ser  .  Leu MET Phe Lys Leu Lys Phe Leu Ser
  Lys Val Phe Leu Arg Ala Arg Arg Asn  .  Cys Leu Asn  .  Asn Phe  .  Val

                                567                                 594
TCG ACG CTC GCG CGA CAT TGG TAT TGA CAT TCT TTA GCG CGC AGT CGG CAG CAC
Ser Thr Leu Ala Arg His Trp Tyr  .  His Ser Leu Ala Arg Ser Arg Gln His
 Arg Arg Ser Arg Asp Ile Gly Ile Asp Ile Leu  .  Arg Ala Val Gly Ser Thr
  Asp Ala Arg Ala Thr Leu Val Leu Thr Phe Phe Ser Ala Gln Ser Ala Ala Gln

                                621                                 648
AGC TGT GGC CAC AAT GTG TTT TTT TCA AAC TGA AGA TCT CGT ATG ACG TTG TTT
Ser Cys Gly His Asn Val Phe Phe Ser Asn  .  Arg Ser Arg MET Thr Leu Phe
 Ala Val Ala Thr MET Cys Phe Phe Gln Thr Glu Asp Leu Val  .  Arg Cys Leu
  Leu Trp Pro Gln Cys Val Phe Phe Lys Leu Lys Ile Ser Tyr Asp Val Val  .

                                675                                 702
AAA GTT TAG GTC GAG TAA AGC GCA AAT CTT TTT TAA ATA ATA GTT TCT AAT TTT
Lys Val  .  Val Glu  .  Ser Ala Asn Leu Phe  .  Ile Ile Val Ser Asn Phe
 Lys Phe Arg Ser Ser Lys Ala Gln Ile Phe Phe Lys  .   .  Phe Leu Ile Phe
  Ser Leu Gly Arg Val Lys Arg Lys Ser Phe Leu Asn Asn Ser Phe  .  Phe Phe

                                729                                 756
TTT ATT ATT CAG CCT GCT GTC GTG AAT ACC GTA TAT CTC AAC GCT GTC TGT GAG
Phe Ile Ile Gln Pro Ala Val Val Asn Thr Val Tyr Leu Asn Ala Val Cys Glu
 Leu Leu Phe Ser Leu Leu Ser  .  Ile Pro Tyr Ile Ser Thr Leu Ser Val Arg
  Tyr Tyr Ser Ala Cys Cys Arg Glu Tyr Arg Ile Ser Gln Arg Cys Leu  .  Asp

                                783                                 810
ATG TCG TAT CTA GGC CTT TTT AGT TTT TCG CTC ATC GAC TTG ATA TTG TCC GGA
MET Ser Tyr Leu Gly Leu Phe Ser Phe Ser Leu Ile Asp Leu Ile Leu Ser Gly
 Cys Arg Ile  .  Ala Phe Leu Val Phe Arg Ser Ser Thr  .  Tyr Cys Pro Asp
  Val Val Ser Arg Pro Phe  .  Phe Phe Ala His Arg Leu Asp Ile Val Arg Thr
```

```
                                405                                 432
AGA AAA CAT GAT TAC GTC TGT ACA CGA ACG CTA TTA ACG CAG AGT GCA AAG TAT
Arg Lys His Asp Tyr Val Cys Thr Arg Thr Leu Leu Thr Gln Ser Ala Lys Tyr
 Glu Asn MET Ile Thr Ser Val His Glu Arg Tyr  .  Arg Arg Val Gln Ser Ile
  Lys Thr  .  Leu Arg Leu Tyr Thr Asn Ala Ile Asn Ala Glu Cys Lys Val  .

                                459                                 486
AAG AGG GTT CAA AAA TAT ATT TTA CGC ACC ATA TAC GCA TCG GGT TGA TAT CGT
Lys Arg Val Gln Lys Tyr Ile Leu Arg Thr Ile Tyr Ala Ser Gly  .  Tyr Arg
 Arg Gly Phe Lys Asn Ile Phe Tyr Ala Pro Tyr Thr His Arg Val Asp Ile Val
  Glu Gly Ser Lys Ile Tyr Phe Thr His His Ile Arg Ile Gly Leu Ile Ser Leu

                                513                                 540
TAA TAT GGA TCA ATT TGA ACA GTT GAT TAA CGT GTC TCT GCT CAA GTC TTT GAT
 .  Tyr Gly Ser Ile  .  Thr Val Asp  .  Arg Val Ser Ala Gln Val Phe Asp
 Asn MET Asp Gln Phe Glu Gln Leu Ile Asn Val Ser Leu Leu Lys Ser Leu Ile
  Ile Trp Ile Asn Leu Asn Ser  .  Leu Thr Cys Leu Cys Ser Ser Leu  .  Ser

                                567                                 594
CAA AAC GCA AAT CGA CGT AAA ATG TGT CCG GAC AAT ATC AAG TCG ATG AGC GAA
Gln Asn Ala Asn Arg Arg Lys MET Cys Pro Asp Asn Ile Lys Ser MET Ser Glu
 Lys Thr Gln Ile Asp Val Lys Cys Val Arg Thr Ile Ser Ser Arg  .  Ala Lys
  Lys Arg Lys Ser Thr  .  Asn Val Ser Gly Gln Tyr Gln Val Asp Glu Arg Lys

                                621                                 648
AAA CTA AAA AGG CCT AGA TAC GAC ATC TCA CAG ACA GCG TTG AGA TAT ACG GTA
Lys Leu Lys Arg Pro Arg Tyr Asp Ile Ser Gln Thr Ala Leu Arg Tyr Thr Val
 Asn  .  Lys Gly Leu Asp Thr Thr Ser His Arg Gln Arg  .  Asp Ile Arg Tyr
  Thr Lys Lys Ala  .  Ile Arg His Leu Thr Asp Ser Val Glu Ile Tyr Gly Ile

                                675                                 702
TTC ACG ACA GCA GGC TGA ATA ATA AAA AAA TTA GAA ACT ATT ATT TAA AAA AGA
Phe Thr Thr Ala Gly  .  Ile Ile Lys Lys Leu Glu Thr Ile Ile  .  Lys Arg
 Ser Arg Gln Gln Ala Glu  .   .  Lys Asn  .  Lys Leu Leu Phe Lys Lys Asp
  His Asp Ser Arg Leu Asn Asn Lys Lys Ile Arg Asn Tyr Tyr Leu Lys Lys Ile

                                729                                 756
TTT GCG CTT TAC TCG ACC TAA ACT TTA AAC AAC GTC ATA CGA GAT CTT CAG TTT
Phe Ala Leu Tyr Ser Thr  .  Thr Leu Asn Asn Val Ile Arg Asp Leu Gln Phe
 Leu Arg Phe Thr Arg Pro Lys Leu  .  Thr Thr Ser Tyr Glu Ile Phe Ser Leu
  Cys Ala Leu Leu Asp Leu Asn Phe Lys Gln Arg His Thr Arg Ser Ser Val

                                783                                 810
GAA AAA AAC ACA TTG TGG CCA CAG CTG TGC TGC CGA CTG CGC GCT AAA GAA TGT
Glu Lys Asn Thr Leu Trp Pro Gln Leu Cys Cys Arg Leu Arg Ala Lys Glu Cys
 Lys Lys Thr His Cys Gly His Ser Cys Ala Ala Asp Cys Ala Leu Lys Asn Val
  Lys Lys His Ile Val Ala Thr Ala Val Leu Pro Thr Ala Arg  .  Arg MET Ser
```

Fig. 4.

```
                                        837                                         864
CAC ATT TTA CGT CGA TTT GCG TTT TGA TCA AAG ACT TGA GCA GAG ACA CGT TAA
His Ile Leu Arg Arg Phe Ala Phe  .  Ser Lys Thr  .  Ala Glu Thr Arg  .
 Thr Phe Tyr Val Asp Leu Arg Phe Asp Gln Arg Leu Glu Gln Arg His Val Asn
  His Phe Thr Ser Ile Cys Val Leu Ile Lys Asp Leu Ser Arg Asp Thr Leu Ile

                                        891                                         918
TCA ACT GTT CAA ATT GAT CCA TAT TAA CGA TAT CAA CCC GAT GCG TAT ATG GTG
Ser Thr Val Gln Ile Asp Pro Tyr  .  Arg Tyr Gln Pro Asp Ala Tyr MET Val
 Gln Leu Phe Lys Leu Ile His Ile Asn Asp Ile Asn Pro MET Arg Ile Trp Cys
  Asn Cys Ser Asn  .  Ser Ile Leu Thr Ile Ser Thr Arg Cys Val Tyr Gly Ala

                                        945                                         972
CGT AAA ATA TAT TTT TGA ACC CTC TTA TAC TTT GCA CTC TGC GTT AAT AGC GTT
Arg Lys Ile Tyr Phe  .  Thr Leu Leu Tyr Phe Ala Leu Cys Val Asn Ser Val
 Val Lys Tyr Ile Phe Glu Pro Ser Tyr Thr Leu His Ser Ala Leu Ile Ala Phe
      Asn Ile Phe Leu Asn Pro Leu Ile Leu Cys Thr Leu Arg  .   .  Arg Ser

                                        999                                         1026
CGT GTA CAG ACG TAA TCA TGT TTT CTT TTT TGG ATA AAA CTC CTA CTG AGT TTG
Arg Val Gln Thr  .  Ser Cys Phe Leu Phe Trp Ile Lys Leu Leu Leu Ser Leu
 Val Tyr Arg Arg Asn His Val Phe Phe Phe Gly  .  Asn Ser Tyr  .  Val  .
  Cys Thr Asp Val Ile MET Phe Ser Phe Leu Asp Lys Thr Pro Thr Glu Phe Glu

                                        1053                                        1080
AGC CTC ATA TTA GAC CCT CAC AAG TTG CAA AAC GTG GCA TTT TTT ACC AAT GAA
Ser Leu Ile Leu Asp Pro His Lys Leu Gln Asn Val Ala Phe Phe Thr Asn Glu
 Ala Ser Tyr  .  Thr Leu Thr Ser Cys Lys Thr Trp His Phe Leu Pro MET Lys
  Pro His Ile Arg Pro Ser Gln Val Ala Lys Arg Gly Ile Phe Tyr Gln  .  Arg

                                        1107                                        1134
GGA ATT TAA AGT TAT TTT AAA AAA TTT CAG TCA CAG ATT TAA AGA AGA ACC AAA
Gly Ile  .  Ser Tyr Phe Lys Lys Phe Gln Ser Gln Ile  .  Arg Arg Thr Lys
 Glu Phe Lys Val Ile Leu Lys Asn Phe Ser His Arg Phe Lys Glu Glu Pro Lys
  Asn Leu Lys Leu Phe  .  Lys Ile Ser Val Thr Asp Leu Lys Lys Asn Gln Lys

                                        1161                                        1188
AAA TCT AAA CTT CAA CAG TTT AAT CGA CCA GTT AAT CAA CGT GTA CAC AGA GCG
Lys Ser Lys Leu Gln Gln Phe Asn Arg Pro Val Asn Gln Arg Val His Arg Ala
 Asn Leu Asn Phe Asn Ser Leu Ile Asp Gln Leu Ile Asn Val Tyr Thr Glu Arg
  Ile     Thr Ser Thr Val  .  Ser Thr Ser  .  Ser Thr Cys Thr Gln Ser Ala
```

```
                                        837                                         864
CAA TAC CAA TGT CGC GCG AGC GTC GAC TTA AAA ATT TTA ATT TAA ACA TTA ATT
Gln Tyr Gln Cys Arg Ala Ser Val Asp Leu Lys Ile Leu Ile  .  Thr Leu Ile
 Asn Thr Asn Val Ala Arg Ala Ser Thr  .  Lys Phe  .  Phe Lys His  .  Leu
  Ile Pro MET Ser Arg Glu Arg Arg Leu Lys Asn Phe Asn Leu Asn Ile Asn Tyr

                                        891                                         918
ACG ACG GGC CCG TAA AAA TAC TTT GTG GCC GCC ACA GGA GCT TTT GCT GCT TAC
Thr Thr Gly Pro  .  Lys Tyr Phe Val Ala Ala Thr Gly Ala Phe Ala Ala Tyr
 Arg Arg Ala Arg Lys Asn Thr Leu Trp Pro Pro Gln Glu Leu Leu Leu Leu Thr
  Asp Gly Pro Val Lys Ile Leu Cys Gly Arg His Arg Ser Phe Cys Cys Leu Leu

                                        945                                         972
TTA AAA AAA CTC GCG ACG CTC TGT TGC GTT TTA CAA ATA TAT TTT CTA CTT TGT
Leu Lys Lys Leu Ala Thr Leu Cys Cys Val Leu Gln Ile Tyr Phe Leu Leu Cys
 .  Lys Asn Ser Arg Arg Ser Val Ala Phe Tyr Lys Tyr Ile Phe Tyr Phe Val
  Lys Lys Thr Arg Asp Ala Leu Leu Arg Phe Thr Asn Ile Phe Ser Thr Leu  .

                                        999                                         1026
AAA AAC TGT TAT GGT GAC GCG ACG AAA AGA GTC GGC TGT TTA TTG TTA AAA TGA
Lys Asn Cys Tyr Gly Asp Ala Thr Lys Arg Val Gly Cys Leu Leu Leu Lys  .
 Lys Thr Val MET Val Thr Arg Arg Lys Glu Ser Ala Val Tyr Cys  .  Asn Glu
  Lys Leu Leu Trp  .  Arg Asp Glu Lys Ser Arg Leu Phe Ile Val Lys MET Ser

                                        1053                                        1080
GCC AAA ATA TCG AAT ACT CTC AAA GCA AGC AGA GAT TAC CGC CTT TTC ATA GCG
Ala Lys Ile Ser Asn Thr Leu Lys Ala Ser Arg Asp Tyr Arg Leu Phe Ile Ala
 Pro Lys Tyr Arg Ile Leu Ser Lys Gln Ala Glu Ile Thr Ala Phe Ser  .  Arg
  Gln Asn Ile Glu Tyr Ser Gln Ser Lys Gln Arg Leu Pro Pro Phe His Ser Asp

                                        1107                                        1134
ATA GTC GAT AAT TTT GAG TCT GAA AAC GAT AGC GAA AAA ATG CTT CAA AAT TTA
Ile Val Asp Asn Phe Glu Ser Glu Asn Asp Ser Glu Lys MET Leu Gln Asn Leu
 .  Ser Ile Ile Leu Ser Leu Lys Thr Ile Ala Lys Lys Cys Phe Lys Ile  .
  Ser Arg  .  Phe  .  Val  .  Lys Arg  .  Arg Lys Asn Ala Ser Lys Phe Asn

                                        1161                                        1188
ATT TAA TTT AAT TAA GAA TTA ATT TTA GTA TCA AGT TGA AAC TCG TAA TGT AAG
Ile  .  Phe Asn  .  Glu Leu Ile Leu Val Ser Ser  .  Asn Ser  .  Cys Lys
 Phe Asn Leu Ile Lys Asn  .  Phe  .  Tyr Gln Val Glu Thr Arg Asn Val Arg
  Leu Ile  .  Leu Arg Ile Asn Phe Ser Ile Lys Leu Lys Leu Val MET  .  Glu
```

```
                                   1215                      1242
CAA AAA ACA CGC AGC CCG ACG TGT TGG CTA AAA TTA TTA AAT CAA CTT GTG TTA
Gln Lys Thr Arg Ser Pro Thr Cys Trp Leu Lys Leu Leu Asn Gln Leu Val Leu
 Lys Lys His Ala Ala Arg Arg Val Gly  .  Asn Tyr  .  Ile Asn Leu Cys Tyr
  Lys Asn Thr Gln Pro Asp Val Leu Ala Lys Ile Ile Lys Ser Thr Cys Val Ile

                                   1269                      1296
TAG TCA CGA TTT GCC GTC CAA CGT GTT CCT CAA AAA GTT GAA GAC CAA CAA GTT
 .  Ser Arg Phe Ala Val Gln Arg Val Pro Gln Lys Val Glu Asp Gln Gln Val
 Ser His Asp Leu Pro Ser Asn Val Phe Leu Lys Lys Leu Lys Thr Asn Lys Phe
  Val Thr Ile Cys Arg Pro Thr Cys Ser Ser Lys Ser  .  Arg Pro Thr Ser Leu

                                   1323                      1350
TAC GGA CAC TAG TTA ATT ATT TGA TTT TGC CCC ACT TCA TTT TGT GGG ATC ACA
Tyr Gly His  .  Leu Ile Ile  .  Phe Cys Pro Thr Ser Phe Cys Gly Ile Thr
 Thr Asp Thr Ser  .  Leu Phe Asp Phe Ala Pro Leu His Phe Val Gly Ser Gln
  Arg Thr Leu Val Asn Tyr Leu Ile Leu Pro His Phe Ile Leu Trp Asp His Asn

                                   1377
ATT TTG TTA TAT TTT AAA CAA AGC TT
Ile Leu Leu Tyr Phe Lys Gln Ser Leu
 Phe Cys Tyr Ile Leu Asn Lys Ala Tyr
  Phe Val Ile Phe  .  Thr Lys Leu
```

```
                                   1215                      1242
AAA AAA AAC ACT AAT ATG GTG TGA TTC TTG CCA GAC GTC CAT TTT ATG TGA TAA
Lys Lys Asn Thr Asn MET Val  .  Phe Leu Pro Asp Val His Phe MET  .   .
 Lys Lys Thr Leu Ile Trp Cys Asp Ser Cys Gln Thr Ser Ile Leu Cys Asp Lys
  Lys Lys His  .  Tyr Gly Val Ile Leu Ala Arg Arg Pro Phe Tyr Val Ile Asn

                                   1269                      1296
ATA AAA CAC AAA CAC AAA GTT TTT ATA TTG TCT TTT ATT GAT ATG TTG CCT AAA
Ile Lys His Lys His Lys Val Phe Ile Leu Ser Phe Ile Asp MET Leu Pro Lys
 .  Asn Thr Asn Thr Lys Phe Leu Tyr Cys Leu Leu Leu Ile Cys Cys Leu Asn
  Lys Thr Gln Thr Gln Ser Phe Tyr Ile Val Phe Tyr  .  Tyr Val Ala  .  Ile

                                   1323                      1350
TAT TAT CAA AAC CTT CCC TAT AAC GGT AAG AGG ATA TTT GAA AAA TTC TAC GAT
Tyr Tyr Gln Asn Leu Pro Tyr Asn Gly Lys Arg Ile Phe Glu Lys Phe Tyr Asp
 Ile Ile Lys Thr Phe Pro Ile Thr Val Arg Gly Tyr Leu Lys Asn Ser Thr Ile
  Leu Ser Lys Pro Ser Leu  .  Arg  .  Glu Asp Ile  .  Lys Ile Leu Arg Ser

                                   1377
CGT AGT TTA CAA AAA TAT AAG TCG AC
Arg Ser Leu Gln Lys Tyr Lys Ser Thr
 Val Val Tyr Lys Asn Ile Ser Arg Pro
  .  Phe Thr Lys Ile  .  Val Asp
```

Fig. 4. Sequence and analysis of open reading frames in the FP locus. Map indicates the relative location of several restriction enzyme sites in the *Hind*III-J fragment of GmMNPV-C3. This fragment differs from the corresponding fragment (*Hind*III-I) of AcMNPV-E2 in the presence of a single *Bam*HI site (not shown) near the *Eco*RI site. (⬡) *Apa*I; (△) *Ava*I; (▽) *Cla*I; (▲) *Eco*RI; (■) *Hind*III; (●) *Sal*I; (○) *Xho*I. Multiple stop codons are encountered in all reading frames from either strand (A or B).

Short Exposure

Long Exposure

GmFP2 Sal/Sal
GmFP2 Sal/Pst (lg)
GmFP2 Sal/Pst (sm)
GmC3 Sal/Sal
GmC3 Sal/Hind
GmFP2 Sal/Sal
GmFP2 Sal/Pst (lg)
GmC3 Sal/Hind

Sf Cyto RNA
Tn Cyto RNA
GmC3 Cyto RNA
GmFP2 Cyto RNA
pUC 18 DNA
Calf Thymus DNA

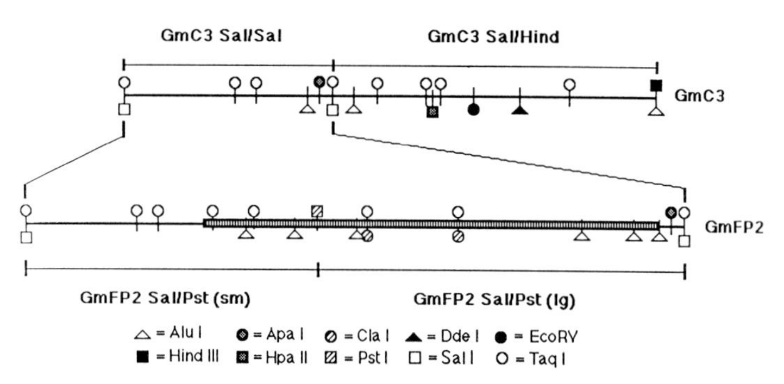

Fig. 5. Analysis of IFP2 homologous RNA transcripts in infected and uninfected cell cultures. Total cytoplasmic RNA was isolated from infected or uninfected cell cultures using the procedure described by Smith *et al.* (1983a) and 5 μg of each RNA species was hybridized to nitrocellulose paper using a dot blot manifold (Shleicher and Schuell) under conditions described by Thomas (1980). One microgram of pUC18 DNA and 10 μg of calf thymus DNA were included in the dot blot array as positive and negative controls, respectively. The restriction map indicates the location of the several DNA probes to which the RNAs were hybridized. The results indicate that transcripts are being produced from the *Sal/Hind* region and to a lesser extent the *Sal/Sal* region of the wild-type virus, GmC3. The levels of RNA homologous to the IFP2 insert-containing probes were similar (relative to the pUC18 standard) in the cells infected with GmFP2 or GmC3, indicating that IFP2 is probably not significantly transcribed in GmFP2 mutant-infected cells. Homologies between TN-368 cytoplasmic RNA and IFP2-containing sequences could be detected after longer exposures, while only scant homology (not significant over the background detected with the wild-type viral sequences) could be detected in uninfected SF21AE cells. These results demonstrate that the IFP2 sequence is transcriptionally active in the TN-368 cell genome, from which it originated.

TABLE I

Structural Features of Some Representative Eukaryotic Retrotransposons

Element	Short inverted terminal repeats (bp)	Long terminal repeats (bp)	Internal domain (kb)	Target site duplication (bp)	Host	Reference[a]
Tyl	None	338	5–5.6	5	Yeast	1, 2
copia	17 (13)	276	4.5	5	*Drosophila*	3, 4
412	10 (8)	481,571	6.5	4	*Drosophila*	5, 6
297	3	400	5.7	4	*Drosophila*	5
B104	5	429	7.9	3	*Drosophila*	7
mdg1	16 (13)	442–444	6.3	4	*Drosophila*	8
TED	10 (7)	270	6.8	4	Lepidoptera	9, 10
THE	None	350	1.6	5	Human	11
mys-1	None	343	2.4	6	Mouse	12
MMTV	6	1192	6.5	6	Mouse	13, 14
MMuSV	11	588	4.3	4	Mouse	15
MMuLV	11	517	7.5	4	Mouse	16

[a]References: 1, Farabaugh and Fink (1980); 2, Gafner and Philippsen (1980); 3, Dunsmuir *et al.* (1980); 4, Levis *et al.* (1980); 5, Potter *et al.* (1979); 6, Will *et al.* (1981); 7, Scherer *et al.* (1982); 8, Kulguskin *et al.* (1981); 9, Miller and Miller (1982); 10, Friesen *et al.* (1986); 11, Paulson *et al.* (1985); 12, Wichman *et al.* (1985); 13, Varmus (1982); 14, Majors and Varmus (1981); 15, Dhar *et al.* (1980); and 16, Van Beveren *et al.* (1980).

TABLE II
Structural Features of Some Eukaryotic P-like Transposable Elements

Element	Short inverted repeats (bp)	Internal domain (kb)	Target site duplication (bp)	Host	Reference[a]
Tc1	54	0.819	(2 or none)	Nematode	1
Ds family	11	Variable	8	Maize	2
Ac	11	4.5	8	Maize	3
Cin1	6	0.679	5	Maize	4
P elements	31	2.9	8	Maize	5,6
Sigma	8	0.325	5	Yeast	7

[a]References: 1, Rosenzweig *et al.* (1983a,b); 2, Doring *et al.* (1984); 3, Pohlman *et al.* (1984); 4, Shepherd *et al.* (1984); 5, O'Hare and Rubin (1983); 6, Karess and Rubin (1984); and 7, Sandmeyer and Olson (1982).

poson. Other structural features which are distinctive for most transposons are the terminal inverted repeats they contain as part of their sequence (Tables I–III). The resulting combination of short direct target site duplications flanking short indirect repeats of transposon sequences is considered a characteristic fingerprint of a transposition event.

Although we can say with certainty that inserted elements exhibiting these types of terminal redundancies must have inserted via transposition, those which do not exhibit such sequence duplications cannot be ruled out. One prokaryotic transposon, Tn554, has been described for which no terminal inverted repeats or target site duplications are found (Murphy and Lofdahl, 1984).

TABLE III
Structural Features of Several Eukaryotic FB-like Transposable Elements

Element	Terminal inverted repeats (bp)	Internal domain (bp)	Target site duplication (bp)	Host	Reference[a]
Mul	235	1350	9	Maize	1
FB family	Variable	Variable	9	*Drosophila*	2,3
TU family	840	Variable	8	Sea urchin	4

[a]References: 1, Freeling (1984); 2, Truett *et al.* (1981); 3, Collins and Rubin (1983); and 4, Hoffmann-Liebermann *et al.* (1985).

VI. FP MUTATION AND TRANSPOSON MUTAGENESIS OF NUCLEAR POLYHEDROSIS VIRUSES

Studies by Miller and Miller (1982) on the FP-D mutant of Potter and Miller (1980) confirmed the supposition that the inserted sequences in the region of the viral genome were derived from host cell DNA. The study further characterized the inserted sequence as a *copia*-like retrotransposon from the host cell genome. The characteristics that led to this determination were that the element was homologous to dispersed repetitive DNA of the host cell and that this repetitive DNA exhibited a different pattern of dispersion between larvae and cell DNAs.

The *copia*-like element was named TED and had terminal inverted repeats 0.27 kb in length, an overall size of 7.3 kb, and frequently deleted most of its length, leaving behind a copy of one terminal direct repeat (Miller and Miller, 1982). The flanking long terminal direct repeats are typical retroviruslike long terminal repeats (LTRs) which contain imperfect inverted repeats (7 of 10 nucleotides) at their termini, and the insertion of TED results in the direct duplication of a four-nucleotide target sequence, AATG (Friesen *et al.*, 1986). There is a typical retrovirus primer binding site located immediately adjacent to the 5′ LTR which is identical to that of the *copia*-like element 297. This primer binding site in 297 is complementary to the 3′ end of the *Drosophila* serine tRNA (Friesen *et al.*, 1986; Saigo *et al.*, 1984). Within the terminal repetition there is a retoviruslike promoter sequence, which in this case is capable of initiating transcription in two directions (Friesen *et al.*, 1986).

The study of Miller and Miller (1982) was important in establishing the phenomenon of transposon-mediated mutagenesis in baculoviruses as a group and NPVs in particular, but it failed to correlate transposon insertion with the FP phenotype. The ability to correlate transposon mutagenesis with an easily identified phenotypic marker such as plaque morphology is critical to development of a useful system for experimental manipulation. As examples of other experimental systems for studying transposition, Tn3 and Tn10 are both associated with antibiotic resistance in *E. coli*, so one may simply follow this trait through successive transfers of these transposons from one genome to the next (for reviews, see Heffron, 1983; Kleckner, 1983). Other examples include the *ADR2* or *his4* loci of yeast, which are used for studying Ty1 insertion (Williamson *et al.*, 1983; Farabaugh and Fink, 1980), the *white* locus of *Drosophila*, which has been utilized to study both *copia* and P element transposition (Karess and Rubin, 1984; see also Rubin, 1983, for review), and the *bronze*, *waxy*, and *shrunken* loci of maize, which were important in initially establishing transposition phenomena (see Fedoroff, 1983, for review).

Fraser *et al.* (1983) showed that a specific region of the AcMNPV or GmMNPV genome was associated with insertions of host DNA sequences and that these insertions could be correlated with the FP phenotype. However, there was no confirmation that any of the repetitive host sequences inserting at this point in the viral genome were transposons. Since none of these host DNA insertions exhibited any restriction site symmetry, it was necessary to resort to sequence analysis to confirm transposition as the mechanism of insertion of these host sequences. We hoped to find the characteristic duplications of a target sequence resident in the wild-type viral genome at the point of insertion and terminal inverted repeats at the ends of the inserted sequences.

Analysis of the mutant GmFP1 revealed no such characteristic features of the transposition event (Fraser *et al.*, 1985). The inserted DNA had an unusual ninefold repetition of TCGC toward the 5′ end, and insertion had occurred with deletion of a single base pair. No target site duplications were evident, and no terminal inverted or direct repeats were found in the inserted element itself. Transposition could not be confirmed as the mechanism of insertion. Homologous recombination could not be implicated either, since no sequence homology between viral DNA and inserted sequences was observed in the terminal regions analyzed.

The analysis of the inserted DNA in GmFP3 was more rewarding. Here we were able to confirm that transposition had occurred. In this case the inserted sequences were flanked by direct duplications of a viral target sequence, TTAA, and the insertion itself had terminal inverted repeats 7 bp in length (Fraser *et al.*, 1985). Of the mutants isolated and characterized, GmFP3 was known to revert spontaneously through loss of most if not all of the inserted host sequences (Fraser *et al.*, 1983). Together, these data confirm that the inserted host sequence in GmFP3 is a transposon, which we called TFP3.

This inserted TFP3 sequence may be an entire transposon, encoding its own transposase, or it may be part of a larger transposon in the host cell genome. Another possibility is that TFP3 represents a defective transposon from the host genome which is incapable of movement on its own. Data arguing against the latter possibility are that the revertants of GmFP3 were isolated from virus which had been propagated in SF21AE cells, and no homology between SF21AE cellular DNA and the inserted sequence in GmFP3 was ever detected (Fraser *et al.*, 1983; M. J.

Fraser, unpublished). However, more data are needed before any firm conclusions can be drawn.

VII. SIGNIFICANCE OF TRANSPOSON-MEDIATED MUTAGENESIS OF BACULOVIRUSES

Our current understanding of eukaryotic transposons suggests some means for movement of these elements between species (Syvanen, 1984). In the case of retroviruslike transposons, these elements may have arisen from either fortuitous or aborted infection of germ line tissues with retroviruses or defective derivatives of retroviruses. Since the replication mechanism of retroviruses involves insertion of the transposonlike DNA proviral genome into the host chromosomes, it is easy to envision subsequent mutation in genes responsible for assembly of virions resulting in defective provirus which cannot produce viable progeny (see Varmus, 1982, for review). Such events occurring in germ line tissues could eventually lead to establishment of defective provirus as part of the genome of the species.

Although transposon shuttling between cells has not yet been confirmed for any system, it is theoretically possible. The most likely mode of transposon movement between cells would be as insertions in viral genomes. Any virus which replicates in the nucleus would be a potential candidate. In fact, recombination between certain viral genomes and the host cell DNA has been documented. Exchanges occur between the SV40 and adenovirus genomes and chromosomal DNA of their host cells (Brockman, 1977; Jones and Shenk, 1978; Papamatheakis *et al.*, 1981; Deuring *et al.*, 1981; Larsen, 1982; Norkin and Tirrell, 1982). These exchanges result in loss of essential viral functions, producing defective particles. It is not difficult to envision such defective particles invading a host and perhaps establishing infection in germ line tissues. These particles could then conceivably deposit the passenger transposon in the nucleus of the host cell. Although this occurrence would be relatively rare, in evolutionary terms it could be quite significant.

The retroviruses are also capable of transporting DNA sequences between cells. Acquisition of cellular sequences occurs during the integrated proviral phase of the retrovirus replication cycle. This is a mechanism for formation of transforming retroviruses through acquisition of cellular oncogenes (see Varmus, 1982, for review). The same mechanism has the potential for vectoring other sequences as well, including cellular mobile elements.

The major drawback to examining transposon mutagenesis in most mammalian virus systems is the inability to detect and isolate many of these defective mutants. The constraints of capsid size significantly limit the amount of DNA that can be accommodated. This means that viral genomes acquiring host cell sequences must often give up essential genes. The resulting defective mutants are incapable of completing their replication cycles in the absence of helper viruses. This in turn makes it relatively difficult to specifically detect and isolate these defective mutants.

While transposon mutagenesis could occur with any DNA viruses that infect the nucleus, the NPVs possess unique features which allow such mutants to be easily detected. The first of these features is the complex replication process which provides superfluous genes related to OB formation. Many late and very late gene functions are geared toward formation of these intranuclear inclusions. Functions such as *de novo* envelope synthesis, envelopment of nucleocapsids in the nucleus, encapsidation of enveloped nucleocapsids in enlarging OBs, and envelopment of OBs in polyhedral envelopes (which are apparently produced by another major intranuclear inclusion, FM) are all nonessential to basic replication of NPVs. Mutation of such genes does not preclude formation of viable progeny virions but could lead to aberrations or reductions in the occlusion process which might result in mutants with a distinctive plaque morphology that could be screened with relative ease.

The other feature of baculoviruses that increases the ability to form and detect insertion mutants is their capacity to accommodate additional DNA sequences in the viral genome. Recent observations of a probable packaging mechanism for viral genomes in the cylindrical capsids suggest a capsid-filling model for genome packaging in NPVs (Bassemir *et al.*, 1983; Fraser, 1987) (Fig. 1). Such a mechanism could allow elongation of the capsid during genome packing. Observations by others have demonstrated that the capsid length can vary substantially. Double-length capsids have been described (Skuratovskaya *et al.*, 1982). The capsid length of viruses carrying insertions of 15 kb is greater than the length of wild-type capsids (Fraser, 1987). All these observations confirm that longer capsids are made around enlarged genomes.

The FP mutants of NPVs offer a unique system for examining transposon-mediated mutagenesis of viruses. Acquisition of cellular sequences by AcMNPV-like viruses in the *Hin*dIII region occurs frequently and results in viable progeny which are capable of independent propagation. Insertion of cellular sequences in this region also results in the distinctive FP plaque phenotype. The fact that this

FP locus is associated with transposon mutagenesis has now been established. Although we cannot demonstrate that transposon insertion is responsible for the FP phenotype in all cases, the majority of inserted sequences that have been characterized are at least movable elements in the cellular genome. We essentially have a mutable locus similar to the *white* locus of *Drosophila* or the *shrunken* locus of maize, providing a useful experimental system for studying this phenomenon.

VIII. BACULOVIRUSES AS VECTORS OF GENETIC ELEMENTS

Miller and Miller (1982) suggested the possible involvement of baculoviruses in vectoring genetic elements between species. This is theoretically possible only if certain criteria are satisfied. From the available experimental evidence it is already apparent that many of these criteria can be satisfied under natural conditions.

First, genetically significant transposon vectoring requires host survival of the virus infection. Investigators have identified several ways in which susceptible hosts might survive a baculovirus infection. One way is through exposure to elevated temperatures either during or following ingestion of virulent OBs (Thompson, 1959; Ignoffo, 1966b; Stairs, 1978; Kobayashi *et al.*, 1981). There is also experimental evidence that many species are less susceptible to lethal infection during later instars (Stairs, 1965; Ignoffo, 1966a; Boucias and Nordin, 1977). It is conceivable that virus could enter and establish a sublethal infection or be "cured" by one of these methods and in the process transfer the transposon to cellular genes in germ line tissues.

There is experimental support for NPV transmission and even replication in nonsusceptible species as well. Sherman and McIntosh (1979) demonstrated that AcMNPV can enter and even begin replication at low levels in cell cultures of *Aedes aegypti*, a nonsusceptible species. The infected cells do not die, but continually produce low levels of virus.

A recent study by Carbonell *et al.* (1985) using the chloramphenicol acetyltransferase (CAT) gene detected transcriptional activity and replication of AcMNPV in *Drosophila* cell cultures. This gene was placed under the control of a constitutive promoter from a retrovirus LTR and recombined into AcMNPV using a polyhedrin gene transfer vector (Pennock *et al.*, 1984). Cultured *Drosophila* cells inoculated with this recombinant virus expressed the CAT gene product, and levels of expression increased for a time. The increased levels of CAT expression were inhibited by the presence of cytosine arabinoside, indicating that increased CAT expression is coupled to viral replication. Similar results were obtained following anal administration of the CAT recombinant ECV in *A. aegypti* larvae. In this case CAT activity could be detected in the midgut cells.

These studies demonstrate that NPVs are capable of invading nonsusceptible host cells. They can conceivably infect any tissue, including germ line, of a nonsusceptible host such as *Drosophila* or *A. aegypti*. In fact, it is probably not necessary that a productive infection be established in the midgut for circulating virus to be found in the hemolymph. Granados and Lawler (1981) have shown that two routes of invasion are possible for baculoviruses in insects. If a virus can penetrate into midgut epithelial cells it could presumably be transported through the cell layer and bud directly into the hemolymph. In theory, it could then be transported to germ line tissue, penetrate, and initiate abortive replication in the nuclei of germ line cells. In this scheme a transposon carried by a mutant virus would be placed in the best possible position to cause significant genetic alterations.

The potential this transposon shuttling might have in terms of speciation has been demonstrated through analysis of transposition bursts in *Drosophila*. Gerasimova *et al.* (1984) present evidence that several known classes of transposons in *Drosophila* may be mobilized by similar control mechanisms. They propose that simultaneous mobilization of several transposon classes (transposition bursts) could be stimulated by the introduction of certain transposons into cells which previously had none of that type. Such transposition bursts would result in simultaneous mutagenesis at several genetic loci, increasing the probability of speciation. Transposon shuttling by baculoviruses could provide such a stimulus.

IX. CONCLUDING REMARKS

A region of the NPV genome which accepts host cell sequences on a frequent basis and results in the FP plaque morphology has been identified. In many cases the inserted host sequences may be transposable elements. This FP locus provides a phenotypic marker much like the *white* locus of *Drosophila* or the *shrunken* locus of maize, both of which have been studied extensively with respect to transposon mutagenesis. The major differences here are the ability to study the potential for vectoring inserted transposons to other species using viruses as vehicles and the possible involvement of such a phenomenon in speciation.

The NPVs have several features which make them ideally suited for this role. First, there are several possible loci in the viral genome which can undergo insertional mutagenesis without destroying its ability to infect and replicate (Burrand and Summers, 1982; Smith *et al.*, 1983a; Miller and Miller, 1982; Fraser *et al.*, 1983; Wang and Fraser, 1987). Although we have detected only two regions associated with host transposons thus far, there could be more which are not associated with a discernible phenotype. Second, NPVs undergo replication in the nucleus, where they are in close proximity to cellular genes. During uncoating and initiation of replication they are vulnerable to mutagenesis of this type. Third, these viruses are able to accommodate large pieces of additional DNA without loss of viability.

The ability to rapidly select and screen FP mutants of NPVs and the availability of numerous insect cell lines provide a unique *in vitro* system for verification of transposon shuttling by viruses. By exploring this phenomenon we may uncover a significant mechanism of speciation through transposon shuttling as well as gain important new insights into transposition mechanisms in eukaryotic organisms.

ACKNOWLEDGMENTS

I thank Dr. Stuart Tsubota for helpful comments and critical reading of the manuscript and Lynne Csiszar Cary for the restriction mapping analysis. Research presented in Fig. 1 was supported by U.S. Department of Agriculture competitive grant GAM 8400211 and that in Figs. 2–5 by U.S. Public Health Service grant 1RO1 AI 22610-01.

REFERENCES

Bassemir, U., Miltenburger, H. G., and David, P. (1983). Morphogenesis of nuclear polyedrosis virus from *Autographa californica* in a cell line from *Mamestra brassica* (cabbage moth). *Cell Tiss. Res.* **228,** 587–597.

Boucias, D. G., and Nordin, G. L. (1977). Interinstar susceptibility of the fall webworm, *Hypantria cunea*, to its nucleopolyhedrosis and granulosis viruses. *J. Invertebr. Pathol.* **30,** 68–75.

Brockman, W. W. (1977). Evolutionary variants of simian virus 40. *Prog. Med. Virol.* **23,** 69–88.

Burand, J. P., and Summers, M. D. (1982). Alteration of *Autographa californica* nuclear polyhedrosis virus DNA upon serial passage in cell culture. *Virology* **101,** 286–290.

Carbonell, L. F., Klowden, M. J., and Miller, L. K. (1985). Baculovirus-mediated expression of bacterial genes in dipteran and mammalian cells. *J. Virol.* **56,** 153–160.

Collins, M., and Rubin, G. M. (1983). High frequency precise excision of the *Drosophila* foldback transposable element. *Nature (London)* **303,** 259–260.

Deuring, R., Klotz, G., and Doerfler, W. (1981). An unusual symmetric recombinant between adenovirus type 12 DNA and human cell DNA. *Proc. Natl. Acad. Sci. U.S.A.* **78,** 3142–3146.

Dhar, R., McClements, W. L., Enquist, L. W., and Vande Woude, G. F. (1980). Nucleotide sequences of integrated Maloney sarcoma provirus long terminal repeats and their host and viral junctions. *Proc. Natl. Acad. Sci. U.S.A.* **77,** 3937–3941.

Doring, H. P., Tillmann, E., and Starlinger, P. (1984). DNA sequence of the maize transposable element *Dissociation*. *Nature (London)* **307,** 127–130.

Dunsmuir, P., Brorein, W. J., Simon, M. A., and Rubin, G. M. (1980). Insertion of the *Drosophila* transposable element *copia* generates a 5 base pair duplication. *Cell (Cambridge, Mass.)* **21,** 575–579.
Farabaugh, P. J., and Fink, G. R. (1980). Insertion of the eucaryotic transposable element Tyl creates a 5-base-pair duplication. *Nature (London)* **286,** 352–355.
Fedoroff, N. V. (1983). Controlling elements in maize. *In* "Mobile Genetic Elements" (J. A. Shapiro, ed.), pp. 1–63. Academic Press, New York.
Fraser, M. J. (1981). The genotypic characterization of MP and FP variants of *Galleria mellonella* nuclear polyhedrosis virus. Dissertation, Ohio State University, Columbus.
Fraser, M. J. (1987). Ultrastructural observations of virion maturation in *Autographa californica* nuclear polyhedrosis virus infected *Spodoptera frugiperda* cell cultures. *J. Ultrastruct. Mol. Struct. Res.* **95,** 189–195.
Fraser, M. J., and Hink, W. F. (1982a). Comparative sensitivity of several plaque assay techniques employing TN-368 and IPLB-SF21AE insect cell lines for the plaque variants of *Galleria mellonella* nuclear polyhedrosis virus. *J. Invertebr. Pathol.* **40,** 89–97.
Fraser, M. J., and Hink, W. F. (1982b). The isolation and characterization of MP and FP plaque variants of *Galleria mellonella* nuclear polyhedrosis virus. *Virology* **117,** 366–378.
Fraser, M. J., and McCarthy, W. J. (1984). The detection of FP plaque variants of *Heliothis zea* nuclear polyhedrosis virus grown in the IPLB-HZ 1075 insect cell line. *J. Invertebr. Pathol.* **43,** 427–429.
Fraser, M. J., Smith, G. E., and Summers, M. D. (1983). The acquisition of host cell DNA sequences by baculoviruses: Relation between host DNA insertions and FP mutants of *Autographa californica* and *Galleria mellonella* nuclear polyhedrosis viruses. *J. Virol.* **47,** 287–300.
Fraser, M. J., Brusca, J. S., Smith, G. E., and Summers, M. D. (1985). Transposon-mediated mutagenesis of a baculovirus. *Virology* **145,** 356–361.
Freeling, M. (1984). Plant transposable elements and insertion sequences. *Annu. Rev. Plant Physiol.* **35,** 277–298.
Friesen, P. D., and Miller, L. K. (1985). Temporal regulation of baculovirus RNA: Overlapping early and late transcripts. *J. Virol.* **54,** 392–400.
Friesen, P. D., Rice, W. C., Miller, D. W., and Miller, L. K. (1986). Bidirectional transcription from a solo long terminal repeat of the retrotransposon TED: Symmetrical RNA start sites. *Mol. Cell. Biol.* **6,** 1599–1607.
Gafner, J., and Philippsen, P. (1980). The yeast transposon Tyl generates duplications of target DNA on insertion. *Nature (London)* **286,** 414–418.
Gerasimova, T. I., Mizrokhi, L. J., and Georgiev, G. P. (1984). Transposition bursts in genetically unstable *Drosophila melanogaster. Nature (London)* **309,** 714–716.
Granados, R. R., and Lawler, K. A. (1981). *In vivo* pathway of *Autographa californica* Baculovirus invasion and infection. *Virology* **108,** 297–308.
Guarino, L. A., and Summers, M. D. (1986). Functional mapping of a trans-activating gene required for expression of a baculovirus delayed-early gene. *J. Virol.* **57,** 563–571.
Heffron, F. (1983). Tn3 and its relatives. *In* "Mobile Genetic Elements" (J. A. Shapiro, ed.), pp. 223–260. Academic Press, New York.
Hink, W. F., and Strauss, E. (1976). Replication and passage of alfalfa looper nuclear polyhedrosis virus plaque variants in cloned cell cultures and larval stages of four host species. *J. Invertebr. Pathol.* **27,** 49–55.

Hink, W. F., and Vail, P. V. (1973). A plaque assay for titration of alfalfa looper nuclear polyhedrosis virus in the cabbage looper (TN-368) cell line. *J. Invertebr. Pathol.* **22,** 168–174.

Hoffmann-Liebermann, B., Liebermann, D., Kedes, L. H., and Cohen, S. N. (1985). Tu elements: A heterogeneous family of modularly structured eucaryotic transposons. *Mol. Cell. Biol.* **5,** 991–1001.

Ignoffo, C. M. (1966a). Effects of age on mortality of *Heliothis zea* and *Heliothis virescens* larvae exposed to a nuclear polyhedrosis virus. *J. Invertebr. Pathol.* **8,** 279–283.

Ignoffo, C. M. (1966b). Effects of temperature on mortality of *Heliothis zea* larvae exposed to sublethal doses of a nuclear polyhedrosis virus. *J. Invertebr. Pathol.* **8,** 290–292.

Jones, N., and Shenk, T. (1978). Isolation of deletion and substitution mutants of adenovirus type 5. *Cell (Cambridge, Mass.)* **13,** 181–188.

Karess, R. E., and Rubin, G. R. (1984). Analysis of P transposable element functions in *Drosophila. Cell (Cambridge, Mass.)* **38,** 135–146.

Kelly, D. C., and Lescott, T. (1981). Baculovirus replication: Protein synthesis in *Spodoptera frugiperda* cells infected with *Trichoplusia ni* nuclear polyhedrosis virus. *Microbiologica (Bologna)* **4,** 35–57.

Kleckner, N. (1983). Transposon Tn10. *In* "Mobile Genetic Elements" (J. A. Shapiro, ed.), pp. 261–294. Academic Press, New York.

Kobayashi, M., Inagaki, S., and Kawase, S. (1981). Effect of high temperature on the development of nuclear polyhedrosis virus in the silkworm, *Bombyx mori. J. Invertebr. Pathol.* **38,** 386–396.

Kulguskin, V. V., Ilyin, Y. V., and Georgiev, G. P. (1981). Mobile dispersed genetic element MDG1 of *Drosophila melanogaster:* Nucleotide sequence of long terminal repeats. *Nucleic Acids Res.* **9,** 3451–3464.

Larsen, S. H. (1982). Evolutionary variants of mouse adenovirus containing cellular DNA sequences. *Virology* **116,** 573–580.

Laski, F. A., Rio, D. C., and Rubin, G. M. (1986). Tissue specificity of *Drosophila* P element transposition is regulated at the level of mRNA splicing. *Cell (Cambridge, Mass.)* **44,** 7–19.

Levis, R., Dunsmuir, P., and Rubin, G. M. (1980). Terminal repeats of the *Drosophila* transposable element *copia:* Nucleotide sequence and genomic organization. *Cell (Cambridge, Mass.)* **21,** 581–588.

Lubbert, H., and Doerfler, W. (1984). Mapping of early and late transcripts encoded by the *Autographa californica* nuclear polyhedrosis virus genome: Is viral RNA spliced? *J. Virol.* **50,** 497–506.

Maeda, S., Kawai, T., Obinata, M., Fujiwara, H., Horiuchi, T., Saeki, Y., Sato, Y., and Furusawa, M., (1985). Production of human a-interferon in silkworm using a baculovirus vector. *Nature (London)* **315,** 592–594.

Majors, J. E., and Varmus, H. E. (1981). Nucleotide sequences at host–proviral junctions for mouse mammary tumour virus. *Nature (London)* **289,** 253–258.

Miller, D. W., and Miller, L. K. (1982). A virus mutant with an insertion of a *copia*-like transposable element. *Nature (London)* **299,** 562–564.

Murphy, E., and Lofdahl, S. (1984). Transposition of Tn554 does not generate a target duplication. *Nature (London)* **307,** 292–294.

Norkin, L. C., and Tirrell, S. M. (1982). Emergence of simian virus 40 variants during serial passage of plaque isolates. *J. Virol.* **42,** 730–733.

O'Hare, K., and Rubin, G. M. (1983). Structures of P transposable elements and their sites of insertion and excision in the *Drosophila melanogaster* genome. *Cell (Cambridge, Mass.)* **34,** 25–35.

Papamatheakis, J., Lee, T. N. H., Thayer, R. E., and Singer, M. F. (1981). Recurring defective variants of simian virus 40 containing monkey DNA segments. *J. Virol.* **18,** 1040–1050.

Paulson, K. E., Deka, N., Schmid, C. W., Misra, R., Schindler, C. W., Rush, M. G., Radyk, L., and Leinwand, L. (1985). A transposon-like element in human DNA. *Nature (London)* **316,** 359–361.

Pennock, G. D., Shoemaker, C., and Miller, L. K. (1984). Strong and regulated expression of *E. coli* B-galactosidase in insect cells using a baculovirus vector. *Mol. Cell. Biol.* **4,** 399–406.

Pohlman, R. F., Fedoroff, N. V., and Messing, J. (1984). The nucleotide sequence of the maize controlling element *Activator. Cell (Cambridge, Mass.)* **37,** 635–643.

Potter, K. N., and Miller, L. K. (1980). Correlating genetic mutations of a baculovirus with the physical map of the DNA genome. *In* "Animal Virus Genetics" (B. N. Fields and R. Jaenisch, eds.), pp. 71–80. Academic Press, New York.

Potter, K. N., Faulkner, P., and MacKinnon, E. A. (1976). Strain selection during serial passage of *Trichoplusia ni* nuclear polyhedrosis virus. *J. Virol.* **18,** 1040–1050.

Potter, S. S., Brorein, W. J., Dunsmuir, P., and Rubin, G. M. (1979). Transposition of elements of the 412, *copia,* and 297 dispersed repeated gene families in *Drosophila. Cell (Cambridge, Mass.)* **17,** 415–427.

Rosenzweig, B., Liao, L. W., and Hirsh, D. (1983a). Sequence of the *C. elegans* transposable element Tcl. *Nucleic Acids Res.* **11,** 4201–4209.

Rosenzweig, B., Liao, L. W., and Hirsh, D. (1983b). Target sequences for the *C. elegans* transposable element Tcl. *Nucleic Acids Res.* **11,** 7137–7140.

Rubin, G. M. (1983). Dispersed repetitive DNAs in *Drosophila. In* "Mobile Genetic Elements" (J. A Shapiro, ed.), pp. 329–361. Academic Press, New York.

Saigo, K., Kugimiya, W., Matasuo, Y., Inouye, S., Yoshioka, K., and Yuki, S. (1984). Identification of the coding sequence for a reverse transcriptase-like enzyme in a transposable genetic element in *Drosophila melanogaster. Nature (London)* **312,** 659–661.

Sandmeyer, S., and Olson, M. V. (1982). Insertion of a repetitive element at the same position in the 5′ flanking regions of two dissimilar yeast tRNA genes. *Proc. Natl. Acad. Sci. U.S.A.* **79,** 7674–7678.

Scherer, G., Tschudi, C., Perera, J., Delius, H., and Pirrotta, V. (1982). B104, a new dispersed repeated gene family in *Drosophila melanogaster* and its analogies with retroviruses. *J. Mol. Biol.* **157,** 435–452.

Shepherd, N. S., Schwarz-Sommer, Z., Blumberg vel Spalve, J., Gupta, M., Weinand, U., and Saedler, H. (1984). Similarity of the Cinl repetitive family of *Zea mays* to eucaryotic transposable elements. *Nature (London)* **307,** 185–187.

Sherman, K. E., and McIntosh, A. H. (1979). Baculovirus replication in a mosquito (dipteran) cell line. *Infect. Immun.* **26,** 232–234.

Skuratovskaya, I. N., Fodor, I., and Strokovskaya, L. I. (1982). Properties of the nuclear polyhedrosis virus of the great wax moth: Oligomeric circular DNA and the characteristics of the genome. *Virology* **120,** 465–471.

Smith, G. E., Vlak, J. M., and Summers, M. D. (1983a). Physical analysis of *Autographa californica* nuclear polyhedrosis virus transcripts for polyhedrin and 10,000-molecular-weight protein. *J. Virol.* **45,** 215–225.

Smith, G. E., Fraser, M. J., and Summers, M. D. (1983b). Molecular engineering of the *Autographa californica* nuclear polyhedrosis virus genome: Deletion mutations within the polyhedrin gene. *J. Virol.* **46,** 584–593.

Smith, G. E., Summers, M. D., and Fraser, M. J. (1983c). Production of human beta interferon in insect cells infected with a baculovirus expression vector. *Mol. Cell. Biol.* **3,** 2156–2165.

Stairs, G. R. (1965). Quantitative differences in susceptibility to nuclear polyhedrosis virus among larval instars of the forest tent caterpillar, *Malacosoma disstria* (Hubner). *J. Invertebr. Pathol.* **7,** 427–429.

Stairs, G. R. (1978). Effects of a wide range of temperatures on the development of *Galleria mellonella* and its specific baculovirus. *Environ. Entomol.* **7,** 297–299.

Stoltz, D. G., Pavan, C., and da Cunha, A. B. (1973). Nuclear polyhedrosis virus: A possible example of de novo intranuclear membrane morphogenesis. *J. Gen. Virol.* **19,** 145–150.

Syvanen, M. (1984). The evolutionary implications of mobile genetic elements. *Annu. Rev. Genet.* **18,** 271–293.

Thomas, P. S. (1980). Hybridization of denatured RNA and small DNA fragments transferred to nitrocellulose. *Proc. Natl. Acad. Sci. U.S.A.* **77,** 5201–5205.

Thompson, C. G. (1959). Thermal inhibition of certain polyhedrosis virus diseases. *J. Insect Pathol.* **1,** 189–192.

Truett, M. A., Jones, R. S., and Potter, S. S. (1981). Unusual structure of the FB family of transposable elements in *Drosophila. Cell (Cambridge, Mass.)* **24,** 753–763.

Van Beveren, C., Goddard, J. G., Berns, A., and Verma, I. M. (1980). Structure of Moloney murine leukemia viral DNA: Nucleotide sequence of the 5′ long terminal repeat and adjacent cellular sequences. *Proc. Natl. Acad. Sci. U.S.A.* **77,** 3307–3311.

Varmus, H. E. (1982). Form and function of retrovirus proviruses. *Science* **216,** 812–820.

Volkman, L. E., and Summers, M. D. (1975). Nuclear polyhedrosis virus detection: Relative capabilities of clones developed from *Trichoplusia ni* ovarian cell line TN-368 to serve as indicator cells in a plaque assay. *J. Virol.* **16,** 1630–1637.

Wang, and Fraser, M. J. (1987). In preparation.

Wichman, H. A., Potter, S. S., and Pine, D. S. (1985). Mys, a family of mammalian transposable elements isolated by phylogenetic screening. *Nature (London)* **317,** 77–81.

Will, B. M., Bayev, A. A., and Finnegan, D. J. (1981). Nucleotide sequence of terminal repeats of 412 transposable elements of *Drosophila melanogaster. J. Mol. Biol.* **153,** 897–915.

Williamson, V. E., Cox, D., Young, E.T., Russell, D. W., and Smith, M. (1983). Characterization of transposable element-associated mutations that alter yeast alcohol dehydrogenase II expression. *Mol. Cell. Biol.* **3,** 20–31.

Wood, H. A. (1980). Isolation and replication of an occlusion body-deficient mutant of the *Autographa californica* nuclear polyhedrosis virus. *Virology* **105,** 338–344.

18
Expression of Foreign Genes in Insect Cells

LOIS K. MILLER

I. INTRODUCTION

Using recombinant DNA technology, it is now possible to insert foreign genes into insect baculoviruses and achieve expression of those genes in insect cells rapidly and efficiently (Maeda *et al.*, 1985; Pennock *et al.*, 1984; Smith *et al.*, 1983). Two applications of baculovirus expression vector technology have immediate commercial applications: production of proteins of biomedical or agricultural importance and improvement of the efficacy of baculoviruses as biological pesticides. These applications of baculoviruses are discussed from the perspective of the advantages and limitations in their use.

II. BACULOVIRUSES FOR COMMERCIAL PRODUCTION OF MEDICALLY OR AGRICULTURALLY IMPORTANT PROTEINS

One of the most exciting potentials in the biotechnology industry is the commercial production of proteins which produce a biological response in a human or animal that protects or improves the health or productivity of the individual. Often proteins that elicit a biological response are produced naturally in only minute quantities in the organism and their isolation is not practical for widespread use in health

BIOTECHNOLOGY IN INVERTEBRATE
PATHOLOGY AND CELL CULTURE

care or farm management. Recombinant DNA technology has afforded a means of producing these proteins outside the organism in which they naturally occur, but production of the proteins on a commercial scale requires a cell-based gene expression system that can "manufacture" the protein in a biologically active form. At one time, many biotechnologists thought that most protein production would be feasible using typical fermentation cell systems such as bacterial or yeast cells. But it has become increasingly clear over the past 5 years that some proteins, particularly those of higher eukaryotes (e.g., plants and animals), cannot easily be expressed in a biologically active form in these prokaryotic or lower eukaryotic cell systems. Specific events or environment during or following translation of the genetic information into protein may be essential for the proper folding, stability, and/or inherent enzymatic activity of the protein. Hence, the biotechnology industry is turning to higher eukaryotic cell systems for the expression of eukaryotic genes, particularly those that are active in the blood. Proteins of current interest include modulators of immune systems such as interleukin 2, growth factors, tumor antagonists, blood clotting or anticlotting factors, protein hormones, antibodies, and vaccines.

The need for a eukaryotic expression system was predicted and the basic prototype for an insect baculovirus host–vector system was outlined at an early stage (Miller, 1981). Development of the system to a practical level required knowledge of the precise location of key viral genes such as the gene encoding the major protein of the occlusion body, the polyhedrin gene. The polyhedrin gene is a key one in the insect baculovirus host–vector system for several reasons. It is a nonessential gene for virus replication in cell culture, and yet it is synthesized in enormous quantities at late times (24–70 hr) in the infection process. Foreign genes inserted in place of the polyhedrin gene, but remaining under the control of the temporally regulated and active polyhedrin promoter, can be expressed abundantly in a regulated fashion. Replacement of the polyhedrin gene with a foreign gene also results in a specific viral phenotype which can be visually selected in a relatively rapid fashion, thus facilitating recombinant virus isolation and propagation. Principles and methods of the insect baculovirus host–vector system have been described (Miller *et al.*, 1986). Numerous successful demonstrations of the insect beculovirus expression vector system have been reported (Maeda *et al.*, 1985; Miller *et al.*, 1986; Miyamoto *et al.*, 1985; Pennock *et al.*, 1984; Smith *et al.*, 1983).

In considering the advantages that insect baculovirus expression vectors offer, it is important to compare the system with other higher eukaryotic expression systems rather than bacterial or yeast systems.

Because inexpensive fermentation systems can be developed for bacteria and yeast, it is preferable to use such fermentation systems if proteins with equivalent biological activity can be produced in such cells. For the cases in which fermentation systems fail to produce active proteins, one must consider and compare higher eukaryotic (plant, invertebrate, and vertebrate animal) expression systems.

Two of the greatest advantages of a baculovirus expression system are the rapidity and simplicity with which expression can be achieved. Starting with a recombinant plasmid construct containing the foreign gene of interest, it takes roughly 2–3 weeks to insert the foreign gene into the virus, select recombinant viruses, produce a substantial virus stock, and obtain high-level foreign gene expression in the appropriate cell culture. This time frame compares very favorably with mammalian cell expression systems based on gene amplification technology, which takes months to nurse to full expression. For industries wishing to develop second-generation biologicals through site-directed mutagenesis of the genes, a rapid and simple expression system becomes essential.

In addition to saving time in obtaining sizable quantities of protein for testing biological parameters, the virus system can reduce costs involved in production since a single cell line can support the production of a variety of different bioproducts. Optimizing cell growth parameters, nutritional requirements, and so forth for a variety of cell lines can be time-consuming and costly. Switching from one cell line to another in a production facility is also time-consuming, costly, and inconvenient.

Developing insect cell cultures to production scale should present no more difficulties than mammalian cell growth, although the conditions for growth will be considerably different. Insect cells, for instance, grow optimally at 27° rather than 37°C. Bovine serum is frequently used as an additive to insect cell media in the laboratory but may be substituted with egg yolk in mass-scale production (see Miller *et al.*, 1986). The use of protein-free media may also be considered during the final stages of viral gene expression. At the laboratory scale, it is possible to remove the medium approximately 18 hr after infection, substitute a phosphate-buffered saline solution, and allow foreign gene expression during the final phase of virus development in the absence of a protein-based medium. Such a production scheme is thought to facilitate the purification of the protein product, a process which can be difficult if high levels of other proteins are present. The biotechnology industry still seems to be divided on the importance of these issues, probably because it is still in the development phase in terms of mass-scale protein production techniques. It may be noteworthy that one suggested production method is the use of insect larvae as protein production machines (Maeda *et al.*, 1985). In this scheme, *Bombyx mori* (silkworm) larvae are injected with recombinant virus and the foreign protein product retrieved from the hemolymph of the infected cell after several days of infection. Whether mass-scale collection of hemolymph or purification from intact larvae will be physically and/or economically feasible remains to be determined. Whether larval-derived biologicals will be acceptable to the Food and Drug Administration will also be a critical question.

Insect baculovirus expression systems have several additional technical advantages, some of which have not yet been fully utilized. The baculoviruses have a large capacity for foreign gene inserts. Although only 14 kb of additional DNA has been inserted to date, it is estimated that 100 kb or more of foreign DNA can be inserted stably into the baculovirus DNA genome, which is packaged in a rod-shaped nucleocapsid. Foreign gene inserts are generally quite stable with little or no pressure for gene deletion; genes carrying direct or inverted repeat sequences, however, may experience deletions or inversions as a result of homologous recombination. The polyhedrin promoter is not activated until after the bulk of infectious virus particles are made, so normal quantities of progeny recombinant virus are usually made. It is possible that this will also be true of viruses producing cytotoxic gene products but this remains to be demonstrated.

In addition to the natural polyhedrin promoter, other promoters may also be used to express more than one gene. The promoter for the abundant late p10 protein, for instance, is activated in similar fashion to the polyhedrin promoter (Friesen and Miller, 1985). Foreign gene insertion at this site as well as the polyhedrin site would allow the simultaneous expression of two genes and would be clearly advantageous for the expression of the peptides of multimeric proteins (proteins composed of two or more polypeptide chains) or possibly proteins involved in a common biosynthetic pathway.

Although baculovirus expression systems have now been shown to be useful for producing high levels of a variety of foreign proteins, there is no guarantee that every protein will be produced at high levels or in a biologically active form. Polyhedrin itself is normally expressed at a level of approximately 0.5–1 mg/ml of cells. The highest production from a recombinant virus was a polyhedrin/β-galactosidase fusion at approximately 200–400 μg/ml of cells (Pennock *et al.*, 1984). Other foreign genes are expressed at levels ranging from 1 to 100 μg/ml (Miller *et al.*, unpublished results; Miyamato *et al.*, 1985; Smith *et al.*, 1983). The factors responsible for low-level expression of some proteins have not been determined. In developing a system for expression of polyoma virus large T antigen, we observed only low levels (approximately 1 μg/ml) of T-antigen production by pulse-labeling proteins with [^{35}S]methionine. Despite the low levels of T antigen observed, the baculovirus expression system compares very favorably with other higher eukaryotic systems that have been described for polyoma T-antigen expression. Thus what baculovirologists consider to be poor expression in their system may be considered significant expression by scientists using other expression systems. Nevertheless, some genes are expressed very poorly by the insect baculovirus expression system or, if expressed, are biologically inactive. This situation is certainly not unique to the baculovirus system and will probably prove to be true of all expression systems.

The question of whether proteins produced in insect cells will have the same structure or posttranslational modifications as those produced in mammalian cells is a multifaceted one. In the cases studied to date, insect cells recognized the signal sequence of the mammalian protein, cleaved the signal sequence appropriately, and secreted the protein (personal communications of D. W. Miller, S. Maeda, and G. E. Smith). However, there are substantial reasons to suspect that insect cells will not glycosylate mammalian proteins in an identical fashion to mammalian cells. Although insect cells are capable of synthesizing *N*-glycosyl-linked carbohydrate chains assembled by tunicamycin-sensitive steps, they do not appear to synthesize complex *N*-glycans and appear to lack the *N*-acetylglucosaminyl-, galactosyl-, and sialyltransferases necessary for the terminal glycosylation of mature mammalian glycoproteins (Butters *et al.*, 1981). In a comparative study of Sindbis virus proteins grown in mosquito cells or vertebrate cells, $Man_3GlcNAc_2$ glycans were found in insect-derived proteins at the two glycosylation sites normally occupied by complex glycans in vertebrate-derived proteins (Hsieh and Robbins, 1984). Thus it appears that insects recognize the same core glycosylation sites as mammalian cells but do not terminally glycosylate in the same manner and may instead reduce the core to simpler glycans such as $Man_3GlcNAc_2$. Clearly the nature of the differences in insect and mammalian glycosylation events needs further study and more examples. However, even if insect glycosylation is different, it is not clear that the exact nature of the glycosyl moieties will be important for the function of the proteins. It is possible that insect-specific glycosylation will improve the stability or activity of some glycoproteins. It must also be noted that different mammalian cells show different types of glycosylation and glycosylation may be tissue-specific. Thus it is not possible to claim, without direct evidence, that proteins produced in mammalian cell cultures are identical to those produced in the mammal itself. Whether insects posttranslationally modify proteins in an identical fashion may become a moot point, but it will be of considerable basic interest to more thoroughly define what the differences are.

Thus an insect baculovirus host–vector expression system appears to have great promise for the production of proteins of use in health and agriculture. The simplicity, rapidity, high production levels, and potential ease of production are unrivaled by other higher eukaryotic expression systems currently available. The limitations of the system are that it is more expensive than fermentation (bacterial and yeast expression systems) and it does not express some proteins in active or stable form. The posttranslational modifications may not be identical to those of mammalian cells, but it is not clear that specific posttranslational modifications will be necessary for active, stable protein production. There is every reason to believe that recombinant baculovirus infection of mass-scale insect cultures or silkworm larvae could be an extremely effective and efficient means of producing biologicals commercially.

III. IMPROVED BACULOVIRUS PESTICIDES

Some baculoviruses are highly pathogenic in their insect hosts and cause natural epizootic diseases within insect host populations. The viruses play a major role in the natural control of certain insect populations and this observation has led to their development as microbial pest control agent (for reviews, see Burges, 1981; Carter, 1984; Miller *et al.*, 1983). Being more expensive to produce than chemicals, the viruses are clearly at a disadvantage compared to chemical insecticides from a cost-competitive standpoint, at least when the chemicals are newly released and the insect pests have not yet developed resistance to them. It is thought that if regulatory pressure were brought to bear on the use of chemical pesticides, viruses could find a prominent role in integrated pest management systems. Several baculoviruses have been registered by the U.S. Environmental Protection Agency for the control of specific insect pests on specific crops or forests.

Although the use of natural viruses has met with considerable success in specific cases, more widespread application of viruses is limited by a variety of features, including the delayed effect and narrow host range of the viruses. Baculoviruses generally do not kill their host until they have undergone multiple cycles of replication within the host over a 5–7-day period from the time of ingestion to morbidity. To a farmer in a crisis, this is unacceptably slow because the insect continues to feed on the crop for an extended period of time. Although such problems might be alleviated with a more sophisticated pest management strategy, including earlier sensing of population dynamics, the delayed efficacy of viral pesticides is a current handicap to their application within the standard agricultural framework. The narrow host range of baculoviruses is perceived to be a problem at the commercial production and marketing level, where it is claimed that it is too expensive to produce and market a different virus for each different pest species. There are additional efficacy problems which are still considered a handicap to the use of the viruses as pesticides; particularly significant is the UV inactivation of viruses applied by conventional spraying techniques to the surface of the plants.

The ability to genetically engineer baculoviruses may provide a means of correcting these perceived defects in the efficacy and marketability of viral pesticides. Some critics may argue that nature has designed the very best viruses possible through millions of years of evolution. But such an argument loses sight of the fact that viruses evolved under the constraint that they maintain both themselves and their hosts during their evolution. If viruses are produced and applied by humans in specific (and artifical) crop production arenas, it should be possible to optimize their genetics so that they can perform specific pesticide functions in a predicted fashion. For instance, a virus's sensitivity to UV light may be reduced by supplying the virus with a more effective DNA repair system. The rapidity with which a virus affects its host's feeding behavior might be increased by expressing an insect-specific neurotoxin gene or the gene for an insect-specific hormone affecting feeding behavior or development. At present, these possibilities remain hypothetical, but there are many avenues to explore in furthering the development of these biological pesticides.

During studies on the regulation of baculovirus gene expression, an interesting observation was made regarding virus host range that could have profound implications for and applications to viral pesticide development (Carbonell *et al.*, 1985). When the bacterial gene encoding chloramphenicol acetyltransferase (CAT) was inserted into a baculovirus genome under the control of a broad-host-range promoter, the Rous sarcoma virus long terminal repeat (RSV-LTR), the CAT gene was expressed relatively early in the infection of permissive lepidopteran cells. When expression of the baculovirus-borne CAT gene was tested in nonpermissive dipteran cells infected with recombinant virus, substantial CAT gene expression was observed but expression from a polyhedrin promoter-controlled gene was not observed. The results indicate that the barrier to infection in dipteran cells is not at the cell surface (e.g., virus entry) but at the level of gene expression (Carbonell *et al.*, 1985). This conclusion is further supported by recent studies of baculovirus transcription in dipteran cells (W. C. Rice and L. K. Miller, unpublished results).

We have proposed from these studies that the effective host range of a virus might be expanded by inserting an insect-specific neurotoxin or behavior-modifying gene into the baculovirus genome so that the gene is under the control of a promoter that is expressed early on virus entry into the cell. This would not actually alter the host range for virus replication but could alter the ability of the virus to influence host behavior and thereby increase the "effective" host range of the virus as a pesticide. A system for testing this hypothesis is currently in the developmental stage. Interestingly and importantly, mammalian cells exhibited a block in baculovirus entry and expression, providing evidence for the relative safety of these viruses with regard to vertebrates (Carbonell *et al.*, 1985).

One of the limiting features of these applications of genetic engineering technology to viral pesticide development may be the current regulations regarding the release of genetically modified microorganisms into the environment. How and when these issues are resolved will certainly affect industry's interest in developing microbial forms of insect pest control. The key questions that should be addressed by the regulatory agencies are whether the host range of the engineered virus has increased to include ecologically valuable species and whether the genetically engineered virus has a competitive advantage for survival outside the agrisystem in which it is to be employed. The ability to more effectively monitor viral entry and gene expression in nontarget host cells using the sensitive CAT assay could be used advantageously in future safety testing of environmentally released viruses.

The future of viral pesticides will depend largely on the ability of scientists to improve the efficacy of these viruses within the framework of existing or future agricultural and silvicultural practices. Since these improved pesticides must be approved by government regulatory agents before release, it will be necessary to consider factors such as host range and survival in product development. The impasse concerning the release of genetically modified organisms into the environment must be resolved so that academia and industry can have a rational framework for developing and improving this form of insect pest control, which holds so much promise for the selective elimination of insect pest species and the reestablishment of predator populations.

ACKNOWLEDGMENTS

The support of U.S. Department of Agriculture competitive research grant 85-CRCR-1-1796 for viral pesticide development is gratefully acknowledged. Research on the development of insect baculovirus vectors for expression of genes of biomedical importance was supported in part by Public Health Service grant AI 23719 from the National Institute of Allergy and Infectious Diseases.

REFERENCES

Burges, H. D., ed. (1981). "Microbial Control of Pests and Plant Diseases 1970–1980." Academic Press, London.

Butters, T. D., Hughes, R. C., and Vischer, P. (1981). Steps in the biosynthesis of mosquito cell membrane glycoproteins and the effects of tunicamycin. *Biochim. Biophys. Acta* **640,** 672–686.

Carbonell, L. F., Klowden, M. J., and Miller, L. K. (1985). Baculovirus-mediated expression of bacterial genes in dipteran and mammalian cells. *J. Virol.* **56,** 153–160.

Carter, J. B. (1984). Viruses as pest-control agents. *Biotechnol. Genet. Eng. Rev.* **1,** 375–419.

Friesen, P. D., and Miller, L. K. (1985). Temporal regulation of baculovirus RNA: Overlapping early and late transcripts. *J. Virol.* **54,** 392–400.

Hsieh, P., and Robbins, P. W. (1984). Regulation of asparagine-linked oligosaccharide processing. *J. Biol. Chem.* **259,** 2375–2382.

Maeda, S., Kawai, T., Obinata, M., Fujiwara, H., Horiuchi, T., Saeki, Y., Sato, Y., and Furusawa, M. (1985). Production of human α-interferon in silkworm using a baculovirus vector. *Nature (London)* **315,** 592–594.

Miller, D. W., Safer, P., and Miller, L. K. (1986). An insect baculovirus host–vector system for high-level expression of foreign genes. *Genet. Eng.* **8.**

Miller, L. K. (1981). A virus vector for genetic engineering in invertebrates. *In* "Genetic Engineering in the Plant Sciences" (N. J. Panopoulos, ed.), pp. 203–224. Praeger, New York.

Miller, L. K., Lingg, A. J., and Bulla, L. A., Jr. (1983). Bacterial, viral and fungal insecticides. *Science* **219,** 715–721.

Miyamoto, C., Smith, G. E., Farrell-Towt, J., Chizzonite, R., Summers, M. D., and Ju, G. (1985). Production of human c-myc protein in insect cells infected with a baculovirus expression vector. *Mol. Cell. Biol.* **5,** 2860–2865.

Pennock, G. D., Shoemaker, C., and Miller, L. K. (1984). Strong and regulated expression of *E. coli* β-galactosidase in insect cells with a baculovirus vector. *Mol. Cell. Biol.* **4,** 399–401.

Smith, G. E., Summers, M. D., and Fraser, M. J. (1983). Production of human beta interferon in insect cells infected with a baculovirus expression vector. *Mol. Cell. Biol.* **3,** 2156–2165.

19

Genotypic Variants in Wild-Type Populations of Baculoviruses

A. H. MCINTOSH, W. C. RICE, AND C. M. IGNOFFO

I. INTRODUCTION

Baculoviruses are insect viruses belonging to the family Baculoviridae, genus *Baculovirus* (Matthews, 1982, 1985). Baculoviruses are large enveloped rod-shaped particles measuring approximately 200–400 × 40–60 nm (Smith, 1976; Matthews, 1982). The viral particles contain circular duplex DNA genomes ranging in molecular weight from 50 to 100 × 10^6 (Burgess, 1977; Kelly, 1977, Smith and Summers, 1978; Matthews, 1982).

Three subgroups (A, B, and C) comprise this genus. Subgroup A consists of the nuclear polyhedrosis viruses (NPVs) that contain a single nucleocapsid per envelope (SNPVs) or multiple nucleocapsids per envelope (MNPVs). Both SNPVs and MNPVs produce polyhedral inclusion bodies (PIBs) in the nuclei of infected cells. Viral particles are occluded singly or multiply into PIBs. Subgroup B viruses consist of the granulosis viruses (GVs), which have predominantly one nu-

BIOTECHNOLOGY IN INVERTEBRATE
PATHOLOGY AND CELL CULTURE

cleocapsid per envelope and mostly a single virion per occlusion. Subgroup C viruses (also known as "Oryctes virus group"), unlike the viruses of subgroups A and B, are not occluded into an inclusion body but, like subgroup A and B viruses, replicate in the nuclei of infected cells.

The morphogenesis of baculoviruses of subgroups A and B is intriguing in the complexity of viral replication. Both *in vitro* and *in vivo* infectivity studies have shown the production of two distinct types of enveloped virions: occluded virions (OVs) and nonoccluded virions (NOVs). It should be pointed out that the term NOV also has been used to describe baculoviruses belonging to subgroup C (Matthews, 1982). In our discussion, the term NOV will be used for viruses of subgroup A or B that have not been occluded in PIBs. It is recognized that the term budded virus (BV) has also been used synonymously with NOV (Volkman, 1983). Budded virus particles acquire their envelopes by budding from the cell membrane, whereas occluded virus particles acquire their envelopes in the cell nuclei. The OV is believed to be responsible for transmission between larvae, whereas the NOV is responsible for the spread of the virus from one tissue to another. The NOV also is commonly employed for initiating infection of cell cultures from infectious hemolymph or from infectious cell culture fluids.

Infection in the host occurs after ingestion of PIB-contaminated food sources and subsequent release of viral particles from PIBs by the alkaline midgut of larvae. It is believed that released viral particles infect midgut columnar cells (Granados, 1980) via fusion of the viral envelope and the columnar cell plasma membrane (Kawanashi *et al.*, 1972; Tanada *et al.*, 1975). Infected columnar cells release NOVs into the hemocoel, resulting in the infection of other tissues and organs within the host. Further release of NOV from infected cells in the hemocoel spreads the infection to other cells and tissues.

The occurrence of genotypic variants in wild-type isolates of baculovirus populations is a well-documented phenomenon (Miller and Dawes, 1978; Lee and Miller, 1978; Smith and Summers, 1978, 1979; Tjia *et al.*, 1979; Knell and Summers, 1981; Gettig and McCarthy, 1982; Kislev and Edelman, 1982; Stiles *et al.*, 1983; Maruniak *et al.*, 1984; Cherry and Summers, 1985; McIntosh and Ignoffo, 1986). Such variants are easily recognized by the presence of submolar fragments in the electrophoretic patterns of restriction endonuclease (REN) digestion products. Plaque purification confirmed that wild-type isolates contained a mixture of variants (Lee and Miller, 1978; Smith and Summers, 1978). The origin of genotypic variants is unknown, but possible sources of variation include recombinants (Summers *et al.*, 1980), host cell DNA insertions (Miller and Miller, 1982; Fraser *et al.*, 1983) and reiterations of the viral genomic sequences (Burand and Summers, 1982). The term geographic isolate recently has come into common usage and refers to the occurrence of a baculovirus in the same host collected from different geographic sites. Variants can occur both within and between geographic isolates.

The term genotypic variants as used here refers to baculoviruses with very similar DNA genomes that can be distinguished only by small changes in their DNA restriction endonuclease profiles.

The objective of the present chapter is to review what is known about the occurrence of genotypic variants of baculoviruses, the possible role of variants in the expression of phenotypic markers in the host (e.g., changes in virulence, host range, size and number of PIBs, host symptoms, and host lysis), and the relatedness of baculoviruses.

II. GENOTYPIC VARIANTS OF THE SINGLE-ENVELOPED NUCLEAR POLYHEDROSIS VIRUSES

Genotypic variants of the SNPVs infecting *Heliothis armigera* (HaSNPV), *Heliothis zea* (HzSNPV), *Trichoplusia ni* (TnSNPV), and *Orgyia pseudotsugata* (OpSNPV) have been reported by Smith and Summers (1978), Miller and Dawes (1978), Rohrmann *et al.* (1978), Jewell and Miller (1980), Gettig and McCarthy (1982), Bilimoria (1983), Williams and Payne (1984), and McIntosh and Ignoffo (1983, 1986). Restriction endonuclease analysis of various baculovirus DNAs clearly distinguishes SNPVs from MNPVs and also identified distinct genotypes among the various SNPVs (Smith and Summers, 1978; Miller and Dawes, 1978; Rohrmann *et al.*, 1978). McIntosh and Ignoffo (1983) observed distinct genotypic variants associated with a GV, an SNPV, and an MNPV as well as submolar bands present in all baculoviruses isolated from species of *Heliothis*. Gettig and McCarthy (1982), in a comparison of geographic isolates of baculoviruses from various *Heliothis* species, observed two major genotypes corresponding to the MNPV or SNPV phenotype. Three isolates, HzS-EL, HzS-US, and HvS-US (Table I), were identical in the pattern of molar fragments after digestion with six restriction endonuclease enzymes. Only isolate HvS-US contained submolar bands, indicating the presence of genotypic variants. Four other SNPV isolates (HaS-US, HpS-US, HaS-SA, and HaS-BU) digested with three restriction endonucleases also had submolar bands in their REN profiles. Often submolar fragments in one isolate corresponded to molar fragments in another and vice versa. In addition, four isolates (HpS-AU, HaS-IN, HaS-SA, and HaS-CH)

TABLE I

Geographic Isolates of *Heliothis* Nuclear Polyhedrosis Viruses[a]

Heliothis host[b]	NPV type[c]	Geographic site	Abbreviation in text
H. armigera	MEV	Tadhzykistan, USSR	HaM-RU
H. armigera	MEV	Shanghai, China	HaM-CH
H. armigera	SEV	Southern India	HaS-IN
H. armigera	SEV	Stellenbosch, South Africa	HaS-SA
H. armigera	SEV	Barolong, Botswana	HaS-BW
H. armigera	SEV	Hupeh Province, China	HaS-CH
H. punctigera	SEV	Queensland, Australia	HpS-AU
H. virescens	SEV	United States	HvS-US
H. zea	SEV	United States	HzS-US
H. zea	SEV	United States[d]	HzS-EL

[a]Modified from Gettig and McCarthy (1982).
[b]Host from which virus was originally isolated.
[c]MEV (multiple-enveloped virion); SEV (single-enveloped virion).
[d]Commercial isolate: Elcar, Sandoz, Inc., San Diego, California.

showed variations in the molar fragments of the *Bam*HI profiles. These alterations were accounted for by the acquisition or loss of cleavage sites; i.e., HpS-AU apparently contains an extra *Eco*RI cleavage site compared to HzS-US (a missing 5.8 MDa fragment is replaced by two additional fragments of 3.5 and 2.21 MDa). Similar relationships were observed between the other SNPV isolates. Of the various isolates, HaS-IN had the most distinctive *Eco*RI and *Hin*dIII patterns.

Hughes *et al.* (1983) used survival time (half-life) values (ST_{50}) to correlate genotype with biological activity for several of these HzSNPV isolates. In general, SNPVs with identical REN patterns showed similar ST_{50} values and those with different REN patterns also differed in their biological activity; e.g., HzS-US and HaS-IN had ST_{50} values of 73.8 ($\pm$3.9) and 110.1 ($\pm$4.9) hr, respectively, as assayed in *H. zea* neonate larvae. Plaque purification of isolates were not performed.

Williams and Payne (1984) described a new isolate of HzSNPV that was genotypically distinct from other SNPVs of *H. armigera* and *H. zea*. When the virulence of this isolate was compared with that of other NPVs, small (twofold) but significant differences were observed.

Baculovirus isolates from *H. zea* (Br, Vh, El) which were separated by different passages and production histories over a 20-year period were compared by McIntosh and Ignoffo (1986). No major differences were observed in the REN patterns of the three isolates generated by three enzymes and submolar fragments were present in all of the isolates. On serial passage of the Vh isolate *in vivo* in both *H. zea* and *H. virescens* larvae, a change in the *Eco*RI pattern was observed at the 20th passage of virus in both hosts. A common *Eco*RI 3-MDa fragment was missing in both Vh Hz-20 and Vh Hv-20 passage isolates while the presence of new submolar fragments was observed. Further examination revealed that this change coincidentally occurred at passage 20 in both hosts. No significant changes in virulence were detected (in either *H. zea* or *H. virescens* larvae) between the original isolate and the 20th serial passage isolate. Plaque purification of the isolates was not performed. Crozier *et al.* (1985) also observed changes in the REN profiles of *Mamestra brassicae* MNPV following twenty-five serial passages in fifth instar larvae of the cabbage armyworm. No change in infectivity of the virus due to passaging was observed.

In a study of SNPVs of plusiine hosts by Bilimoria (1983), host range was predicted better by the genotypic patterns of the isolates than by the host from which the virus was isolated. Five SNPV isolates were analyzed: three from *T. ni* [(TNY, New York), (TAR, Arkansas), (TAL, Alabama)], one from *Syngrapha selecta* (SMI, Michigan), and one from *Pseudoplusia includens* (PGT, Guatamala). Genotypic variants with respect to major molar fragments existed between the TAL, TAR, and TNY isolates. Estimates of genomic divergence using the equation of Upholt (1977) yielded 0.2–0.4% divergence between TAL and TAR and 1.1–1.3% divergence between TAL/TAR and TNY. Isolate SMI showed the same relationship compared to TAL/TAR and TNY. The PGT isolate had apparently diverged by 8–11% from the other isolates. PGT infects both *T. ni* and *P. includens* larvae (Harper, 1976), but TAL does not appear to infect *P. includens* larvae. Unpublished observations of Young (Bilimoria, 1983) have shown that PGT is moderately infectious in *T. ni*. Neonatal larvae of *P. includens* are susceptible to TAL at doses 100- to 1000-fold higher than those reported by Harper (1976).

III. GENOTYPIC VARIANTS OF THE MULTIPLE-ENVELOPED NUCLEAR POLYHEDROSIS VIRUSES

Andrews *et al.* (1980) tested the virulence (LD_{50} values) of five different genotypic variants of *Autographa californica* MNPV (AcMNPV) in larvae of *T. ni*. The LD_{50} values of the five genotypes (L-1, L-6, L-7, L-9, L-10) and uncloned parent AcMNPV ranged from 10 to 21 polyhedra per larva and the genotypes were not statistically different from each other. Although several of the variants (L-6, L-7, L-10) had small deletions of DNA, these deletions did not alter virulence for *T. ni*. It was

TABLE II
Virulence of Genotypic Variants of Selected Multiple Nuclear Polyhedrosis Viruses (MNPVs)

Baculovirus (MNPV)	Genotypic variant	Virulence[a]	Host	Reference
Autographa californica	1, 6, 7, 9, 10	NS	*T. ni*	Andrews *et al.*, 1980
	E-2, M-3	NS	*T. ni, H. virescens*	Vail *et al.*, 1982
	S1	−S(9X)	*T. ni*	Vail *et al.*, 1982
	S1	NS	*H. virescens*	Vail *et al.*, 1982
	I-1	NT		Miller *et al.*, 1980
Galleria mellonella	G-1	−S(3X)	*T. ni*	Vail *et al.*, 1982
		NS	*H. virescens*	Vail *et al.*, 1982
Trichoplusia ni	T-1	+S(4X)	*T. ni*	Vail *et al.*, 1982
		−S(3X)	*H. virescens*	Vail *et al.*, 1982
Rachiplusia ou	R-1	NS	*T. ni*	Vail *et al.*, 1982
		NS	*H. virescens*	Vail *et al.*, 1982
Spodoptera frugiperda	B1, B2, D5, D7	$-S^{M}$	*S. frugiperda*	Hamm and Styer, 1985
	D6	NT	*S. frugiperda*	Hamm and Styer, 1985

[a]NS, not statistically significant from that of wild-type parent isolate based on LD_{50} values; −S, statistically less virulent than wild-type parent isolate based on LD_{50} values; +S, statistically more virulent than wild-type parent isolate based on LD_{50} values; NT, not tested; M, based on percent mortality.

concluded that plaque purification of the virus was neither deleterious nor advantageous with respect to virulence for *T. ni* larvae (Table II).

Vail *et al.* (1982) attempted to relate *in vivo* infectivity of baculovirus isolates, variants, and recombinants to genomes of physically mapped baculoviruses. Four closely related variants (Smith and Summers, 1979) were studied: AcMNPV, *Galleria mellonella* MNPV (GmMNPV), *T. ni* MNPV (TnMNPV), and *Rachiplusia ou* MNPV (RoMNPV). Plaque-purified isolates of AcMNPV (E-2, M-3, S-1) and TnMNPV-RoMNPV (AR-1, AR-66) were tested for their virulence in *T. ni* and *Heliothis virescens* larvae and compared with the activities of the wild-type standard (AcMNPV-13). The plaque-purified isolate T-1 from TnMNPV was the only one that was more virulent (4 times) for *T. ni* larvae than the wild-type isolate AcMNPV-13. However, it proved to be less virulent (3 times) for *H. virescens* larvae (Table II). Variants E-2, M-3, AR-1 (recombinant), and R-1 were similar in their LC_{50} activities for both *T. ni* and *H. virescens* larvae. Variant G-1 plaque purified from the wild isolate GmMNPV was 2.5 times less virulent than AcMNPV-13 for *T. ni* larvae but showed equal virulence when tested against *H. virescens*. Variant S-1 (with an insertion of DNA) was approximately 9 times less virulent to *T. ni* larvae than AcMNPV-13 (Table II). AR-66 was not bioassayed because of insufficient PIB production. The LC_{50} values of the standard, AcMNPV-13, ranged from 0.32 to 0.51 PIB/mm^2 of diet for *T. ni* and from 0.12 to 0.23 for *H. virescens* larvae. The LC_{50} values of isolates ranged from 0.09 to 3.69 for *T. ni* and 0.08 to 0.62 for *H. virescens*. Other variants of AcMNPV that have been reported include the AcMNPVI-1 isolated from *A. californica* larvae from a field epizootic in Idaho (Miller *et al.*, 1980). Although this isolate was genotypically homogeneous (plaque purification confirmed this observation), it was genotypically different from previously characterized variants. Another variant of AcMNPV was isolated from *Spodoptera exempta* and a physical map was generated by Brown *et al.* (1984). Virulence studies on the latter two variants of AcMNPV were not reported.

In a study of sequence homology between AcMNPV and OpMNPV (Rohrmann *et al.*, 1982b), 13–25% homology was found between the two viruses under varying conditions of stringency. The relatedness of these two viruses was further supported by the earlier findings of Martignoni *et al.* (1980) and Smith and Summers (1981). No variants were reported for OpMNPV and no virulence studies were performed. However, in another study, Martignoni and Iwai (1986) showed an increase in virulence (10×) of OpMNPV for its natural host *O. pseudotsugata* following several passages in an alternate host *T. ni* and one reverse passage in its natural host.

Knell and Summers (1981) obtained four isolates (designated A, B, C, D) of *Spodoptera frugiperda* MNPV (SfMNPV) from different sources and were able to distinguish them by minor differences in their *Eco*RI patterns. Submolar fragments, detected in all isolates indicating the presence of genotypic variants, were confirmed by plaque purification of isolates B and D. Of 13 clones, 5 showed distinct differences in their *Eco*RI patterns and 3 of the 5 clones (SfMNPV-B1, B2, D6) had comigrating fragments similar to those of wild-type isolates. A specific fragment, denoted as *Eco*RI b, occurred in submolar concentrations in isolates A, B, and C but in molar concentrations in cloned variants SfMNPV-B1 and B2. It was concluded that isolates A, B, and C contained subpopulations of cloned isolate SfMNPV B1 or B2. In a more recent investigation, Hamm and Styer (1985) employed isolates and variants described by Knell and Summers (1981) to determine their virulence for neonatal larvae of *S. frugiperda* and *S. exigua* (Table II). Isolate A and variant SfMNPV-B2 were the most pathogenic, whereas isolate B and variants SfMNPV-D5 and D7 appeared to be slightly less virulent for *S. frugiperda* larvae. An interesting finding was a loss in virulence of D5 and D7 for *S. exigua*. Polyhedral inclusion bodies of isolate A and variant B2 were highly virulent for both *S. frugiperda* and *S. exigua* when the PIBs were produced in *S. frugiperda*. On the other hand, PIBs of A and B2 produced in *S. exigua* were nonvirulent for either host.

Dramatic differences in phenotypic expressions of comparative pathology and morphogenesis between original isolates (wild type) and variants were observed by Hamm and Styer (1985). Wild-type isolate A showed typical NPV symptomology (Summers and Arnott, 1969) in its homologous host *S. frugiperda*, including production of a well-developed virogenic stroma, production of large numbers of PIBs, and "melting" or liquefaction of larvae. In contrast, infection of *S. frugiperda* with variant D7 was delayed, with fewer infected cells, absence of well-developed virogenic stroma, less liquefaction of larvae, and reduced numbers and size of PIBs. Isolate A was infectious for its heterologous host *S. exigua* but showed a reduction and delay in infectivity as well as abnormal morphogenesis. Variant D7 produced neither nucleocapsids nor PIBs in *S. exigua*. Abnormal PIB production has also been reported for MNPVs in *in vitro* systems (MacKinnon *et al.*, 1974; Hirumi *et al.*, 1975; McIntosh *et al.*, 1979; Andrews *et al.*, 1980). Reduced PIB production could be related to a selective advantage of some variants during *in vitro* passage similar to that observed for SfMNPV-D7 (Knell and Summers, 1981).

Additionally, Loh *et al.* (1982) studied four geographical isolates of

SfMNPV from Georgia (GA), Mississippi (MS), North Carolina (NC), and Ohio (OH) by REN with *Eco*RI, *Bam*HI, and *Hin*dIII. Results showed NC and OH to be identical whereas MS and GA could be distinguished from each other as well as from NC and OH. The EcoRI digest of OH DNA had a submolar fragment in its cleavage pattern. No virulence studies on the isolates were reported.

Maruniak *et al.* (1984) isolated seven unique variants from wild-type SfMNPV. Physical maps were generated for the predominant variant and other variants but virulence studies were not conducted.

Genotypic variants also were found to occur in geographic isolates of *S. littoralis* MNPV (Kislev and Edelman, 1982; Cherry and Summers, 1985) from different regions in Israel. Two distinct viral types occurred with equal frequency but no assessment of virulence of isolates and variants was performed.

Brown *et al.* (1981) studied five European isolates of *Mamestra brassicae* MNPV (Oxford, Germany, France, Netherlands, Denmark) for their biological, serological, and biochemical properties. Two distinct variants were identified with seven restriction endonucleases; however, they were not tested for pathogenicity. However, isolates were tested but no statistical differences in virulence (LD_{50}) between the five isolates were found. Vlak and Groner (1980) also studied two MbMNPV isolates from Germany and the Netherlands and found them to be unique genotypes. Submolar fragments indicative of the presence of genotypic variants were not described. The LD_{50} values of both isolates were similar.

Genotypic variants as evidenced by the presence of submolar fragments have also been reported for *H. armigera* NPV (HaMNPV) (Smith and Summers, 1978; Gettig and McCarthy, 1982; McIntosh and Ignoffo, 1983; Williams and Payne, 1984). No attempt was made to plaque-purify variants and therefore no information on virulence or host specificity of such variants is available. Williams and Payne (1984), however, did compare HaMNPV with HaSNPV and HzSNPV and found HaMNPV to be approximately two-fold less virulent for *H. armigera* 6-day-old larvae than the other two isolates. Ignoffo *et al.* (1983) demonstrated that HzSNPV was equally virulent against 24-hr-old larvae of *H. zea*, *H. virescens*, and *H. armigera* and that HaMNPV was about 5 to 6 times more virulent against *H. zea* than against *H. virescens* or *H. armigera*. In contrast, HaGV was 18 to 33 times more virulent against *H. virescens* than against *H. armigera* or *H. zea*. The HzSNPV was about 5 to 15 times more virulent than HaMNPV against all three species of *Heliothis* and 2 to 4 log times more virulent than HaGV against the same three species of *Heliothis*. Hughes *et al.* (1983), employing a time–mortality response (ST_{50}), found that two wild-type isolates of HaMNPV were slower acting than other wild-type baculoviruses of *Heliothis*. No cloned variants were employed in these studies.

Dougherty (1983) studied two isolates (LDP-67, VIRIN-ENSh) of the gypsy moth baculovirus *Lymantria dispar* (LdMNPV) by analysis with four different restriction enzymes. The VIRIN-ENSh (Russian isolate) and LDP-67 (United States isolate) gave different REN patterns. The presence of submolar bands in both isolates indicated the occurrence of genotypic variants. Neither isolate was plaque-purified. Shapiro (1983) bioassayed the same isolates in *L. dispar* larvae and found that LDP-67 was significantly (38 times) more virulent. However, following three passages of the VIRIN-ENSh isolate in *L. dispar* larvae, its virulence was equivalent to that of LDP-67. This indicated that the differences were probably due to production artifacts rather than real differences in virulence. Stiles *et al.* (1983) reported many genomic variants in REN patterns of two isolates (LdMNPV HA and LdMNPV IT) of LdMNPV. Isolates were not cloned and no virulence studies were reported.

Shapiro *et al.* (1984) tested the biological activities of nineteen geographical isolates of *L. dispar* NPV. LC_{50} values ranged from 1.7×10^3 PIB/ml to 5×10^6 PIB/ml. Twelve statistically different activity levels were observed. No REN were performed on the isolates.

IV. GENOTYPIC VARIANTS OF GRANULOSIS VIRUSES

Genome variants from the same host insect species, based on the few reported studies of REN profiles of granulosis virus DNA, have demonstrated homologies ranging from 88 to over 99% (Harvey and Volkman, 1983; Harvey and Tanada, 1985). In contrast, different GVs (REN patterns) inhabiting the same host insect species were less than 10% homologous (Harvey and Tanada, 1985) or over 95% homologous (Crook, 1981). If there are any conserved DNA sequences coding for host specificity in the two GVs isolated from *Pseudaletia unipuncta* (Harvey and Tanada, 1985), they could not be detected, nor have they been detected in any other baculovirus isolated from the same host insect species (Smith and Summers, 1982). *Eco*RI digests of the DNA of GV have produced 15–27 fragments or an average molecular weight of $4.8–5.2 \times 10^6$ per fragment (Table III).

Harvey and Volkman (1983) compared isolates of the GV of the codling moth *Cydia pomonella* (CpGV) obtained from three different sources. In their studies REN profiles were obtained for an isolate from Mexico propagated for 9 years at Berkeley, California (CpGV-MB) and

TABLE III
Number of Restriction Endonuclease–DNA Fragments from *Eco*RI Digests and Molecular Weights of Genomes of Selected Insect Granulosis Viruses

Host species	Number of fragments	Molecular weight (× 10^6)	Reference
Cydia pomonella	12–14	74.5–80.6[a]	Burgess, 1977; Harvey and Volkman, 1983, Crook *et al.*, 1985
Heliothis armigera	15–18	86.6	Burgess, 1983; McIntosh and Ignoffo, 1983
Phthorimaea operculella	12	68.3	Burgess, 1983
Pieris brassicae	—	74.6[b]	Brown *et al.*, 1977; Crook, 1981
Pieris rapae	15	72.7	Tweeten *et al.*, 1980; Crook, 1981
Plodia interpunctella	27	69.9	Tweeten *et al.*, 1980
Pseudaletia unipuncta (H)	19	85.0	Harvey and Tanada, 1985; also personal communication, 1986
Pseudaletia unipuncta (O)	14	67.0	Harvey and Tanada, 1985; also personal communication, 1986
Scotogramma trifolii (S_1, S_2)	18	83.0	Harvey and Tanada, 1985; also personal communication, 1980
Spodoptera frugiperda	15	77.3	Smith and Summers, 1978
Trichoplusia ni	23	111.1	Smith and Summers, 1978

[a]Burgess, 1977; electron microscopy measurement of DNA length was 71.2 × 10^6.
[b]Estimated by reassociation kinetics.

Darmstadt, Germany (CpGV-MD) and an isolate from Kazakstay, Russia (CpGV-R). Inclusion bodies of GVs normally contain a single-enveloped virion per inclusion body. However, the three GV isolates from *C. pomonella* (possibly genotypic variants) contained as many as 22 virions per envelope per inclusion body (Falcon and Hess, 1985). Isolates CpGV-MB and CpGV-MD were different in only one of six enzymatic digests. Of the 68 fragments (from the six enzymes) from CpGV-MB and CpGV-MD, 67 comigrated and only 1 was unique. Isolates CpGV-MB and CpGV-MD were calculated to be more than 99% homologous (Upholt, 1977). The isolate from Russia (CpGV-R), in contrast, was different from the CpGV-MB and CpGV-MD isolates in five of the six restriction profiles. Of the 68 fragments, 60 comigrated and 8 were unique. Isolate

CpGV-R had 88% of the fragments in common with either CpGV-MB or CpGV-MD. All the isolates were homogeneous, i.e., free of submolar bands. The average molecular weight of the CpGV-MB and CpGV-MD was 76.2×10^6; the average molecular weight of CpGV-R was 74.5×10^6. There was little, if any, difference in virulence between CpGV-MB and CpGV-MD isolates, but both isolates were more virulent (36 to 73 times more for 1st instar larvae and 5 to 14 times more for 5th instar larvae) than the CpGV-R isolate. These differences in virulence, however, were not confirmed when two of the same isolates subsequently were rebioassayed (Crook *et al.*, 1985). Crook *et al.* (1985) bioassayed two (CpGV-MD and CpGV-R) of the three isolates that Harvey and Volkman (1983) used as well as another isolate from England (CpGV-E). Differences in virulence could not be detected between any of the three isolates. In addition, there was little genotypic variation between any of seven isolates studied by Crook *et al.* (1985). Restriction enzyme profiles were analyzed for geographic isolates from Canada, England, Italy, Mexico, New Zealand, Russia, and Switzerland. Small genotypic differences were obtained in the Russian and English isolates, and the English isolate turned out to be a mixture of two variants.

Crook (1981) utilized immunodiffusion techniques with capsules, ELISA of capsules and viral particles, electrophoresis of viral polypeptides (capsules, virus particles, nucleocapsids), buoyant densities of DNA, restriction endonuclease analysis, and pathogenicity to differentiate between GVs from two closely related insect hosts, *Pieris brassicae* (PbGV) and *Pieris rapae* (PrGV). Major distinctions between the PbGV and PrGV viral isolates were obtained only with REN analysis and pathogenicity tests (Crook, 1981). Crook (1981) indicated that at least "three unique bands were generated from either virus using *Eco*-RI, *Hin*dIII, and *Bam*HI" and about "one-third of the total number of fragments . . . for each virus were unique to each virus." Of 43 fragments, 29 comigrated and 14 were unique to each virus. The two viruses were calculated to be 97.7% homologous to each other (Upholt, 1977). Submolar fragments were detected only in a *Bam*HI digest of the PbGV isolate. No attempts were made to clone these variants since *in vitro* techniques were not available. In pathogenicity studies, larvae of *P. rapae* were equally susceptible to either the PrGV or PbGV isolate, with an average LD_{50} of 200 or 300 capsules/larva, respectively. In contrast, larvae of *P. brassicae* were over 1000 times more susceptible to the PbGV isolate than to the PrGV isolate.

Submolar fragments of REN-digested DNA, considered genomic variants (Smith and Summers, 1978), have been found in many wild-type isolates of GV. It is possible that some submolar bands are products of partial digestion, since Harvey and Volkman (1983) reported that submolar bands of CpGV were eliminated after redigestion with enzymes. The significance of these variants, however, has not been determined. When *Eco*RI digests of the genomes of five GVs (*P. interpunctella, P. rapae, H. armigera, T. ni,* and *S. frugiperda*) were examined, two-thirds of the submolar DNA fragments had molecular weights in the range 1–5 $\times 10^6$.

V. GENOTYPIC VARIANTS OF NONOCCLUDED VIRUSES

The *Oryctes* baculovirus (Huger, 1966) is the type species of the subgroup C baculoviruses (Matthews, 1982). This subgroup comprises approximately 15 baculoviruses isolated from a wide variety of arthropods including arachnids and crustaceans. One of these baculoviruses, IMC-Hz-1-NOV, was isolated from a persistently infected *H. zea* cell line (Hink and Ignoffo, 1970). Both of the IMC-Hz-1 and the *Oryctes* baculoviruses have been characterized by REN (Langridge, 1981; Crawford *et al.*, 1985, 1986). Genomic variants of *Oryctes* baculoviruses were reported for 12 geographical isolates by Crawford *et al.* (1986). In addition, a physical map of a clonal isolate of *Oryctes* baculovirus has been reported (Crawford *et al.*, 1985).

VI. RELATEDNESS AMONG BACULOVIRUSES

Within the past 5 years there have been few published reports evaluating relatedness of baculoviruses. The approaches utilized included restriction endonuclease digestion, hybridization, a combination of REN and hybridization, and a comparison of polyhedrin amino acid and nucleotide sequences (Jewell and Miller, 1980; Rohrmann *et al.*, 1981, 1982a; Rohrmann, 1986; Smith and Summers, 1982; Knell and Summers, 1984). Rohrmann (1986) constructed a phylogenetic tree based on nucleotide and amino acid sequence data for six polyhedrin and granulin genes. On the basis of polyhedrin serological and sequence data Rohrmann (1986) was able to identify 4 major divisions within the occluded baculoviruses which did not always correspond to the classically defined morphological divisions (NPV and GV). The major divisions are (a) the dipteran NPV group, (b) the hymenopteran NPV group, (c) the granulosis virus group, and (d) the lepidopteran NPV group. The lepidopteran polyhedrins are the most conserved group (85–90% amino acid homology) while the granulins contain an N-terminal insertion, and additional cysteine residues along with a variety of the changes. Hymenopteran NPV apparently arose via a cross infection from the lepidopteran NPV lineage. The dipteran polyhedrin (*Tipula paludosa* SNPV) showed no serological or *N*-terminal amino acid sequence relatedness to the other polyhedrins (Guelpa *et al.*, 1977; Rohrmann *et al.*, 1981) and thus is more distantly related.

The DNA homology among 18 baculoviruses from subgroups A, B, and C was compared using REN analysis, DNA hybridization, and Southern blot hybridization. Smith and Summers (1982) detected a 2- to 10-fold difference in DNA homology in the 18 viruses. At the least stringent conditions viral DNA from subgroups A, B, and C had detectable sequence homology. Using Southern blots, Smith and Summers (1982) showed that the viral DNA of all subgroups (A, B, and C) hybridized to AcMNPV *Eco*RI I and that the viral DNA of A and B also hybridized to *Eco*RI fragments L and M of AcMNPV. Presumably polyhedrin or polyhedrin-related sequences are responsible for the hybridization to the *Eco*RI I fragment. The *Eco*RI L and M fragments may code for gene products that play an essential or highly conserved function. Several NPVs [e.g., OpMNPV, RoMNPV, and *Choristoneura fumiferana* MNPV (CfMNPV)] showed a high degree of homology to *Eco*RI blots of AcMNPV (Smith and Summers, 1982). A physical map of HzSNPV was constructed by Knell and Summers (1984) using six restriction endonucleases. The degree of homology between AcMNPV and HzSNPV was then assessed. The polyhedrin gene of HzSNPV (using an AcMNPV polyhedrin probe) mapped to the *Xho*I F fragment and the physical maps of AcMNPV or HzSNPV were orientated accordingly. Six *Eco*RI fragments (C, G, F, J, M, P) and two *Hin*dIII fragments (F, V) of AcMNPV hybridized to various HzSNPV *Hin*dIII fragments. Of the eight AcMNPV fragments sharing sequences with HzSNPV, six shared homology at about the same map positions (suggesting some similarities in the physical and functional organization of these two genomes).

Perhaps the most striking example of conservation of genome organization between apparently unrelated baculoviruses was that between AcMNPV and OpMNPV. Although the REN patterns were quite distinct (Jewell and Miller, 1980), Rhormann *et al.* (1982b) demonstrated 24% duplex formation between AcMNPV and OpMNPV. Using cosmid cloning and mapping, Leisy *et al.* (1984) established a physical map for OpMNPV. Selected cosmid clones were used to probe Southern blots of OpMNPV and AcMNPV DNA under conditions of lower stringency (30% formamide, 42°C). A high degree of colinearity was observed between the two viruses. Interestingly, the three regions of the AcMNPV genome (*Hin*dIII fragments X/J, S, and K) that did not hybridize to any of the OpMNPV DNA probes closely corresponded to regions that were not hybridized by labeled DNA of the *Anticarsica gemmatalis* MNPV (Smith and Summers, 1982). Repeated sequences found in the genomes of AcMNPV and CfMNPV (Arif and Doerfler, 1984) may also occur in OpMNPV. The presence of an inversion was suggested in that the polyhedrin genes of OpMNPV (Rohrmann *et al.*, 1982a) and AcMNPV (Smith *et al.*, 1983) are transcribed in opposite directions with respect to their aligned maps.

Brown *et al.* (1984) characterized variants of *S. exempta* MNPV (SeMNPV-25) that is a REN DNA map variant similar to AcMNPV. Fourteen REN variants were identified out of 71 isolates, and subsequently were analyzed for insertions, deletions, and new restriction sites. A comparison of the physical maps of the SeMNPV variants revealed that, although distinctive, they were clearly genomic AcMNPV variants. Baculovirus genomic variation was found to be restricted to defined regions and three separate mechanisms for generating variation was proposed. The five regions of the SeMNPV-25 genome to which all detectable SeMNPV genomic variants mapped were 8.5–11.0 map units (m.u.), 18.3–20.0 m.u., 34.5–37.1 m.u., 76.2–77.3 m.u., and 88.6–89.5 m.u. Genomic variations detected in AcMNPV isolates were found to map at 8.5–11.0 m.u., 18.3–20.0 m.u., 26.8 m.u., 52.7–53.4 m.u., 58.9–63.7 m.u., 70.0–71.5 m.u., 76.2–77.3 m.u., and 84.1–86.2 m.u. Although SeMNPV and AcMNPV isolates were derived from separate hosts, genotypic variants of the hr (associated with homologous viral sequences) and v (associated with the insertion of cellular DNA sequences) types appear to be associated with specific regions of the genome.

In another study directed toward evaluating conservation of genome organization, Brown *et al.* (1987) looked for conserved homology between SfMNPV-2 and SeMNPV-25. Using reciprocal hybridization studies employing genomic DNA and cloned viral fragments as hybridization probes, they established that a strong degree of colinearity existed between the genomes of SfMNPV-2 and SeMNPV-25 even though they share less than 33% sequence homology. Interestingly the hr 2 region of SeMNPV-25 when viewed on a circular map has moved approximately 180° with respect to its position of SfMNPV-2.

VII. SUMMARY AND CONCLUSIONS

The natural occurrence of genotypic variants in wild-type populations of subgroup A, B, and C baculoviruses appears to be ubiquitous.

Genotypic variants are usually recognized by the presence of submolar bands in REN patterns. Since submolar bands could be products of partial digestion (Harvey and Volkman, 1983), it is advisable to take precautions such as use of extended enzymatic digestion, redigestion of samples, or plaque purification of the parental isolate. The origin of genotypic variants is not known, but they could result from recombination (Summers *et al.*, 1980), acquisition of host cell DNA (Miller and Miller, 1982; Fraser *et al.*, 1983). Other possible sources of genomic variants are mutations resulting in sequence deletion and/or reiteration within the viral genome. Both deletions and reiterations have been reported for baculoviruses propagated *in vitro* (Burand and Summers, 1982) and *in vivo* (McIntosh and Ignoffo, 1986). In addition to changes in virulence and specificity, other possible expressions of genotypic variants include changes in number of virions per PIB, in size and number of PIB, host symptoms, and host lysis. Although genotypic variants may have significance with regard to expression of virulence and specificity (two very important characteristics of biological control agents) the virulence of most variants (recovered through plaque purification from the wild-type parent) was equal to or less than that of the wild-type isolate from which they were cloned (Table II). One exception is the TnMNPV variant of AcMNPV, which was 4 times more virulent than the wild type against *T. ni* and 3 times less virulent against *H. virescens*. Another variant was 9 times less virulent against *T. ni*. Andrews *et al.* (1980) concluded from their studies on cloned variants "that cloning is neither deleterious nor advantageous with regard to virulence in *T. ni*." Whether this statement holds true for other baculoviruses and hosts remains to be determined by further studies.

The relationship between specificity (host range) and variant genotype requires more concerted study. Although McIntosh and Ignoffo (1986) reported a change in REN patterns after the 20th passage of HzSNPV in *H. zea*, there was no difference in its virulence when compared with the parent strain. Furthermore, no change in host specificity was observed when the 20th passage virus was tested in nonhomologous hosts (*T. ni, S. frugiperda, A. gemmatalis, Plutella xylostella*, and *P. rapae*). Conserved DNA sequences coding for host specificity or virulence could not be detected in two GVs isolated from *P. unipuncta* (Harvey and Tanada, 1985).

Variants occurring in the subgroup B baculoviruses, namely the GVs, have not been studied to the same extent as those in subgroup A baculoviruses, primarily because of lack of *in vitro* systems for plaque purification. The recent successful *in vitro* cell cultivation of *C.*

pomonella GV (Naser *et al.*, 1984) and *T. ni* GV (Granados *et al.*, 1986) should lead to the cultivation and plaque purification of other GVs.

Clearly additional studies are needed to further establish the relationships of the baculovirus subgroups, as well as isolates and variants within each subgroup.

REFERENCES

Andrews, R. E., Spence, K. D., and Miller, L. K. (1980). Virulence of cloned variants of *Autographa californica* nuclear polyhedrosis virus. *Appl. Environ. Microbiol.* **39,** 932–933.

Arif, B. M., and Doerfler, W. (1984). Identification and location of reiterated sequences in the *Choristoneura fumiferana* MNPV genome. *EMBO J.* **3,** 525–529.

Bilimoria, S. L. (1983). Genomic divergence among single-nucleocapsid nuclear polyhedrosis viruses of plusiine hosts. *Virology* **127,** 15–23.

Brown, D. A., Bud, H. M., and Kelly, D. C. (1977). Biophysical properties of the structural components of a granulosis virus isolated from the cabbage white butterfly (Pieris brassicae). *Virology* **81,** 317–327.

Brown, D. A., Evans. H. F., Allen, C. J., and Kelly, D.C. (1981). Biological and biochemical investigations of five European isolates of *Mamestra brassicae* nuclear polyhedrosis virus. *Arch. Virol.* **69,** 209–217.

Brown, S. E., Maruniak, J. E., and Knudson, D. L. (1984). Physical map of SeMNPV baculovirus DNA: An AcMNPV genomic variant. *Virology* **136,** 235–240.

Brown, S. E., Maruniak, J. E., and Knudson, D. L. (1987). Conserved homologous regions between two baculovirus DNA's, *J. Gen. Virol.* **68,** 207–212.

Burand, J. P., and Summers, M. D. (1982). Alteration of *Autographa californica* nuclear polyhedrosis virus DNA upon serial passage in cell culture. *Virology* **119,** 223–229.

Burgess, S. (1977). Molecular weights of lepidopteran baculovirus DNAs: Derivation by electron microscopy. *J. Gen. Virol.* **37,** 1–10.

Burgess, S. (1983). *Eco*RI restriction endonuclease fragment patterns of eight lepidopteran baculoviruses. *J. Invertebr. Pathol.* **42,** 401–404.

Cherry, C. L., and Summers, M. D. (1985). Genotypic variation among wild isolates of two nuclear polyhedrosis viruses isolated from *Spodoptera littoralis. J. Invertebr. Pathol.* **46,** 289–295.

Crawford, A. M., Ashbridge, K., Sheehan, C., and Faulkner, P. (1985). A physical map of the *Oryctes* baculovirus genome. *J. Gen. Virol.* **66,** 2649–2658.

Crawford, A. M., Zelazny, B., and Alfiler, A. R. (1986). Genotypic variation in geographical isolates of *Oryctes baculovirus. J. Gen Virol.* **67,** 949–952.

Croizier, G., Croizier, L., Biache, G., and Chaufaux, J. (1985). Evolution de la composition genetique et du pouvoir infectieux du baculovirus de *Mamestra brassicae* L. au cours de 25 multiplications successives sur les larves de la noctuelle du chou. *Entomophaga* **30,** 365–374.

Crook, N. E. (1981). A comparison of the granulosis viruses from *Pieris brassicae* and *Pieris rapae. Virology* **115,** 173–181.

Crook, N. E., Spencer, R. A., Payne, C. C., and Leisy, D. J. (1985). Variation in *Cydia pomonella* granulosis virus isolates and physical maps of the DNA from three variants. *J. Gen. Virol.* **66,** 2423–2430.

Dougherty, E. M. (1983). A comparison of the Gypchek and VIRIN-ENSh preparations of a multiple embedded nuclear polyhedrosis virus of the gypsy moth *Lymantria dispar* utilizing restriction endonuclease analysis. *In* "A Comparison of the US (Gypchek) and USSR (VIRIN-ENSh) Preparations of the Nuclear Polyhedrosis Virus of the Gypsy Moth, *Lymantria dispar*" (C. M. Ignoffo, M. E. Martignoni, and J. L. Vaughn, eds.), results of research conducted under Project V Insect Pests, US/USSR Joint Working Group on the Production of Substances by Microbiological Means, pp. 21–30. Am. Soc. Microbiol., Washington, D.C.

Falcon, L. A., and Hess R. T. (1985). Electron microscope observations of multiple occluded virions in the granulosis virus of the codling moth, *Cydia pomonella*. *J. Invertebr. Pathol.* **45,** 356–359.

Fraser, M. J., Smith, G. E., and Summers, M. D. (1983). Acquisition of host cell DNA sequences by baculoviruses: Relationship between host DNA insertions and FP mutants of *Autographa californica* and *Galleria mellonella* nuclear polyhedrosis viruses. *J. Virol.* **47,** 287–300.

Gettig, R. R., and McCarthy, W. J. (1982). Genotypic variation among wild isolates of *Heliothis* spp. nuclear polyhedrosis viruses from different geographical regions. *Virology* **117,** 245–252.

Granados, R. R. (1980). Infectivity and mode of action of baculoviruses. *Biotechnol. Bioeng.* **22,** 1377–1405.

Granados, R. R., Derksen, A. C. G., and Dwyer, K. G. (1986). Replication of the *Trichoplusia ni* granulosis and nuclear polyhedrosis viruses in cell cultures. *Virology* **152,** 472–476.

Guelpa, B., Bergoin, M., and Croizier, G. (1977). La proteine d'inclusion et les proteines du virion du baculovirus du diptèra *Tipula paludosa*. *C. R. Hebd. Seances Acad. Sci. D* **284,** 779–782.

Hamm, J. J., and Styer, E. L. (1985). Comparative pathology of isolates of *Spodoptera frugiperda* nuclear polyhedrosis virus in *S. frugiperda* and *S. exigua*. *J. Gen. Virol.* **66,** 1249–1261.

Harper, J. D. (1976). Cross-infectivity of six plusiine nuclear polyhedrosis virus isolates to plusiine hosts. *J. Invertebr. Pathol.* **27,** 275–277.

Harvey, J., and Tanada, Y. (1985). Characterization of the DNAs of five baculoviruses pathogenic for the armyworm, *Pseudaletia unipuncta*. *J. Invertebr. Pathol.* **46,** 174–179.

Harvey, J. P., and Volkman, L. E. (1983). Biochemical and biological variation of *Cydia pomonella* (codling moth) granulosis virus. *Virology* **124,** 21–34.

Hink, W. F., and Ignoffo, C. M. (1970). Establishment of a new cell line (IMC-Hz-1) from ovaries of cotton bollworm moths, *Heliothis zea* (Boddie). *Exp. Cell Res.* **60,** 307–309.

Hirumi, H., Hirumi, K., and McIntosh, A. H. (1975). Morphogenesis of a nuclear polyhedrosis virus of the alfalfa looper in a continuous cabbage looper cell line. *Ann. N.Y. Acad. Sci.* **266,** 302–326.

Huger, A. M. (1966). A virus disease of the Indian rhinoceros beetle *Oryctes rhinoceros* (Linnaeus), caused by a new type of insect virus, *Rhabdionvirus oryctes* gen. n., sp. n. *J. Invertebr. Pathol.* **8,** 38–51.

Hughes, P. R., Gettig, R. R., and McCarthy, W. J. (1983). Comparison of the time–mortality response of *Heliothis zea* to 14 isolates of *Heliothis* nuclear polyhedrosis virus. *J. Invertebr. Pathol.* **41,** 256–261.

Ignoffo, C. M., McIntosh, A. H., and Garcia, C. (1983). Susceptibility of larvae of *Heliothis zea*, *H. virescens*, and *H. armigera* (Lepidoptera: Noctuidae) to three baculoviruses. *Entomophaga* **28,** 1–8.

Jewell, J. E., and Miller, L. K. (1980). DNA sequence relationships among six lepidopteran nuclear polyhedrosis viruses. *J. Gen. Virol.* **48,** 161–175.

Kawanashi, C. Y., Summers, M. D., Stoltz, D. B., and Arnott, H. J. (1972). Entry of an insect virus *in vivo* by fusion of viral envelope and microvillus membrane. *J. Invertebr. Pathol.* **20,** 104–108.

Kelly, D. C. (1977). The DNA contained by nuclear polyhedrosis viruses isolated from four *Spodoptera* sp. (Lepidoptera, Noctuidae): Genome size and homology assessed by DNA reassociation kinetics. *Virology* **76,** 468–471.

Kislev, N., and Edelman, M. (1982). DNA restriction-pattern differences from geographical isolates of *Spodoptera littoralis* nuclear polyhedrosis virus. *Virology* **119,** 219–222.

Knell, J. D., and Summers, M. D. (1981). Investigation of genetic heterogeneity in wild isolates of *Spodoptera frugiperda* nuclear polyhedrosis virus by restriction endonuclease analysis of plaque-purified variants. *Virology* **112,** 190–197.

Knell, J. D., and Summers, M. D. (1984). A physical map for the *Heliothis zea* SNPV genome. *J. Gen. Virol.* **65,** 445–450.

Langridge, W. H. R. (1981). Biochemical properties of a persistent nonoccluded baculovirus isolated from *Heliothis zea* cells. *Virology* **112,** 770–774.

Lee, H. H., and Miller, L. K. (1978). Isolation of genotypic variants of *Autographa californica* nuclear polyhedrosis virus. *J. Virol.* **27,** 754–767.

Leisy, D. J., Rohrmann, G. F., and Beaudreau, G. S. (1984). Conservation of genome organization in two multicapsid nuclear polyhedrosis viruses. *J. Virol.* **52,** 699–702.

Loh, L. C., Hamm, J. J., Kawanishi, C., and Huang, E. (1982). Analysis of the *Spodoptera frugiperda* nuclear polyhedrosis virus genome by restriction endonucleases and electron microscopy. *J. Virol.* **44,** 747–751.

McIntosh, A. H., and Ignoffo, C. M. (1983). Restriction endonuclease patterns of three baculoviruses isolated from species of *Heliothis*. *J. Invertebr. Pathol.* **41,** 27–32.

McIntosh, A. H., and Ignoffo, C. M. (1986). Restriction endonuclease cleavage patterns of commercial and serially passaged isolates of *Heliothis* baculovirus. *Intervirology* **41,** 172–176.

McIntosh, A. H., Shamy, R., and Ilsley, C. (1979). Interference with polyhedral inclusion body (PIB) production in *Trichoplusia ni* cells infected with a high passage strain of *Autographa californica* nuclear polyhedrosis virus (NPV). *Arch. Virol.* **60,** 353–358.

MacKinnon, E. A., Henderson, J. F., Stoltz, D. B., and Faulkner, P. (1974). Morphogenesis of nuclear polyhedrosis virus under conditions of prolonged passage *in vitro*. *J. Ultrastruct. Res.* **49,** 419–435.

Martignoni, M. E., and Iwai, P. J. (1986). Propagation of multicapsid nuclear polyhedrosis virus of *Orgyia pseudotsugata* in larvae of *Trichoplusia ni*. *J. Invertebr. Pathol.* **47,** 32–41.

Martignoni, M. E., Iwai, P., and Rohrmann, G. (1980). Serum neutralization of nucleopolyhedrosis viruses (baculovirus subgroup A) pathogenic for *Orgyia pseudotsugata*. *J. Invertebr. Pathol.* **36,** 12–20.

Maruniak, J. E., Brown, S. E., and Knudson, D. L. (1984). Physical maps of SfMNPV baculovirus DNA and its genomic variants. *Virology* **136,** 221–234.

Matthews, R. E. F. (1982). Classification and nomenclature of viruses. Fourth report of the International Committee on Taxonomy of Viruses. *Intervirology* **17,** 1–200.

Matthews, R. E. F. (1985). Viral taxonomy for the nonvirologist. *Annu. Rev. Microbiol.* **39,** 451–474.

Miller, D. W., and Miller, L. K. (1982). A virus mutant with an insertion of a *copia*-like transposable element. *Nature (London)* **299,** 562–564.

Miller, L. K., and Dawes, K. P. (1978). Restriction endonuclease analysis to distinguish two closely related nuclear polyhedrosis viruses; *Autographa californica* MNPV and *Trichoplusia ni* MNPV. *Appl. Environ. Microbiol.* **35,** 1206–1210.

Miller, L. K., Franzblau, S. G., Homan, H. W., and Kish, L. P. (1980). A new variant of *Autographa californica* nuclear polyhedrosis virus. *J. Invertebr. Pathol.* **36,** 159–165.

Naser, W. L., Miltenburger, H. G., Harvey, J. P., Huber, J., and Huger, A. M. (1984). In vitro replication of the *Cydia pomonella* (codling moth) granulosis virus. *FEMS Microbiol. Lett.* **24,** 117–121.

Rohrmann, G. F. (1986). Polyhedrin structure. *J. Gen. Virol.* **67,** 1499–1513.

Rohrmann, G. F., McParland, R. H., Martignoni, M. E., and Beaudreau, G. S. (1978). Genetic relatedness of two nuclear polyhedrosis viruses pathogenic for *Orgyia pseudotsugata*. *Virology* **84,** 213–217.

Rhormann, G. F., Pearson, M. N., Bailey, T. J., Becker, R. R., and Beaudreau, G. S. (1981). N-terminal polyhedrin sequences and occluded baculovirus evolution. *J. Mol. Evol.* **17,** 329–333.

Rohrmann, G. F., Leisy, D. J., Chow, K. C., Pearson, G. D., and Beaudreau, G. S. (1982a). Identification, cloning, and R-loop mapping of the polyhedrin gene from the multicapsid nuclear polyhedrosis virus of *Orgyia pseudotsugata*. *Virology* **121,** 51–60.

Rohrmann, G. F., Martignoni, M. E., and Beaudreau, G. S. (1982b). DNA sequence homology between *Autographa californica* and *Orgyia pseudotsugata* nuclear polyhedrosis viruses. *J. Gen. Virol.* **62,** 137–143.

Shapiro, M. (1983). Comparative infectivity of Gypchek L-79 and VIRIN-ENSh to *Lymantria dispar. In* "A Comparison of the US (Gypchek) and USSR (VIRIN-ENSh) Preparations of the Nuclear Polyhedrosis Virus of the Gypsy Moth, *Lymantria dispar*" (C. M. Ignoffo, M. E. Martignoni, and J. L. Vaughn, eds.), results of research conducted under Project V Insect Pests, US/USSR Joint Working Group on the Production of Substances by Microbiological Means, pp. 38–42. Am. Soc. Microbiol., Washington, D.C.

Shapiro, M., Robertson, J. L., Injac, M. G., Katagiri, K., and Bell, R. A. (1984). Comparative infectivities of gypsy moth (Lepidoptera: Lymantriidae) nuclear polyhedrosis virus isolates from North America, Europe, and Asia. *J. Econ. Entomol.* **77,** 153–156.

Smith, G. E., and Summers, M. D. (1978). Analysis of baculovirus genomes with restriction endonucleases. *Virology* **89,** 517–527.

Smith, G. E., and Summers, M. D. (1979). Restriction maps of five *Autographa californica* MNPV, variants, *Trichoplusia ni* MNPV and *Galleria mellonella* MNPV DNAs with endonucleases *Sma*I, *Kpn*I, *Bam*HI, *Xho*I and *Eco*RI. *J. Virol.* **30,** 828–838.

Smith, G. E., and Summers, M. D. (1981). Application of a novel radioimmunoassay to identify baculovirus structural antigens that share interspecies antigenic determinants. *J. Virol.* **39,** 125–137.

Smith, G. E., and Summers, M. D. (1982). DNA homology among subgroup A, B, and C baculoviruses. *Virology* **123,** 393–406.

Smith, G. E., Vlak, J. M., and Summers, M. D. (1983). Physical analysis of *Autographa californica* nuclear polyhedrosis virus transcripts for polyhedrin and 10,000 molecular weight protein. *J. Virol.* **45,** 215–225.

Smith, K. M. (1976). "Virus–Insect Relationships." Longmans, Green, New York.

Stiles, B., Burand, J. P., Meda, M., and Wood, H. A. (1983). Characterization of gypsy moth (*Lymantria dispar*) nuclear polyhedrosis virus. *Appl. Environ. Microbiol.* **46,** 297–303.

Summers, M. D., and Arnott, H. J. (1969). Ultrastructural studies on inclusion formation and virus occlusion in nuclear polyhedrosis and granulosis virus-infected cells of *Trichoplusia ni* (Hubner). *J. Ultrastruct. Res.* **28,** 462–480.

Summers, M. D., Smith, G. E., Knell, J. D., and Burand, J. P. (1980). Physical maps of *Autographa californica* and *Rachiplusia ou* nuclear polyhedrosis virus recombinants. *J. Virol.* **34,** 693–703.

Tanada, Y., Hess, R. T., and Omi, E. M. (1975). Invasion of a nuclear polyhedrosis virus in midgut of the armyworm *Pseudaletia unipuncta* and the enhancement of the synergistic enzyme. *J. Invertebr. Pathol.* **267,** 99–104.

Tjia, S. T., Carstens, E. B., and Doerfler, W. (1979). Infection of *Spodoptera frugiperda* cells with *Autographa californica* nuclear polyhedrosis virus. *Virology* **99,** 399–409.

Tweeten, K. A., Bulla, L. A., Jr., and Consigli, R. A. (1980). Restriction enzyme analysis of the genomes of *Plodia interpunctella* and *Pieris rapae* granulosis viruses, *Virology* **104,** 514–519.

Upholt, W. B. (1977). Estimation of DNA sequence divergence from comparison of restriction endonuclease digests. *Nucleic Acids Res.* **4,** 1257–1265.

Vail, P. V., Knell, J. D., Summers, M. D., and Cowan, D. K. (1982). *In vivo* infectivity of baculovirus isolates, variants and natural recombinants in alternate hosts. *Environ. Entomol.* **11,** 1187–1192.

Vlak, J. M., and Groner, A. (1980). Identification of two nuclear polyhedrosis viruses from the cabbage moth *Mamestra brassicae* (Lepidoptera: Noctuidae). *J. Invertebr. Pathol.* **35,** 269–278.

Volkman, L. E. (1983). Occluded and budded *Autographa californica* nuclear polyhedrosis virus: Immunological relatedness of structural proteins. *J. Virol.* **46,** 221–229.

Williams, C. F., and Payne, C. C. (1984). The susceptibility of *Heliothis armigera* larvae to three nuclear polyhedrosis viruses. *Ann. Appl. Biol.* **104,** 405–412.

20

Biotechnological Application of Invertebrate Cell Culture to the Development of Microsporidian Insecticides

TIMOTHY J. KURTTI AND ULRIKE G. MUNDERLOH

I. INTRODUCTION

Since the epochal work of Pasteur on the pebrine disease of silkworms, the Protozoa have occupied a prominent position in the history of insect pathology. Though they are widespread among beneficial and pest insects, major gaps remain in our understanding of the impact these parasites make on insect populations. Generally, they produce chronic rather than acute diseases. Few of the entomopathogenic protozoans of pest insects are virulent or fast-acting enough for use in the control of insect populations (Pramer and Al-Rabiai, 1973). One group, the Microsporida, is potentially of commercial importance as it contains organisms possessing characteristics desired of an effective microbial insecticide.

Data from the ecology of animal populations and invertebrate pathology have been used to formulate models on the dynamics of host–parasite interactions (Anderson and May, 1981). Theses models define

BIOTECHNOLOGY IN INVERTEBRATE
PATHOLOGY AND CELL CULTURE

the characteristics a pathogen should possess if it is to be used effectively in the control of a pest insect and the quantities of infectious stages needed to effect this control. Candidate microbial insecticides should be highly virulent, have short latent periods, demonstrate good survival in the environment, and produce large numbers of long-lived infectious stages. These requirements are especially met by some of the insect-pathogenic microsporidia. The prospects for the use of microspordians for microbial control of insects have been reviewed (Brooks, 1980; Canning, 1982), but the molecular biology and genetics of these parasites are poorly understood, and additional knowledge is needed before their biotechnological exploitation is feasible. Cell culture systems for the propagation, selection, and manipulation of the microsporidia would facilitate such studies. The potential for developing and scaling up invertebrate cell cultures for propagating Microsporida for experimental or commercial purposes is addressed in this chapter.

The microsporidians are obligate intracellular protozoans that form resistant spores for transmission. The spores are highly differentiated cells adapted to persist in the environment until they are ingested by a potential host. Within the trilaminar spore case are enclosed the sporoplasm and a coiled tubular filament that are extruded during germination. The filament serves to penetrate the gut of the host insect and to inoculate the sporoplasm into a host cell. In a simplified life cycle, the development of the parasite can be subdivided into three phases: infection, growth, and sporulation. Depending on the species, sporulation may involve the formation of one (monomorphic) or two (dimorphic) types of spores. Of the two groups, the dimorphic microsporidians are considered to have the greatest potential for development as biological control agents (Maddox *et al.*, 1981; Canning, 1982). These included the *Vairimorpha* species of Lepidoptera and the *Amblyospora* and related species that infect mosquitoes. The *Vairimorpha* species (e.g., *Vairimorpha necatrix*) are very pathogenic and can cause high levels or larval mortality. Field trials with *V. necatrix* have demonstrated that, with appropriate application technology, this parasite can be as efficacious as the commercially produced bacterial insecticide Dipel (Fuxa and Brooks, 1979a). The dimorphic species that infect mosquitoes are also effective because their high rates of vertical (transovarial) transmission (up to 100%) can lead to massive infections and the elimination of males, quickly reducing mosquito populations. In contrast, most of the monomorphic species, e.g., the *Nosema*, are much less pathogenic and mortality is enhanced by stress. Consequently, these are often considered to function more as natural regulators of host populations.

II. CELL CULTURE

Living cells are an absolute requisite for the growth and development of the microsporidia. To mass-produce spores, the microsporidians are currently propagated in insect hosts. The usual method involves per os inoculation or intrahemocoelic injection of the parasites into their normal or an alternative host, generally laboratory-reared lepidopteran larvae. Frequently, the use of alternative hosts is advantageous, due to improved yields of spores, larger size, resistance to bacterial septicemia, and ease of surface disinfection. The progeny spores are subsequently harvested and separated from the host tissues by homogenization, filtration, and differential centrifugation. Spore yields have ranged from 9×10^5 to 2×10^{10} spores per host. Nevertheless, the production of entomopathogens *in vivo* is felt to be inherently less well controlled though not more expensive than their production *in vitro* (Brooks, 1980). Few of the entomophilic microsporidia have been grown *in vitro* using cell culture systems (Table I). Use of fermentation technology for the mass culture of insect cells to produce spores is not presently feasible. Not only is the technology for economical large-scale culture of insect cells lacking, but also the yield of spores from cell cultures is too low (Table I) for practical purposes if we consider the numbers needed for field applications. In a recent study, spores of *V. necatrix* were applied to tobacco at a level of 2.5×10^{12} spores/ha, an amount produced in 100 *Heliothis zea* larvae (Fuxa and Brooks, 1979b). In other trials 2.5×10^9 spores of *Nosema locustae* per hectare (Henry and Oma, 1981) and 2.2×10^9 spores of *Nosema algerae* per square meter (Anthony *et al.*, 1978) were applied. Currently, 10^6–10^7 spores/ml may be harvested from cell culture systems; thus 10^2–10^3 liters are needed to produce the spores applied to 1 ha. Other problems are low rates of infection, replication, and spore formation and loss of spore-forming ability with *in vitro* passage. Another shortcoming of *in vitro* systems is the absence of culture systems for most species of microsporidia, especially the commercially more promising dimorphic species. Most of the published work has focused on one genus, *Nosema*, with emphasis on (1) the development of procedures for establishing the parasite in culture; (2) elucidation of the microsporidian life cycle and determination of how the parasite moves from cell to cell; (3) the quantitation of parasite growth with observations on the percentage of cells that are infected and the number of parasites per cell; (4) the ultrastructure of the parasite and the host cell; and (5) the effects of chemotherapeutic agents on the microsporidian.

Cell lines persistently infected with microsporidians can be isolated

TABLE I

Comparison of Yields of Microsporidia from Insects and Insect Cell Cultures[a]

Microsporidium	*In vivo*		*In vitro*		References
	Host insect	Spores/host	Cell line	Spores/ml	
Nosema algerae	*Anopheles stephensi*	8.9×10^5	*Mamestra brassicae* (IZD-Mb-0503)	2×10^6	Undeen and Maddox, 1973; Streett *et al.*, 1980
	Heliothis zea	1.8×10^9	*Heliothis zea* (IPLB-1075)	4×10^6	
			Trichoplusia ni (TN 368)	Negligible	
Nosema heliothidis	*Heliothis zea*	2.0×10^9	*Heliothis zea* (IPLB-1075)	$2\text{–}6 \times 10^7$	McLaughlin and Bell, 1970; Kurtti and Brooks, 1976b
Nosema locustae	*Melanoplus bivittatus*	3.9×10^9	n.d.[a]	n.d.	Henry and Oma, 1981
Nosema disstriae	*Malacosoma disstria*	n.d.	*Malacosoma disstria* (UMN-MDH-1)	8×10^6	Kurtti *et al.*, 1983
			Heliothis zea (IPLB-1075)	4×10^6	Kurtti *et al.*, 1983
Nosema gastroideae	*Leptinotarsa decemlineata*	$6\text{–}9 \times 10^7$	n.d.	n.d.	Hostounsky, 1984
Nosema equestris	*Leptinotarsa decemlineata*	$1\text{–}2 \times 10^8$	n.d.	n.d.	Hostounsky, 1984
Vairimorpha necatrix	*Heliothis zea*	1.7×10^{10}	n.d.	n.d.	Fuxa and Brooks, 1979b
Vairimorpha sp. 696	*Heliothis virescens*	4.9×10^9	n.d.	n.d.	Sedlacek *et al.*, 1985

[a]n.d., not determined.

by simply explanting infected cells from the insect, but their isolation and spore production are erratic. Sohi and Wilson (1976) isolated a line (Md-66) from the hemolymph of *Malacosoma disstria* larvae infected with *Nosema disstriae*. The line remained infected for 6 years (199 transfers). Although there was a reduction of spore formation with repeated transfer, the spores remained infectious for tent caterpillars. With continued passage, the level of infected cells ranged from 17 to 80% and the average number of parasites per infected cell remained fairly constant at 21–24. Later studies by Wilson and Sohi (1977) revealed that the optimum temperature for maintaining the cell line was 28°C. While temperatures ranging from 20° to 30°C did not influence the number of protozoans per cell or the percentage of infected cells, a temperature of 35°C was detrimental to the parasite and could be used to cure the line of its microsporidian infection. The drugs benomyl and fumagillin, but not gentamicin, could also be used to cure the infection (Sohi and Wilson, 1979).

A. Use of Spores as Inocula

Spores are generally used to infect healthy cell cultures, particularly if synchronous parasite development is desired. The success of this approach depends on the availability of sufficient numbers of uncontaminated spores and their controlled germination in the presence of suitable host cells. Because spores are generally obtained from the insect, two procedures have been used to avoid or eliminate microbial contaminants. One is based on the aseptic removal of heavily infected tissues or hemolymph from surface-disinfected hosts (Kurtti and Brooks, 1977; Streett *et al.*, 1980) and the other on density gradient centrifugation using Ludox HS-40, Percoll, or Urografin to separate contaminating microorganisms and host debris from the spores (Streett *et al.*, 1980; Jouvenaz, 1981; Kawarabata and Ishihara, 1984). Several stimuli for controlled germination of spores have been identified (see Jaronski, 1984, for a listing). Shifting the spore suspension to an alkaline pH in the presence of sodium or potassium ions is particularly effective with the microsporidians that infect lepidopteran insects (Ishihara and Sohi, 1966). In some cases, e.g., *N. algerae*, centrifugation of the spores in Ludox HS-40 gradients followed by a rapid increase in the concentration of potassium ions provides important stimuli. Methods for germinating the spores of the dimorphic species *in vitro* have also been developed but have not been applied to the *in vitro* infection of cultured cells. The effectors again include pH or osmotic pressure changes, rehydration, or inorganic ions (Undeen, 1978, 1983; Undeen and Avery, 1984; Malone, 1984b).

TABLE II

Influence of Spore : Cell Ratio on *in Vitro* Infection Rates by Microsporidia

Microsporidium	Host cells	Ratio of spore to cell	Percentage of cells infected		References
			24 hr	3–7 day	
Nosema algerae	*Mamestra brassicae* (IZD-Mb-0503)	5 : 1	29	n.d.	Jaronski, 1984
		10 : 1	46	n.d.	Jaronski, 1984
		20 : 1	63	n.d.	Jaronski, 1984
	Mamestra brassicae (IZD-Mb-0503)	10 : 1	30.4	22.6	Streett *et al.*, 1980
	Heliothis zea (IPLB-1075)	10 : 1	27.1	32.3	Streett *et al.*, 1980
	Trichoplusia ni (TN 368)	10 : 1	27.0	26.8	Streett *et al.*, 1980
Nosema bombycis	*Antheraea eucalypti* (Grace's)	10 : 1	1.5	34	Kawarabata and Ishihara (1984)
Nosema disstriae	*Heliothis zea* (IPLB-1075)	10 : 1	1	25	Kurtti and Brooks, 1977
Nosema heliothidis	*Heliothis zea* (IPLB-1075)	10 : 1	28	n.d.	Kurtti and Brooks, 1976a

[a]n.d., not determined.

The key to the successful infection of cells with spores is to ensure maximum contact between spores and host cells during germination. It is especially important to consider the length of the polar filament (usually 50–150 μm), as infection can occur only when the polar filament strikes a cell and transplants the sporoplasm into the cytoplasm. There is no evidence for passive infection of cells by sporoplasms released into the extracellular environment. Because the extrusion of the polar filament is randomly oriented, maximum infection is achieved by addition of activated spores to concentrated host cell suspensions rather than to cell layers. The optimal conditions have not been completely defined but, in general, suspensions of 10^6–10^7 host cells/ml and 10^7–10^8 spores/ml are successfully used. The degree of infection (percentage of cells infected) is influenced by the ratio of spores to host cells (Table II). The data of Jaronski (1984) for *Mamestra brassicae* cells and *N. algerae* spores demonstrate that, as the ratio of spores to cells is increased, the proportion of infected cells increases. A multiplicity of infection of 20 : 1 is needed to obtain an infection rate of 63%, but this is not an efficient way to produce spores, especially as only 20–30 spores are often produced per infected cell. In one study, Kurtti and Brooks (1976b) estimated that 90% of the spore germinations are abortive. The type of host cell does not seem to influence the initial rate of infection but does influence the subsequent growth rate of the parasite and spore formation (Table III, p. 337). Streett *et al.* (1980) found that *N. algerae* produced similar levels of infection in three different lepidopteran cell lines from *Trichoplusia ni*, *H. zea*, or *M. brassicae*, but parasite growth and sporulation were poor in the *T. ni* cell line. Such findings are not unexpected as various insect tissues differ in their ability to support microsporidian growth *in vivo*.

Even when spores are used at high multiplicities of infection, the synchronous development of parasites is not ensured and the yield of infected cells after 1 week of culture rarely exceeds 30%. Kawarabata and Ishihara (1984), using a multiplicity of infection of 10, observed that the level of *Nosema bombycis*-infected *Antheraea eucalypti* cells remained stationary at 1–2% for the first 36 hr, after which there was an increase in the proportion of infected cells to 35% by day 6 and to 80% by day 10 postinoculation. Similar observations were made by Kurtti and Brooks (1977) using cultures of *H. zea* cells infected with *N. disstriae*. These results may reflect, in part, difficulties in detecting the presence of retarded sporoplasms in the cytoplasm of the host cell, and this can be alleviated by the use of lactoacetoorcein stains (Hazard *et al.*, 1981). Extracellular parasites labeled as secondary infective forms have been detected (Ishihara, 1969; Kawarabata and Ishihara, 1984) and may play a role in the spread of the parasite from cell to cell within a cell culture system. It has not been critically demonstrated how much the asynchronous development of sporoplasms or the occurrence of secondary infective forms contributes to the observed increase in the number of infected cells.

Controlled infection of cell cultures with spores has been particularly useful for the analysis of the sequence of developmental steps in the life cycle of *Nosema* organisms. *In vitro*, the development of *Nosema* can be separated into three stages (Kawarabata and Ishihara, 1984): (1) the infection process, during which the sporoplasm is inoculated into a host cell; (2) the additive or growth phase, when the sporoplasm or the secondary infective form differentiates into a schizont capable of multiplication by repeated binary fission; and (3) the sporulation phase and the differentiation of the putative secondary infective forms. Growth and development take place in the cytoplasm of the host cell, and these parasites are binucleated throughout most of the life cycle with diplokaryotic spores being formed by binary fission of the sporont. Mononucleated stages appear shortly (12 hr) after the sporoplasm is inoculated into the host cell, indicating that autogamy occurs *in vitro* (Streett and Lynn, 1984). Mature spores can be harvested from cell cultures 5–6 days after the cultures were inoculated with activated spores. Kawarabata and Ishihara (1984) observed mature progeny spores in *A. eucalypti* cultures 6 days postinoculation, but the proportion of parasites in the various stages of development was not stated. Six days was also required for the development of *N. disstriae* spores in *H. zea* cell cultures and 40–50 spores were recovered from each infected cell(Kurtti and Brooks, 1977).

B. Use of Infected Cells as Inocula

Infected tissues from the host or infected cells from cultures can also be used to infect healthy cell layers. This procedure is useful when methods for germinating, purifying, and activating the spores are not known or contaminants are difficult to remove from the spore preparation. Lines of parasites can be maintained simply by mixing healthy with infected cells at the time of subculture. Streett *et al.* (1980) maintained *N. algerae* in *M. brassicae* cell cultures by mixing healthy with infected cells in the ratio of 2 : 1 or 4 : 1. The shortcomings of this method are that parasite development is highly asynchronous and infection spreads slowly within the culture. Also, spore production declines with repeated transfer; e.g., after six subcultures by this method, spore production in *N. algerae* ceased (Streett *et al.*, 1985).

The mechanism(s) whereby the parasite is transferred to the healthy cell is not known but several hypotheses have been advanced. These include the existence of secondary infectious forms that actively migrate from cell to cell (Ishihara, 1969); the spontaneous germination of spores, shed from infected cells, that inoculate their sporoplasm into the cytoplasm of a neighboring cell; and the fusion of parasite-bearing sloughed cell vesicles from infected cells with uninfected cells (Tsang *et al.*, 1982). Kurtti *et al.* (1983) identified some of the factors that regulate movement of the parasite in the cell layer. These include the age of the culture, i.e., time postinoculation of infected cells; seeding density, i.e., number of cells per milliliter or per square millimeter; and ratio of infected to uninfected cells at the time of subculture. When uninfected *M. disstria* cells (UMN-MDH-1) were mixed with infected cells to achieve an initial infection level of 4% and a seeding density of 4.3×10^5 cells/ml, there was a 2-day lag period in cell growth and a 5-day lag period in the rise of infected cells. By the end of the second week of culture, the population density was 1.1×10^6 cells/ml and 32% of the cells were infected. At a cell density below 1×10^5 cells/ml, there was little or no transfer. Thus, spread of the parasite was promoted by culture of the cells at high concentrations. Maximum yields were obtained using confluent cell layers. The movement of the parasite was not dependent on cell growth, and the greatest overall increases were obtained when cultures were seeded with very high cell densities (more than 1×10^6 cells/ml), with 18 infected cells harvested for each infected cell seeded. Streett *et al.* (1980) also observed that the spread of *N. algerae* was dependent on population density, and the lines that grew to higher densities, e.g., *M. brassicae,* yielded more parasites. Parasite yield was also influenced by the initial ratio of infected to uninfected cells. In one study the seeding density was held constant at 5.6×10^5 *H. zea* cells (IPLB-1075) per milliliter while the percentage of cells infected with *N. disstriae* varied from 5 to 27%. While there was little growth of the host cells in these high-density cultures (final concentration was 6.3×10^5 cells/ml), the parasites still grew and cross-infected the healthy cells. The greatest overall increase in infected cells was obtained when the initial infection rate ranged from 5 to 10%, and 10 infected cells were harvested for each infected cell initially seeded into the culture. With this method we were able to achieve higher yields in shorter times than with the use of spores to infect the cells. For example, infection rates approaching 100% within 1 week were realized when we mixed *N. disstriae*-infected cells with uninfected cells in the ratio of 3 to 7.

III. BIOTECHNOLOGICAL CONSIDERATIONS

Little research has been done on the factors that influence the quantity or quality of the microsporidia grown in cell cultures. From the practical consideration of economy, the dictum of Hostounsky (1984) for the production of entomopathogens in insects is also applicable to cell cultures. The parasite should be produced within the shortest possible time using the lowest infectious dose and with the smallest loss of host cells. A comparison of the yield of spores from cell cultures with those from insects is given in Table I. In general, the yields of fully differentiated spores from cell cultures are quite low, precluding the use of fermentation technology at this time. Considering the size of the inocula and the proportion, as well as the number, of mature and infective spores harvested, the cell culture system is presently not as efficient as the insect. In fact, it takes 1 liter of cell cultures to produce the number of spores that can generally be harvested from one insect. For example, Kurtti and Brooks (1977) harvested 1–4 × 10^7 spores/ml from cultures of *M. disstria* (UMN-MDH-1) or *H. zea* (IPLB-1075) cells infected with *N. disstriae*. Furthermore, only 10–30% of the spores were refractile and less than 50% of the refractile spores germinated, indicating that sporulation in most parasites was incomplete. Also, the prolonged serial passage of microsporidia in cell culture may result in reduction of spore yields due to the amplification of aberrant non-spore-forming parasites and the spontaneous germination of spores in the culture medium. Several researchers demonstrated that the spores recovered from the cultures were infective for insects but did not quantitate the level of infectivity. Streett *et al.* (1980) showed that spores of *N. algerae* produced *in vitro* were infective for *H. zea* larvae. Spores of *N. bombycis* from cultures of infected *A. eucalypti* cells were infectious for and killed second instar *B. mori* larvae (Kawarabata and Ishihara, 1984). Cultures of *M. disstria* cells persistently infected with *N. disstriae* produced spores that were infective for *M. disstria* larvae, but the spores were less refractile and produced lighter infections in caterpillars than spores harvested from the insect (Sohi and Wilson, 1976).

A limited number of studies have evaluated the use of fermentation technology to grow insect cells as substrates for entomopathogens. The growth of lepidopteran cells (*Spodoptera frugiperda*, IPL-21 AE III) (Weiss *et al.*, 1980, 1981, 1982) and mosquito cells (Singh's *Aedes albopictus*) (Hilwig, 1981; Hilwig and Alapatt, 1981) in large roller bottle culture vessels has been analyzed. The growth of lepidopteran cells (*T. ni*, TN 368) in spinner culture was evaluated by Hink and Strauss (1976). These studies demonstrate the technological feasibility

TABLE III
Influence of Host Cell on Growth in Microsporidia

Microsporidium	Cell line	Average no. of parasites per cell	References
Nosema algerae	*Trichoplusia ni* (TN-368)	9	Streett *et al.*, 1980
	Heliothis zea (IPLB-1075)	33	Streett *et al.*, 1980
	Mamestra brassicae (IZD-Mb-0503)	28	Streett *et al.*, 1980
Nosema bombycis	*Antheraea eucalypti* (Grace's)	68	Kawarabata and Ishihara, 1984
Nosema disstriae	*Heliothis zea* (IPLB-1075)	29	Kurtti and Brooks, 1977
	Malacosoma disstria (Md-66)	24	Sohi and Wilson, 1976

of such systems and suggest that more sophisticated means of controlling the culture parameters, e.g., medium flow rates, pH, oxygen tension, and glucose utilization, are needed. However, the major hurdle is the high serum requirement of cells grown in this system and the expensive media that are used. The studies by Kurtti *et al.* (1983) reveal that there are likely to be serious shortcomings in the use of suspension cell cultures for the production of microsporidian spores. It was demonstrated that parasite yields were highest in cultures having high cell densities. This indicates that systems which allow the cells to grow to near-tissue densities, such as hollow fiber technology, are likely to find utility in the large-scale production of the microsporidia.

The influence of temperature on the *in vitro* development of the microsporidia is another area in need of research as it plays an important role in the developmental genetics of protozoan parasites. The type of spore that is formed by the dimorphic microsporidia is influenced by temperature, and it is conceivable that this dimorphism can be directed *in vitro* by manipulating the temperature. There is evidence that the critical event is meiosis, which has a narrow temperature requirement (Jouvenaz and Lofgren, 1984). With *Vairimorpha* species the binucleated spores are formed during the early stages of infection and the rapid development that occurs at higher temperatures (25° to 31°C). Uninucleated spores are formed later and when parasite growth is retarded by lower temperatures (20°–21°C) (Pilley, 1976; Malone, 1984a). Several tests have demonstrated that the two types of spores are from a single species and are not the result of a mixed infection (Maddox and Sprenkel, 1978).

Some studies have characterized the gene products, mainly proteins, of the microsporidia, but none has compared cultured forms with those produced in the insect at the molecular level. This is due in part to the absence of procedures for separating and concentrating the parasites from the cultured cells. Jaronski (1984) described a method for the release of parasites from infected cells by rupturing the cells with shear forces produced by passing them through a 30-gauge needle. The various stages of parasite development were then separated using discontinuous gradients of Percoll. This and similar procedures promise to be important tools in the biochemical and serological characterization of the Microsporida. Streett and Briggs (1982a,b, 1984) used sodium dodecyl sulfate–polyacrylamide gel electrophoresis (SDS-PAGE) to characterize the polypeptides of spores isolated from insects and found that the polypeptide profiles were unique, reproducible, and uninfluenced by host species or rearing temperature. The tests detected 40–45 polypeptides ranging in molecular weight from 100,000 to 200,000. Serology is also a useful technique but has the disadvantage that a large number of parasites (6×10^{10}) are currently needed to elicit usable antibody titers in animals. Unless methods for boosting the yield of parasites from cell cultures are found, cultures are unsuitable sources of parasites for antigens for serological studies. Knell and Zam (1978) used a double-immunodiffusion technique to investigate the taxonomic relationship of six different species. The enzyme-linked immunosorbent assay (ELISA) developed by Greenstone (1983) for a microsporidian detected as little as 2 ng of spore protein in a homogenate containing as few as 2000 spores.

In vitro systems will aid in the genetic manipulation and engineering of the microsporidia, but the basic techniques for *in vitro* culture and genetic analyses of many species with commerical potential are presently lacking (Table I). A study of their genetics would provide an understanding of the phenotypic stability and variation found in a population of microsporidians. Phenotypic characteristics that determine the usefulness of microbial insecticides are of particular interest. They are pathogenicity (virulence), intensity of spore germination, spore resistance, developmental or growth rate, temperature, and ultraviolet light sensitivity (DNA repair mechanisms). Other useful markers might include protein (isozyme) variation and drug sensitivity. The acquisition of pure strains and their characterization will be an important first step in the biotechnological exploitation of the microsporidia.

To evaluate the degree of heterogeneity in existing strains and new isolates, techniques for cloning the parasite and analyzing the clones are needed. Cloning could be achieved by end point dilution or plaque assays. The various methods available for plaque assay of the insect pathogenic baculoviruses *in vitro* (Fraser, 1982) should be evaluated for their applicability to the microsporidia. Cell culture-based assays have the advantage that cells are easier to handle and are more homogeneous than insects. They also readily permit detection and harvesting of variants or mutants for genetic studies. Recent evidence for sexual stages in some of the microsporidia indicates that these organisms could be analyzed by conventional genetic methodology, but this requires the development of techniques for hybridization and selection of the progeny from such crosses. An understanding of the genetic processes such as sexual recombination and mutation rates by which variant forms of microsporidians arise will also be of value. Because the morphological characteristics of these protozoans display considerable ontogenetic and/or environmental variation, characterization of their genome by DNA technology promises to play an important role. To this end it will be necessary to work with clones because genotypic heterogeneity would, e.g., during endonuclease analyses, result in the formation of restriction fragments that could be misleading in the determination of linkage relationships when physical maps of the genome are constructed.

To better understand the molecular basis of aberrant sporulation *in vitro* it would be useful to plaque-isolate some of these parasites after various numbers of serial passage and analyze the changes that have taken place in the genome. Because developmental abnormalities in other protozoans serially passaged *in vivo* or *in vitro* have been linked with changes in the genome, it might be useful to evaluate the DNA of microsporidians maintained in culture. Malarial parasites, for example, lose their ability to form gametocytes infective for mosquitoes with repeated *in vitro* passage (Burkot *et al.*, 1984). A decline in mosquito infectivity has been correlated with a decrease in the amount of repetitive sequences in the parasite's genome (Birago *et al.*, 1982; Casaglia *et al.*, 1985).

Success in genetic manipulation of the microsporidia depends on understanding their chromosomal and genomal organization and control. Central to this is the identification and localization of the meiotic and syngamic steps in the life cycle. The formation of uninucleate and binucleate spores by some of the microsporidia indicates that sexual and asexual phases of development occur in these parasites. If confirmed, these promise to be important developmental events for the genetic manipulation of the microsporidia. It will be necessary to establish the degree of replication of the parasite's genome prior to meiotic division to be able to predict the outcome of genetic crossing experiments. The cytogenetic studies on the *Amblyospora* and related microsporidians that infect mosquitoes provide important information in this regard. The binucleated stages of the parasites are transmitted transovarially and subsequent development is determined by the sex of the host. In male larvae, the parasite develops rapidly in the fat body and thoracic tissues, leading to a fatal infection and the formation of uninucleate spores (Andreadis and Hall, 1979). Evidence for karyogamy and a meiosislike process to form uninucleate spores has been presented (Hazard and Brookbank, 1984). This included the presence of synaptonemal complexes in the nuclei, the meiotic configuration of the chromosomes, and the DNA content of individual nuclei. The process, however, lacked important features of meiosis (chiasmata or metaphase II) and the genetic functions of such events remain to be elucidated. Until recently, the role of the haploid spores remained enigmatic as they are not infective for mosquito larvae. The spores are now known to be infective for an intermediate host, the copepod, and spores formed in the copepod are infectious for mosquito larvae (Sweeney *et al.*, 1985). The early stages in horizontal transmission to the mosquito have been examined in some detail (Andreadis, 1985). They involve the formation of gametes and plasmogamy that lead to the formation of the binucleated parasite (Hazard *et al.*, 1985; Sweeney *et al.*, 1985; Andreadis, 1985). Similar studies to unravel the role of the uninucleate spores of the *Vairimorpha* that infect caterpillars are needed as well. A major difficulty in working with these species that infect mosquitoes is that frequently the host–parasite complex cannot be maintained in the laboratory. *In vitro* systems may allow us to overcome this hurdle.

IV. SAFETY CONSIDERATIONS

Little effort has been made to assess the environmental safety of the microsporidia to be used as microbial insecticides. The limited information that is available indicates that the microsporidia are safe and pose little threat to the health of humans or domestic animals. Their safety for beneficial insects, however, requires closer scrutiny (Huger, 1984). Because of the close taxonomic relationship between the insect and the mammalian pathogenic microsporidians and the fact that several different orders of insects can be infected by a given parasite, a more thorough testing program is warranted than has been executed to date. Even mammalian cells have been infected by *Nosema*. Ishihara (1968) infected primary cultures of rodent cells with *N. bombycis*, and Undeen (1975) found that *N. algerae* can infect porcine kidney cells. In both of these cases, the microsporidians could not multiply at 37°C and it was necessary to incubate the cells at lower temperatures. Using recombinant DNA technologies, it might be possible to construct microsporidians with narrower temperature tolerances. The high pathogenicity of *Vairimorpha* species for caterpillars results from disruption of the gut by the polar filaments extruded during germination and the ensuing septicemia caused by the gut bacteria introduced into the hemocoel, rather than replication of the parasite within the tissues (primarily fat body) of the host insect (Maddox *et al.*, 1981). Spore germination and sporoplasm viability (i.e., ability to initiate vegetative growth) have been shown to be two separate processes (Undeen, 1983). Parasites that can germinate and disrupt the gut function of a wide spectrum of insects but do not replicate in the insect tissues could be designed.

ACKNOWLEDGMENTS

This report is Paper No. 14929, Scientific Journal Series, Minnesota Agricultural Experiment Station.

REFERENCES

Anderson, R. M., and May, R. M. (1981). The population dynamics of microparasites and their invertebrate hosts. *Philos. Trans. R. Soc. London, Ser. B.* **291,** 451–524.

Andreadis, T. G. (1985). Life cycle, epizootiology, and horizontal transmission of *Amblyospora* (Microspora: Amblyosporidae) in an univoltine mosquito, *Aedes stimulans. J. Invertebr. Pathol.* **46,** 31–46.

Andreadis. T. G., and Hall, D. W. (1979). Development, ultrastructure, and mode of transmission of *Amblyospora* sp. (Microspora) in the mosquito. *J. Protozool.* **26,** 444–452.

Anthony, D. W., Savage, K. E., Hazard, E. I., Avery, S. W., Boston, M. D., and Oldacre, S. W. (1978). Field tests with *Nosema algerae* Vavra and Undeen (Microsporida, Nosematidae) against *Anopheles albimanus* Wiedmann in Panama. *Misc. Publ. Entomol. Soc. Am.* **11,** 17–27.

Birago, C., Bucci, A., Dore, E., Frontali, C., and Zenobi, P. (1982). Mosquito infectivity is directly related to the proportion of repetitive DNA in *Plasmodium berghei. Mol. Biochem. Parasitol.* **6,** 1–12.

Brooks, W. M. (1980). Production and efficacy of Protozoa. *Biotechnol. Bioeng.* **22,** 1415–1440.

Burkot, T. R., Williams, J. L., and Schneider, I. (1984). Infectivity to mosquitoes of *Plas-*

modium falciparum clones grown *in vitro* from the same isolate. *Trans. R. Soc. Trop. Med. Hyg.* **78,** 339–341.

Canning, E. U. (1982). An evaluation of protozoal characteristics in relation to biological control of pests. *Parasitology* **84,** 119–149.

Casaglia, O., Dore, E., Frontali, C., Zenobi, P., and Walliker, D. (1985). Re-examination of earlier work on repetitive DNA and mosquito infectivity in rodent malaria. *Mol. Biochem. Parasitol.* **16,** 35–42.

Fraser, M. J. (1982). Simplified agarose overlay plaque assay for insect cell lines and insect nuclear polyhedrosis viruses. *J. Tissue Culture Methods* **7,** 43–46.

Fuxa, J. R., and Brooks, W. M. (1979a). Effects of *Vairimorpha necatrix* in sprays and corn meal on *Heliothis* species in tobacco, soybeans, and sorghum. *J. Econ. Entomol.* **72,** 462–467.

Fuxa, J. R., and Brooks, W. M. (1979b). Mass production and storage of *Vairimorpha necatrix* (Protozoa: Microsporida). *J. Invertebr. Pathol.* **33,** 86–94.

Greenstone, M. H. (1983). An enzyme-linked immunosorbent assay for the *Amblyospora* sp. of *Culex salinarius* (Microspora: Amblyosporidae). *J. Invertebr. Pathol.* **41,** 250–255.

Hazard, E. I., and Brookbank, J. W. (1984). Karyogamy and meiosis in an *Amblyospora* sp. (Microspora) in the mosquito *Culex salinarius. J. Invertebr. Pathol.* **44,** 3–11.

Hazard, E. I., Ellis, E. A., and Joslyn, D. J. (1981). Identification of microsporidia. *In* "Microbial Control of Pests and Plant Diseases 1970–1980" (H. D. Burges, ed.), pp. 163–182. Academic Press, London.

Hazard, E. I., Fukuda, T., and Becnel, J. J. (1985). Gametogenesis and plasmogamy in certain species of Microspora. *J. Invertebr. Pathol.* **46,** 63–69.

Henry, J. E., and Oma, E. A. (1981). Pest control by *Nosema locustae*, a pathogen of grasshoppers and crickets. *In* "Microbial Control of Pests and Plant Diseases 1970–1980" (H. D. Burges, ed.), pp. 573–586. Academic Press, London.

Hilwig, I. (1981). Manipulation with *Aedes albopictus* cell line (Aa). *Z. Angew. Entomol.* **91,** 480–486.

Hilwig, I., and Alapatt, F. (1981). Insect cell lines in suspension, cultivated in roller bottles. *Z. Angew. Entomol.* **91,** 1–7.

Hink, W. F., and Strauss, E. (1976). Growth of the *Trichoplusia ni* (TN-368) cell line in suspension culture. *In* "Invertebrate Tissue Culture: Applications in Medicine, Biology and Agriculture" (E. Kurstak and K. Maramorosch, eds.), pp. 297–300. Academic Press, New York.

Hostounsky, Z. (1984). Production of microsporidia pathogenic to the Colorado potato beetle (*Leptinotarsa decemlineata*) in alternate hosts. *J. Invertebr. Pathol.* **44,** 166–171.

Huger, A. M. 1984. Susceptibility of the egg parasitoid *Trichogramma evanescens* to the microsporidium *Nosema pyrausta* and its impact on fecundity. *J. Invertebr. Pathol.* **44,** 228–229.

Ishihara, R. (1968). Growth of *Nosema bombycis* in primary cell cultures of mammalian and chicken embryos. *J. Invertebr. Pathol.* **11,** 328–329.

Ishihara, R. (1969). The life cycle of *Nosema bombycis* as revealed in tissue culture cells of *Bombyx mori. J. Invertebr. Pathol.* **14,** 316–320.

Ishihara, R., and Sohi, S. S. (1966). Infection of ovarian tissue culture of *Bombyx mori* by *Nosema bombycis* spores. *J. Invertebr. Pathol.* **8,** 538–540.

Jaronski, J. (1984). Microsporida in cell culture. *Adv. Cell Cult.* **3,** 183–229.

Jouvenaz, D. P. (1981). Percoll: An effective medium for cleaning microsporidian spores. *J. Invertebr. Pathol.* **37,** 319.

Jouvenaz, D. P., and Lofgren, C. S. (1984). Temperature-dependent spore dimorphism in *Burnella dimorpha* (Microspora: Microsporida). *J. Protozool.* **31,** 175–177.

Kawarabata, T. and Ishihara, R. (1984). Infection and development of *Nosema bombycis* (Microsporida: Protozoa) in a cell line of *Antheraea eucalypti. J. Invertebr. Pathol.* **44,** 52–62.

Knell, J. D., and Zam, S. G. (1978). A serological comparison of some species of Microsporida. *J. Invertebr. Pathol.* **31,** 280–288.

Kurtti, T. J., and Brooks, M. A. (1976a). Propagation of a microsporidan in a moth cell line. *In* "Invertebrate Tissue Culture: Applications in Medicine, Biology and Agriculture" (E. Kurstak and K. Maramorosch, eds.), pp. 395–398. Academic Press, New York.

Kurtti, T. J., and Brooks, M. A. (1976b). Propagation of microsporidia in invertebrate cell culture. *Proc. Int. Colloq. Invertebr. Pathol., 1st,* 1976, pp. 123–127.

Kurtti, T. J., and Brooks, M. A. (1977). The rate of development of a microsporidan in moth cell culture. *J. Invertebr. Pathol.* **29,** 126–132.

Kurtti, T. J., Tsang, K. R., and Brooks, M. A. (1983). The spread of infection by the microsporidian, *Nosema disstriae,* in insect cell lines. *J. Protozool.* **30,** 652-657.

McLaughlin, R. E., and Bell, M. R. (1970). Mass production *in vivo* of two protozoan pathogens, *Mattesia grandis* and *Glugea gasti* of the boll weevil, *Anthonomus grandis. J. Invertebr. Pathol.* **16,** 84–88.

Maddox, J. V., and Sprenkel, R. K. (1978). Some enigmatic microsporidia of the genus *Nosema. Misc. Publ. Entomol. Soc. Am.* **11,** 65–84.

Maddox, J. V., Brooks, W. M., and Fuxa, J. R. (1981). *Vairimorpha necatrix,* a pathogen of agricultural pests: Potential for pest control. *In* "Microbial Control of Pests and Plant Diseases 1970–1980" (H. D. Burges, ed.), pp. 587–594. Academic Press, London.

Malone, L. A. (1984a). A comparison of the development of *Vairimorpha plodiae* and *Vairimorpha necatrix* in the Indian meal moth, *Plodia interpunctella. J. Invertebr. Pathol.* **43,** 140–149.

Malone, L. A. (1984b). Factors controlling *in vitro* hatching of *Vairimorpha plodiae* (Microspora) spores and their infectivity to *Plodia interpunctella, Heliothis virescens,* and *Pieris brassicae. J. Invertebr. Pathol.* **44,** 192–197.

Pilley, B. (1976). A new genus, *Vairimorpha* (Protozoa: Microsporida), for *Nosema necatrix* Kramer 1965: Pathogenicity and life cycle in *Spodoptera exempta* (Lepidoptera: Noctuidae). *J. Invertebr. Pathol.* **28,** 177–183.

Pramer, D., and Al-Rabiai, S. (1973). Regulation of insect populations by protozoa and nematodes. *Ann. N. Y. Acad. Sci.* **217,** 85–92.

Sedlacek, J. D., Dintenfass, L. P., Nordin, G. L., and Ajlan, A. A. (1985). Effects of temperature and dosage on *Vairimorpha* sp. 696 spore morphometrics, spore yield, and tissue specificity in *Heliothis virescens. J. Invertebr. Pathol.* **46,** 320–324.

Sohi, S. S., and Wilson, G. G. (1976). Persistent infection of *Malacosoma disstria* (Lepidoptera: Lasiocampidae) cell cultures with *Nosema (Glugea) disstriae* (Microsporida: Nosematidae). *Can. J. Zool.* **54,** 336–342.

Sohi, S. S. and Wilson, G. G. (1979). Effect of antimicrosporidian and antibacterial drugs on *Nosema disstriae* (Microsporida) infection in *Malacosoma disstria* (Lepidoptera: Lasiocampidae) cell cultures. *Can. J. Zool.* **57,** 1222–1225.

Streett, D. A., and Briggs, J. D. (1982a). An evaluation of sodium dodecyl sulfate–polyacrylamide gel electrophoresis for the identification of microsporidia. *J. Invertebr. Pathol.* **40,** 159–165.

Streett, D. A., and Briggs, J. D. (1982b). Variation in spore polypeptides from four species of *Vairimorpha*. *Biol. Syst. Ecol.* **10,** 161–165.

Streett, D. A., and Briggs, J. D. (1984). Separation of spore polypeptides from an *Amblyospora* sp. infecting *Culex salinarius*. *J. Invertebr. Pathol.* **43,** 128–129.

Streett, D. A., and Lynn, D. E. (1984). *Nosema bombycis* replication in a *Manduca sexta* cell line. *J. Parasitol.* **70,** 452–454.

Streett, D. A., Ralph, D., and Hink, W. F. (1980). Replication of *Nosema algerae* in three insect cell lines. *J. Protozool.* **27,** 113–117.

Sweeney, A. W., Hazard, E. I., and Graham, M. F. (1985). Intermediate host for an *Amblyospora* sp. (Microspora) infecting the mosquito, *Culex annulirostris*. *J. Invertebr. Pathol.* **46,** 98–102.

Tsang, K. R., Brooks, M. A., and Kurtti, T. J. (1982). Culture conditions regulating the infection of cells by an intracellular microorganism. *In* "Invertebrate Cell Culture Applications" (K. Maramorosch and J. Mitsuhashi, eds.), pp. 125–156. Academic Press, New York.

Undeen, A. (1975). Growth of *Nosema algerae* in pig kidney cell cultures. *J. Protozool.* **22,** 107–110.

Undeen, A. (1978). Spore-hatching processes in some *Nosema* species with particular reference to *N. algerae* Vavra and Undeen. *Misc. Publ. Entomol. Soc. Am.* **11,** 29–49.

Undeen, A. H. (1983). The germination of *Vavraia culicis* spores. *J. Protozool.* **30,** 274–277.

Undeen, A. H. and Avery, S. W. (1984). Germination of experimentally nontransmissible microsporidia. *J. Invertebr. Pathol.* **43,** 229–301.

Undeen, A. H., and Maddox, J. V. (1973). The infection of nonmosquito hosts by injection with spores of the microsporidan *Nosema algerae*. *J. Invertebr. Pathol.* **22,** 258–265.

Weiss, S. A., Kalter, S. S., Vaughn, J. L., and Dougherty, E. (1980). Effect of nutritional, biological and biophysical parameters on insect cell culture of large scale production. *In Vitro* **16,** 222–223.

Weiss, S. A., Smith, G. C., Kalter, S. S., and Vaughn, J. L. (1981). Improved method for the production of insect cell cultures in large volume. *In Vitro* **17,** 495–502.

Weiss, S. A., Smith, G. C., Vaughn, J. L., Dougherty, E. M., and Tompkins, G. J. (1982). Effect of aluminum chloride and zinc sulfate on *Autographa californica* nuclear polyhedrosis virus (AcNPV) replication in cell culture. *In Vitro* **18,** 937–944.

Wilson, G. G., and Sohi, S. S. (1977). Effect of temperature on healthy and microsporidia-infected continuous cultures of *Malacosoma disstria* hemocytes. *Can. J. Zool.* **55,** 713–717.

Undeen, A. H., and Maddox, J. V. (1973). The infection of nonmosquito hosts by injection with spores of the microsporidan *Nosema algerae*. *J. Invertebr. Pathol.* **22,** 258–265.

21

Grasshopper and Locust Control Using Microsporidian Insecticides

S. K. RAINA, M. M. RAI, AND A. M. KHURAD

I. INTRODUCTION

Plagues of locusts and grasshoppers have been known in India from early times. In March 1983, locust activity was sighted at Bassi of Pokaran Tehsil in the Jaisalmer district. From June onward, migration of desert locusts continued from the west and there was a sudden explosion in their population at Raura of Phalodi Tehsil in the Jodhpur district. In July, due to widespread rainfall and adequate ecological conditions, a countless number of locusts was recorded at Mayani of Sheo Tehsil in the Barmer district. The first exotic desert locust swarm was reported in August at Ghantiyali village in the Jaisalmer district. This swarm was small and uneven in maturity. Later, several swarms entered India from the Indo-Pak border through the Barmer and Jai-

BIOTECHNOLOGY IN INVERTEBRATE PATHOLOGY AND CELL CULTURE

salmer districts of Rajasthan and the Kutch (Bhuj) district of Gujarat. Two waves of swarm incursions occured, the first in August and the second in October. In the former 21 swarms and in the latter 10 swarms entered India from the west. In September 1986, three small swarms ranging from 3 to 8 sq km entered across the Indo-Pak border in the Jaisalmer district (Chandra, 1986). Control of these swarms with chemical pesticides was undertaken by the Locust Warning and Control Organization, Directorate of Plant Protection, Quarantine and Storage, of the government of India. While this organization is usually involved in taking preventive measures to forestall locust activities, the grasshopper problem is rarely recognized until the economically tolerable threshold has been exceeded. In both cases, successful orthopteran pest management in previous years has given way to misgivings with the recognition of pesticide resistance and toxicity. Many of the chemicals used have proved hazardous both ecologically and economically. Bacterial and viral biological pesticides, used effectively against lepidopteran pests in many countries, proved ineffective against orthopterans (Canning, 1982). However, a protozoan microsporidian, *Nosema locustae*, a pathogen of grasshoppers and locusts, has been developed and used successfully to suppress the population of many species of grasshoppers in the United States (Henry, 1971; Henry and Oma, 1981; Henry and Onsager, 1984). Its application has been further extended to Canada, Argentina, and Australia (Ewen and Mukerji, 1980; Luna *et al.*, 1981; Moulden, 1981).

The pathogen was first noted by Canning (1953) in the field-collected African migratory locust *Locusta migratoria migratorioides*. Its geographic range has recently been recorded in 10 species of grasshoppers in Saskatchewan, Canada (Ewen, 1983), in the desert locust *Schistocerca gregaria* in Rajasthan (Srivastava and Bhanotar, 1985), and in the rice pest *Hieroglyphus* sp. in Vidarbha, India, (Raina, 1985). There now seems to be real promise that this pathogen could be introduced effectively against grasshoppers and locusts in management programs in India. Any new biotechnology aimed at insect pest management in India requires as a prerequisite thorough screening against useful insects, particularly silkworms and honeybees, which are the basis of important cottage industries in India. The honeybees, *Apis mellifera*, are not susceptible to infection by *N. locustae* (Menapage *et al.*, 1978), but the silkworm, *Bombyx mori*, has not been tested. Hence, a Biological Pest Management Project, funded by the International Development Research Center, has been initiated to carry out *in vivo* and *in vitro* investigations on the microsporidian pathogen *N. locustae* for its use against grasshoppers and locusts in India. This chapter describes the first phase of the project, including the isolation, *in vivo* production, and pathogenesis of the pathogen and its screening against the silkworm.

II. MATERIALS AND METHODS

Locust and grasshopper populations and the incidence of pathogens in these populations have been monitored in the following sites chosen for this work: Bikaner, Jaisalmer, Barmer, and Jodhpur (Rajasthan); Rann of Kutch (Gujarat); Balaghat (Madhya Pradesh) and Bhandara, Nerla, Tumsar, Gondia, Umrer, Pauni, and adjoining rice-growing areas (Vidarbha). Laboratory rearing of the desert locust *S. gregaria* and the migratory locust *L. migratoria migratorioides* (both gregaria phase) has been undertaken in specially designed cages, using fresh cabbage and *Sorghum* leaves as feed. Similarly, a culture of NB_4 D_2 × pure Mysore race of the silkworm *B. mori* has also been maintained for screening of the developed pathogen *N. locustae*. This microsporidian was isolated in the laboratory from field-collected grasshoppers, *Hieroglyphus* sp., by the centrifugation method (Cantwell, 1970). It was produced on a large scale using *L. migratoria* and *S. gregaria* as host insects, since *Hieroglyphus* sp. is rather difficult to rear and maintain in the laboratory because it is an egg diapause grasshopper.

To determine the ideal dose rate of *N. locustae* spores for infection, locusts were fed with various concentrations (1.5×10^5,1.5×10^6, 1.5×10^7, and 1.5×10^8 spores/ml of distilled water) of pathogen by using single, double, and triple inoculations for large-scale production. Mortality and host tissue infections were recorded at regular intervals. The harvested spores were passed twice and their virulence tested using the same host.

To examine pathogenesis in the host tissue, the inoculated insects were sacrificed after every second day and various tissues (midgut, fat body, nerve tissue, muscle, and ovary) were fixed in Bouin's fixative, dehydrated, cleared in xylene, and embedded in paraffin wax. The sections were cut at 5 μm thickness and stained with Azan. A hemolymph smear was prepared before every sacrifice and stained with Giemsa (Humason, 1962). Screening of *N. locustae* against the silkworm *B. mori* was carried out by spraying 1.5×10^6 spores/ml on mulberry leaves. The treated leaves were fed once a day for up to 7 days to developing silkworm larvae (3rd to 5th instars). To examine the specificity of *N. locustae* another microsporidian, *Nosema* sp., was isolated from diseased tasar silkmoth, *Antheraea* spp., and simultaneously fed to the other batch of silkworm larvae. The life cycle, mortality, effect on internal organs, growth of silk gland, cocoon weight, fecundity, egg hatching, and effects on the F_1 generation were recorded.

III. RESULTS AND DISCUSSION

A. Incidence of Pathogens in Field-Collected Locusts and Grasshoppers

Swarms of locusts were not reported in 1984 and 1985 in Rajasthan and Gujarat; cadavers of locusts (probably from the 1983 swarm), the dominant population being *S. gregaria*, were observed in the millions at the long seashore of Kunwarbet near the Indian bridge. A few cadavers were also located near the Surwari cotton fields. In 1985 the number of locusts recorded in Bikaner and Jaisalmer districts of Rajasthan was insignificant. Specimens were brought to the laboratory and tissues were examined but the incidence of *N. locustae* was not observed. In Vidarbha and Madhya Pradesh, weekly or biweekly surveys were conducted during 1984 and 1985 in the rice fields under study and the population of the grasshopper *Hieroglyphus* was monitored by sweeping and visual counting methods. A low percentage incidence of *N. locustae* was observed in field-collected specimens (Table I).

B. Isolation of *Nosema locustae* Spores

The field-collected grasshoppers were crushed and kept in a flask containing distilled water at room temperature for 1–2 days. The

TABLE I
Occurrence of *Nosema locustae* Spores in Field-Collected Grasshoppers, *Hieroglyphus* sp., and Locust, *Schistocerca gregaria*

Species	Survey site[a]	Infection (%) 1984	Infection (%) 1985
Hieroglyphus sp.	Ambajhari	3–4	4–5
	Dhurkheda	1–2	1–2
	Mounda	nil	1–2
	Nerla	1–2	1–2
	Pandhrabodee	1–2	1–2
Schistocerca gregaria	Gujarat	nil	nil
	Rajasthan	nil	nil

[a]All listed villages are in Vidarbha province.

crushed material was homogenized in a mortar to make a paste. The homogenate was filtered through at least four layers of cheesecloth. Spores of *N. locustae* were isolated by the centrifugation method developed by Cantwell (1970). The small quantity of isolated spores was fed to laboratory-reared *L. migratoria* and *S. gregaria*. Both species of locusts became infected and, in the process, a sufficient quantity of *N. locustae* spores was obtained.

C. *In Vivo* Production of *Nosema locustae* Spores in *Locusta migratoria*

Since a large number of *L. migratoria* can easily be reared in an insectary, this species is preferred for *in vivo* production of *N. locustae* spores.

Third instars were used for the initial inoculations with various concentrations of spores. The nymphs were kept individually in plastic vials covered with muslin cloth at both ends. The nymphs were starved for 24 hr before being fed a known spore inoculum. The vials were separated into three groups coded S (single inoculation), D (double inoculation), and T (triple inoculation).

A single dose of spores was given to each of four batches of the S group by feeding an approximately 1-cm piece of *Sorghum* leaf dipped in 1.5×10^5, 1.5×10^6, 1.5×10^7, or 1.5×10^8 spores/ml. The D group was given the first inoculation on the same day as the S group, followed by a second inoculation after 48 hr. The T group was treated similarly to groups S and D for the first and second inoculations, followed by a third inoculation 48 hr after the second inoculation. After inoculation, the nymphs were distributed into separate cages and served a regular feed of fresh *Sorghum* leaves.

It was observed (Table II) that triple inoculation with 1.5×10^5 spores/ml was effective, producing about 1.3×10^9 spores per infected adult, whereas single and double inoculations of the same concentration yielded fewer spores. Three inoculations of 1.5×10^6 spores/ml also proved effective and yielded appreciable quantities of spores. The insects inoculated with 1.5×10^7 spores/ml three times gave overall good performance, but the yield remained close to that obtained with the previous dose (1.5×10^6). A single inoculum of 1.5×10^8 spores/ml yielded only 1.3×10^9 spores per insect. Moreover, double and triple applications were lethal to early stages and disappointing in terms of spore production.

The mortality record (Table III) indicates that there was an increase in nymphal mortality at the 4th and 5th nymphal stages with an in-

TABLE II

Effect of Initial *Nosema locustae* Spore Inoculations in Third Instar on Final Spore Production in Adult *Locusta migratoria*

	Inoculum concentration (spores/ml) and spore production			
Inoculation	1.5×10^5	1.5×10^6	1.5×10^7	1.5×10^8
Single	$6,234,375 \times 10^2$	$12,122,880 \times 10^2$	$11,955,600 \times 10^2$	$12,556,800 \times 10^2$
	6.2×10^8	1.2×10^9	1.2×10^9	1.3×10^9
Double	$7,176,000 \times 10^2$	$14,158,800 \times 10^2$	$16,500,000 \times 10^2$	$5,825,520 \times 10^2$
	7.2×10^8	1.4×10^9	1.7×10^9	5.8×10^8
Triple	$12,589,200 \times 10^2$	$20,958,240 \times 10^2$	$20,838,400 \times 10^2$	$5,481,600 \times 10^2$
	1.3×10^9	2.1×10^9	2.1×10^9	5.5×10^8

TABLE III

Effects of Various Concentrations of *Nosema locustae* Spores on Mortality of *Locusta migratoria*

Inoculation	Concentration (spores/ml)	No. of insects tested	Mortality					Mortality (%)		Mortality (days)	
			Nymph								
			3rd instar	4th instar	5th instar	Total	Adult	Nymph	Adult	50%	100%
Control	0	43	0	2	3	5	4	11.6	9.3	—	—
Single	1.5×10^5	43	1	2	10	13	30	30.2	69.8	31	63
	1.5×10^6	43	0	2	9	11	32	25.6	74.4	30	58
	1.5×10^7	44	1	5	20	26	18	59.1	40.9	29	44
	1.5×10^8	45	2	13	25	40	5	88.9	11.1	24	37
Double	1.5×10^5	44	1	3	11	15	29	34.1	65.9	27	56
	1.5×10^6	44	3	4	8	15	29	34.1	65.9	24	55
	1.5×10^7	43	2	5	28	35	8	81.4	18.6	23	35
	1.5×10^8	45	5	19	19	43	2	95.6	4.4	17	32
Triple	1.5×10^5	43	2	4	12	18	25	41.9	48.1	20	53
	1.5×10^6	43	4	8	14	26	17	60.5	39.5	19	51
	1.5×10^7	44	6	14	21	41	3	93.2	6.8	19	31
	1.5×10^8	46	16	15	13	44	2	95.7	4.3	12	27

crease in spore concentrations of single, double, and triple inoculums. At high concentrations (1.5×10^7 and 1.5×10^8 spores/ml) in single, double, and triple inoculations the nymphal stages succumbed to the infection more rapidly and died prior to reaching the adult stage. This was also confirmed by examining stained hemolymph smears containing microsporidian stages in the infected hemocytes. On the other hand, at low concentrations the nymphal infection proceeded comparatively slowly, and more than 65% of the nymphs reached the adult stage, except for those receiving the triple inoculation. However, mortality at the adult stage reached 100% by 51–63 days after the initial inoculations. The numbers of days required for the pathogen to cause 50% mortality in the infected nymphs and adults were 24–31, 17–27, and 12–20 at single, double, and triple inoculations, respectively, and 100% mortality was observed within 37–63, 32–56, and 27–53 days at single, double, and triple inoculations, respectively.

The data indicated that although a desirable number of spores could be produced effectively with a single inoculation of 1.5×10^6 spores/ml, double inoculation of the same concentration was preferable to ensure 99% infection of nymphs. Triple inoculation of this concentration enhanced spore production but caused high nymphal mortality, reducing the total yield. Similarly, 1.5×10^7 and 1.5×10^8 spores/ml resulted in a higher rate of premature deaths during nymphal stages.

It has been observed that production of spores is higher in adult *L. migratoria* than in nymphs. Henry and Oma (1981) also obtained a greater yield in adults of the grasshopper *Melanoplus bivittatus* using the fifth instar as a host. In spruce budworm, *Choristoneura fumiferana* (Wilson, 1976), the maximum mean spore count was 1.36×10^8 spores per larva when second instar larvae were inoculated with a concentration of 2.0×10^5 *Nosema fumiferanae* spores per ml. McLaughlin and Bell (1970) inoculated second instar boll weevils, *Anthonomus grandis*, with 10^6 *Glugea gasti* spores/ml, achieving approximately 6.8×10^7 spores per weevil. Hostounsky and Weiser (1972) noted wide variability in the spore count when fourth instar caterpillars of *Mamestra brassicae* were fed with a spore suspension of *Nosema plodiae*. They concluded that low initial doses were more efficient in mass production of spores than high doses. In *L. migratoria*, the maximum mean spore count was 2.1×10^9 spores per adult with a triple inoculation of 1.5×10^6 or 1.5×10^7 spores/ml. However, this resulted in high nymphal mortality, hence it was not considered for mass production. Double inoculation of the 1.5×10^6 concentration, even though yielding 1.2×10^9 spores per adult, resulted in more infected adults and improved total yield.

D. Pathogenesis of *Nosema locustae* in Host Tissue

1. Midgut

The ingested spores germinate in the lumen of the midgut, possibly due to change in the pH. The sporoplasms ooze out and enter the midgut epithelium. This process in *L. migratoria* may occur in two ways: (a) the sporoplasms are absorbed during the process of digestion, and (b) the spores that come in close contact with the brush border directly inject sporoplasm into the midgut epithelium.

There has been confusion regarding the transfer of microsporidians from the lumen of the gut into the cells. Some authors (Weiser, 1961; Steche, 1965) have proposed that the sporoplasm is capable of ameboid movement and actively penetrates the host cell, but most workers think that the sporoplasm is injected into the host cell by the evaginating polar tube (Ishihara, 1967; Weidner, 1972; Vavra, 1965). Lom and Vavra (1963) suggested that sporoplasms are pushed out by the expanded posterior vacuole; this does not apply to all microsporidians because of the absence of posterior vacuoles in some of them (Vinckier *et al.*, 1970; Canning and Sinden, 1973).

In the midgut epithelium of *L. migratoria* the sporoplasm may or may not initiate the developmental process and most sporoplasms pass through the epithelium and enter into the hemolymph. However, a few sporoplasms begin their development in the midgut epithelium, around which they form cysts. This process damages the midgut and exposes the hemolymph to lumen bacteria, causing premature septicemial death of nymphs.

2. Hemocyte

Many of the sporoplasms that enter the hemolymph are encapsulated by the phagocytes and in about 4 to 8 days postinoculation all the phagocytes become infected with various developmental stages of the pathogen. Other types of hemocytes are not infected with the pathogen. In infected phagocytes the nuclei are disrupted and the whole cytoplasm is replaced by developing spores (Fig. 1). The cells appeared swollen and larger because of the pathogen invasion. In *Melanoplus sanguinipes* spore-filled phagocytes were observed during the course of the infection (Raina and Ewen, 1980).

3. Fat Body

In *L. migratoria* the fat body is a translucent bright yellow structure forming a complete sheath around the intestine and gonads. Histo-

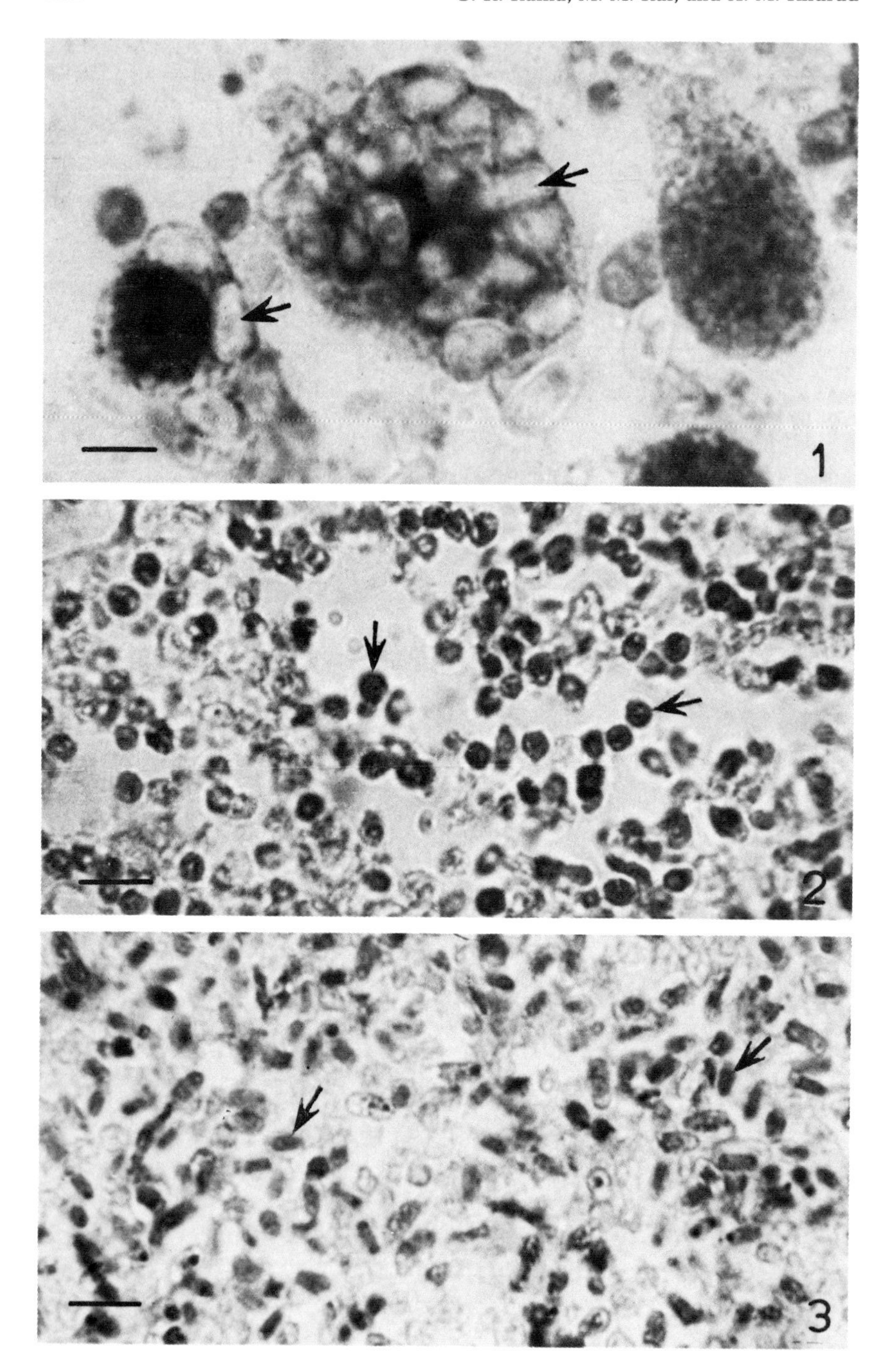
1
2
3

logically, the fat body is composed of a network of large cells with centrally situated nuclei. The fat body begins to lose translucence with the initiation of infection and becomes opaque and creamy 10–12 days postinoculation. In a fully infected host, the fat body cells are filled with mature spores.

Histopathological examinations revealed that sporoplasms enter fat body cells through the hemolymph. Sporoplasms divide and give rise to schizonts. During schizogony the pathogen divides two or three times, producing sporonts at its final division (Fig. 2). The sporont has a thicker outer wall than the schizonts and is comparatively smaller. The sporogony gives rise to sporoblasts, which organize themselves into spores (Fig. 3). The immature spores do not show chitinization and appear like sporoblasts except that they are comparatively smaller. The chitinization process begins soon after division and lasts 24–48 hr, when sporogony is complete and the fat body becomes a mass of mature spores.

The percent infection in fat body tissue (Table IV) indicates that *N. locustae* invades 80–90% of the fat body 15–18 days after the inoculation of the host. The pathogen utilizes most of the fat body reserve, probably reducing fecundity and longevity of the host. Other workers have also reported the deleterious effect on host development caused by disruption of the gut wall and depletion of fat body reserves by microsporidians (Canning and Hulls, 1970; Hazard and Lofgren, 1971; Hulls, 1971; Raina and Ewen, 1980; Henry, 1981).

4. Muscles and Nerve Tissues

Nosema locustae has no affinity for nerve and muscle tissues and no trace of infection was observed in the present study. However, the pathogen may indirectly inhibit the development of these tissues by utilizing their nourishment. Henry (1967) reported that *Nosema acridophagus* attacked nerve tissue, pericardial cells, and ovarian tissue.

Fig. 1. Heavily infected phagocytes of *Locusta migratoria* filled with *Nosema locustae* spores (arrows) in cytoplasm. Giemsa stain; magnification, ×1350. Bar, 5 μm.

Fig. 2. Section through infected fat body of *Locusta migratoria* showing sporont (arrows) stage of *Nosema locustae*. The sporonts are identified by their round dense structure. Azan stain; magnification, ×900. Bar, 10 μm.

Fig. 3. Sporoblast (arrows) stage of *Nosema locustae* in the heavily infected fat body of *Locusta migratoria*. Azan stain; magnification, ×900. Bar, 10 μm.

TABLE IV

Intensity of Infection in Fat Body Tissue of *Locusta migratoria* Postinoculated with Various Concentrations of *Nosema locustae* Spores at the Third Nymphal Stage

Inoculation	Concentration (spores/ml)	Infection (%) after inoculation (days)[a]						
		1	3	6	9	12	15	18
Single	1.5×10^5	−	−	−	±	+	++	++++
	1.5×10^6	−	±	+	+	++	+++	+++
	1.5×10^7	−	−	−	+	+	++	+++++
	1.5×10^8	−	−	±	++	+++	+++++	+++++
Double	1.5×10^5	−	−	−	±	++	+++	+++
	1.5×10^6	−	−	−	±	+++	++++	+++++
	1.5×10^7	−	−	−	±	++	+++++	+++++
	1.5×10^8	−	−	−	+	++	*	*
Triple	1.5×10^5	−	−	−	+	++	++	+++
	1.5×10^6	−	−	±	+++	++++	+++++	+++++
	1.5×10^7	−	+	++	++	++++	+++	*
	1.5×10^8	−	−	+	++	++++	*	*

[a]Infection (%): −, nil; ±, 0–5; +, 5–20; ++, 20–40; +++, 40–60; ++++, 60–80; +++++, 80–100.

*, Mortality.

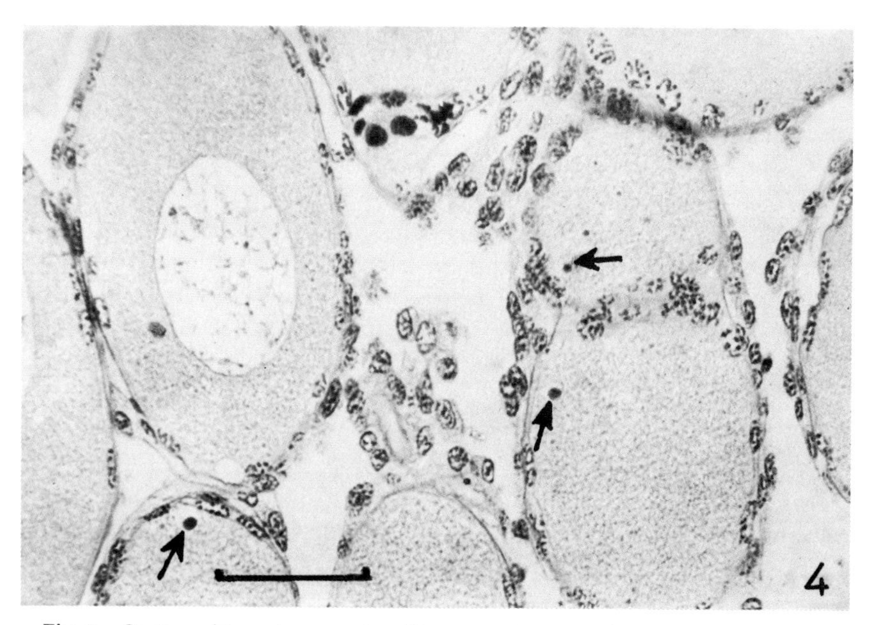

Fig. 4. Section of immature oocytes of *Locusta migratoria* showing developing stages of *Nosema locustae*. Azan stain; magnification, ×500. Bar, 0.1 mm.

TABLE V
Effect of Various Concentrations of *Nosema locustae* Spores on the Ovarian Weight of *Locusta migratoria* When Inoculated as Third Instar

Inoculation	Concentration (spores/ml)	Ovarian weight (mg) after inoculation (days)						
		3	6	9	12	15	18	21
Control	0	2.0	3.9	6.8	10.8	16.2	18.0	19.5
Single	1.5×10^5	1.4	2.1	2.7	3.7	4.3	7.8	9.7
	1.5×10^6	1.3	2.0	4.8	5.9	8.0	12.0	16.1
	1.5×10^7	1.3	3.2	6.1	7.5	9.0	12.3	*
	1.5×10^8	1.2	3.3	4.5	4.9	5.2	*	*
Control	0	—[a]	3.3	5.0	9.5	14.4	17.4	19.0
Double	1.5×10^5	—	2.0	4.9	6.5	9.7	11.7	14.5
	1.5×10^6	—	1.9	3.8	5.6	8.5	12.2	16.5
	1.5×10^7	1.1	2.6	4.4	5.5	8.0	10.0	13.0
	1.5×10^8	1.0	1.6	3.6	5.6	7.5	*	*
Control	0	2.6	4.9	8.1	12.6	16.8	18.5	19.5
Triple	1.5×10^5	2.2	3.9	4.6	7.7	10.0	10.7	12.3
	1.5×10^6	2.1	3.8	5.5	9.6	15.5	16.8	17.9
	1.5×10^7	1.5	2.0	2.4	3.0	2.7	2.5	*
	1.5×10^8	1.4	2.1	2.3	2.2	*	*	*

[a]—, Not recorded.
*, Mortality.

5. Ovary

The ovary was not affected directly by *N. locustae* except in a few cases, where oocytes showed slight infection in the follicular epithelium and ooplasm (Fig. 4). The overall effect on the ovary of four different concentrations of spores revealed a decline in ovarian weight compared to that of controls (Table V). A significant decrease in ovarian weight was observed with 1.5×10^7 and 1.5×10^8 spores/ml in all inoculations. This decline in ovarian weight indicates arrested development of the ovary due to nonavailability of yolk material during oocyte development as the fat body is completely infected. We could not establish unequivocally transovarial transmission of microsporidia in *L. migratoria*. In spruce budworm, *C. fumiferana*, transovarial transmission of *N. fumiferanae* has been demonstrated (Thomson, 1958; Wilson, 1982) in infected females. A few locusts in our insectary became naturally infected with *N. locustae*. The parents of these insects had been given a spore inoculum of lower concentration and a few laid eggs.

E. Screening of *Nosema locustae* against the Silkworm *Bombyx mori*

In the screening tests, the overall emphasis was directed toward the yield of silk because in India, where sericulture is of considerable importance, the mulberry trees grow in close proximity to other important agricultural crops, which are frequently damaged by grasshoppers and locusts. Drifts of microbial pesticides are hazardous to silkworms feeding on mulberry leaves if the pesticides prove deleterious to any of the life stages of the developing moth. Therefore it was imperative to test *Nosema* microbial pesticides for possible activity against the silkworm. To establish the specificity of *N. locustae* as an orthopteran pathogen, another *Nosema* species from the tasar silkworm (Lepidoptera) was also tested simultaneously.

The life cycle of the silkworm *B. mori* (egg to egg) was usually completed in 47–52 days. The egg stage took 9–11 days for hatching in a multivoltine race. Larval life lasted for about 27 ± 2 days, followed by a pupal stage of 10–12 days in a silk cocoon and final emergence as an adult, which lasted for 2–4 days. This cycle runs smoothly at temperatures ranging from 24° to 28°C at a relative humidity of 70–90%. Any fluctuation in temperature and relative humidity brings about a change in the duration of life stages which is reflected in the production of silk.

About 1500 third instar larvae were divided into three groups. The first group, coded SA, was fed with an *N. locustae* spore suspension (1.5×10^6 spores/ml) sprayed on the mulberry leaves. The second group was coded SB for *Nosema* sp.-fed silkworm (1.5×10^6 spores/ml). The third group was kept as the control (SC). Groups SA and SB were given seven inoculations of respective spores over a period of 13 days. The life cycle, internal organs, cocoon weight, egg laying and hatching, and F_1 generation were compared.

1. Life Cycle

The molting of 3rd, 4th, and 5th instar larvae of *N. locustae*-treated groups was quite normal and there was no delay in the completion of the life cycle. However, *Nosema* sp. of tasar silkmoth slightly delayed molting at the 4th and 5th instar stages and consequently the life cycle was prolonged by 4–5 days compared to that of the controls and *N. locustae*-treated larvae (Table VI). Previous work has also shown that infection with protozoan parasites can delay molts (Canning, 1962) and reduce the time period of larval growth (Ashford, 1967). The mortality of 3rd, 4th, and 5th instar stages was insignificant in all instances.

TABLE VI

Effect of *Nosema locustae* and *Nosema* sp. (Tasar) Spore Inoculations on Life Cycle of Silkworm *Bombyx mori*

	Period required for development (days)							
	Larval development					Pupal development		
Group[a]	3rd instar	Molt	4th instar	Molt	5th instar	Spinning	Adult emergence	Total
SA	3–4	1–2	4	1–2	10	3–4	10	31–36
SB	4	1–2	4–5	2–3	9–11	4–5	10–12	34–42
SC	3–4	1–2	4	1–2	9	3–4	9–10	30–35

[a]SA, *Nosema locustae* inoculated; SB, *Nosema* sp. (tasar) inoculated; SC, control.

However, a slight infection of *Nosema* sp. was evident in some SB larvae.

The spinning was complete in 3–4 days in almost all caterpillars. The mortality that occurred at the time of cocoon spinning was comparatively higher in group SB than in group SA and the controls (Table VII).

2. Internal Organs

a. Midgut

The midgut in silkworm larvae is composed of long columnar cells with a brush border and centrally situated nuclei. Histopathological examination of SA and SB caterpillars did not reveal pathogen development in the midgut epithelium.

TABLE VII

Effect of *Nosema locustae* and *Nosema* sp. (Tasar) Inoculation on Mortality in Silkworm *Bombyx mori* When Inoculated as Third Instar

		Mortality at different stages				
Group[a]	No. of larvae tested	3rd instar	4th instar	5th instar	Spinning	Mortality (%)
SA	460	0	0	4	20	5.2
SB	460	2	4	13	44	13.7
SC	460	0	0	2	31	7.4

[a]SA, *Nosema locustae* inoculated; SB, *Nosema* sp. (tasar) inoculated; SC, control.

b. Hemocytes

Any infection in the body of the silkworm is exhibited in the hemolymph, where the invading pathogens are encapsulated by phagocytes. Hemolymph of the silkworms from groups SA, SB, and SC was examined at various intervals. Only the SB larvae at a late stage of development contained *Nosema* sp., and SA and SC larvae showed no trace of infection throughout their life cycle.

c. Silk Gland

The silk gland is composed of three parts: anterior, middle, and posterior. These portions assimilate the mulberry protein in the course of larval development and grow to a considerable size and weight (0.500 g/g body weight) to provide protein material during the spinning of the cocoon. The processes of protein assimilation and growth of the silk gland depend mainly on the physiological condition of the caterpillar during larval growth. The weights of the silk glands of SA, SB, and SC larvae were compared to observe any effect due to feeding of pathogens.

The data (Fig. 5) indicated that the silk glands of the final instar of SA larvae were unaffected by the pathogen and the per gram body weight (0.516 g) remained very close to that of group SC (0.500 g), whereas a decrease in the weight of the silk gland was observed in group SB (0.344 g). This suggests that *N. locustae* does not interfere with the normal physiology of the silkworm, whereas *Nosema* sp. of tasar silkmoth affects the weight of the silk gland, possibly by reducing the assimilation of proteins.

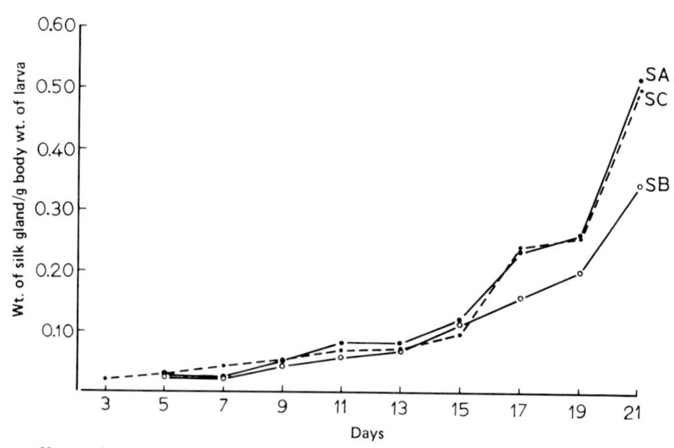

Fig. 5. Effect of *Nosema locustae* of grasshopper and *Nosema* sp. tasar silkmoth on weight of the silk gland of *Bombyx mori*. SA, *Nosema locustae* spores; SB, *Nosema* sp. of silkmoth; SC, Controls.

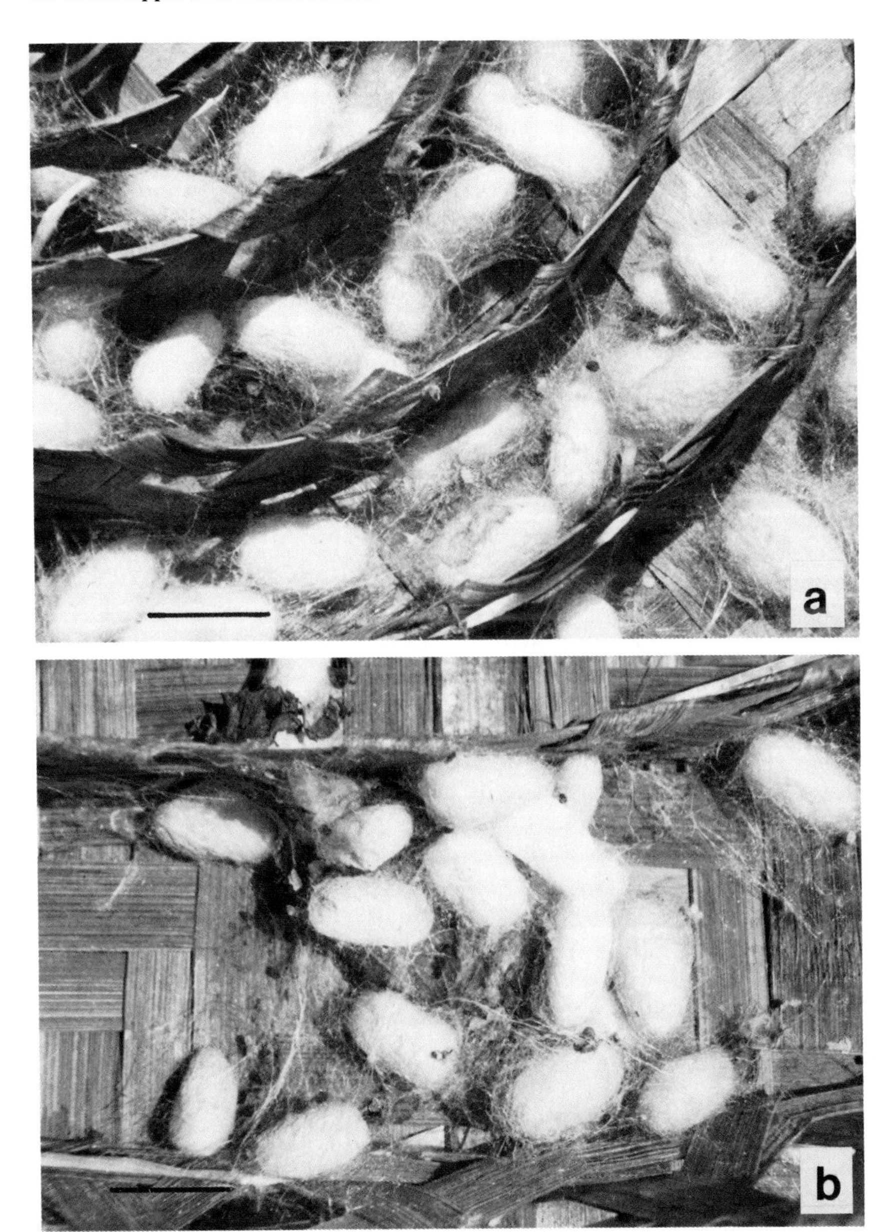

Fig. 6. (a) Cocoons produced by the silkworm *Bombyx mori* (group SC, controls). Bar, 3.5 cm. (b) Cocoons produced by the silkworm *Bombyx mori* when fed with *Nosema locustae* (group SA)-sprayed mulberry leaves at the third instar stage. Bar, 3.5 cm.

3. Cocoon Weight

The silk cocoons produced by SA, SB, and SC caterpillars differed very slightly in weight. Their partial dry weight was recorded on the 11th day after spinning and the average weight of a single cocoon was 0.69 g in SA, 0.60 g in SB, and 0.79 g in SC. The cocoons produced by a few infected SB larvae were flimsy and of poor quality. However, the cocoons of groups SA and SC were of good quality (Figs. 6a and 6b).

4. Egg Laying and Hatching

Comparison of egg laying by adult females of groups SA, SB, and SC showed that a reduced number of eggs and low percentage of hatching were predominant in group SB, whereas 97% hatching occurred in groups SA and SC (Table VIII). The unhatched eggs of group SB contained more black and yellow heads than those of SA and SC.

The low hatching percentage in group SB may have been due to transovarial transmission of *Nosema* sp. of tasar silkmoth. Wilson (1982) reported that *N. fumiferanae* can be transmitted transovarially within a host population.

5. Effect on F_1 Generation

In groups SA and SC, the larval period ranged from 28 to 29 days and mortality was less than 5%. In SB the mortality was greater than 30%; most of the dead larvae showed brown spots on the body surface and became almost weightless during drying. This mortality in SB may be attributed to a rise in the environmental temperature at the end of February, which F_1 larvae of SB could not withstand because of the presence of transmitted pathogens in their body. On the other hand, SA

TABLE VIII

Effect of *Nosema locustae* and *Nosema* sp. (Tasar) Spore Inoculations on Egg Laying and Hatching of Silkworm *Bombyx mori* When Inoculated as Third Instar

Group[a]	No. of eggs	Eggs hatched	Hatching (%)	Egg mortality		
				Blackhead	Yellow	Total
SA	316 ± 28.4	308 ± 27.0	97.4	6	2	8
SB	219 ± 29.6	170 ± 21.6	78.1	20	29	49
SC	304 ± 25.4	297 ± 27.4	97.7	3	4	7

[a]SA, *Nosema locustae* inoculated; SB, *Nosema* sp. (tasar) inoculated; SC, control.

and SC could resist this change indicating the least possibility of *N. locustae* transmission.

The results obtained in this study indicate that the microsporidian genus *Nosema* is very diversified and its species are specific to various insect groups. The silkworm *B. mori* can be affected by *Nosema* sp. isolated from lepidopterans, whereas *N. locustae* of Orthoptera does not affect the silkworm and has no deleterious effects on its life stages. Thus, the use of *N. locustae* against grasshoppers and locusts should present no hazard to sericulture.

IV. CONCLUSION

An effort has been made to present a new approach to grasshopper and locust control in India through the use of microsporidians. It is hoped that hazardous and uneconomical chemical pesticides may be replaced by, or integrated with, inexpensive and ecologically safe biocides of protozoan origin. The protozoan *N. locustae* has been isolated locally and developed in an insectary, using locusts. It has been successfully screened against the silkworm to make it acceptable for field use. An effort is now being made by the Biological Pest Management Project of Nagpur University to develop this biocide by *in vitro* techniques.

ACKNOWLEDGMENTS

We wish to thank the International Development Research Centre, Canada, for their research grant to implement the Biological Pest Management Project at Nagpur University. The authors are grateful to Dr. Kenneth W. Riley, Mr. Balraj Aher, Dr. V. K. Thakare, Dr. P. D. Prasad Rao, and Dr. R. K. Downey for facilities and timely help and to Dr. Karl Maramorosch for helpful suggestions. We express our appreciation to Ms. Supriya Das and Mr. V. V. Adolkar for excellent technical assistance, C. M. Sarodey for photography, and P. S. Mahulikar and Diane Allen for typing the manuscript.

REFERENCES

Ashford, R. W. (1967). A study of protozoan parasites of stored products Coleoptera. Ph.D. Thesis, University of London.

Canning, E. U. (1953). New microsporidian, *Nosema locustae, Nosema* sp., from the fat body of the African migratory locust, *Locusta migratoria migratorioides* R&F, *Parasitology* **43,** 287–290.

Canning, E. U. (1962). A pathogenicity of *Nosema locustae. J. Insect Pathol.* **4,** 248–256.

Canning, E. U. (1982). An evaluation of protozoal characteristics in relation to biological control of pests. *Parasitology* **84,** 119–149.

Canning, E. U., and Hulls, R. M. (1970). A microsporidian infection of *Anopheles gambiae* Giles, from Tanzania; interpretation of its mode of transmission and notes on *Nosema* infections in mosquitoes. *J. Protozool.* **17,** 531–539.

Canning, E. U., and Sinden, R. E. (1973). Ultrastructural observations on the development of *Nosema algerae* Vavra & Undeen (Microsporida, Nosematidae) in mosquito, *Anopheles stephensi* Liston. *Protistologica* **9,** 405–425.

Cantwell, G. E. (1970). Standard methods for counting *Nosema* spores. *Am. Bee J.* **110,** 222–223.

Chandra, S. (1986). Locust Situation Bulletin, India **38**(18), 1–4.

Ewen, A. B. (1983). Extension of geographic range of *Nosema locustae* (Microsporida) in grasshoppers (Orthoptera: Acrididae). *Can. Entomol.* **115**(8), 1049–1050.

Ewen, A. B., and Mukerji, M. K. (1980). Evaluation of *Nosema locustae* (Microsporida) as a control agent of grasshopper populations in Saskatchewan. *J. Invertebr. Pathol.* **35,** 295–303.

Hazard, E. I., and Lofgren, C. S. (1971). Tissue specificity and systematics of a *Nosema* in some species of *Aedes, Anopheles* and *Culex. J. Invertebr. Pathol.* **18,** 16–24.

Henry, J. E. (1967). *Nosema acridophagus* sp. N., a microsporidian isolated from grasshoppers. *J. Invertebr. Pathol.* **9,** 331–341.

Henry, J. E. (1971). Experimental application of *Nosema locustae* for control of grasshoppers. *J. Invertebr. Pathol.* **18,** 389–394.

Henry, J. E. (1981). Natural and applied control of insects by Protozoa. *Annu. Rev. Entomol.* **26,** 49–73.

Henry, J. E., and Oma, E.A. (1981). Protozoa: Pest control by *Nosema locustae,* a pathogen of grasshoppers and crickets. *In* "Microbial Control of Pests and Plant Disease 1970–1980" (H. D. Burges, ed.), pp. 573–586. Academic Press, New York.

Henry, J. E., and Onsager, J. A. (1984). Experimental control of Mormon cricket, *Anabrus simplex,* by *Nosema locustae* (Microspora: Microsporida), a protozoan parasite of grasshoppers (Orthoptera: Acrididae). *Entomophaga* **27**(2), 197–201.

Hostounsky, Z., and Weiser, J. (1972). Production of spores of *Nosema plodiae* Kellen et Lindegren in *Mamestra brassicae* L. after different infective dosages. I. Vestr. Cesk. Spol. Zool. **36,** 97–100.

Hulls, R. H. (1971). The adverse effects of a microsporidian on sporogony and infectivity of *Plasmodium berghei. Trans. R. Soc. Trop. Med. Hyg.* **65,** 421–422.

Humason, L. H. (1962). "Animal Tissue Technique," 3rd ed. Freeman, San Francisco, California.

Ishihara, R. (1967). Stimuli causing extrusion of polar filaments of *Glugea fumiferanae* spores. *Can. J. Microbiol.* **13,** 1321–1332.

Lom, J. and Vavra, J. (1963). The mode of sporoplasm extrusion in microsporidian spores. *Acta Protozool.* **1,** 81–90.

Luna, G. C., Henry, J. E., and Ronderos, R. A. (1981). Experimental and natural infections with pathogenic protozoa in acridids from Argentina. *Rev. Soc. Entomol. Argent.* **40**(1/4), 243–247.

McLaughlin, R. E., and Bell, M. R. (1970). Mass production *in vivo* of two protozoan pathogens, *Mattesia grandis* and *Glugea gasti,* of the boll weevil, *Anthonomus grandis. J. Invertebr. Pathol.* **16,** 84–88.

Menapage, D. M., Sackett, R. R., and Wilson, W. T. (1978). Adult honey bees *(Apis mellifera)* are not susceptible to infection by *Nosema locustae. J. Econ. Entomol.* **71**(2), 304–306.

Moulden, J. (1981). Disease could control grasshoppers. *J. Agric., West. Aust.* **22**(2), 53–54.

Raina, S. K. (1985). "*In Vitro* Propagation of the Microsporidians Potentially Useful as Biological Control Agents of Insect Pests on Semi-arid Crops in India," UGC Proj. I and II, Tech. Rep. Dep. Zool. Entomol., Nagpur University.

Raina, S. K., and Ewen, A. B. (1980). Morphology of primary grasshopper fat body cell cultures isolated from *Melanoplus sanguinipes* and their subsequent infection by the microsporidian *Nosema locustae*. *Proc. 2nd Trienn. Meet., Pan Am. Acridol. Soc.*, 203.

Srivastava, Y. N., and Bhanotar, R. K. (1985). A new record of a protozoan pathogen, *Nosema locustae* canning infecting desert locust from Rajasthan. *Indian J. Entomol.* **45**(4), 500–501.

Steche, W. (1965). Zur Ontogenie von *Nosema apis* Zander im Mitteldarm der Arbeitsbiene. *Bull. Apic. Doc. Sci. Tech. Inf.* **8,** 181–183.

Thomson, H. M. (1958). The effect of a microsporidian parasite on the development, reproduction and mortality of the spruce budworm, *Choristoneura fumiferana* (Clem). *Can. J. Zool.*, **36,** 499–511.

Vavra, J. (1965). Etude au microscope électronique da la morphologie et du développement de quelques microsporidies. *C. R. Hebd. Seances Acad. Sci.* **261,** 3467–3470.

Vinckier, D., Devauchelle, G., and Pres, G. (1970). *Nosema vivieri* n. sp. (Microspora: Nosematidae) hyperparasite d'une gré vivant dans le Coelome d'une Nes. *C. R. Hebd. Seances Acad. Sci.* **270,** 821–823.

Weidner, E. (1972). Ultrastructural study of microsporidian invasion into cells. *Z. Parasitenkd.* **40,** 227–242.

Weiser, J. (1961). Die Mikrosporidian als Parasiten der Insekten. *Monogr. Angew Entomol.* **17,** 149.

Wilson, G. G. (1976). A method for mass producing spores of the microsporidian *Nosema fumiferanae* in its host, the spruce budworm, *Choristoneura fumiferana* (Lepidoptera: Tortricidae), *Can. Entomol.* **108,** 383–386.

Wilson, G. G. (1982). Transmission of *Nosema fumiferanae* (Microsporida) to its host *Choristoneura fumiferana* (Clem). *Z. Parasitenkd.* **68,** 47–51.

22

Establishment of Embryonic Cell Lines from the Brown Ear Tick Rhipicephalus appendiculatus and Their Immunogenicity in Rabbits

M. NYINDO, L. R. S. AWITI, AND T. S. DHADIALLA

I. INTRODUCTION

The brown ear tick, *Rhipicephalus appendiculatus*, is the main vector of *Theileria parva*, the causative agent for East Coast fever or theileriosis in cattle. Because of the economic importance of *R. appendiculatus* as a disease carrier and the effect on cattle due to the insect's feeding, various approaches have been taken to understand the biology of this tick and the vector–parasite relationships. One of the approaches has been to establish tick cell lines, because of the tremendous potential they hold for the study of viruses, tick-borne pathogens of veterinary and medical importance, and genetics. There has been considerable effort in developing the cultivation of tick cells *in vitro* (Rehacek, 1958; Martin and Vidler, 1962; Pudney *et al.*, 1973; Varma *et*

BIOTECHNOLOGY IN INVERTEBRATE PATHOLOGY AND CELL CULTURE

al., 1975). As a result of work by these and other investigators, the needed background information on tick cell cultures and the techniques required in their establishment was published in a very comprehensive review by Kurtti and Buscher (1979).

Recently, interest in antigens from tick tissues and tick cells in culture as potential vaccine candidates for the immunization of livestock against tick infestations has been generated (Wikel, 1982). At the International Centre of Insect Physiology and Ecology (ICIPE), one of the research priorities is in the field of acquired resistance to tick infestation and the development of artificial ways to immunize livestock against ticks. It is hoped that tick-immune animals will suffer less from tick-borne diseases because of a reduced tick burden. To this end efforts are being made at ICIPE to identify and purify tick tissue antigens for use as vaccine material. We have also established cell lines from *R. appendiculatus* embryos in the hope that these cells will provide a cleaner source (free from host blood proteins) of tick cell antigens as well as an opportunity to manipulate the enhanced expression of desirable gene products.

II. METHODS

A. Establishment of Cell Lines

Tick cell lines from embryonating *R. appendiculatus* (and *R. evertsi evertsi*) 14–21 days postoviposition were established, based primarily on techniques outlined by Kurtti and Buscher (1979). Mated females were surface-sterilized and individually placed in sterile tubes stoppered with cotton wool. Pooled egg batches from four ovipositing females were surface-sterilized, washed three times with Leibovitz L-15 medium (Leibovitz, 1963), and crushed in a tissue grinder to release cells. From our experience the eggshells must be removed by a series of washings with L-15 medium to achieve satisfactory cell growth since contamination of cells with eggshells retarded cell growth. Dispersing reagents such as trypsin were not used in the procedure to disperse cells. The crushed eggs were washed four times in L-15 medium by centrifugation and resuspension of the pellet. The cell pellet was finally resuspended in 12 ml of L-15 medium supplemented with 20% heat-inactivated fetal bovine serum, 10% tryptose phosphate broth, 1% lactalbumin hydrolyzate, 100 units/ml penicillin, 100 μg/ml streptomycin, and 50 μg/ml mycostatin. The cell suspension was dispersed into 3 × 25 cm^2 tissue culture flasks. Cultures were incubated at 37°C. Spent medium was replaced once a week with fresh medium. It usually took 10–14 days for adherent cells to reach confluence. Culture vessels that showed cell confluence were passaged once a week at a split ratio of 1 : 3 by slight agitation of the flasks to detach the cells.

B. Electron Microscopy

Cells were detached from culture vessels by slight agitation. The detached cells were spun down and, after decanting the medium, were fixed in 2.5% glutaraldehyde in 0.5 *M* cacodylate buffer. After postfixing in 1% osmium tetroxide, they were dehydrated in a graded ethanol series, infiltrated, and embedded in Araldite. Thin sections were stained with uranyl acetate and lead acetate before examination under the electron microscope.

C. Immunochemical Techniques

The cells were detached from culture flasks and collected as mentioned above. They were washed four or five times in 0.1 *M* phosphate-buffered saline (PBS), pH 7.4. The pellet was resuspended in 5 ml of PBS. Cells in suspension were subjected to sonication and the sonicate spun at 10,000 g. The supernatant consisting of soluble proteins was stored at -20°C until required for use. Protein concentration of the soluble cell protein extract was determined by the method of Lowry *et al.* (1951). Rabbits were immunized with 8 μg of protein extract in Freund's complete adjuvant by the intramuscular route. Ten days later they were given a booster of the same antigen in Freund's incomplete adjuvant. Rabbits were bled for antiserum by day 30. The immune serum was tested against protein extracts from cultured cells, larvae, nymphs, salivary glands, midguts, and whole fully engorged female ticks by Ouchterlony's double-immunodiffusion test. Immune serum from cattle rendered tick-resistant by tick infestation was also tested against the cell culture antigens.

Soluble protein extracts of tick cells were also subjected to polyacrylamide gel electrophoresis (PAGE) to resolve them into different protein bands. Electrophoresis under denaturing conditions [sodium dodecyl sulfate (SDS)-PAGE] was performed on 3–15% gradient gel slabs (1.5 × 18 × 20 cm). Electrophoresis was performed according to Laemmli (1970). Protein bands were visualized by staining the gel with Coomassie Blue R-250.

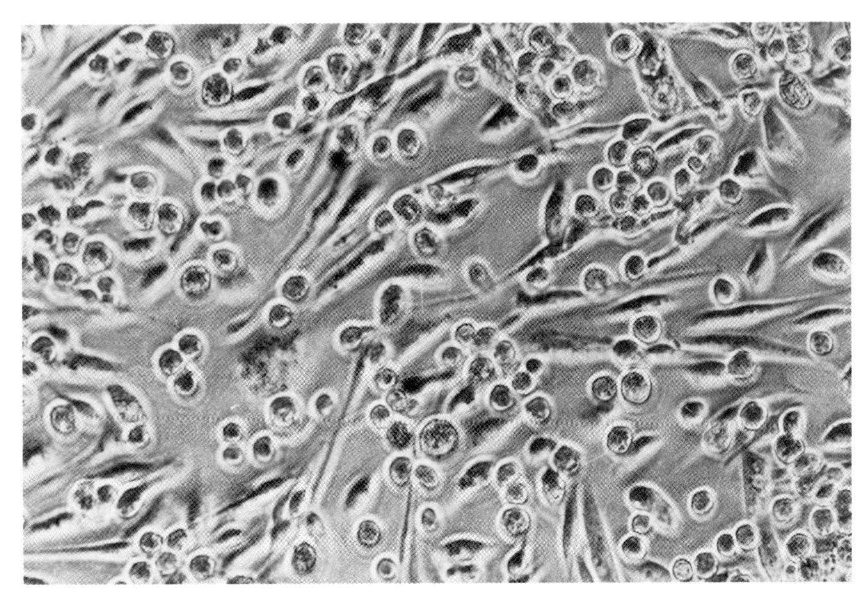

Fig. 1. Light micrograph of tick embryonic cells on day 425 of cultivation. Epithelial cells (round) and spindle-shaped cells are shown. Magnification, ×1600.

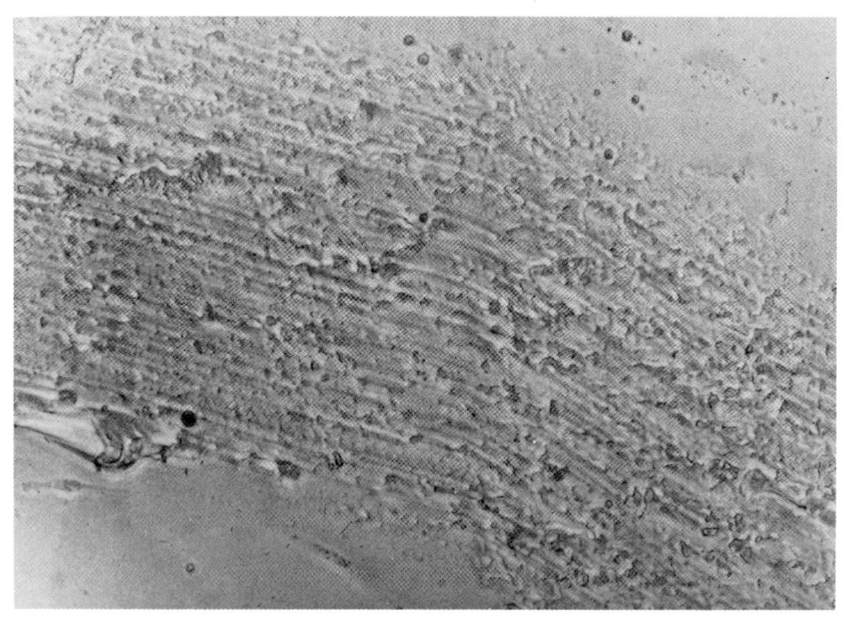

Fig. 2. Light micrograph of muscle fiber cells of *R. appendiculatus* embryonic cells in culture on day 425 of cultivation. Magnification, ×1700.

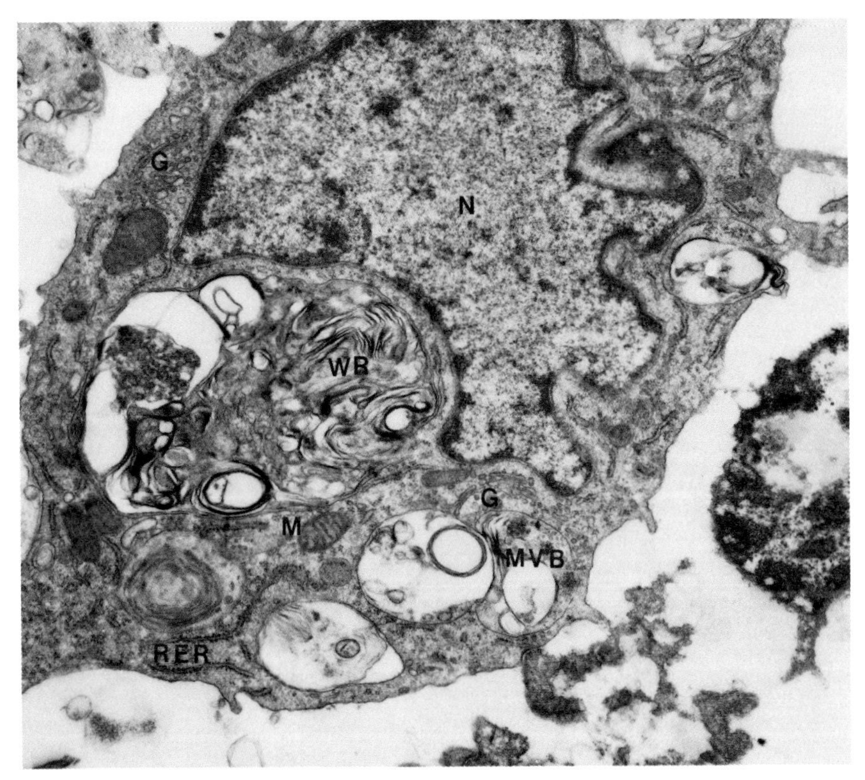

Fig. 3. Electron micrograph of an epithelial-type cell on day 96 of cultivation. Cell nucleus (N), Golgi apparatus (G), multivesicular body (MVB), mitochondria (M), rough endoplasmic reticulum (RER), and whorled rings (WR) are shown. Magnification, ×22,500.

III. OBSERVATIONS

Three types of adherent cells were observed to have grown from the tick embryos. These were epithelial-like, spindle-shaped, and muscle fibers. The latter cell type was very scarce; the other types were predominant. The population doubling time interval was estimated to be 5–6 days. Figure 1 shows the epithelial cells and spindle-shaped cells of *R. appendiculatus* on day 425 of cultivation, Fig. 2 shows the rare cell type consisting of muscle fibers, and Fig. 3 is an electron micrograph of an epithelial cell showing the presence of Golgi apparatus, rough endoplasmic reticulum, mitochondria, whorled rings, multivesicular body, and nucleus. Vacuolation was a common feature of the cells. When the soluble proteins were electrophoresed on SDS-PAGE (Fig. 4) they resolved into several protein bands ranging in molecular

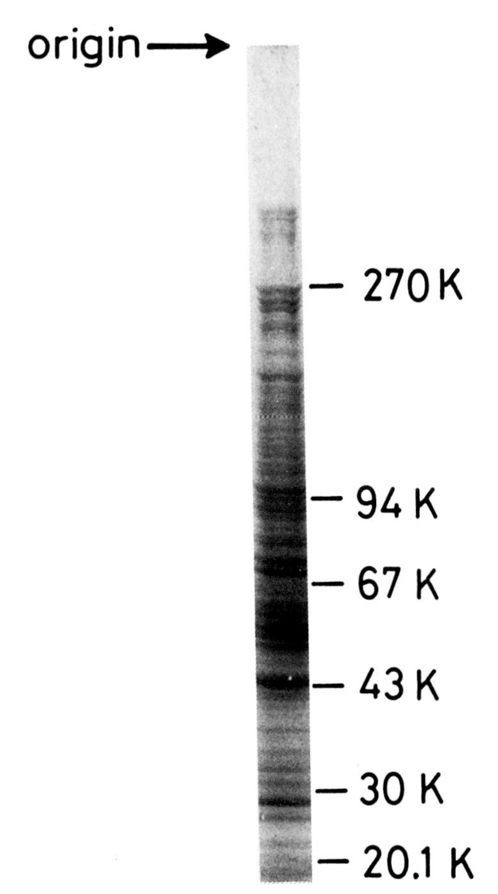

Fig. 4. SDS-PAGE of soluble protein extract of tick embryonic cells in culture on 3–15% gradient slab gel. On the right of the electrophretogram the relative mobility and molecular weights of marker proteins are indicated.

weight from 300,000 to 20,000. Antiserum from rabbits immunized with the cell culture soluble antigens reacted with antigens from midgut, salivary glands, and larvae to produced precipitin lines (Fig. 5). Serum from cattle made resistant to ticks by deliberate tick infestation also produced precipitin lines when reacted against the soluble antigen from the cell cultures (results not shown).

IV. SUMMARY

Tick cells have been established from embryonating eggs of *R. appendiculatus* 14–21 days postoviposition. It was observed that eggshells

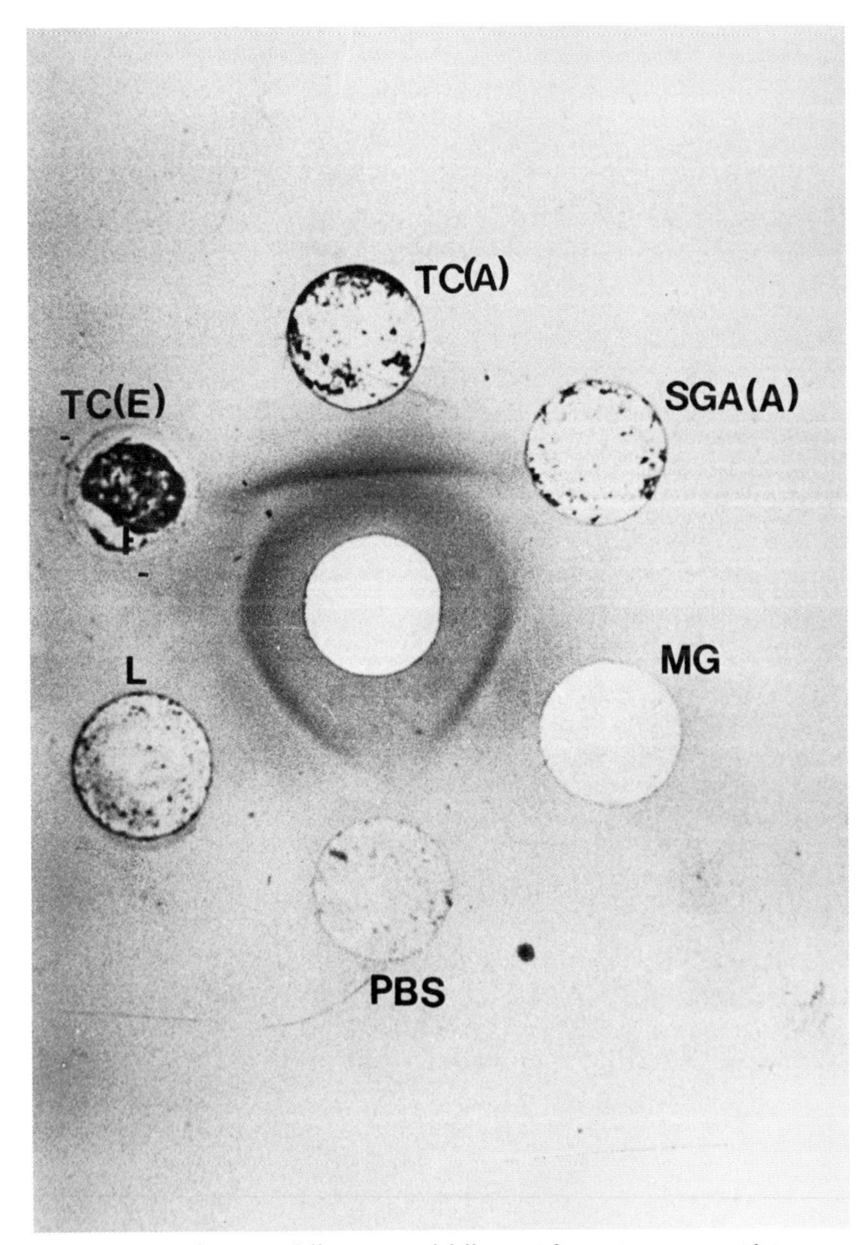

Fig. 5. Agar gel immunodiffusion test of different tick protein extracts with immune serum to soluble protein extracts of embryonic cell lines of *R. appendiculatus* raised in rabbits. The outer wells contain *R. appendiculatus* protein extracts from TC(A), tissue culture; SGA(A), salivary glands from both sexes; MG, midgut; L, larvae; TC(E), cell culture antigen derived from *R. evertsi evertsi;* PBS, phosphate-buffered saline. The center well contains the antiserum.

were detrimental to the successful establishment of adherent cells, and they must be removed after the eggs have been crushed. Epithelial cells and spindle-shaped cells are the predominant cell types. A cell type resembling muscle fibers was occasionally encountered. Soluble proteins from the cell culture have molecular weights ranging from 300,000 to 20,000. The soluble proteins are immunogenic in rabbits and the immune serum arising from inoculation of the proteins can recognize proteins from tick midguts, salivary glands, and larval and nymphal extracts. Serum from cattle made resistant to ticks recognizes antigens in the cell culture soluble fraction.

REFERENCES

Kurtti, T. J., and Buscher, G. (1979). *In* "Practical Tissue Culture Applications " (K. Maramorosch and H. Hirumi, eds.), pp. 351–371. Academic Press, New York.
Laemmli, U. K. (1970). *Nature (London)* **227,** 680.
Leibovitz, A. (1963). *Am. J. Hyg.* **78,** 173–180.
Lowry, O. H., Roserbrough, N. J., Ford, A. L., and Randall, R. J. (1951). *J. Biol. Chem.* **193,** 265–275.
Martin, H. M., and Vidler, B. O. (1962). *Exp. Parasitol.* **12,** 192–193.
Pudney, M., Varma, M. G. R., and Leake, C. J. (1973). *J. Med. Entomol.* **10,** 493–496.
Rehacek, J. (1958). *Acta Virol.* **2,** 253–254.
Varma, M. G. R., Pudney, M. and Leake, C. J. (1975). *J. Med. Entomol.* **11,** 698–706.
Wikel, S.K. (1982). *Annu. Rev. Entomol.* **27,** 21–48.

23

Establishment of an Ovarian Cell Line in the Cotton Bollworm Heliothis armigera and in Vitro Replication of Its Cytoplasmic Polyhedrosis Virus

DE-MING SU, ZHONG-JIAN SHEN,
AND YUN-XIAN YUE

I. INTRODUCTION

The cytoplasmic polyhedrosis viruses (CPVs) constitute the second largest group of insect viruses in terms of the number of isolates found in their hosts. Their importance in the biological control of insect pests has long attracted the attention of insect virologists (Payne, 1982). Thus it is surprising that fewer than a dozen isolates of CPVs have been studied *in vitro* (Payne and Mertens, 1983). Apparently, more insect virus–cell culture systems are needed for the investigation of various aspects of insect CPV replication (Granados, 1976).

In China, both the nuclear polyhedrosis virus (NPV) and CPV of the cotton bollworm, *Heliothis armigera*, were described in the late 1970s

BIOTECHNOLOGY IN INVERTEBRATE
PATHOLOGY AND CELL CULTURE

(Yue *et al.*, 1978; Su *et al.*, 1978). Since then, *Heliothis* NPV has been used extensively as a viral pesticide for the control of this economically important pest (Su, 1982). Several cell lines of *Heliothis* were established and are now available for replication study of the NPV (Yang and Xie, 1982–1983). The *Heliothis* CPV is common among field and laboratory populations of the cotton bollworm. Its *in vivo* morphogenesis had been studied in this laboratory by electron microscopy (Yue *et al.*, 1981). Cell lines need to be developed for the in-depth study of CPV *in vitro*.

This chapter will deal with the establishment of an ovarian cell line of the cotton bollworm, designated as SFE-HA-831, and the *in vitro* replication of its cytoplasmic polyhedrosis virus.

II. MATERIALS AND METHODS

The virus stock to be used was isolated in this laboratory (Su *et al.*, 1978).

Culture vessels and media. Cells were grown in 25 ml glass flasks. BML-TC10 medium (Gardiner and Stockdale, 1975) was used throughout the study. The medium was supplemented with 20% calf serum, 0.2% yeastolate (Difco), 100 IU/ml penicillin, and 100 μg/ml streptomycin. After 30 passages of the cell line only 15% calf serum was used to supplement the medium.

Primary culture and cell lines. Ovaries were dissected out of 10 *Heliothis* adults 24 hr after their eclosion. They were rinsed three times with sterile Rinaldini solution, and the fat bodies and tracheas attached to the ovaries were removed carefully with forceps. Then the ovarioles were cut into small pieces, pipetted into the flask with BML-TC10 medium, and incubated at 28°C. Cell growth was examined and the medium partly replaced once a week. The first passage of the cells was performed after the stationary stage of the cell culture.

Karyology. Cells were prepared for chromosome analysis as described by Schneider (1973) with slight modifications.

Serology. Antisera were produced in rabbits separately from SFE-HA-831 and ovaries of the *Heliothis* adults. SFE-HA-831 was compared to two other insect cell lines. One was also an ovarian cell line from the cotton bollworm, established in this laboratory along with SFE-HA-831. The other was an ovarian cell line of the geometrid *Ectropis obliqua*, designated as SIE-EO-803 (Liu *et al.*, 1981). Extracts of ovaries of the *Heliothis* adults and the greater wax moth, *Galleria*, were used as antigens. Immunodiffusion and immunoelectrophoresis were performed in characterizing these cell lines.

Isozyme analysis. Isozyme analyses of SFE-HA-831 and the geometrid cell line SIE-EO-803 were performed as described by Tabachnik and Knudson (1980) and the results were compared. Isozyme patterns of esterase, glucose-6-phosphate dehydrogenase, malate dehydrogenase, and lactate dehydrogenase were analyzed.

Virus replication in vitro. CPV replication *in vitro* was studied by light and electron microscopy and the virus produced was checked again by sodium dodecyl sulfate–polyacrylamide gel electrophoresis (SDS-PAGE) of the viral genome.

CPV particles prepared from alkaline-treated polyhedra and the supernatant of the extracts of infected *Heliothis* midguts were used as inocula for infecting the cells. Then the cells were incubated at 28°C. After adsorption of the virus for 2 hr, samples were taken at 40 min postinfection and every 2 hr thereafter. The cells were fixed and processed for electron microscopy as in the *in vivo* study of CPV morphogenesis (Yue *et al.*, 1981).

III. RESULTS

A. Establishment of the *Heliothis* Cell Line SFE-HA-831

1. Primary Culture and Cell Lines

Cells were seen migrating out of the explants 12 hr after incubation. Then these cells settled down and adhered to the bottom of the flask. Cell migration was most active in the first 48 hr of incubation. It continued for 2–3 weeks and then stopped. The stationary stage of cell culture began and lasted for 3 months.

After the stationary stage the cells were again activated and resumed their mitotic activities. Colonies of cells in which various cell types were present appeared in the culture. Then these colonies came into confluence, and the monolayer was formed.

The first passage of cells was performed when stacked cells were formed in patches in the monolayer. In the first few passages it took 12–18 hr for the cells to adhere to the bottom of the flask. The growth was slow. Each passage took 10–15 days before the next passage could be started. After the sixth passage, however, the cell growth accelerated and each passage was completed within 1 week.

The primary culture was started in January 1983, and the *Heliothis* cell line was in its 148th passage as of March 1986. It has been characterized and designated as SFE-HA-831 (Fig. 1).

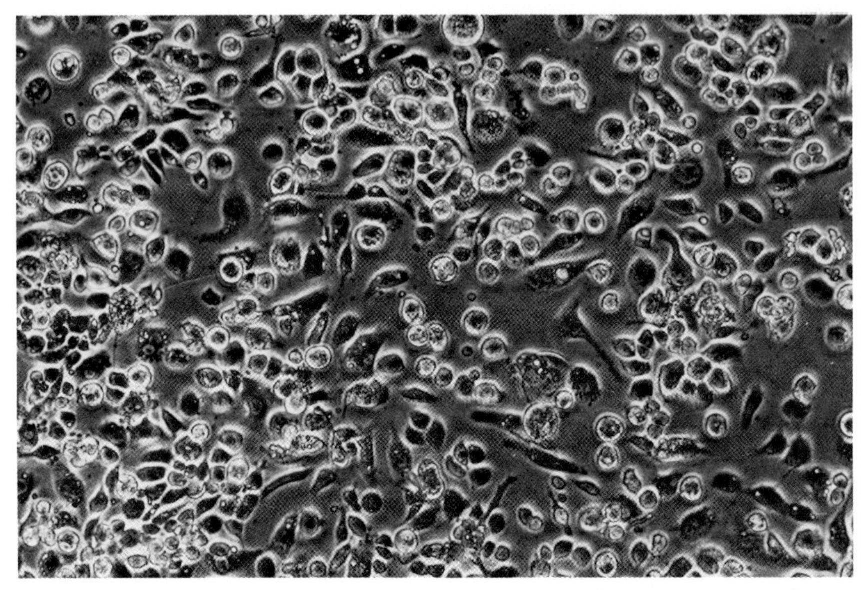

Fig. 1. *Heliothis* ovarian cell line, SFE-HA-831. Cells of various types are shown. Magnification, ×425.

2. Morphology and Growth

Three cell types of different morphology were present in SFE-HA-831: round cells, epithelial-like cells, and spindle cells.

Some growth features of the cells were followed. The first passage started from the primary culture at a 1:2 split. As mentioned above, cell growth was slow at this time. After the 6th passage cell growth was greatly accelerated, and each passage was done at a 1:10 split.

Three phases of the cell cycle were noted as shown in the growth curve (Fig. 2): the lag phase, during the first 24 hr; the logarithmic growth phase, from 24 to 120 hr; and the plateau phase, from 120 hr to 9 days. After that, cell growth decreased drastically.

Other parameters calculated during the cell cycle were the optimum cell density for inoculation, 150×10^3 cells/ml; the cell doubling time during the logarithmic growth phase, 24 hr; and the maximum cell density attained, 1.75×10^6 cells/ml after 5 days of growth at 28°C.

3. Karyology

Chromosomes of the cell line were round or short rods and very varied in number, as is typical of lepidopteran cells (Schneider, 1973).

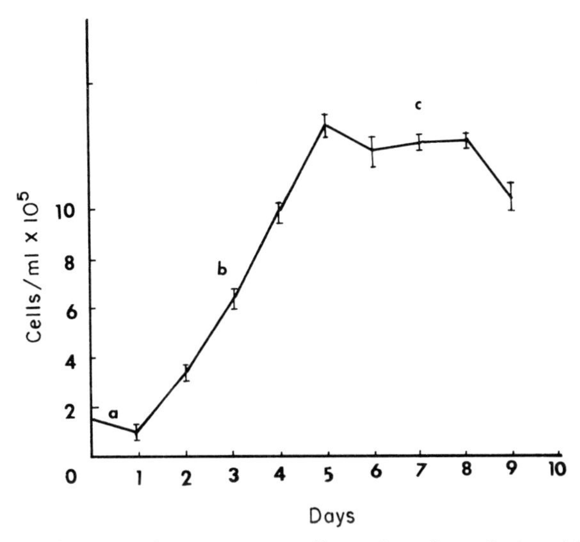

Fig. 2. Growth curve of SFE-HA-831 cells. a, Lag phase; b, logarithmic phase; c, plateau phase.

4. Serology

With double immunodiffusion, antisera prepared separately against SFE-HA-831 and the pupal and adult ovaries of the cotton bollworm reacted strongly with their homologous antigens but only weakly with extracts of the geometrid cell line and *Galleria* ovaries. This distinction was even more pronounced in experiments performed by immunoelectrophoresis. Antisera prepared against the extracts of *Heliothis* ovaries reacted with the homologous antigens from *Heliothis* ovaries and SFE-HA-831. In both these cases five or six precipitation arcs were formed, and their shapes and positions in the gel were also quite similar. On the other hand, in the heterologous reaction systems, only two or three precipitation arcs were formed and these arcs differed in their shapes and positions in the gel.

5. Isozyme Analysis

The esterase pattern of SFE-HA-831 in SDS-PAGE was completely different from that of the SIE-EO-803. The former showed only three light bands, whereas as many as nine heavy bands could be resolved in the latter. In the case of glucose-6-phosphate dehydrogenase, malate dehydrogenase, and lactate dehydrogenase the isozyme analysis also showed differences between these two insect cell lines.

B. Virus Replication *in Vitro*

The general picture of CPV morphogenesis as demonstrated by light and electron microscopy was the same in *in vivo* and *in vitro* systems. As expected, there were some differences in the duration of events important in morphogenesis of the virus.

1. Light Microscopy

Only a few bright particles were observed 18 hr postinfection in the cytoplasm, although the nature of these particles is not yet clear. No other cytopathic effects were detected in infected cells. Polyhedra were formed 48 hr postinfection. By 72 hr postinfection the infected cells rounded up. After 120 hr postinfection they began to aggregate and became detached from the bottom of the flask. These floating cells did not disintegrate; therefore no polyhedra were ever seen in the medium. However, after long incubation the infected cells eventually died and completely disintegrated, leaving masses of polyhedra in the medium.

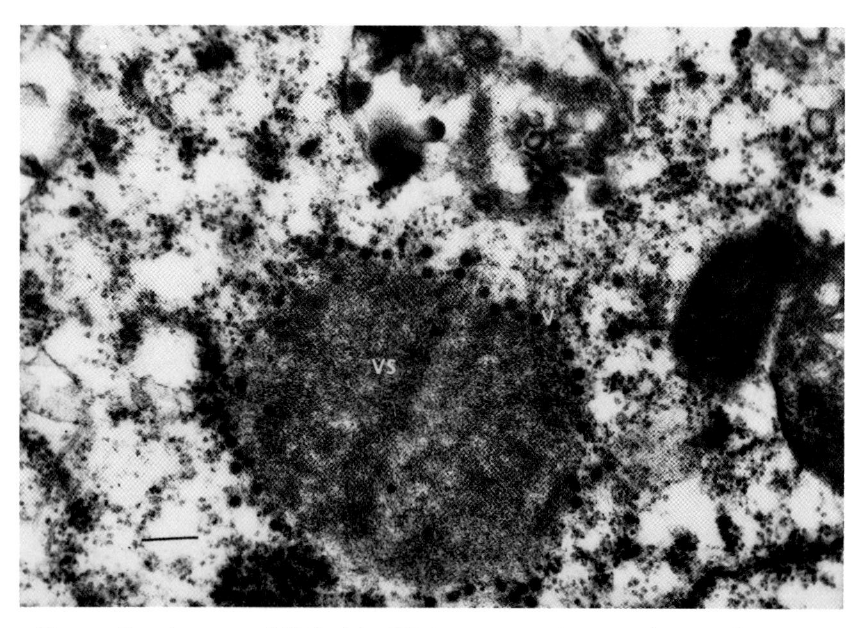

Fig. 3. Development of *Heliothis* CPV *in vitro* at 2 hr postinfection, showing the virogenic stroma (vs) and the associated virus particles (v) in the cell cytoplasm. Magnification, ×40,000. Bar, 250 nm.

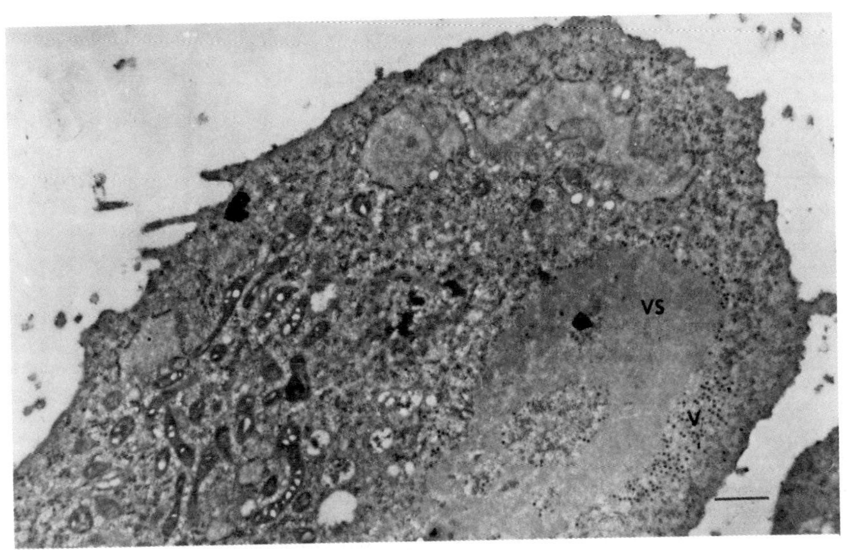

Fig. 4. Development of *Heliothis* CPV *in vitro* at 12 hr postinfection. More virogenic stroma (vs) and a large number of virus particles (v) were formed. Magnification, ×10,000. Bar, 1000 nm.

2. Electron Microscopy

The virogenic stroma and the virus particles of *Heliothis* CPV were detected early, i.e., 2 hr postinfection, in the cytoplasm of the infected cells (Fig. 3). The virogenic stroma increased in volume 12 hr postinfection and the number of virus particles free in the cytoplasm also increased (Fig. 4). Polyhedra were formed 30 hr postinfection, and accumulated in mass in the cytoplasm (Fig. 5). Their shapes were the same as those of the polyhedra produced *in vivo*. Large amounts of free virus particles were again present. It is interesting that no cytopathic lesions were evident in various organelles at this stage of infection.

IV. DISCUSSION AND CONCLUSIONS

In this chapter we reported the establishment of an ovarian cell line of the cotton bollworm, *Heliothis armigera*. The cell line, SFE-HA-831, is permissive for the replication of its homologous CPV, which was isolated in this laboratory in 1978 (Su *et al.*, 1978). Basically, the *in vitro* replication of *Heliothis* CPV is the same as its replication *in vivo* (Yue *et al.*, 1978), as demonstrated by both light and electron microscopy. Polyhedra and free virus particles produced *in vitro* infect both

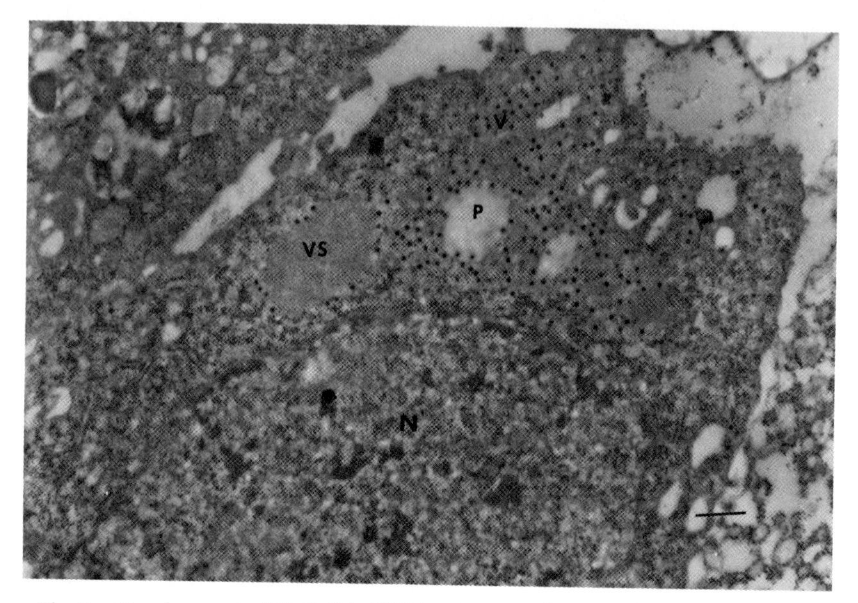

Fig. 5. Development of *Heliothis* CPV *in vitro* at 30 hr postinfection, showing complete formation of the polyhedra (P). Virogenic stroma (vs) and nucleus (N) are also shown. Magnification, ×20,000. Bar, 500 nm.

the host larvae and the cell line (unpublished observations). In addition, polyhedra produced *in vivo* and *in vitro* are the same shape—most of them are spherical. This is in sharp contrast to the case of *Euxoa* CPV in a *Lymantria* cell line (Belloncik and Bellemare, 1980; Quiot and Belloncik, 1977) and the *Heliothis* cell line, SFE-HA-831 (Belloncik *et al.*, 1985), in which a high percentage of cuboidal polyhedra are produced in infected cells. The puzzle of what governs the shape of CPV polyhedra remains unsolved (cf. Payne and Mertens, 1983). In conclusion, the establishment of the present cell line permissive to the replication of *Heliothis* CPV is a step toward understanding the molecular biology of this important CPV.

ACKNOWLEDGMENTS

We thank Drs. Yu-Jian, Hu, Qi-Gan, Liu, and Li-Mei Shen, Shanghai Institute of Entomology, Shanghai Academy of Sciences, for donating the geometrid cell line, SEI-EO-803. The technical assistance of Mr. Tong-Run Cai and Miss Hui-Ling Zhao, Laboratory of Electron Microscopy, Fudan University, for electron microscopy of infected cells is also acknowledged.

REFERENCES

Belloncik, S., and Bellemare, N. (1980). Polyèdres du CPV d'*Euxoa scandens* (Lep.: Noctuidae) produits in vivo et sur cellules cultivées in vitro: Etudes comparatives. *Entomophaga* **25,** 199–207.

Belloncik, S., Rocheleau, H., Su, D.-M., and Arella, M. (1985). Replication of a cytoplasmic polyhedrosis virus (CPV) in cultured insect cell. *Pap., Int. Cell Cult. Congr., 3rd, 1985.*

Gardiner, G. R., and Stockdale, H. (1975). Two tissue culture media for production of lepidopteran cells and nuclear polyhedrosis viruses. *J. Invertebr. Pathol.* **25,** 363–370.

Granados, R. R. (1976) Infection and replication of insect pathogenic viruses in tissue culture. *Adv. Virus Res.* **20,** 189–236

Liu, Q.-G., Hu, Y.-J., and Shen, L.-M. (1981) Establishment of two cell lines from pupal ovary of *Ectropis obliqua* Warren. *Contrib. Shanghai Inst. Entomol.* **2,** 123–128.

Payne, C. C. (1982). Insect viruses as control agents. *In* "Parasites as Biological Control Agents" (R. M. Anderson and E. U. Canning eds.).

Payne, C. C., and Mertens, P. P. C. (1983). Cytoplasmic polyhedrosis viruses. *In* "The Reoviridae" (W. K. Joklik, ed.), pp. 425–504. Plenum, New York.

Quiot, J. M., and Belloncik, S. (1977). Caractérisation d'une polyédrose cytoplasmique chez le lepidoptère *Euxoa scandens* Riley (Noctuidae, Agrotinae). Etudes in vivo et in vitro. *Arch. Virol.* **55,** 145–153.

Schneider, I. (1973). Karyology of cells in culture. F. Characteristics of insect cells. *In* "Tissue Culture: Methods and Applications" (P. F. Kruse, Jr. and M. K. Patterson, Jr., eds.), pp. 788–790. Academic Press, New York.

Su, T.-M. (1982). Use of bacteria and other pathogens to control insect pests in China. *In* "Microbial and Viral Pesticides" (E. Kurstak, ed.), pp. 317–332. Dekker, New York.

Su, T.-M., Yue, Y.-X., Chen, M.-C., and Yang, J.-Y. (1978). Studies on a cytoplasmic polyhedrosis of *Heliothis armigera* (Hübner). *J. Fudan Univ. (Nat. Sci.)* **17,** 74–78, and 85.

Tabachnik, W. K., and Knudson, D. L. (1980). Characterization of invertebrate cell lines. II. Isozyme analysis employing starch gel electrophoresis. *In Vitro* **16,** 392–398.

Yang, S.-Y., and Xie, R.-D. (1982–1983). Establishment of SIE-Ha-798 and SIE-Ha-806 cell lines of *Heliothis armigera* Hübner and their characteristics. *Contrib. Shanghai Inst. Entomol.* **3,** 129–136.

Yue, Y.-X., Zhan, H.-M., Ciang, Z.-Q., Chen, M.-C., Shu, Y.-M., Yao, H.-Z., and Su, T.-M. (1978) Studies on the nuclear polyhedrosis of *Heliothis armigera* (Hübner). I. Symptoms and the pathogen. *J. Fudan Univ. (Nat. Sci.)* **17,** 79–85.

Yue, Y.-X., Wu, Y.-L., Chen, Z.-Y., Chen, M.-Q., and Su, T.-M. (1981). Morphogenesis of the cytoplasmic polyhedrosis virus of the cotton bollworm, *Heliothis armigera* (Hübner). *Acta Entomol. Sin.* **24,** 475–476.

IV

Cell Fusion

24

Fusion of Insect Cells

JUN MITSUHASHI

I. INTRODUCTION

In some plant and animal cells, cell fusion is becoming a routine technique, although there are still many other cells in which fusion has not yet been successful.

In insect cell cultures, cell fusion has been tried in only a limited number of cell lines, especially those of *Drosophila*, and a technique applicable to all types of cells has not yet been established.

In this chapter I review insect cell fusion previously performed and also show the results of my own experiments.

II. METHODS FOR CELL FUSION AND ISOLATION OF HYBRIDS

There are two kinds of cell fusion methods: fusion with fusogens and fusion due to electric stimuli.

Some viruses and chemicals are used as fusogens. The hemagglutinating virus of Japan (HVJ), which is also called the Sendai virus, was the first fusogen used. In addition to HVJ, simian parainfluenza SV_5 virus, Aujeski herpes virus, Semliki Forest virus, caprine retrovirus, hemorrhagic fever with renal syndrome virus, Epstein–Barr virus, iridoviruses, and densonucleosis viruses have been known to

BIOTECHNOLOGY IN INVERTEBRATE
PATHOLOGY AND CELL CULTURE

cause cell fusion. These viruses are inactivated by ultraviolet irradiation or ß-propiolactone treatment before use. The treated viruses, however, retain their ability to fuse cells. The chemical fusogen that is most commonly used is polyethylene glycol (PEG). Dextran, lysolecithin, concanavalin A (Con A), and wheat germ agglutinin (WGA) are also used as fusogens.

When a solution of a fusogen is added to a mixture of cells, the cells begin to fuse randomly. This results in various cell combinations. Besides hybrid cells, homologous cells also fuse. Polykaryons are also formed, although they usually do not multiply. Following heterokaryon formation, synkaryons are formed. The resulting cell population consists of heterologously fused cells, homologously fused cells, polykaryons, and parent cells which did not fuse. From such cell populations, we have to select only heterologously fused cells. Various methods have been devised for this purpose.

Electric fusion is usually accomplished in two steps. The first step is achieved by the application of alternating current. A suitable current frequency of about 600 kHz and a voltage of 10–30 V/mm are required. When the voltage is too low, cells do not show any movement. With an increase in voltage, cells begin to move and attach to each other. This phenomenon is called dielectrophoresis. A further increase in voltage causes rotation of cells, and the cells burst when the voltage becomes too high. After the first cell contact, the cells are pushed against each other, if the voltage is increased slightly. This state is suitable for applying a direct current pulse to fuse them. Prolonged application of alternating current causes lines of cells which are called pearl chains. The pearl chain will produce polykaryons when a direct current pulse is given. A direct current pulse of several hundred volts per millimeter and several microseconds duration is usually applied. If the voltage is too high or the duration is too long, the cells burst. The appropriate pulse causes fusion of cells.

To induce successful cell fusion, the composition of the fluid, in which a mixture of cells is suspended, is important. Usually 0.3 *M* mannitol, sorbitol, or glucose is used. Some investigators add proteinase or calcium chloride to these solutions.

Two types of electrodes have been devised: a microelectrode and a parallel electrode. The former is small and induces fusion of cells in a limited area. This is suitable for observation of the fusion process and also for verification of occurrence of fusion. The latter is large and can produce many fusants at one time. This is convenient for separation of the fusants desired.

As in the case of cell fusion with fusogens, fusion occurs randomly, resulting in heterologous cell fusion, homologous cell fusion, and formation of polykaryons. Synkaryon formation follows and the cell population becomes a mixture of homologously fused cells, heterologously fused cell, polykaryons, and parent cells which did not fuse. From such a cell population, we must select only heterologously fused cells, and various methods have been devised for this purpose.

The mechanical method for selecting a desired fusant is the same as a single cell clone method. However, it is well known that a singly isolated cell dies in most cases. Therefore, it is necessary to use a conditioned medium, feeder layer, microcapillary, or microdrops to prevent cells from deteriorating. Using this method we can pick up only the cells we want, but we cannot select many cells at one time, because the procedure takes time and is laborious.

The most sophisticated method for selecting fusants is to use selection medium. However, this method can be applied only to special cells. The parent cells should be mutants lacking some enzymes for DNA biosynthesis in order to make fusants that complement the functional deficiency. Usually, the $HGPRT^-$ mutant lacks hypoxanthine-guanine phosphoribosyltransferase in the purine salvage pathway, and the TK^- mutant lacks thymidine kinase in the pyrimidine salvage pathway. To obtain such mutants, cells are first treated with X-rays, γ-rays, or mutagens such as *N*-methyl-*N*′-nitrosoguanidine (MNNG) or ethylmethane sulfonate (EMS) and are then selected with 8-azaguanine or 6-thioguanine for the $HGPRT^-$ mutant and with 5-BrdU for the TK^- mutant. Since both mutants lack one enzyme in the salvage system, their DNA biosynthesis depends solely on *de novo* biosynthesis. When such cells are fused to each other, the fusants recover salvage biosynthesis by complementation from both parent cells. Therefore fusants have both pathways: *de novo* and salvage biosynthesis of DNA. When such fusants are placed in media that contain an inhibitor of *de novo* DNA biosynthesis such as aminopterin and substrates for the salvage pathway such as hypoxanthine and thymidine, they can survive with their DNA biosynthesized in the salvage pathway. However, the parent cells, which have a defective salvage pathway, cannot survive in such media. Hypoxanthine–aminopterin–thymidine (HAT) supplemented medium is an example of such a selection medium. By the use of such a system, hybrid cells can be selected efficiently.

Temperature and nutritional requirements may be used for selection of hybrid cells if appropriate temperature-sensitive mutants or mutants that require special substances for their growth are used as parents.

III. HISTORICAL REVIEW OF INSECT CELL FUSION

The first insect cell fusion was attempted between a human cancer cell line, HeLa, and a mosquito (*Aedes aegypti*) cell line (Zepp *et al.*, 1971; Conover *et al.*, 1971), which was later proved to be a moth (*Antheraea eucalypti*) cell (Greene *et al.*, 1972). The HeLa cells were labeled with [^{3}H]thymidine. The uniformly labeled HeLa cells were suspended in UV- inactivated HVJ. After thorough mixing, the suspension was used to resuspend mosquito cells. The mixture was placed in a 4°C water bath on an automatic shaker, shaken at low speed for 20 min, and then shaken for 60 min at 37°C. Autoradiography of direct smears immediately after these procedures gave evidence of HeLa–mosquito heterokaryon formation. This was also evident from the morphology of the cells because the nuclei of both parents were morphologically distinguishable. The HeLa cells had been cultured in Eagle's minimum essential medium (MEM) at 37°C. Little or no growth was obtained when HeLa cells were incubated at 26–28°C. The mosquito cells could not be maintained in MEM, and they cannot withstand the high temperature of 37°C. When a mixture of equal parts of complete MEM and mosquito culture medium was made (EPM), HeLa cells grew in it at 37°C. The HVJ-treated cell population could be subcultured 10 times in EPM. The population contained hybrid cells and residual HeLa cells. Evidence of synkaryon formation was obtained from chromosome analysis of 36–40-hr first-passage cultures. The presence of both HeLa and mosquito chromosomes, which apparently share the same spindle apparatus, was observed. However, there was no instance in which complete summation of parental complements was observed. Chromosome analysis at each passage also showed both parental chromosome types present within single cells; however, it also demonstrated a predominance of the HeLa chromosomes with progressively smaller numbers of the mosquito chromosomes. This might be due to the fact that the culture conditions were more suited to HeLa cells than mosquito cells.

Cell fusion between insect cells has been studied mostly with *Drosophila* cell lines. Becker (1972) reported the fusion of *Drosophila melanogaster* cells with Con A. One of the parent cell lines was labeled with [^{3}H]thymidine. Mixtures of labeled and unlabeled cell suspensions were added to Con A at a final concentration of 100 μg/ml in the presence of $CaCl_2$ and $MnSO_4$ at 10^{-4} *M*. After treatment for 30 min, the cells were seeded in a Leighton tube and rinsed with fresh culture medium when they attached to the vessel. Within 1 hr after Con A treatment, cytoplasmic bridges were formed between attached cells. During the next 10 hr heterokaryons were formed, and most of them formed synkaryons within 24 hr. Hybrids were demonstrated by autoradiographs showing labeled and unlabeled nuclei in a single cell. The hybrids were not isolated.

Rizki *et al.* (1975) conducted experiments similar to Becker's using another plant lectin WGA. Exposure of *Drosophila* cells to high concentrations of WGA (100–200 μg/ml) appeared simply to shrivel the cell surface, while exposure to low concentrations of WGA (5–10 μg/ml) caused fusion of cells as observed with a scanning electron microscope. The cell fusion was evident following treatment for 3 min with WGA. It was not ascertained whether synkaryons were formed.

Halfer and Petrella (1976) compared the fusion-inducing capacity and cytotoxicity of Con A and lysolecithin on *Drosophila* cells. The latter was found to be more rapid and drastic than the former in fusion capacity. Lysolecithin was also more cytotoxic to *Drosophila* cells. They confirmed the formation of heterokaryons with autoradiography but did not examine the formation of synkaryons.

Bernhard (1976) showed that PEG was also effective for the fusion of *Drosophila* cells. He fused [^{3}H]thymidine-labeled Kc cells with unlabeled Kc cells or imaginal disc cells isolated from the 3rd instar larvae by incubating the mixed cells in 50 m*M* PEG (molecular weight 4000) for 10 min at 25°C. The fusion was proved with autoradiography. The estimated frequency of the fusion was 5%. The resulting heterokaryons did not survive.

Nakajima and Miyake (1978) isolated fused hybrids of *Drosophila* cells. They used two temperature-sensitive mutants as parent cell lines. In order to get *ts* mutants, the GM_1 cell line was treated with the mutagen ethylmethane sulfonate (500 μg/ml). Mutants that did not grow at a high temperature were selected by poisoning colonies growing at 30°C with 5-FldU (25 μg/ml). The 5-FldU treatment was repeated four times. Finally two clones, ts-15 and ts-58, were selected as parent cell lines for fusion. They formed colonies at 23°C about one-third as well as the wild-type cells when inoculated at a density of 10^3 cells per dish (60 × 15 mm glass dish), but they formed no colonies at 30°C. These mutants were fused by treatment with 50% PEG (molecular weight 6000) for 1 min. In some experiments, 15% dimethyl sulfoxide was added to the PEG solution. When PEG-treated cells were seeded at a density of 10^4 per dish a few colonies were formed at 30°C. These colony-forming cells showed plating efficiency comparable to that of the wild-type cell line at 30°C. Karyotype analysis showed that the PEG-treated and selected cell lines were nearly tetraploid, while both parent *ts* mutants were nearly diploid. These results supported the hypothesis that fusion of these two distinct *ts* mutants produced hybrids that complemented the functional deficiency, permitting growth at a high temperature.

Becker (1974) and Moisenko and Kakpakov (1974) demonstrated that *Drosophila* cells have no functional HGPRT. Then Wyss (1979) developed a process analogous to the HAT selection system. This system was called the TAM (thymidine–adenine–methotrexate) selection system. He selected a clone (MDR3) resistant to 6-methylpurine and diaminopurine from the Kc line of *Drosophila*. The clone was an adenine salvage-deficient variant and was sensitive to TAM selection medium, which was medium ZH1% containing thymidine, adenine, and methotrexate. Two wild-type (TAM-resistant) cell lines, Schneider's line 3 (S3) and Dübendorfer's line 1 (D1), were unable to proliferate in medium ZH1% used for MDR3. This allowed the selection of hybrids between MDR3 and either D1 or S3 in TAM cloning medium after treatment with PEG. A pellet of mixed parent cells was treated with 45% PEG 1000 for 30 sec. Cell fusion was obvious 5–10 min after beginning the treatment. Hybrids were selected in the above-mentioned TAM selection medium and were confirmed by isozyme analysis of isocitrate dehydrogenase bands. MDR3 and S3 did not proliferate in the presence of 20-hydroxyecdysone. However, some ecdysone-resistant clones were obtained from MDR3 and were designated MDER. A great majority of all hybrids formed between ecdysone-sensitive parents were also sensitive to 20-hydroxyecdysone. In contrast, most of hybrids resulting from fusion of any MDER-clones to S3 proved to be resistant to 20-hydroxyecdysone (Wyss, 1980a). When MDR3 was hybridized with cells derived from wild-type *Drosophila* embryos, the resulting hybrids were able to proliferate in the presence of 20-hydroxyecdysone (Wyss, 1980a). On the other hand, no hybrids resulting from fusion of MDR3 to primary cells from eye–antennal discs of third-instar larva, were resistant to 20-hydroxyecdysone (Wyss, 1980b).

Hybridization between *Drosophila* cell lines with contrasting phenotypes and different responses to 20-hydroxyecdysone has been studied (Berger and Wyss, 1980). *Drosophila* S3 line cells maintained a high basal level of acetylcholinesterase (AChE), which was lost and then reinduced following exposure to 20-hydroxyecdysone; MDR line cells had a lower basal AChE activity, which was induced to a modest level by the hormone. MDER line cells, an ecdysone-resistant strain, also showed a low basal AChE level, which was not elevated by the hormone. S3/MDR and S3/MDER hybrids were produced by treatment with PEG and selected by cloning in a soft agar medium containing ZH1% + TAM. In S3/MDR hybrids the high basal level phenotype was extinguished and the hormone-induced AChE level was modest. In

S3/MDER hybrids two of the hybrid clones, F1 and F6, showed a phenotype and response to the hormone similar to those of the MDER parent. However, one hybrid clone, F7, was ecdysone-sensitive and was induced by the hormone to a specific higher level of AChE production.

With the same experimental system, another effect of 20-hydroxyecdysone in *Drosophila* cells was measured by examining the rate of polypeptide synthesis (Berger *et al.*, 1980). Proteins were labeled with [^{35}S]methionine, and two-dimensional polyacrylamide gel electrophoresis of the peptides was compared before and after the hormone treatment. A set of 10 of the more than 300 resolvable spots were selected as internal standards for subsequent quantitative work. Five peptide spots appeared to either increase or decrease in relative labeling intensity following hormone treatment. Significant changes occurred in the rate of synthesis of several peptides in two different *Drosophila* cell lines following exposure to the hormone. For peptides 7/8, which could not be excised separately because of their close proximity, the synthesis rate increased by a factor of 10 in line S3 and by a factor of 1.3 in line MDR. In hormone-sensitive hybrids S3/MDR and S3/MDER-F7 the rate increased by a factor of 3–4.5. For peptide 10 the synthesis rate decreased by a factor of 8 in line S3 but increased by a factor of 2 in line MDR. In hormone-sensitive hybrids, the synthesis rate increased by a factor of 2–3. In hormone-insensitive cell lines MDER, S3/MDER-F1, and S3/MDER-F6, no change in the rate of peptide (both 7/8 and 10) synthesis occurred in the presence of the hormone.

Recently, fusion of a lepidopteran cell line was reported by Miltenburger *et al.* (1985). Instead of using HAT or TAM selection system, they employed irreversible inhibition of cell growth with drugs. A cell line (IZD-Cp-2202) from *Cydia pomonella* hemocytes was treated with actinomycin D (0.25 μg/ml) and puromycin (2×10^{-4} *M*) for 3 hr at 27°C. About 35% of the cells were alive at the time of fusion. However, only a small percentage (0.1%) of the cells recovered after the blocking. The cells, which survived but were unable to proliferate, were used as parent cells. Other parent cells were prepared directly from *C. pomonella* embryos. About 200 2- to 3-day old eggs were crushed after surface sterilization and passed through a steel mesh. The homogenate, containing single cells and small cell aggregates, was used for fusion after washing.

The mixed cell pellet was treated with PEG for 1 min at 37°C. After washing, the cells were seeded into the wells of a hybridoma tissue culture tray. About 2 weeks later some colonies were obtained. The cells of the colonies obtained differed in morphology from the permanent parental cells, IZD-Cp-2202. Isozyme analysis for esterases was particularly indicative of the hybrid nature of the resulting cells, because several enzymes with esterase activity found either in the primary cell line or in IZD-Cp-2202 cells were present together in the resulting cells. The hybrid cells were also found to be much less susceptible to *C. murinana* NPV than Cp-2202.

IV. ATTEMPTS TO ESTABLISH CELL FUSION METHODS FOR INSECT CELLS OTHER THAN *DROSOPHILA*

As stated in the preceding section, fusion of insect cells has been done mostly with *Drosophila* cell lines, although there have been one or two exceptions. It cannot be said that a general method for obtaining hybrid cell lines of insects has been established. I have conducted some experiments on cell fusion with dipteran and lepidopteran cell lines.

Since karyotypes are quite different between lepidopteran cells and dipteran cells, the first attempt was made to fuse a butterfly (*Papilio xuthus*) cell line, NIAS-PX-58, with a mosquito (*Aedes albopictus*) cell line, NIAS-AeAl-2. Both parental cells were maintained in Mitsuhashi and Maramorosch's medium (MM) containing 3% fetal bovine serum (FBS) (Mitsuhashi, 1982). Suspensions of both cell lines were mixed and centrifuged at 1500 rpm for 5 min. The supernatant was discarded, and the cell pellet was treated with 50% PEG for 3 min. After being washed, the cells were suspended in MM with 3% FBS and an aliquot of the cell suspension was used for examination of cell fusion. Many cells were attached to each other. Homologous combinations, PX-58 to PX-58 or AeAl-2 to AeAl-2, were predominant (Figs. 1A and 1B), al-

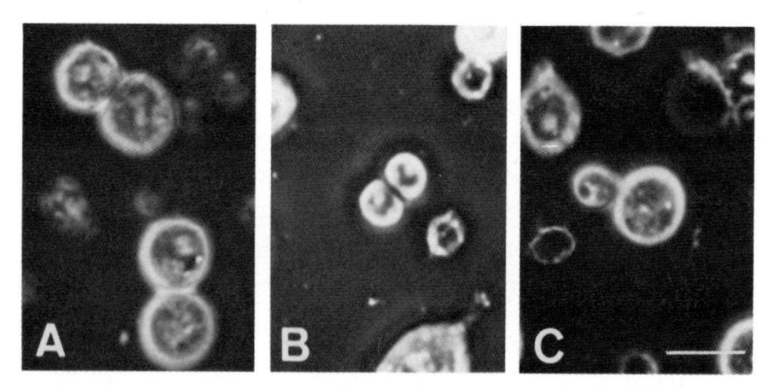

Fig. 1. Attachment of cells after PEG treatment. (A) Homologous attachment of NIAS-PX-58 (a butterfly cell line) cells; (B) homologous attachment of NIAS-AeAl-2 (a mosquito cell line) cells; (C) attachment of an NIAS-PX-58 (large cell) to an NIAS-AeAl-2 (small cell). Bar, 10 μm.

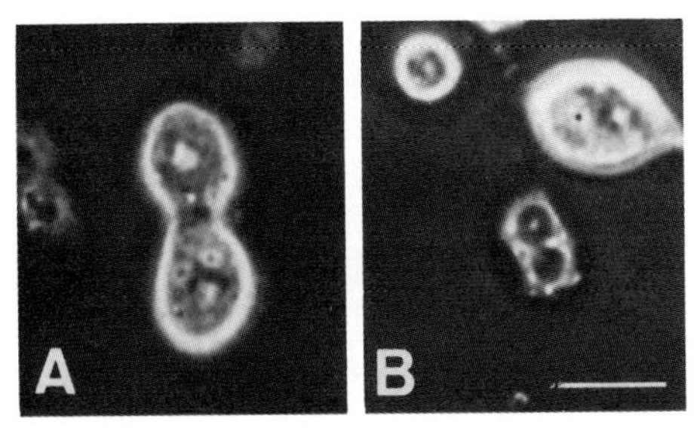

Fig. 2. Fusion of cells immediately after PEG treatment. (A) Homologous fusion of NIAS-PX-58 cells; (B) homologous fusion of NIAS-AeAl-2 cells. Bar, 10 μm.

though heterologous cell combinations, PX-58 to AeAl-2, were sometimes observed (Fig. 1C). These cells attached to each other but were not fused; i.e., both cells were separated by cell membranes. There were also cells which appeared to be fused. However, the ratio of such cells was very low. In this case, homologous cell fusion was observed more frequently than heterologous fusion (Fig. 2).

The remaining cells treated with PEG were cultured for several days, and their karyotypes were examined. Mitoses were stopped at metaphase by treatment with 0.5 μg/ml colchicine for 4 hr. After hypotonic treatment with 0.5–0.6% KCl, these cells were fixed with a mixture of ethanol and acetic acid (3 : 1), spread on cover glasses by the air-dry method, and stained with Giemsa. Many attached cells, some of which were at the mitotic metaphase, were present. Homologously attached cells were common, but there were not too many heterologous ones (Fig. 3). Attached cells, both of which were in the mitotic metaphase,

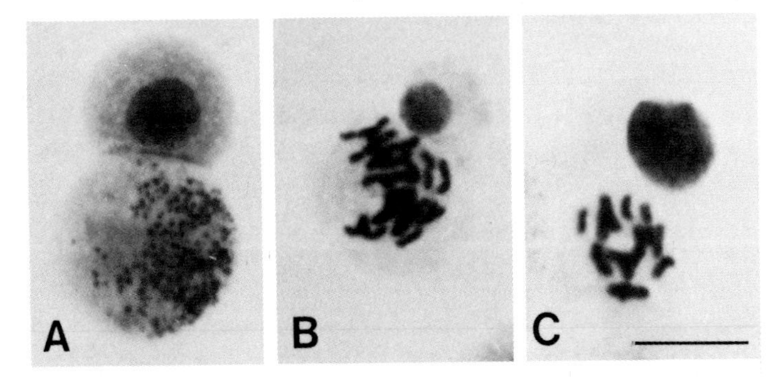

Fig. 3. Attached cells showing unsynchronized mitosis. One of the paired cells is at the mitotic metaphase. (A) Homologous attachment of NIAS-PX-58 cells; (B) homologous attachment of NIAS-AeAl-2 cells; (C) attachment of an NIAS-PX-58 cell to an NIAS-AeAl-2 cell. The latter cell is at the mitotic metaphase. Bar, 5 μm.

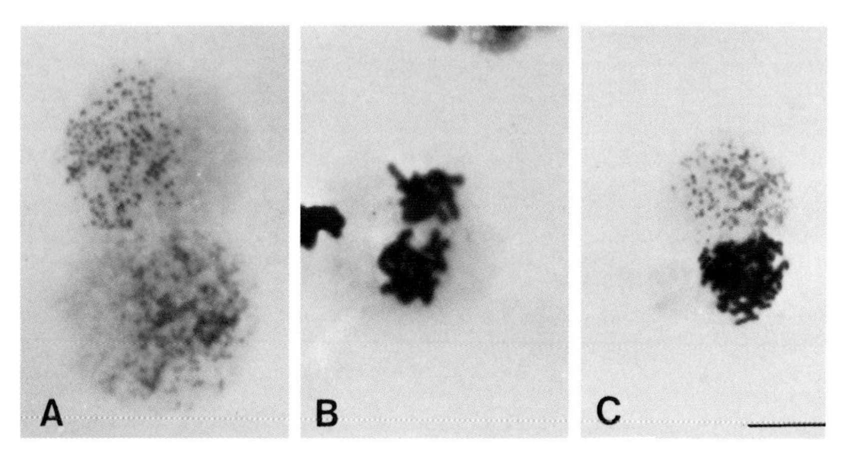

Fig. 4. Simultaneous mitosis of attached cells. (A) homologous attachment of NIAS-PX-58 cells; (B) homologous attachment of NIAS-AeAl-2 cells; (C) heterologous attachment of an NIAS-PX-58 cell and an NIAS-AeAl-2 cell. Bar, 5 μm.

could be seen (Fig. 4); most of them were homologously attached cells, and heterologous ones were rather rare. Very few cells seemed to be fusants. The mosquito cells were predominantly tetraploid (the chromosome number was 12) (Fig. 5A). However, cells with many chromosomes were also seen (Fig. 5B). These cells might be homologously fused mosquito cells, because such highly polyploid cells could not be seen in the parent cell population. Homologously fused butterfly cells might be present. However, they could not be detected with certainty because butterfly cells had many microchromosomes and highly polyploid cells were common in the parent cell population. Chromosomes of both parent cell lines in a metaphase plate have never been observed.

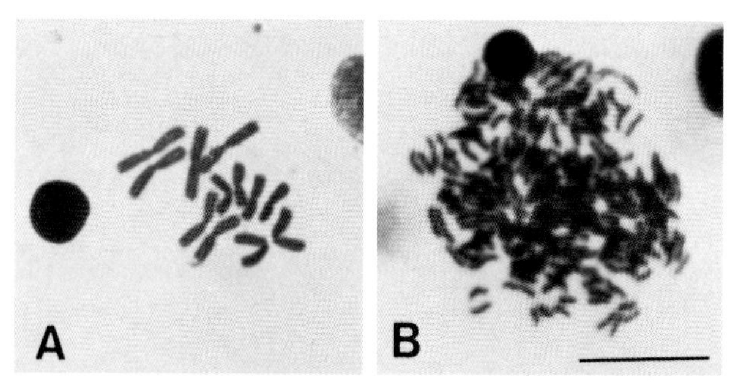

Fig. 5. (A) Typical metaphase plate of a mosquito cell line (NIAS-AeAl-2). The cell is tetraploid ($n = 3$). (B) Metaphase plate of a highly polyploid cell of the NIAS-AeAl-2 line, seen after PEG treatment. Bar, 5 μm.

From the above experiments, definite evidence of cell fusion could not be obtained. Attempts were then made to obtain direct evidence of cell fusion. Since it seemed easier to carry out homologous cell fusion, the following experiments were conducted within the same cell lines.

At first, vital staining of one of the parent cells was attempted. A butterfly cell line (NIAS-PX-58), cabbage armyworm, *Mamestra brassicae*, cell lines (NIAS-MB-32 and NIAS-MaBr-85), a mosquito cell line (NIAS-AeAl-2), and a flesh fly, *Sarcophaga peregrina*, cell line (NIH-SaPe-4) were used. One parent cell population was stained with neutral red and fused to unstained parent cells by the use of PEG. However, this method proved unsuccessful because the dye was rapidly lost during the washing procedure after PEG treatment. Then vital staining of nuclei with fluorescent dye, Hoechst 33258 or ethidium bromide, was applied. Cells of one parent were stained with Hoechst 33258 and cells of the other with ethidium bromide before cell fusion. This technique was also unsuccessful because Hoechst 33258 stained only part of the cell population and ethidium bromide was very toxic to the cells and its fluorescence was weak. Marking by phagocytosis of fluorescent beads was also tried. Parent cell lines were given greenish fluorescent beads (Polysciences 9847) and purplish fluorescent beads (Polysciences 7769), respectively. The cells phagocytized these beads well. However, when the cells were examined under a fluorescence microscope after PEG treatment, cell fusion was difficult to judge by this method. The fluorescence of Polysciences 7769 was weak and free beads, derived from unphagocytized beads or from cells that disintegrated during treatment, were phagocytized after PEG treatment.

In the next set of experiments, karyotype analysis was applied to the cell populations 1 week after fusion treatment with PEG. The results are shown in Fig. 6. The mosquito cells were mostly tetraploid (Fig. 5A). After PEG treatment, the percentage of octaploid cells increased slightly. This suggested that some cells fused and that the hybrid cells were growing. The flesh fly cells were predominantly diploid (Fig. 7). The percentage of tetraploid cells was progressively increased with prolongation of PEG treatment. However, treatment for more than 3 min was detrimental to the cells. This also suggested the occurrence of cell fusion, although the ratio of tetraploid cells was small.

Other methods for cell fusion were also examined. HVJ was inactivated with ultraviolet irradiation. The virus suspension was added to cell pellets and Carlson's fluid (Carlson, 1946), a physiological saline, was then added. After being maintained at 4°C for 10 min, the cells were incubated at 37°C for 30 min. The virus-treated cells were washed by centrifugation and cultured. With this method, the mosquito cell

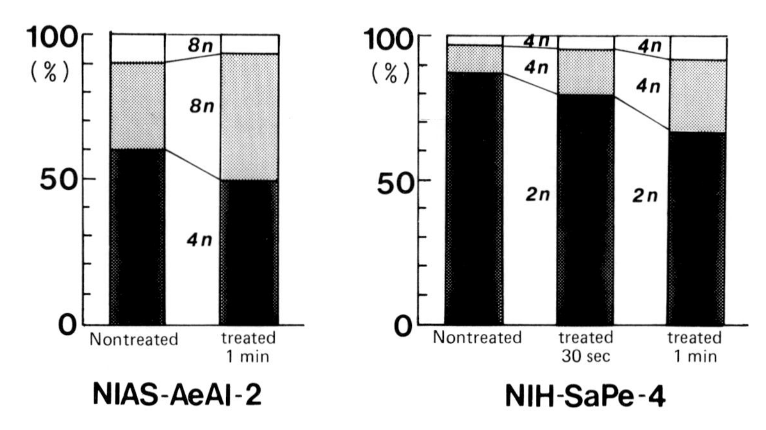

Fig. 6. Rate of ploidy in the cell population treated with PEG. NIAS-AeAl-2, a mosquito cell line; NIH-SaPe-4, a flesh fly cell line.

line or the flesh fly cell line was treated. However, no evidence of cell fusion was obtained. Probably, like the *Drosophila* cell line (Echalier, 1976), these dipteran cell lines lack a receptor for HVJ.

Becker's Con A method (Becker, 1972) was also examined. Cells of the mosquito line or the flesh fly line were incubated for 30 min in medium containing Con A (100 μg/ml) and $CaCl_2$ and $MgSO_4$ (10^{-4} *M*). However, this method proved unsuccessful for the mosquito or the flesh fly cells.

Cell fusion by electric stimuli was attempted. In these experiments a butterfly, *P. xuthus*, cell line (NIAS-PX-64) was used. When the cell suspension in 0.3 *M* mannitol was placed in an alternating current field with frequency 600 kHz at 10–20 V/mm, the cells were dielectrophoresed and attached to each other. If the voltage was low, movement of the cells was very slow but some pairing cells were ob-

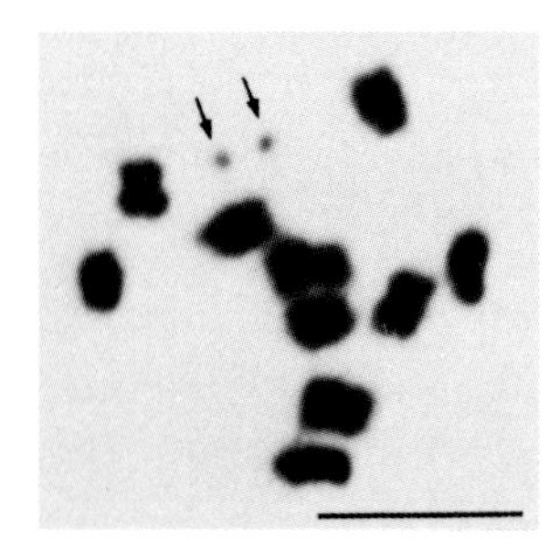

Fig. 7. Typical metaphase plate for a flesh fly cell line, NIH-SaPe-4. The cell is diploid ($n = 6$). Arrows indicate microchromosomes. Bar, 5 μm.

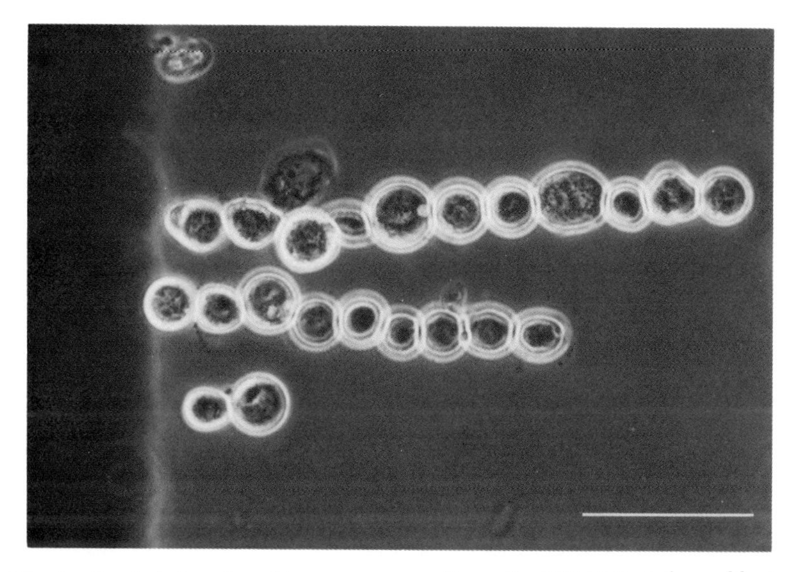

Fig. 8. Pearl chains of a cabbage armyworm line cells, SES-MaBr-4, formed by treatment with alternating current at 600 kHz and 15 V/mm. Bar, 10 μm.

tained. Prolongation of the treatment caused lining up of cells in pearl chains (Fig. 8). If the voltage was too high, cells moved very quickly and many cells started to rotate. When many pairing cells or pearl chains were formed, a direct current pulse at 100–300 V/mm for 2–10 μsec was applied. If the voltage was low, most cells seemed unchanged. If the voltage was high, a considerable number of cells burst. It was very difficult to find appropriate conditions for effective cell fusion. Occasionally cell fusion occurred, but fused cells died soon after the fusion. The composition of the medium used to suspend cells for electrically stimulated fusion might be an important factor. Mannitol, sorbitol, and glucose at 0.3 *M* did not differ in effectiveness. Addition of enzymes such as dispase, trypsin, and pancreatine or addition of $CaCl_2$ to the medium did not improve the effect of electric stimuli. With the same electric treatment, protoplasts of the entomogenous fungi *Entomophaga aulicae* fused quite easily. The membrane of insect cells may be somewhat different from that of plants or higher animals.

V. EPILOGUE

A cell fusion method applicable to various insect cells has not yet been established. PEG has been most widely used, although the rate of hybrid formation is not high. The rate of hybrid formation may not be important if an efficient selection system for hybrids is provided. Therefore, selection system should be developed before cell fusion.

An electric method will become a useful tool in future if the application conditions are specified. If an electric method can be used, the process of cell fusion will be simplified and the time required for cell fusion will be shortened.

Cell fusion is certainly a useful tool for cytogenetics, especially for such insects as *Drosophila* which have been studied in depth genetically. Cell fusion, especially the hybridoma technique, will help in the study of tissue specificity of virus replication. Also, this technique will contribute to the development of host cells which support efficient virus production. Finally, fusion of cells from established lines to dissociated tissue cells may be used to establish cell lines with differentiated functions from cell types that may not proliferate *in vitro*.

REFERENCES

Becker, J. L. (1972). *C. R. Hebd. Seances Acad. Sci.* **275,** 2969–2973.

Becker, J. L. (1974). *Biochimie* **56,** 779–781.

Berger, E., and Wyss, C. (1980). *Somatic Cell Genet.* **6,** 631–640.

Berger, E., Ireland, R., and Wyss, C. (1980). *Somatic Cell Genet.* **6** 719–729.

Bernhard, H. P. (1976). *Experientia* **32,** 786 (abstr.).

Carlson, J. G. (1946). *Biol. Bull. (Woods Hole, Mass.)* **90,** 109–121.

Conover, J. H., Zepp, H. D., Hirshhorn, K., and Hodes, H. L. (1971). *Curr. Top. Microbiol. Immunol.* **55,** 85–92.

Echalier, G. (1976). *In* "Invertebrate Tissue Culture: Applications in Medicine, Biology and Agriculture" (E. Kurstak and K. Maramorosch, eds.). pp. 131–150, Academic Press, New York.

Greene, A. E., Charney, J., Nickols, W. W., and Coriell, L. (1972). *In Vitro* **7,** 313–322.

Halfer, C., and Petrella, L. (1976). *Exp. Cell Res.* **100,** 399–404.

Miltenburger, H. G., Naser, W. L., and Schliermann, M. G. (1985). *In Vitro* **21,** 433–438.

Mitsuhashi, J. (1982). *Adv. Cell Cult.* **2,** 133–196.

Moisenko, E. V., and Kakpakov, V. T. (1974). *Drosophila Inf. Serv.* **51,** 44–45.

Nakajima, S., and Miyake, T. (1978). *Somatic Cell Genet.* **4,** 131–141.

Rizki, R. M., Rizki, T. M., and Andrews, C. A. (1975). *J. Cell Sci.* **18,** 113-121.

Wyss, C. (1979). *Somatic Cell Genet.* **5,** 29–37.

Wyss, C. (1980a). *Exp. Cell Res.* **125,** 121–126.

Wyss, C. (1980b). *In* "Invertebrate Systems in Vitro" (E. Kurstak, K. Maramorosch, and A. Dübendorfer, eds.), pp. 279–289. Elsevier/North Holland Bimedical Press, Amsterdam.

Zepp, H. D., Conover, J. H., Hirshhorn, K., and Hodes, H. L. (1971). *Nature (London), New Biol.* **229,** 119–121.

25

Protoplast Fusion of Insect Pathogenic Fungi

S. SHIMIZU

I. INTRODUCTION

Since the frequency of isolation of *Beauveria bassiana* is higher than that of any other species of entomogenous fungi, this species was selected for the present work on utilization of entomogenous fungi in the microbial control of insect pests. *Beauveria brongniartii* is currently investigated as a microbial control agent for the yellow-spotted longicorn beetle, *Psacothea hiralis*, in Japan (Kawakami, 1978). Most investigations on the genus *Beauveria* concern its impact on the host population, stability and persistence in nature, and interaction with various factors in the agroecosystem (Doderski, 1981; Ignoffo *et al.*, 1983; Ramoska, 1984). Primary requirements for the use of an entomogenous fungus as a microbial control agent are the susceptibility of the insect on the one hand and the virulence of the fungus on the other. The latter depends on the selection of a strain with stable, specific efficacy for a target host (Samišiňáková and Kálalová, 1983). There is, nevertheless, considerable potential for their genetic improvement for microbial control purposes.

There are many reports concerning intra- and interspecific protoplast fusion between mutants of filamentous fungi including *Aspergillus* and *Penicillium* (Ferenczy *et al.*, 1975a,b; Kevei and Peberdy, 1984). In the aspergilli, when taxonomically distantly related species were hybridized, unstable complementation products were recovered at a very low frequency and protoplast fusion between the closely related species resulted in interspecific heterokaryons as primary fusion products (Kevei and Peberdy, 1984). Production of protoplasts of the genus *Beauveria* will allow genetic investigations, and protoplast fusion will provide a possible means for the selection of fungal hybrid strains having desired properties (Riba, 1978; Pendland and Boucias, 1984). However, techniques for protoplast formation or protoplast fusion in *Beauveria* have not been established.

This chapter reports the preparation of protoplasts and protoplast fusion between auxotrophic mutants of the genus *Beauveria*. The results obtained in the protoplast fusion experiments suggested the existence of two types in *B. brongniartii,* and it was found that *B. brongniartii* type 2 and *Beauveria amorpha* were closely related species based on their complementation frequencies. In addition, it was possible to obtain hybrids by protoplast fusion between some combinations of the genus *Beauveria*.

II. MATERIALS AND METHODS

A. Organisms and Growth Conditions

Isolates of the genus *Beauveria* used are listed in Table I. Some isolates were provided by M. Shimazu, Forestry and Forest Products Research Institute. Cultures were grown and maintained on a complete medium (CM) containing 2% sucrose, 1% peptone, 0.5% NaCl, 0.3% yeast extract, 2% agar. A minimal medium (MM) containing 3% glucose, 0.3% $NaNO_3$, 0.1% K_2HPO_4, 0.5% $MgSO_4$, 0.5% KCl, 0.001% $FeSO_4 \cdot 7H_2O$, 2% agar was used for the selection of fusion products. In the protoplast regeneration and fusion experiments both CMS and MMS media were supplemented with osmotic stabilizer, 0.6 *M* KCl.

TABLE I

Isolates of *Beauveria bassiana*, *Beauveria brongniartii*, and *Beauveria amorpha*

Species	Isolate no.	Original host	Place	Serotype[a]	Zymogram type[b]
B. bassiana	F1		Japan	B_1	
	F54	*Bombyx mori*	Japan	B_2	
	K1	*Psacothea hiralis*	Japan	B_3	
B. brongniartii	Bt5	*Psacothea hiralis*	Japan		1
	Bt6	*Psacothea hiralis*	Japan		1
	F16	Scarabaeidae	France		1
	F77	*Anomala costate*	Japan		2
	F47	*Melolontha melolontha*	France		2
	384-4	*Anomala costate*	Japan		2
B. amorpha	357-11	*Anomala cuprea*	Japan		

[a]Shimizu and Aizawa (1983).

[b]Isolates of *B. brongniartii* were classified into two types based on the electrophoretic patterns of three enzymes (esterase, acid phosphatase, and malate dehydrogenase) (S. Shimizu, unpublished data).

B. Isolation of Auxotrophic Mutants

Conidium suspensions were treated with 800 μg/ml *N*-methyl-*N*′-nitro-*N*-nitrosoguanidine at 25°C for 1 hr. The filtration enrichment technique (S. Shimizu, unpublished data) was used to increase the frequency of isolation of mutants.

C. Protoplast Isolation

Conidium suspension was poured into 100 ml of liquid complete medium in a 500-ml Sakaguchi flask. After 72 hr of shaking cultures at 28°C, 10 ml of culture was transferred to 100 ml of the same medium. This was repeated several times, and finally 4 ml of the seed culture was transferred to 100 ml of the same medium. These serial transfers of the seed culture appeared to be very important for the efficient production of protoplasts. Young mycelia of 18-hr cultivation were collected on Toyo No. 2 filter paper and washed with filter sterilized stabilizing medium (0.02 *M* phosphate buffer, pH 7.2, containing 0.6 *M* KCl and 2 m*M* $MgCl_2$).

One gram (fresh weight) of mycelium was transferred to 50 ml of a lytic enzyme solution consisting of the stabilizing medium supplemented with 1 mg/ml Zymolyase-20T (Kirin Brewery Co., Ltd.). Digestion of the cell wall was allowed to proceed at 30°C with gentle shaking. The culture was filtered through a 100- μm steel sieve to remove mycelial fragments, and the filtrate was centrifuged at least twice at 1000 g for 5 min to remove the enzyme. The pellet was resuspended in the stabilizing medium and taken as the protoplast preparation.

D. Protoplast Regeneration

For the liquid culture, protoplasts were suspended in a regeneration medium containing 2% sucrose, 1% peptone, 0.5% NaCl, 0.3% yeast extract, and 0.6 *M* KCl in 0.02 *M* phosphate buffer, pH 7.0, and incubated at 28°C. For the culture on agar medium, small amount of protoplast suspension was plated on CMS or MMS medium and incubated at 28°C for 7 days.

E. Protoplast Fusion

Protoplast fusions were done by a modification of the method of Ferenczy *et al.* (1975b). Protoplasts obtained from the two different auxotrophic mutants were mixed (5×10^6 to 10^7 protoplasts of each auxotroph) and centrifuged. The pelleted protoplasts were resuspended in 1 ml of a solution containing various concentration of polyethylene glycol (molecular weight 3350) (PEG) and 10 m*M* $CaCl_2$. After incubation at 30°C for 30 min, the suspension was centrifuged for 5 min at 1000 g. Fused protoplasts were plated at a suitable dilution on MMS and CMS media. Fusion frequencies were calculated from the ratio of colony numbers on MMS to colony numbers on CMS medium.

F. Nuclear Staining

Protoplasts were fixed by adding 1% glutaraldehyde to the stabilizing medium for 1 hr and stained by 4′, 6-diamido-2-phenylindole (DAPI) as described by Miyakawa *et al.* (1984).

G. Application of Calcofluor White and FITC-Labeled Lectins

Calcofluor white (CAW) (commerical name, Kayaphor conc; Nihon Kayaku Co., Ltd.) was applied at a concentration of 1 μg/ml. The FITC-labeled lectins concanavalin A (Con A), soybean agglutinin (SBA), and peanut agglutinin (PNA) were purchased from Maruzen Oil Biochemical. Before use, 50-μl portions of Con A, SBA, and PNA were diluted, respectively, with 1 ml of 0.01 *M* phosphate buffer (ph 7.5) supplemented with 0.001 *M* $CaCl_2$. To apply them to protoplast-associated structures, these solutions were further supplemented with 0.6 *M* KCl to adjust osmotic pressure. For the assay for binding to mycelia or protoplast-associated structures grown in liquid medium, specimens washed as above were placed on glass slides and dipped in a 50–100-μl drop of CAW or lectin solution. Specimens were allowed to react for 15 min at room temperature. All observations were made with an Olympus BH2-RFL epifluorescence microscope equipped with a high-pressure mercury lamp.

CAW binds β-1,4-linked polymers, either chitin or cellulose. Receptors of SBA are α-N-acetyl-D-galactosamine and α- or β-galactose. Those of Con A and PNA are α-D-glucose or α-D-mannose and galactose-β(1 → 3)-N-acetyl-D-galactosamine, respectively.

III. RESULTS

A. Protoplast Isolation

A high yield of protoplasts from young mycelia of the genus *Beauveria* was obtained by 1- to 2-hr incubation with Zymolyase-20T (Fig.

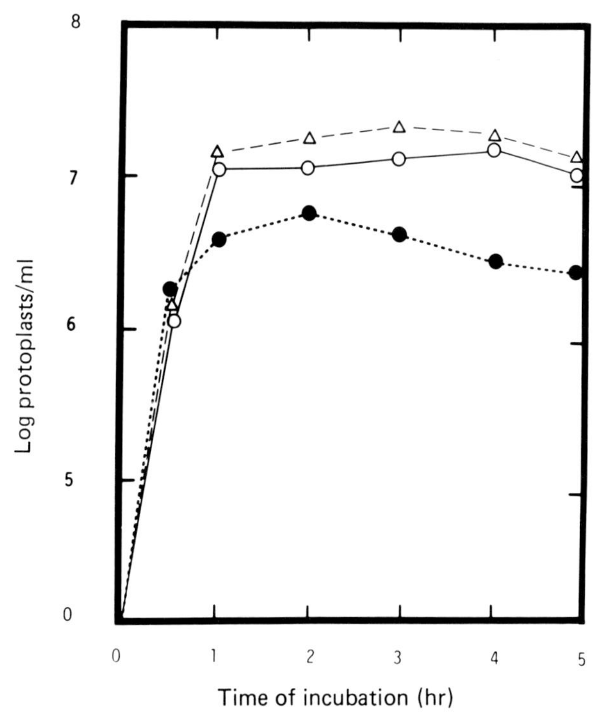

Fig. 1. Time course of release of protoplasts from mycelia of (○) *B. bassiana* F1, (●) *B. brongniartii* Bt6, and (△) *B. amorpha* 357-11. One gram of mycelia was transferred to 50 ml of the lytic enzyme solution consisting of the stabilizing medium supplemented with 1 mg/ml Zymolyase-20T. Digestion of cell wall was allowed to proceed at 30°C.

1). Prolonged treatment for more than 2 hr did not increase the yield of protoplasts. Most of the mycelia disappeared during the incubation, and nearly all cells in mycelia were converted into protoplasts. Protoplasts showed a perfectly sherical form with a diameter of 2.5–6 μm (Fig. 2).

Increasing the volume of the lytic enzyme solution from 50 to 100 ml for 1 g (fresh weight) of mycelia did not influence the yield of protoplasts. The age of the mycelia strongly affected the protoplast yield. Yields of protoplasts prepared from 3- to 4-day-old mycelia were very low.

Nuclei of protoplasts were stained by DAPI. More than 20% of the protoplasts were without a nucleus, 50–70% contained a single nucleus and other protoplasts contained two or more nuclei.

Mycelia and protoplasts reacted with Con A, and the former reacted with CAW, SBA, and PNA. Septa of mycelia stained with CAW but not

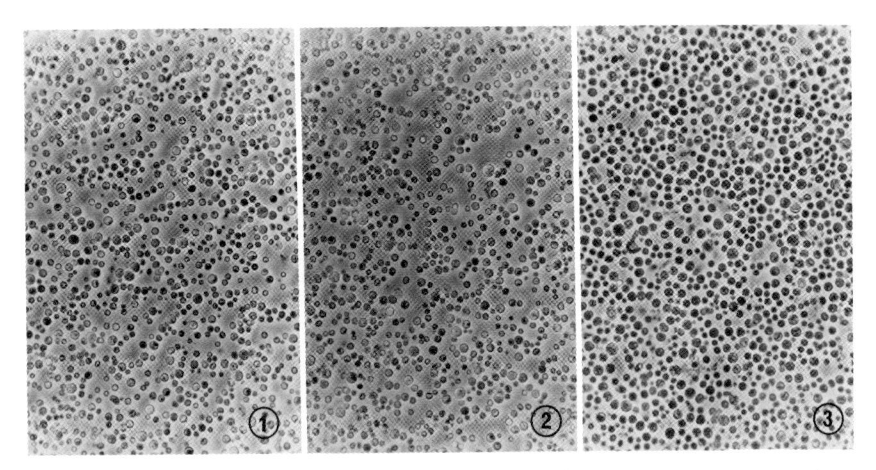

Fig. 2. Phase-contrast photographs of protoplasts of the genus *Beauveria*. 1, *B. bassiana* F1; 2, *B. brongniartii* Bt6; 3, *B. amorpha* 357-11.

with FITC-labeled lectins. However, only septa located near an open cut in mycelia always gave an intensely positive response to FITC-labeled Con A, SBA, and PNA. Mycelia and protoplasts did not react with *Dolichos biflorus* agglutinin (DBA) and *Ricinus communis* agglutinin-1 (RCA-1), which bind α-*N*-acetyl-D-galactosamine and β-galactose, respectively.

B. Protoplast Regeneration

The mode of regeneration of protoplasts of the genus *Beauveria* was similar to that of protoplasts of other filamentous fungi (Garcia Acha *et al.*, 1966; Kunoh, 1983). Three basic patterns of regeneration were observed under a phase-contrast microscope. In type 1, protoplasts produced yeastlike buds and developed into irregularly shaped chains of buds. These were still sensitive to osmotic shock and changed to ghosts by dilution of the suspension. Cytoplasmic content became lucent under a phase-contrast microscope and buds autolysed without hyphal development. Type 2 was similar to type 1, but after the development of an irregularly shaped chain of buds, a germ tube-like hypha protruded from the bud distal to the original protoplast, and a normal apical type of hyphal growth was restored. A very interesting mode of regeneration was observed in type 3. The regeneration protoplasts remained spherical for a long time (usually more than 8 hr), and they could be distinguished from nonregenerating protoplasts under a phase-contrast microscope (Fig. 3). However, they were stained with

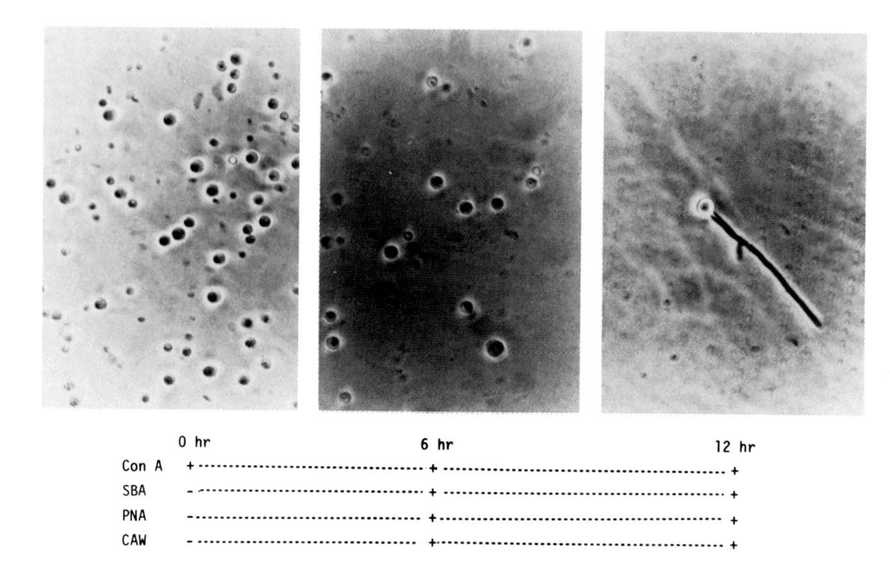

Fig. 3. Phase-contrast photographs of protoplast regeneration (type 3) of *B. bassiana* F1 and their responses to Con A, SBA, PNA, and CAW. +, Positive reaction; −, negative reaction.

CAW, SBA, and PNA and nuclear division was not observed. A germ tube-like hypha protuded directly from the spherical cell and was not sensitive to osmotic shock.

Protoplasts of the genus *Beauveria* were not stained by CAW. However, regenerated cells, yeastlike buds, and reversional hyphae were more intensely stained than intact mycelia by CAW.

The regeneration was tested on CMS medium and approximately 50% of protoplasts regenerated and formed colonies.

C. Protoplast Fusion

Protoplasts of *B. brongniartii* Bt5-16 (methionine^{-}) and *B. brongniartii* Bt6-13 (adenine^{-}) were mixed, centrifuged, and then treated with PEG solutions at various concentrations containing 10 m*M* $CaCl_2$ to find the optimal concentration of the fusion agent. High frequencies of prototrophic colonies in these experiments were obtained at concentrations of 20–50% PEG (Fig. 4). However, incubation of protoplasts with PEG at concentration below 20% caused the lysis of protoplasts.

Mixtures of the protoplasts were treated with 30% solutions of PEG having a molecular weight of 1540, 3350, and 8000, respectively, and

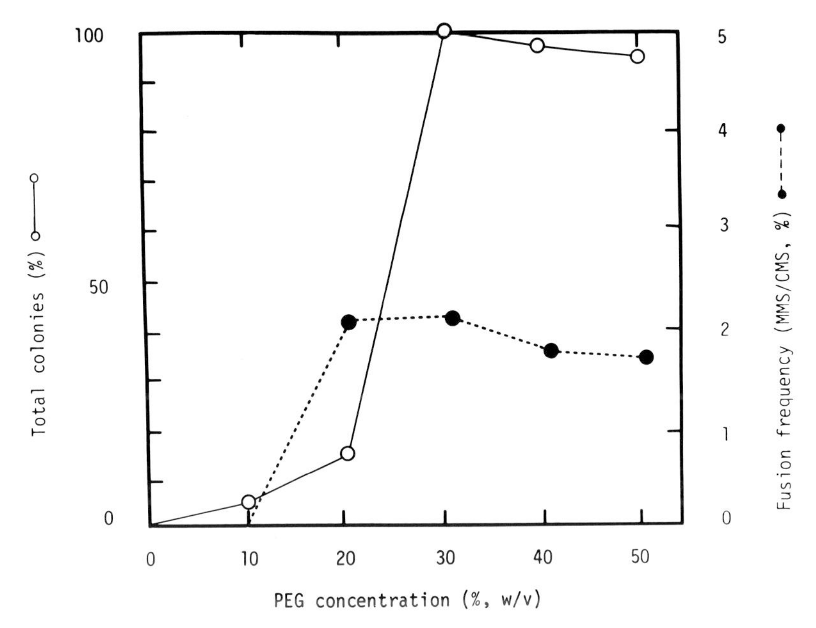

Fig. 4. Effect of PEG concentration on frequency of protoplast fusion between *B. brongniartii* Bt5-16 (methionine$^-$) and Bt6-13 (adenine$^-$). Fusion was performed by suspending the protoplast in a solution containing various concentrations of PEG 3350 and 10 m*M* $CaCl_2$ at 30°C for 30 min. Diluted protoplast suspensions were plated on MMS and CMS media. Colony counts obtained at 30% PEG were taken as 100%.

various concentrations (5–100 mM) of $CaCl_2$. Similar recoveries of prototrophic colonies were obtained.

Based on the results described above, intraspecific protoplast fusion of *B. brongniartii* Bt6 was investigated using histidine-requiring and lysine-requiring mutants (Table II). High frequencies (about 5%) of prototrophic colonies were obtained. Two control experiments were set up with plating on CMS and MMS media. Protoplasts of the parental strains were treated with PEG and plated separately. When mixtures of protoplasts of the parental strains without PEG treatment were used, prototrophic colonies were not obtained.

Results for the fusion between *B. brongniartii* and *amonpha* are shown in Table III. Complementation frequency between *B. brongniartii* Bt5 and Bt6 (same type of zymogram) was 2.1%. However, in the protoplast fusion between *B. brongniartii* Bt5 and F47 (different type of zymogram), it was lower than 0.0048%. In addition, in interspecific protoplast fusion between *B. brongniartii* F47 (type 2) and *B. amorpha*, complementation frequency was 1.1%.

TABLE II

Fusion Frequency between Protoplasts of Auxotrophic Mutants in *Beauveria brongniartii* Bt6

	Colony counts[a]							
	4-3-2 + 4-3-3 + PEG		4-3-2 + 4-3-3 − PEG		4-3-2 + PEG		4-3-3 + PEG	
Dilution	MMS	CMS	MMS	CMS	MMS	CMS	MMS	CMS
10^0	+++	+++	0	+++	0	+++	0	+++
10^{-1}	+++	+++	0	+++	0	+++	0	+++
10^{-2}	+++	+++	0	+++	0	+++	0	+++
10^{-3}	30	+++	0	+++	0	+++	0	+++
10^{-4}	1	71	0	262	0	145	0	110

[a]Strain 4-3-2, histidine$^-$. Strain 4-3-3, lysine$^-$. Fusion was performed by suspending the protoplasts in 30% (w/v) PEG containing 0.01 *M* $CaCl_2$ at 30°C for 30 min. Diluted protoplast suspensions were plated on MMS and CMS media.

Figure 5 shows the consequences of protoplast fusion in the genus *Beauveria*. In protoplast fusion of the same strains in a given species, for example, *B. bassiana* serotype B_1 and serotype B_1 or *B. amorpha* and *B. amorpha*, complementation frequencies were relatively high, mostly around 1–5%.

Intraspecific protoplast fusion of *B. bassiana* serotype B_1 and

TABLE III

Fusion Frequency between Protoplasts of Auxotrophic Mutants of *Beauveria brongniartii* and *Beauveria amorpha*[a]

Protoplast pairs	Fusion frequency (%)
B. brongniartii Bt5 and *B. brongniartii* Bt6	
Bt5-16 (methionine$^-$) + Bt6-13 (adenine$^-$)	2.1
B. brongniartii Bt5 and *B. brongniartii* F47	
Bt5-16 (methionine$^-$) + F47-18 (proline$^-$)	<0.0048
B. brongniartii F77 and *B. brongniartii* F47	
F77-21 (methionine$^-$) + F47-18 (proline$^-$)	2.8
B. brongniartii Bt5 and *B. amorpha* 357-11	
Bt5-16 (methionine$^-$) + 357-11-8 (adenine$^-$)	<0.0013
B. brongniartii F47 and *B. amorpha* 357-11	
F47-18 (proline$^-$) + 357-11-8 (adenine$^-$)	1.1

[a]Fusion was performed by suspending the protoplasts in 30% (w/v) PEG containing 0.01 *M* $CaCl_2$ at 30°C for 30 min. Diluted protoplast suspensions were plated on MMS and CMS media.

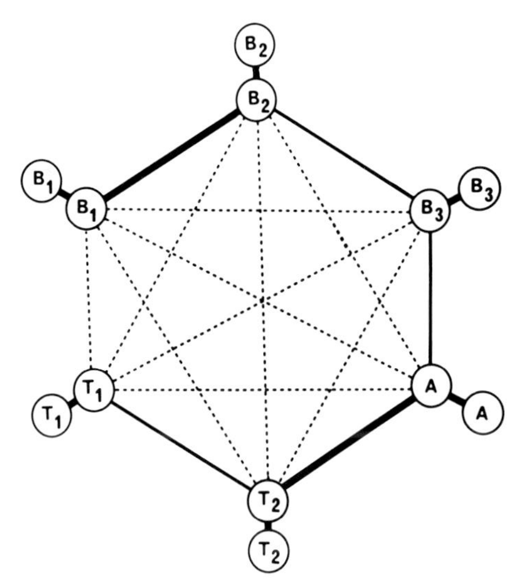

Fig. 5. Consequence of protoplast fusion in the genus *Beauveria*. B_1, *B. bassiana* serotype B_1; B_2, *B. bassiana* serotype B_2; B_3, *B. bassiana* serotype B_3; T_1, *B. brongniartii* type 1; T_2, *B. brongniartii* type 2; A, *B. amorpha*. (▬▬▬) Complementation frequency 10^{-2}–10^{-3}; (———) complementation frequency 10^{-5}–10^{-6}; (- - - - - - - - -) Complementation frequency lower than 10^{-6}.

serotype B_2 gave prototrophic colonies with a frequency of 1–2%. However, in protoplast fusion between serotype B_1 and serotype B_3 or serotype B_2 and serotype B_3, complementation frequencies were lower than 10^{-5}. In the protoplast fusion between type 1 and type 2 of *B. brongniartii*, the complementation frequency was on the order of 10^{-5}, whereas in fusions involving the same type of zymogram a high frequency was obtained. For interspecific protoplast fusion between *B. brongniartii* type 2 and *B. amorpha*, the complementation frequency was on the order of 10^{-2}.

In crosses with a high complementation frequency, colonies formed by protoplast fusion could be cultured, in most cases, by frequent subculturing of young mycelia on MM.

IV. DISCUSSION

Protoplast formation from blastospores of *B. bassiana* using lytic enzymes was reported by Kawamoto and Aizawa (1983) and Kawula and Lingg (1984). In the present study, high yields of protoplasts from young mycelia of the genus *Beauveria* were obtained by the treatment with Zymolyase-20T. Most of the protoplasts had one nucleus.

The results obtained by using various fluorescent markers indicated the presence of α-D-glucose or α-D-mannose, α-D-galactose, and galactose-β(1 → 3)-*N*-acetyl-D-galactosamine on the surface of mycelia; and β-1,4-linked polymer, either chitin or cellulose, is probably the major structural component of the cell wall in the genus *Beauveria*. Septa of mycelia were stained with CAW but not with FITC-labeled lectins. This suggested that the former permeates the cell wall but the latter binds the receptors on the cell wall surface.

The regenerating protoplast of type 3 is very interesting. Nuclear division and synthesis of cytoplasmic contents were somehow suppressed, and only cell wall was actively synthesized. It is possible that the synthesis of cell wall continued until the protoplast established the conditions for apical growth. Protrusion of a germ tube like hypha occurred after the apical conditions were acquired. Further analysis of this situation will reveal some important features on basic cellular structures.

For osmotic stabilization and high-frequency protoplast fusion, a PEG concentration of 30% was needed. At 20–50% PEG, the complementation data were similarly high. The low-frequency complementation below 20% PEG was due to the protoplast bursting rather than to infrequent aggregation. When an osmotic stabilizer was added to prevent protoplast bursting, a higher fusion frequency was obtained.

The effect of calcium ions on fusion frequency was significant. Without calcium ions, the complementation frequency was very low. Addition of less than 5m*M* $CaCl_2$ was effective for protoplast fusion. Interestingly, addition of 5–100 m*M* $CaCl_2$ had a stimulating effect similar to the process of fusion.

Shimizu and Aizawa (1983) reported that isolates of *B. bassiana* could be classified into three serotypes (B_1, B_2, and B_3) and that most isolates belonging to serotypes B_1 and B_2 were more virulent than those belonging to serotype B_3 against the silkworm *Bombyx mori*. In intraspecific protoplast fusion between *B. bassiana* serotypes B_1 and B_3 or serotypes B_2 and B_3 the fusion frequency was lower than 10^{-5}, whereas in the fusion between serotypes B_1 and B_2 a high frequency was obtained. This suggested that there are two types in *B. bassiana* based on complementation frequencies and that serotypes B_1 and B_2 are closely related strains.

Isolates of *B. brongniartii* were classified into two types (1 and 2) based on the electrophoretic patterns of three enzymes (esterase, acid phosphatase, and malate dehydrogenase) (S. Shimizu, unpublished data). In protoplast fusion between type 1 and type 2 of *B. brongniartii* the complementation frequency was 10^{-6}, whereas in fusion involving the same type a high frequency of 10^{-2}–10^{-3} was obtained. The results for the electrophoretic patterns of three enzymes coincided with the results for the complementation frequencies in protoplast fusion. In the interspecific protoplast fusion, the complementation frequency between *B. brongniartii* type 2 and *B. amorpha* was on the order of 10^{-2}, but the frequencies for the other pairing combinations were lower than 10^{-2}. The results suggested the existence of two types in *B. brongniartii*, and it was found that *B. brongniartii* type 2 and *B. amorpha* are closely related species based on their complementation frequencies.

In conclusion, it is possible to obtain hybrids by protoplast fusion between *B. bassiana* strains of serotype B_1 and B_2 and between *B. amorpha* and *B. brongniartii* type 2. Protoplast fusion will provide a useful procedure for breeding of fungi for microbial control.

V. SUMMARY

Protoplast fusion with polyethylene glycol and complementation between auxotrophic mutants of *B. bassiana*, *B. brongniartii*, and *B. amorpha* were investigated. High yields of protoplasts from young mycelia were obtained by treatment with Zymolyase-20T. Most of the protoplasts had one nucleus. With 6 hr of incubation at 28°C, cell wall regeneration on a regenerating medium could be observed under a fluorescence microscope by staining with calcofluor white or two FITC-labeled lectins (soybean agglutinin and peanut agglutinin). Isolates of *B. brongniartii* were classified into two types (1 and 2) based on the electrophoretic patterns of three enzymes (esterase, acid phosphatase, and malate dehydrogenase). In protoplast fusion between types 1 and 2 of *B. brongniartii* the complementation frequency was 10^{-6}, whereas in fusion involving the same type a high frequency of 10^{-2}–10^{-3} was obtained. In interspecific protoplast fusion, the complementation frequency between *B. brongniartii* (type 2) and *B. amorpha* was 10^{-2}–10^{-3}, but the frequencies for the other pairing combinations were lower than 10^{-5}.

REFERENCES

Doderski, J. M. (1981). Comparative laboratory studies on three fungal pathogens of the elm bark beetle *Scolytus scolytus*; Effect of temperature and humidity on infection by *Beauveria bassiana*, *Metarhizium anisopliae*, and *Paecilomyces farinosus*. *J. Invertebr. Pathol.* **37**, 195–200.

Ferenczy, L., Kevei, F., and Szegedi, M. (1975a). Increased fusion frequency of *Aspergillus nidulans* protoplasts. *Experientia* **31,** 50–52.

Ferenczy, L., Kevei, F., and Szegedi, M. (1975b). High-frequency fusion of fungal protoplasts. *Experientia* **31,** 1028–1030.

Garcia Acha, I., Lopez-Belmonte, F., and Villanueva, J. R. (1966). Regeneration of mycelial protoplasts of *Fusarium culmorum. J. Gen. Microbiol.* **45,** 515–523.

Ignoffo, C. M., Garcia, C., Kroha, M., Samšiňáková, A., and Kálalová, S. (1983). A leaf surface treatment bioassay for determining the activity of conidia of *Beauveria bassiana* against *Leptinotarsa decemlineata. J. Invertebr. Pathol.* **41,** 385–386.

Kawakami, K. (1978). On an entomogenous fungus *Beauveria tenella* (Delacroix) Siemaszko isolated from the yellow-spotted longicorn beetle, *Psacothea hilaris* Pascoe. *Bull. Seric. Exp. Stn.* **27,** 445–467 (in Japanese, with English summary).

Kawamoto, H., and Aizawa, K. (1983). Protoplast fusion of *Beauveria bassiana. Abst., Int. Mycol. Cong., 3rd,* p. 139.

Kawula, T. H., and Lingg, A. J. (1984). Production of protoplasts from *Beauveria bassiana* blastospores. *J. Invertebr. Pathol.* **43** 282–284.

Kevei, F., and Peberdy. J. F. (1984). Further studies on protoplast fusion and interspecific hybridization with the *Aspergillus nidulans* group. *J. Gen. Microbiol.* **130,** 2229–2236.

Kunoh, H. (1983). Fungal protoplasts—isolation, regeneration, reversion, fusion, and applications. *Trans. Mycol. Soc. Jp.* **24,** 341–356 (in Japanese, with English summary).

Miyakawa, I., Aoi, H., Sando, N., and Kuroiwa, T. (1984). Fluorescence microscopic studies of mitochondrial nucleoids during meiosis and sporulation in the yeast, *Saccaromyces cerevisiae. J. Cell Sci.* **66,** 21–38.

Pendland, J. C., and Boucias, D. G. (1984). Production and regeneration of protoplasts in the entomogenous hyphomycete *Nomuraea rileyi. J. Invertebr. Pathol.* **43,** 285–287.

Ramoska, W. A. (1984). The influence of relative humidity on *Beauveria bassiana* infectivity and replication in the chich bug, *Blissus leucopterus. J. Invertebr. Pathol.* **43,** 389–394.

Riba, G. (1978). Recombination après hétérocaryose chez le champignon entomopathogène *Paecilomyces fumosoroseus* (Deutéromycète). *Entomophaga* **23,** 417–421.

Samšiňáková, A., and Kálalová, S. (1983). The influence of a single-spore isolate and repeated subculturing on the pathogenicity of conidia of the entomophagous fungus *Beauveria bassiana. J. Invertebr. Pathol.* **42,** 156–161.

Shimizu, S., and Aizawa, K. (1983). Serological properties of *Beauveria bassiana. Abst., Int. Mycol. Congr., 3rd,* p. 288.

V

Future Perspectives

26

University–Industry Perspectives

GEORGE M. GOULD

I. INTRODUCTION

The potential which helps drive technology transfer across the university–industry interface is the amount of grant money available from nonindustry sources. As the amount of grant money from U.S. government sources has dropped, the greater has been the pressure on the technology transfer process to step in and help fill the financial void. To many schools, this has meant an in-depth review and ultimate redrafting of long-standing policies on patents, consultantships, and grants. Tensions have arisen between faculty members and their administrations on issues ranging from esoteric concerns about the role of academia in a technology-driven society, the sanctity of academic freedom, and the right to publish to much more immediate pocketbook issues such as the split of royalty income derived from the successful licensing of a university invention between the inventor(s) and the university administration.

As a result of the recent relatively consistent trend toward less federal funding for university research in areas not involved with na-

BIOTECHNOLOGY IN INVERTEBRATE
PATHOLOGY AND CELL CULTURE

tional defense, many universities have professionalized their technology transfer activities. They have designated or brought in staff to handle this function. Public attention has been focused on particularly successful licensing programs, such as the licensing of the Cohen–Boyer U.S. patent rights covering the basic recombinant DNA process by Stanford University and the University of California. Over 70 companies have licensed these rights and the comparatively modest annual minimum royalties have already brought in several millions of dollars to support research at these two institutions (the two inventors having waived their personal share of the royalty stream). Considering that such income was generated well before there was any appreciable commercial activity with any recombinant DNA products that would produce a royalty stream, this should stimulate attention.

Obviously, the future for substantial income from the licensing of this basic invention is extremely bright. Revenues, even with the 0.5 to 1.0% royalty rates which have been granted, may well run into the tens of millions of dollars per year by the last few years of the patent life if market projections for some "blockbuster" recombinant products, presently in advanced clinical trials, make it to market in timely fashion. Even more institutions will be stimulated to emulate such successful examples.

Not all licensing projects will be winners. Harvard University misjudged the marketplace with an overzealous attempt to emulate the Stanford–California success and tried to raise a large front-end royalty stream with a less than optimum biotechnology patent right. What is interesting is not the lesson learned from one failed licensing program but rather the incredible change in policy that program represented compared to the situation at Harvard hardly a decade ago. Up to the early 1970s, the patent policy at Harvard Medical School required staff members to dedicate to the public any inventions they might make in the health field. The result of such policy was decades of potentially valuable medical discoveries sitting fallow in publications, as the lack of patent protection acted as a disincentive to potential developers to invest the needed developmental scientific resources and monies.

As one looks to the future, it is evident that if the present withdrawal of the public sector from supporting research at universities continues, there will be a greater need to come up with imaginative technology transfer mechanisms to make up at least part of the shortfall from the private sector. Present expectations are that biotechnology will be a particularly fertile area for such transfer as more and more companies will be generating commercial revenues from activities in this field. In order to ensure efficient interactions between the respective entities as we move into this phase, it is important to identify areas of possible difficulty and friction. Once identified, strategies can be developed to minimize their negative effects. The intent of this discussion is to identify some of these potential trouble zones and to discuss some approaches to resolve or at least to close the area of difference. Serious and fundamental as some of these issues appear, they can be resolved by application of reason, knowledge of the needs of the other party, and the mutual self-interests which are served by making beneficial use of the technology transfer process.

However, before examining the substantive issues and problems facing the transfer process it might first be useful to briefly review the various types of arrangements which are commonly used to effectuate technology transfer between universities and industry. The first such arrangement is the consultantship. It is usually a contract for personal services involving an individual scientist which may be limited to a specified field or specific subject matter. The scientist may be called on to work on the technology of the industrial sponsor or the consultantship may be used as a device to assist the process of transfer of a university-based technology to the industrial concern. Generally, it should not encompass carrying out anything other than intellectual activities at the university facilities, nor should it infringe on the scientist's employment commitments with the university. To be safe, each agreement should be checked with the printed policies on consultantships which virtually every university has established or else a university administrator should be contacted to ensure compliance.

Usually when there is a need to carry out research activities at the university facilities, the appropriate arrangement would involve a grant from the industrial concern to the university. The grant can designate support of the specific investigator, a laboratory, or a department and may be restricted or unrestricted. The former type of arrangement implies that the sponsor will obtain a proprietary position if any inventions are made while the latter type usually means that no rights are retained in the research by the sponsor. Custom usually dictates that the sponsor be provided with the research results before publication. Some universities will also give the sponsor an unwritten right of first negotiation to license any inventions coming out of the unrestricted grant-supported research.

A further type of arrangement is a license agreement pursuant to which commercial rights under intellectual property rights (such as patents, trade secrets, and copyrights) or tangible property rights (cell lines, organisms, vectors, and the like) are transferred from the university to an industrial concern for a consideration. Such consideration usually includes some form of front-end payments (license fee or minimum royalties) and a royalty on sales. Obviously, the parties are not limited to any one of these arrangements. In practice, hybrid arrangements involving elements of some or all of these types of agreements can be employed to tailor fit the legal terms to the specific interaction desired to be protected.

We turn now to the substantive issues which can arise during a university–industry technology interaction.

II. ACADEMIC FREEDOM VS. NEEDS FOR CONFIDENTIALITY

Perhaps the single most polarizing issue affecting the university side is the question of academic freedom and whether the technology transfer process is so contrary to its principles that the two cannot coexist in the same institution. Obviously, most universities do not subscribe to this extreme view or else the technology transfer process would not be experiencing its present rapid growth. However, there are still many in academia who express such views in a rather vocal manner and administrators do keep these views in mind during negotiations. In some technology transfer situations, the package of separate issues which form the generic complex known as "academic freedom" can rise to the deal-breaker status and thus warrant careful review and sensitive handling in practice.

Individual subissues which may be part of the "academic freedom" complex include the following: (1) the right to publish; (2) the type of restrictions, if any, that can be placed on students working on sponsored research projects; and (3) whether tangible research results are the property of the investigator, the university, or the sponsor or, as a corollary, whether university researchers can take their research results and materials with them if they leave the university.

Many other variations on these themes can, of course, come readily to mind. It is useful, however, to understand what specific aspects are being questioned under the general academic freedom banner in order to produce an appropriate arrangement that establishes a proper framework to address and solve problems which arise from that aspect. An institution may be committed to a policy that requires that no restrictions be imposed on students with respect to publication or to the students' use of materials generated by reserach activities in which they participated. If the industrial concern is very sensitive to the need of maintaining a proprietary position on their sponsored research program, it might be necessary to specifically exclude student participation from grant-supported activities as a provision of the agreement. Alternatively, if areas of greater or lesser sensitivity can be defined, then the exclusion of student participation might be limited to the specific defined areas of high sensitivity. The lesson from such an exercise is that there are many gradations of agreement possible even if one is facing seemingly irreconcilable policy positions on either side. What is usually needed to achieve a successful resolution of a potential impasse is an analysis of what the real problems are for each party at the operating levels of the arrangement and whether there is room between the respective positions for give in at least part of the program to allow an intermediate accommodation that still is reasonably responsive to each party's critical issues.

Turning specifically to the problems generated by the publication issue, analysis of the various considerations which underlie these issues can be of great assistance in reaching positions of accommodation. A useful place to start the analysis is by asking the question: "Freedom to publish what?" In many cases, the area of greatest sensitivity might lie in the so-called background information—that is, the information independently developed by the university scientists prior to the date of any technology transfer agreement. Conversely, this area might be of lesser importance to the industrial concern for several reasons. It may already have been published in whole or in part, it may be covered by patent applications, or it may be of a general nature and not directed specifically to the scientific problem which is the subject of the arrangement. In such circumstance, the parties could readily agree that no restrictions be placed on publications based on the university's background technology.

A similar analytical procedure may be carried out with respect to activities to be performed pursuant to the arrangement. Will there be substantial inputs of proprietary information from the industrial concern? For example, the industrial concern may provide information and samples of novel compounds which are not yet known to the public. In such circumstances, the industrial concern may have a legitimate need for confidentiality covering not only its inputted proprietary information but also the data and research results which are generated by the sponsored research. Arguments in support of interpreting academic freedom broadly in this circumstance are somewhat muted, since no one can reasonably urge that the university researcher has the unilateral right to destroy the industrial concern's right to maintain its proprietary position on its prior independently developed technology. Accepting this posture, the scientific value of any publication on the research results achieved on this technology is minimal if the university researcher treats the identity of the compounds used as a "black box." Therefore, the university is conceding little by accepting restrictions on publication under this fact pattern.

Somewhat more difficult to solve are publication rights with respect to research which involves more equivalent inputs from the parties, more in the manner of a collaboration than a sponsored research program. Here both parties can provide reasonable arguments to support their positions on whether or not restrictions on publications should be allowed. In such circumstances, it still may be possible to negotiate a working compromise where the university's desire for freedom to publish is directed to the research results achieved by its scientists while the industrial concern is focusing on proprietary materials which it provided to the collaboration. It should be possible to allow the industrial concern to promptly review all manuscripts before submission for publication with the right to remove all of the industrial concern's background information. The review could also include a determination of whether any of the disclosure warranted patent protection. If that were the case, then the agreement could provide for a reasonable delay in publication for a time certain to allow the patent preparation to be effected before submission of the manuscript.

As before, the precise mechanism for carrying out this comprise is extremely flexible and can be typed to take into consideration many variables in the fact patterns. It should be noted that the more the university presses for liberal publication rights, the greater will be its reliance on the patent rights it has established for any substantial revenue under the technology transfer agreement. By allowing early publication, either all trade secrets and lead time advantages are lost or their value is substantially reduced. Any sharing in the commercial rewards of the project would, by necessity, be based on the availability and scope of patents having claims covering the commercially developed product or process.

Developments in biotechnology have also put a premium on maintaining exclusivity using certain tangible research results arising in this field. For example, a possible endpoint for a collaborative research arrangement could be the development of a recombinant microorganism capable of expressing substantial amounts of a novel, therapeutically valuable protein. Alternatively, such an endpoint could be a hybridoma cell line which secretes a new, diagnostically important monoclonal antibody. Unlike the situation with a conventional organic or biochemical project, where access to the end product poses no real breach of a proprietary position once patent filings have been initiated, unrestricted access to the living factories which recombinant organisms and hybridoma cell lines can represent may well mean irrevocably losing technological lead time, bypassing trade secret barriers, and diminishing the threshold barrier of competition should a competitor obtain possession of these materials. The risk of these severe adverse economic consequences to a commercial entity, in a business which is already distinguished by high risks even where all proprietary rights are strongly protected, usually causes the industrial concern to dig in its heels and seek all possible protections.

Weighing the risk-to-benefit ratio of unrestricted access to commercially valuable biological materials, it would appear that the scales tilt more to the side of industry on this issue. There is no reason why this has to be an all-or-nothing issue for the parties. As a minimum, the industrial concern should be able to provide reasonable research quantities, at no cost to the university researchers, of the expressed protein produced by the recombinant microorganism (perhaps even in purified form, a result which might not have been readily achievable by the university researchers on their own) or a monoclonal antibody produced by the hybridoma cell lines if that is the research product. In this way, the university should be better able to carry out its academic research with the needed proprietary reagents in a facile manner. At the same time the university can provide the needed reagents to other laboratories and collaborators so that the scientific reputations of the university scientists and the research commitments of the university are not jeopardized. This is especially important if the materials in question were initially prepared at the university and the university scientists are the principal authors on the resulting papers.

Should it not be possible to satisfy the research needs of the university scientists by providing the product protein or monoclonal antibody and it is necessary that access be provided to the critical organisms or cells, then it may be possible to consider restricted access by the senior university investigator. One continuous problem for the industrial concern in such circumstances arises from the fact that security at most university research laboratories is generally not up to industry standards. In return for a commitment to restrict access to the biological materials to a single indicated laboratory and/or to research personnel specified by the university, the industrial concern might agree to provide suitable means of ensuring secure storage, such as a lockable freezer or refrigerator to allow restricted access to the designated biological materials. Only the specified personnel would have keys to the storage area and the materials could be kept segregated from cold rooms or refrigerators where there is usually unrestricted access.

III. PRODUCT LIABILITY

If one had carried out a poll of the 10 problem areas in technology transfer arrangements just 10 years ago, it is possible that the issue of product liability would not have appeared in a single response at either the industry or the university level. It started to be an issue in the early 1980s when legal writers began to comment on the aggressiveness of the product liability bar. Some began wondering who might become possible targets of the continued push to find the legal theories that might spread liability risk to new classes of defendants. One relatively still virgin class of defendants at the time was the university acting as licensor of technology. Pundits wondered whether a party injured by some new product which was manufactured on the basis of technology licensed from a university could reach back through the licensing agreement to impose liability on the university. In such circumstances the industrial concern, to whom one might ordinarily look for loss due to commercialization activities, might be a start-up company with limited assets and limited insurance coverage. A university with a major endowment fund and operating budgets running in the hundreds of millions of dollars per year could be considered a very tempting target even if it played a relatively passive role in the development and commercialization of the licensed product.

A few years ago, concerns about product liability exposure could be resolved by obtaining product liability insurance with the cost apportioned between the parties as determined by negotiation. However, the recent crisis in product liability insurance has made it virtually impossible for even well-established major industrial enterprises to get reasonable levels of protection on well-established product lines. How then can one reasonably expect to insure a university against risks arising from biotechnology products whose potential for causing injury or damage is still not fully appreciated or documented? The industrial concern can justify a posture of letting the university sink or swim on its own by pointing out that the risk of product liability exposure is merely a cost of being in the technology transfer business. On the other hand, the university trustees may not wish to jeopardize the viability of a major educational and research institution should a major misfortune hit the product and a megaverdict be brought in against the university, especially if the benefit from the license is a royalty revenue stream.

Until the product liability crisis abates and insurance is available for reasonable premiums, the burden of providing protection will, by default, fall on the industrial concern. The mechanism for this will be to provide the university with an indemnity and hold it harmless in the license agreement. This will mean that the costs associated with the project may increase as the industrial concern will have to amortize the added costs of such assumption of risk with the result that marginal programs might be avoided as not being worth the risk. As a matter of policy, most companies would not extend their immunity to cover acts of negligence by university personnel. Thus, some residual risk would remain for the university in most cases.

IV. DIVISION OF ROYALTY INCOME

As the role of technology transfer in the provision of research funds to universities expands, so simultaneously will technology transfer produce a growing proportion of the compensation provided to university research employees. While the policies of universities vary widely with respect to the percentage of the royalty stream which is reserved for the inventors, many do provide very substantial proportions to the individual scientists. Some range as high as 50%.

The intent in liberalizing the sharing of license revenue, which recently has been extended to federal employee inventors of government-owned inventions, was to stimulate the innovation process. The desire was to help ensure the disclosure of more inventions to the university and to ensure the active participation of the inventors in the patenting procedure. The concerns which arise from liberalizing the monetary awards to university inventors involve stimulating highly competitive postures among the staff scientists, with increasing secretiveness and inhibition of free and open interchanges and collaborations. Such behavior could distort the process by which inventorship is determined, particularly in cases where senior level researchers are contesting more junior members' rights to be named as inventors or coinventors.

Not only is there the possibility of tensions and pertubations of normal scientific interactions at the internal university level but also, if the potential for major rewards exists for individual scientists, it could also cause problems in establishing a smooth working collaboration between the university and industrial scientists. Obviously, if two groups of researchers are working on a common problem and one group can benefit directly and personally by being included as inventors on commercially important inventions while the other group receives only less direct benefits, the more base instincts of human nature might emerge. The groups might tend to be less than fully candid and open in discussions. The exchange of needed data and materials could be affected.

Once the nature of the potential problems which can arise from a distorted distribution of economic benefits from a research project are recognized, steps can be taken to assure all researchers that they will be treated equitably. Industry employees, who do not normally share directly in the commercial benefits derived from a specific project, can be advised that contributions to the success of the collaborative program will be fully and timely recognized and appropriately compensated. The most preferable endpoint for maximizing the potential for success from the collaboration would be to establish an environment in both research groups that would make irrelevant the issue of which group or which individual scientist(s) within a group was the named inventor(s). This could be workable if the sizes of the research groups are reasonably manageable, but it becomes somewhat idealistic if the number of people involved is substantial. The moral, if one can be drawn in such circumstances, is to try to prevent the erection of institutional barriers to success. There is little that one can do in an affirmative way to ensure the successful conduct of any research program.

V. TITLE TO BIOLOGICAL MATERIALS

Some of the problems regarding disclosure and restricted access to biological materials, as specific examples of tangible research results, were touched on above in the discussion of academic freedom. However, other areas of consideration should be noted in reviewing these basic chattels of biotechnology. The recent trend in many universities has been to estabilsh a strong policy position that all biological materials produced as research products by university employees are the sole and exclusive property of the university. The federal government has encouraged this trend and has issued guidelines supporting the university's attempts to establish proprietary positions by use of patents or trade secret law.

Free and open exchange of biological materials, which was the generally followed practice in the biological and biochemical acts up to the mid-1970s, has given way to a far more cautious and documented interchange, which is the hallmark of the 1980s. University scientists who would like to enjoy their academic freedom by distributing a cell line, recombinant organism, or other tangible research product to any third party without restriction run the risk that they will be violating their employment agreement with the university by giving away university property without authorization.

Most universities will not wish to interfere with the normal scientific interchanges between researchers. They have established rather simple form letters on which the recipient scientist can acknowledge receiving the identified biological material and agree to use the subject material solely for research purposes. It is also common to require an undertaking on the recipient's part not to pass the material on to any other scientist without the granting scientist's permission.

If the transaction involves scientists at a commercial concern the situation becomes a bit more complicated. Usually, if the intended use by the commercial scientist is clearly and unequivocally noncommercial research, then the same type of letter sent to noncommercially employed scientists should suffice. However, determining what activities in a commercial organization would constitute noncommercial research is not a simple task. A prudent scientist in the commercial organization would do well to consult the legal department before signing off on such a commitment. Examples of troublesome areas include cell lines which are used in screening assays for developing new drugs, in quality control procedures, in toxicology, or in other activities which may not be directly linked to production of product for sale but rather used in an ancillary role which does provide substantial commercial benefit to the user.

Obviously, if there is going to be direct commercial use of the biological sample, then an exchange between scientists is not appropriate to convey an authorized right to use for commercial purposes. This right to use involves a license grant under the university's proprietary rights, which may include patent rights, ownership rights to the tangible property as embodied by the cell line, and/or trade secret rights to any know-how regarding the production and use of the material. Drafting and negotiating such licenses involves application of business and legal skills. This should be left to the university administration and legal staffs to carry out.

The commercially employed scientist should be careful about receiving biological material for commercial or quasi-commercial purposes from university scientists even if they give express permission for such use. Such permission may be null and void as the university scientists may not have the authority to bind the university in such matters, no matter how well intentioned the whole transaction may be.

The transition from the unrestricted exchanges of the early 1970s to the more careful, prudent, and official interactions of today has been catalyzed in large measure by the notoriety generated by the controversy surrounding the accession and use in its interferon project of a cell line by the pharmaceutical firm Hoffmann-La Roche Inc. The cell line in question was developed by two medical scientists at the University of California and was the subject of a prior scientific publication. The line was provided to a number of researchers around the world, including scientists at the National Cancer Institute. Subsequently, a sample of the line was transferred to scientists at Roche to evaluate for interferon production. The facts and understandings surrounding the transfer of the line between the various parties were in dispute. What is clear is that after development in the hands of the Roche scientists, the line became a high producer of leukocyte interferon. Because of this induced property, the line became the source of messenger RNA. This was used in the collaboration between Roche and Genentech as a template to produce complementary DNA for use in the cloning and expression of a number of mature recombinant human leukocyte interferons.

The university scientists, after they learned about the use of the subject cell line in the successful cloning effort, made claims on Roche and Genentech, which claims were subsequently taken over by the University of California. The university took the position that all proprietary rights to tangible research results produced by university employees at university facilities were the property of the university and that included title to cell lines created from human sources. As the controversy escalated into the lay press, Roche brought suit to establish clear title to the fruits of their interferon project including clear title to the recombinant organisms and product proteins as well as the relevant intellectual property rights, including the patent rights.

It was expected that a legal decision in that case would establish legal precedent for determining what property rights, if any, could be established on living materials, particularly human derivative materials, what types of restrictions can be placed on such materials when they are provided to other scientists, and the legal requirements for imposing such restrictions. However, the parties settled the suit before trial and no legal precedents were established. In any case, the fact that the claims were made served to energize the scientific community to change its normal behavior patterns when it came to exchanging biological materials. The process was further catalyzed by the publicity generated by the start-up of a number of biotechnology companies. Scientists could literally become millionaires overnight. Access to proprietary biological materials such as a recombinant microorganism which had the capability of expressing an important protein, for example, growth hormone, interferon, or insulin, could make or break such enterprises. Within a very short period, universities established guidelines for the protection of their proprietary interest in such materials. Many scientists were leaving university positions, attracted by the powerful twin lures of interesting projects and big money offered by the biotechnology companies. To protect their interests, universities encouraged patenting wherever possible. Form letters were provided covering the transfer of biological materials to other researchers. Different provisions were used for researchers in academic institutions as opposed to researchers located at commercial organizations. The government also became aggressive in protecting research results funded by federal research grants and established a mechanism through the National Technology and Information Service (NTIS) of the Department of Commerce to administer the technology transfer of such rights to commercial concerns. There is now a steady stream of inventions and proprietary materials being offered for license by that agency. Offers of such licenses are published regularly in the Federal Register and in the Official Gazette of the U.S. Patent and Trademark Office.

In a somewhat ironic twist to the Roche–University of California matter, one of the university scientists who was a principal in the subject dispute was sued by a patient claiming title to a cell line produced from biopsy material taken from his body.The suit challenged the sufficiency of the consent given for the procedure and cited the lack of candor of the scientist in not telling the patient about the potential value of the sample when proferring the consent for signature (the scientist was alleged to have granted a license to the product cell line as consideration for a package of benefits from a commercial organization which included stock options in the company, substantial consulting fees, and future royalties on any commercial products that arose from use of the line). Again, many of the legal issues initially raised in Roche–University of California are before a court of law but from a somewhat different perspective.

The initial stages of this litigation have resulted in a number of complaints being successfully challenged by the lawyers representing the scientist. This tactic has required the filing of amended complaints to set forth new legal theories of recovery. Should the plaintiff patient ultimately prevail, it will require extensive rethinking of how the sampling for establishing the source of human biological materials should be carried out.

At present, many lawyers are urging their client organizations, whether commercial or academic, to be careful in obtaining human body samples. If it is known that such samples may be used to establish lines or to extract useful reagents such as DNA or RNA, then the release forms should reflect these facts and should clearly indicate that the samples may be used for commercial purposes. If the patient is being brought in specifically for the purpose of providing a sample and not for diagnostic or therapeutic treatment, then reimbursement of all incidental expenses including the cost of the medical services as well as more than a nominal consideration for the release should be considered.

Academic and commercial organizations have started to use reasonable care in investigating the pedigree of biological materials to be used in their commercially interesting projects in order to avoid the problems associated with claims which surface after the project has been completed and the use of the biological material in question has become an integral part of the product and/or the intellectual property rights which protect such a product. The best protection against future surprise liability exposure in this area will be a well-prepared history of each line employed in the project augmented by appropriate releases and licenses as needed.

VI. UNIVERSITY–INDUSTRY AGREEMENTS

Generally, the two extremes in type of arrangements between a university and a commercial concern are a license agreement under preexisting intellectual and tangible property rights developed by the university and an industry-sponsored research agreement. As pointed out above, the reduced financing of non-defense-oriented research taken together with the increasing costs of industrial research has substantially increased the number of industry-sponsored university projects, including specifically the biotechnology field. Simultaneously, the number of university-owned patents and inventions available for licensing in this field has also increased manyfold.

There are a number of reasons why industry would turn to a university to carry out research besides the lower cost factor alluded to previously. Most important is the access to specialized expertise which may only be available in academia. In addition, contracting out provides more flexibility as projects can be taken on without incurring the long-term obligations associated with taking on added employees. If the research does not produce desired results within a predetermined time frame, then the agreement can be terminated without further commitment on the part of the company—a highly desirable benefit.

From the university point of view, the primary goal of undertaking research programs is to advance basic scientific knowledge. University scientists are usually interested in pursuing basic research projects instead of the focused projects directed to meeting specific product needs or improving existing processes which in large measure are the goals of commercially employed scientists.

An important aim in drafting these various types of arrangements is that both parties be in a position to achieve their respective needs and objectives. This can be accomplished by use of certain contract elements which have been negotiated to contain terms that reflect a balance between the policy needs of the individual parties. The most common elements will be reviewed briefly to illustrate the flexibility available in this process.

Definition of the Field. This element is of particular importance in consultantship arrangements, especially if there is a clause restricting the right of the scientist to consult for other companies within the defined field. Obviously, the scientist who desires to retain the opportunity to engage in consulting services with other commercial organizations would wish to narrow the scope of the defined field. The sponsor would wish to keep a broad field definition to ensure the broadest possible use of the consultant's expertise. Another advantage to a broad field definition is the lessening of the chance of potential conflict with respect to disclosure of confidential information to a competitor. If fields are close or overlapping, then the scientist who consults for multiple clients faces an almost impossible task of segregating information that can be used freely from that which has been learned under restriction.

In sponsored research projects, broad field definitions may sweep in additional background patent rights under which the university may have to grant licenses to the sponsoring company. Conversely, such broad definition could result in a broader base for establishing a royalty obligation from the commercial entity. Obviously, it is not possible to generalize as to whether a broad or narrow definition is favorable to a given party. It is important to consider the effect of varying the scope of the field definition on the most important terms of the agreement.

Duration and Scope of Exclusivity. A provision important for most commercial organizations operating in the biotechnology field concerns the duration and scope of exclusivity to be granted under the license to the proprietary rights arising from the research agreement. This question is of particular concern for industries, such as the pharmaceutical industry, where product development involves large early investments with long lag times to market approval.These investments involve very high risk money as only a small percentage of pharmaceutical research projects reach the market. Present estimates for the cost of developing a pharmaceutical ethical product to market launch run between $60 million and $80 million with a total elapsed time from laboratory to druggist's shelf of some 7 to 9 years. Absent exclusivity in the marketplace, it would be virtually impossible for any commercial organization to successfully amortize this immense investment and provide sufficient return to investors to compensate for the time and resources expended as well as reimburse the research costs expended on unsuccessful projects. Thus, from the commercial entity's view, there is little flexibility on the question of exclusivity.

From the other point of view, the university in the licensor's position is also faced with a potentially long period between the time the research project (and presumably the research cost reimbursements) ends and the initiation of license royalties from commercial activities by the licensee. This is particularly true when the activities are in the biotechnology or allied pharmaceutical or agronomic fields. The combination of long latency to payoff and strict requirement for exclusivity in the marketplace requires that the university licensor use care in the selection of the licensee as, by necessity, there must be an extended period of trust and good faith during the development and regulatory approval process. Experience further shows that communication between the parties will be limited and fragmentary during this period.

Although overriding economic policy considerations dictate the posture on the issue of exclusivity, it is not impossible to work out suitable provisions to assure the university that its partner will employ suitable diligence and expend needed resources in the commercialization of the product of the subject joint project.

One device employed by licensors in technology transfer agreements, to help ensure diligence on the part of the licensee and at the same time help relieve the cash flow drought between initiation of the arrangement and the start of commercial operations, is to provide for up-front payments. Unfortunately, these types of payments, particularly if they are substantial, are usually the least desirable from the viewpoint of the commercial concern as they represent high risk dollars and are expensive when considered in terms of a time-discounted value of money. In some circumstances, the university can further justify its request for substantial up-front payments beyond the policy consideration discussed above. Thus, if the university has background assets which predated the investment by the industrial sponsor in the joint research program, these can be made part of the license. These background assets can be quite varied in nature, including, for example, patent rights or proprietary reagents needed to carry out the project such as cell lines, antibodies, catalysts, assay systems, and synthetic processes. There may be ways to structure the front-end payments as payments for the purchase of these assets, which, depending on the tax status of the commercial entity, could provide tax benefits that might offset some of the cited disadvantages of these payments.

An alternative to substantial up-front payments is to use milestone payments. There are disadvantages to using schedules which require substantial payments to be made on certain dates. The university may look on this type of arrangement as providing the "earnest money" needed as an incentive for the commercial partner to invest the resources required to help ensure successful development of the product. However, the practical effect may be to the contrary. Each point in time when a prescheduled payment is due is a decision point for the management of the industrial entity. Thus, at relatively early stages of the development phase of the project, management is given the choice of either continuing the funding or terminating it, releasing monies and resources to support other programs. If the development project has not been achieving the desired objectives, an adverse decision is possible. It appears that better planning would produce an agreement that is constituted to provide as little opportunity as possible to get a wrong answer to an untimely question.

Should there still be a need to have front-end payments, some of the negative impact can be diluted by crediting all or part of these payments against future royalties. In fact, Stanford and the University of California used this technique in their licensing program and did it with a somewhat novel twist. They offered their licensees a multiple credit scheme for front-end payments which the licensees would have to make for a number of years before any commercialization could be expected. The fivefold credit for each dollar paid was a powerful inducement to entering the license early and provided a substantial cash flow to the two universities even with no licensed products being brought to market. It should be noted, however, that such types of provisions are rarely used in exclusive license situations.

As an alternative to up-front payments on predetermined calendar dates, it is possible to tie such payments to specific events, preferably positive development events which could be characterized as confidence builders. The achievement of a positive milestone should not cause resistance to continued investment on the part of the commercial concern's management. Another version of this strategy would involve stepping up the size of the payments as the probability of success of the project increases when the positive milestones are met. Philosophically, these types of payments no longer can properly be considered earnest money to ensure diligence by the commercial partner since, during the development phase, they are keyed to the commercial entity's own progress. If progress is slow, the company's payments are extended.

Another wrinkle can be employed to maintain the diligence-ensuring purpose of these payments. This would involve linking milestone payments with temporal events in the following manner. For example, a triggering event is selected to start the clock. The date on which the university licensor provides to the licensee a required proprietary reagent would be suitable. Then a reasonable time period is set within which the licensee is expected to achieve a significant, well-defined development result. In the example this could be 6 months to open a company Investigational New Drug (IND) application on the licensed product as the time zero event. If the licensee achieves the desired diligence-establishing result within the agreed on time frame, the milestone payment is waived.

This hybrid method uses the carrot-and-stick approach. Diligence is rewarded without burdening the commercial concern with the undesired early high-risk payments. Moreover, payment is triggered, even if the industrial partner misses the date, by a research achievement of the university. This should provide the needed confidence builder to management that the project's chances for success were moving in the right direction.

Conversion of Exclusive to Nonexclusive Rights. The threat of conversion of a license grant from exclusive to nonexclusive was once the principal mode by which licensors enforced diligence. Thus, agreements provided that the licensor had the right—if, for example, the licensee had not commercialized a product within a certain time—to convert the license from exclusive to nonexclusive. In practice, this remedy proved not to be a particularly helpful one to the licensors who invoked it. Remarketing a nonexclusive license after an exclusive license has failed, even if the failure can be clearly attributed to the exclusive licensee's mis- or nonfeasance, is not an easy prospect. Even if it were accomplished, the financial terms that could be expected from the second license would be far more modest than those provided in the initial exclusive license. Moreover, the nonperforming licensee would be in a position to benefit from the efforts of the second licensee to commercialize the target product as the first licensee would still have entry to the proprietary rights of the license.

For most industries the loss of exclusivity is an extremely grave remedy. Thus, licensors tend not to "pull the trigger" as it would result in essential loss of the first licensee's activities while, as pointed out above, the situation facing the licensor after the clause was invoked would not be very attractive.

If the clause is used in the future, great care should be exercised in drafting to ensure that it is absolutely clear who has the responsibility for making the determination that a diligence milestone has not been met—i.e., the licensor unilaterally, a sponsoring agency of the U.S. government such as by implementation of a march-in right clause, or a third party such as a selected arbitrator.

Defining what specific point, in terms of time and/or event, would constitute a failure of diligence is also quite difficult and requires careful consideration. If the product which is the subject of the license agreement requires governmental approvals prior to marketing, such as from the Food and Drug Administration, the Environmental Protection Agency, or the Department of Agriculture, then the commercial concern should be prudent during negotiation to link its diligence to the last event under its sole control. For pharmaceutical products it would be preferable to link diligence to the last event under its sole control. This would be the filing of a New Drug Application (NDA) with the FDA. It would be dangerous to have diligence tied to NDA approval, a process noted for its unpredictability and arbitrariness.

Best Efforts. Many exclusive licenses now contain a provision which obligates the licensee to employ its "best efforts" to develop the licensed product to commercial utility or to use "all reasonable efforts" to maximize the benefits to be realized by the licensor. It is not apparent whether such provisions provide anything but optical value to the agreement since many states impose such obligations on an exclusive licensee as a matter of law. In fact, there may be more protection for the licensor from the rights afforded by act of law then one can provide through negotiation and drafting of a relevant provision.

Should a best-efforts provision be pursued by the university, the commercial entity would most likely wish to modify the language to introduce some limitations on the quantity and quality of the efforts to be expended, by introducing concepts such as "all efforts which in the reasonable business and technical judgment of licensee are needed to . . ." or by defining the effort to be employed in relation to those usually expended by licensees on projects of similar scope and at the same state of development and risk.

VII. FEDERAL FUNDING

Congress has recently defined the government's rights in inventions made with federal funding in the provisions of 35 U.S.C. §200 through 212. For inventions made between December 12, 1980 and November 8, 1984, earlier legislation identified as Public Law 96-517 and administrative rulings such as Office of Management and Budget Circular A-124 issued in February 1982 by the Office of Federal Procurement Policy control the rights of administering organizations to grant licenses.

An important consideration in any licensing arrangement is whether any of the underlying inventions to be included in the licensed rights were made with federal funding. The requirements imposed by the provisions of Title 35 U.S.C. §200 et seq. are reasonably direct. However, if such requirements are not observed there can be some serious consequences, particularly for the licensėe. In addition, some of the specific requirements of Title 35 (such as §204) can influence the commercial value of the license. Thus, if the agreement encompasses inventions made with federal funding one should take the following points into consideration.

1. The date of invention can be important as it determines which law is in effect. If the invention was made prior to July 1, 1981, the provisions of Public Law 96-517 (enacted in 1981) as enabled by OMB Circular A-124 are in effect. Public Law 98-620, which was effective November 8, 1984, amended the earlier legislation and enhanced the administration rights of nonprofit organizations to government-funded inventions. It is interesting to note that Circular A-124 urges federal agencies to treat inventions made under funding agreements made prior to July 1, 1981 in substantially the same manner as contemplated by Public Law 96-517. The liberalized provisions of the later laws may be applied retroactively to inventions made before their effective date, depending on the policy of the funding agency.
2. Another important question is whether the university (or other appropriate nonprofit institution) performed the needed acts to perfect its title to the invention. The university has the right to retain title to inventions made with federal funding. To accomplish this the university must disclose the invention to the funding agency, provide a written notice to the funding agency that it elects to retain title, and then file or cause to be filed through a nonprofit patent organization (e.g., Research Corporation or University Patents) a patent application covering the invention [35 U.S.C. §202(C)]. Should the university fail to make the required disclosure, make the election in the time provided, or file the patent application, title can be lost [35 U.S.C. §202 (S)].
3. Public Law 98-620 allows the university to grant an exclusive license for the full term of any licensed patent which is based on an invention funded by the government. In contrast, the previous law limited the period of exclusivity to 5 years from first commercial sale or 8 years from the date of the license grant, whichever is earlier. All licenses granted under either the earlier or the current law are, however, subject to a nonexclusive, nontransferable, irrevocable, paid-up license reserved for or on behalf of the United States [35 U.S.C. §202(C) (4)].

4. Additional limitations and obligations which may be imposed on licenses subject to federal patent policy are summarized below.
   a. The sponsoring federal agency may require periodic reporting on the development of the invention to commercialization [35 U.S.C. §202(C) (5)].
   b. The issued patent must include a statement that the invention was made in the course of a federal grant and the government has rights to the inventions.
   c. Patents subject to the government's interest cannot be assigned without approval of the funding agency.
   d. The administering institution must share royalties with the inventors.
   e. Net royalties after expenses and inventors' share must be used to support scientific research or education.
   f. The law provides a preference to small businesses as licensees.
   g. The federal government retains march-in rights under which the licensor or the assignee or exclusive licensee may be required to grant a nonexclusive, partially exclusive, or exclusive license in any field of use on reasonable terms if a determination is made that the federally funded invention is not meeting certain elucidated developmental or public need criteria.

VIII. CONCLUSION

The discussion of the perspectives of the university–industry relationship has focused on some of the problem areas that might serve as barriers to successful interactions between these entities with special emphasis on issues related specifically to biotechnology. Recent history suggests that universities and industry have managed to overcome these obstacles and have forged many arrangements that have proved to be of mutual benefit.

In the early 1980s, when a number of the more prominent of these arrangements in the life sciences were being forged, there were serious questions raised as to whether the parties could overcome these many problems without permanent distortions of the roles of the universities and the involved research scientists in the educational and research worlds. It now appears evident that there has been an increase in the number of arrangements between universities and industrial partners, although the nature of the average arrangement is on a far more modest scale than the megaprojects that dominated the headlines just a few years ago. Industry's desires for access to state-of-the-art technology, specialized skills, and reasonable research costs remain undiminished.

The same is true of the university's needs for a more constant source of research funding, outlets for university-based inventions, and dealing with an increased entrepreneurial spirit among faculty members. If anything, an increase in such arrangements in the biotechnology field can be expected as more and more of the genetic engineering houses build up sufficient capital to take more of their projects on to commercialization without seeking sponsored research arrangements with the larger, established pharmaceutical companies. This may free resources in those large companies for increased support of research projects at the universities.

Another trend which is beginning to become more evident involves tripartite arrangements between universities, industry, and government. The mobilization of these three sectors to focus on common problems is already well advanced in Japan and also has started in Europe. Such interactions are developing in the United States in areas such as research on HTLV-III (the AIDS virus) and research directed toward developing a vaccine for malaria. Obviously, such multi-institutional programs are better suited to complex scientific problems which require these interdisciplinary approaches than to problems which can be handled by single laboratories.

The experience developed from the successful arrangements between university and industry will be of great value in meeting the even more complex challenges presented by the tripartite arrangements. To balance the needs of all parties while remaining responsive to the overriding public interest requires that all concerned have astute negotiating skills. Such arrangements may well become the wave of the future, particularly if the initial role models are successful in achieving the desired results. From a national viewpoint, we cannot afford to fail in mobilizing the best talents from the three research spheres if the United States is to remain competitive with our friends across both oceans.

FURTHER READINGS

Battenburg, J. B. (1980). *Soc. Res. Admin. J.* **11,** 3.

Crittendon, A. (1981). *N.Y. Times* July 22, D1.

Lepkowski, W. (1984). *Chem. Eng. News* June 25, pp. 7–11.

Lippert, N. T., and Gould, G. M. (1985). "Trends in Biotechnology and Chemical Patent Law." Practicing Law Institute, New York.

National Science Board (1982). "University–Industry Research Relationships: Myths, Realities and Potentials." Natl. Sci. Found., Washington, D.C.

Olsen, S. (1986). "Biotechnology: An Industry Comes of Age." Natl. Acad. Press, Washington, D.C.

Tatel, D. S., and Guthrie, R. C. (1983). *Educ. Res.* **64,** 2.

Varrin, R. D., and Kukich, D. S. (1985). *Science* **227,** 385–388.

27

Control of Invertebrate Pests through the Chitin Pathway

H. M. MAZZONE

I. INTRODUCTION

Chitin is the major component of the cuticule of invertebrates. It is almost totally restricted to invertebrates, including insects and fungi. During the molting cycle of arthropods, chitin is hydrolyzed and synthesized again. In the regulation of populations of pest insects, either of these stages offers a target for chitin inhibition. Disease-causing fungi may also be inhibited by compounds which affect the chitin pathway. In choosing control measures for pest insects and disease-causing fungi, substances such as antibiotics, with a selective spectrum of action against the chitin pathway, would ensure that there would be no toxicity to humans. Recent biotechnological advances in drug delivery techniques would guarantee protection of these agents against the degradative effects of environmental or host factors.

The antibiotics polyoxin D and nikkomycin are known to inhibit chitin synthase, an enzyme involved in the metabolism of chitin. These substances also adversely affect invertebrate pests and cell cultures of pest insects. Moreover, the antibiotic tropolone and chitosan, a compound intimately associated with chitin, may indirectly promote deleterious effects on chitin-bearing hosts. The action of these compounds as well as that of thuringiensin, a bacterial toxin, on the causal fungus of Dutch elm disease is discussed in this chapter.

In the control of invertebrate pests the chitin pathway may be further exploited through genetic engineering. In this connection, the breakdown of chitin by various procedures is currently a topic of great interest to investigators in the United States and Japan. Isolation of genes that are responsible for chitinase activity is being realized for prokaryotic organisms. This research approach may be used to introduce chitinase activity in other organisms such as the insect-infecting baculoviruses. Such investigations are bound to have beneficial applications in controlling pest insects and disease-causing fungi.

The spread of invertebrate pests throughout the world and the threat of further crop destruction has been augmented by increased resistance of such pests to control agents now in use. Restructuring current strategies to handle this crisis necessarily involves vigorous employment of new methodologics. Recent advances in biotechnology have indicated that experiments can now be designed which could have far-reaching importance in controlling invertebrate pests. Thus, experiments designed to exploit the expression of gene products, the breaking of antibiotic resistance of microorganisms, and progress in drug delivery techniques through microencapsulation procedures represent areas that may be adapted in control efforts.

This chapter considers a number of biotechnological advances which may be employed against destructive insects and disease-causing fungi of trees. For each type of pest, the principal point of concern is the chitin pathway, chitin being a common structural element of the insect cuticle and of fungal cell walls (Neville, 1975). In control programs for invertebrate pests, chitin skeletal structures represent logical targets for compounds with a selective spectrum of activity, e.g., certain antibiotics and enzymes, and baculoviruses that have been genetically altered so as to contain within their genome the gene for expressing chitinase. Presumably, organisms lacking chitin would not be afected by such measures.

II. SUBSTANCES INHIBITING THE CHITIN PATHWAY

Chitin is a homopolymer of β-1,4-*N*-acetylglucosamine (Muzzarelli, 1977). A number of compounds have been shown to inhibit chitin synthesis (Post *et al.*, 1974; Marks and Sowa, 1976; Marx, 1977) and offer some promise to act as antimetabolites of pest insects (Marks and Sowa, 1976; Vincent, 1978) and disease-causing fungi (Misato and Kakiki, 1977; Gooday, 1977). A selected area of the chitin pathway affected by some chitin inhibitors is the enzyme chitin synthase (uridine diphosphate-*N*-acetylglucosamine : chitin *N*-acetylglucosaminyltransferase, EC 2.4.1.16).

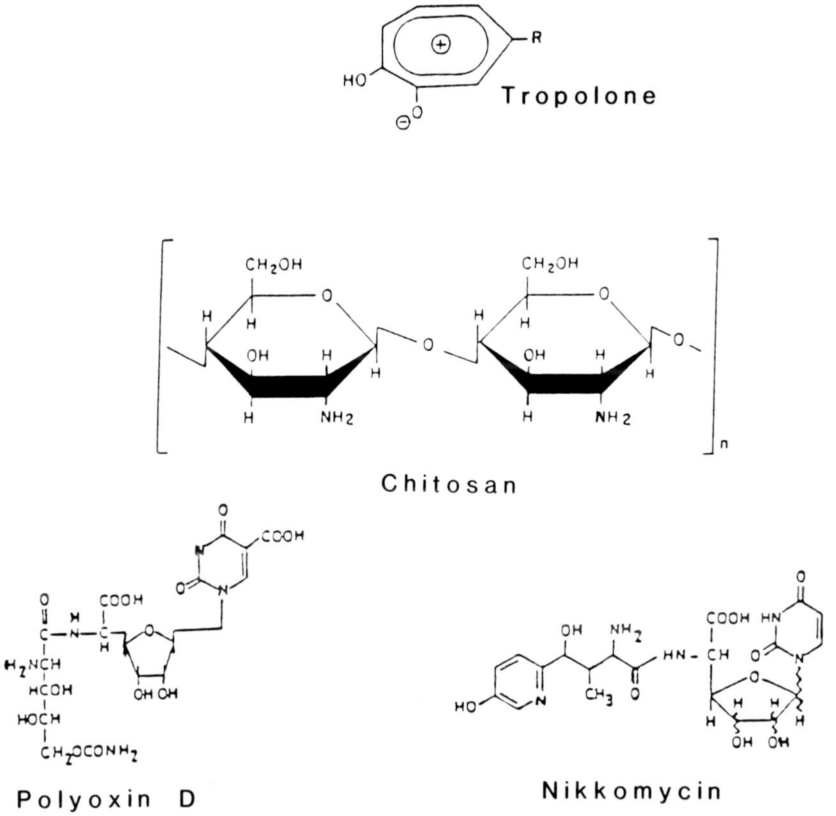

Tropolone

Chitosan

Polyoxin D

Nikkomycin

Fig. 1. Chemical structure of various substances tested against the Dutch elm disease fungus.

Polyoxin D (Isono *et al.*, 1967) and nikkomycin (Höhne, 1974; Dähn *et al.*, 1976), nucleoside peptide antibiotics (Fig. 1), are known to inhibit chitin synthase. In 1976 Marks and Sowa reported incorporation experiments with radiolabeled *N*-acetylglucosamine (NAG) in insect organ cultures. They observed that polyoxin D was capable of inhibiting chitin synthesis in cockroach cells, as it does in fungi (Endo *et al.*, 1970). Leighton *et al.* (1981) extended the study to show that polyoxin D inhibited chitin synthase in a fungus (*Phycomyces*) and in explant cultures of the cockroach. Nikkomycin has been shown to affect chitin synthesis in insects and mites by specifically inhibiting chitin synthase (Brilinger, 1979). Sites other than the citin pathway, such as the osmoregulatory organs of mites, may also be affected by nikkomycin (Mothes-Wagner and Seitz, 1984).

TABLE I
Effect of Substances on *C. ulmi* in Plate Cultures[a]

Substance	Concentration required for complete inhibition[b]
Nikkomycin	4 mg (100 μg/ml)
Nikkomycin + tropolone	1 mg of each (50 μg/ml)
Chitosan	6 mg (150 μg/ml)
Chitosan + tropolone	1 mg of each (50 μg/ml)
Thuringiensin	Noninhibitory at 10 mg
Thuringiensin + tropolone	1 mg of each (50 μg/ml)
Polyoxin D[c]	10 mg (250 μg/ml)
Polyoxin D + tropolone[c,d]	0.1 mg of each (5 μg/ml)
Tropolone[c]	2 mg (50 μg/ml)

[a]Substances were added as 5 ml of agar overlays with fungus (test plate). The agar overlay was of the following composition: 4.0 ml of 0.7% agar, 0.5 ml of stock fungus suspension, 0.5 ml of test substance solution. Agar overlays for control plates contained 4.5 ml of 0.7% agar and 0.5 ml of stock fungus suspension.

[b]Concentrations are per plate culture, 40-ml volume.

[c]Mazzone (1985).

[d]Of the substances tested, singly and in combinations (Mazzone, 1985, and this study), the combination of polyoxin D + tropolone was the most effective in terms of concentration required to completely inhibit fungal growth.

These chitin-inhibiting antibiotics have been tested against *Ceratocystis ulmi*, the causal fungus of Dutch elm disease (DED). Polyoxin D was obtained from CalBiochem-Behring Corp. (La Jolla, California) and nikkomycin was a gift from Prof. H. Zähner (Institute für Biologie, University of Tübingen, Federal Republic of Germany). In a previous report (Mazzone, 1985) the efficacy of polyoxin D, which was low when it was used singly against the DED fungus, was found to be significantly enhanced when it was combined with the antibiotic tropolone. In the present study, nikkomycin also was observed to be more effective in combination with tropolone than when used singly against the DED fungus (Table I).

Chitosan, 2-amino-2-deoxy-β-(1→4)-D-glucan (Fig. 1), a hydrophilic polyelectrolyte obtained by deacetylation of chitin (Muzzarelli, 1973), was tested against the DED fungus. It was a gift from Dr. L. A. Hadwiger (Department of Plant Pathology, Wahington State University, Pullman, Washington) and had been prepared from crab shells by Bioshell, Inc. (Albany, Oregon). Hadwiger and co-workers (Hadwiger and Beckman,

1980; Alan and Hadwiger, 1979) demonstrated that chitosan, occurring naturally in fungi, has antifungal properties. Preliminary data (Hadwiger *et al.*, 1984) indicated that the mode of action by which chitosan inhibits fungal growth is by blockage of the synthesis and accumulation of RNA. Moreover, based on experiments involving the cloning and expression of disease resistance genes, Hadwinger *et al.* (1984) proposed that chitosan may also induce a disease resistance response in plants that are attacked by fungi. The chitin pathway is affected by such a response. Briefly, the disease resistance response is hypothesized as follows. When a fungal cell makes contact with a plant host cell, endochitinase and endo-β-1,3-glucanase present in the plant progressively release fungal wall fragments. Some of the fragments are chitosan segments which enter the plant cell and activate certain genes. The mRNAs from the activated genes produce proteins which enhance the activity of the phenol–propanoid pathway, which in turn produces phenolics potentially adverse to the pathogen, such as phytoalexins (Bailey, 1982) or ligninlike compounds (Vance *et al.*, 1980). Chitosan, tested against the DED fungus, was more effective when used in combination with tropolone (Table I).

The enhancement effect of tropolone in binary combination with a number of substances tested against the DED fungus has been demonstrated (Mazzone, 1985). Tropolone, a catechol compound (Fig. 1), is known to have marked phenolic properties (Belleau and Burba, 1961). In this respect tropolone may be phytoalexinlike in its action. The phytoalexins are antifungal phenolic compounds associated with plants (Coxon, 1982). Tropolone, produced by species of *Pseudomonas* (Lindberg *et al.*, 1980), is lethal to some fungi (Lindberg, 1981). It is available commercially and may be purchased from Aldrich Chemical Co., Milwaukee, Wisconsin.

The mode of action of phytoalexins suggests two lines of activity (Smith, 1982): (1) phytoalexins may represent multisite toxicants, and (2) dysfunction of membrane systems, particularly the plasmalemma, is inherent in their toxicity. A possible mode of action of tropolone involves its ability to affect membrane permeability, perhaps through its well-known chelating action (Muetterties *et al.*, 1966). A substance used in combination with tropolone is then able to enter the target organism and exert its effect. This feature was implied in the present study in the case of thuringiensin, a β exotoxin (McConnell and Richards, 1959) of *Bacillus thuringiensis*. Thuringiensin, a nucleotidic ATP analog, is an inhibitor of RNA biosynthesis and acts by competing with ATP for binding sites (Lecadet and de Barjac, 1981). However, thuringiensin is not a contact poison; it must be ingested to be effective

(Šebesta *et al.*, 1981). Thuringiensin was obtained from Dr. N. R. Dubois, U.S. Department of Agriculture–Forest Service, Hamden, Connecticut. It had been received from Abbott Laboratories, Chicago, Illinois. At a concentration of 10 mg per plate culture of DED fungus, thuringiensin was ineffective in inhibiting fungal growth. However, when thuringiensin was combined with tropolone, at 1 mg of each substance, no fungal growth occurred (Table I). Similar results were obtained in the case of puromycin and 4-deoxypyridoxine (Mazzone, 1985). Puromycin is an inhibitor of RNA synthesis (Yarmolinsky and de la Haba, 1959) and 4-deoxypyridoxine is a potent antimetabolite of vitamine B6 (Wooley, 1952). When used singly at a concentration of 10 mg per plate culture of DED fungus, neither compound had any effect on fungal growth. In binary combination of each compound with tropolone, at 1 mg of puromycin or 1 mg of 4-deoxypyridoxine plus 1 mg of tropolone, no growth of the fungus occurred.

These studies demonstrate the potential of antibiotics and other substances in controlling the DED fungus. In this connection, inhibition of the chitin pathway can be an achievable objective, provided that antichitin compounds are able to penetrate the fungus. The action of tropolone when used with such substances would appear to ensure this prerequisite.

The information presented here will be of value in ongoing experiments designed to break the resistance of the DED fungus to antibiotics. Even should such a breakthrough occur, the results of using antibiotics to retard tree diseases have not produced optimism. As a systemic disease, DED is treated systemically by injecting fungicides into infected trees. However, conditions within the tree may not ensure the functional integrity of substances such as antibiotics. Recently, major strides have been made biotechnologically in drug delivery techniques. Microencapsulation procedures (Lim, 1984) provide protection of drugs while allowing dispersal throughout the organism. Moreover, confined within biodegradable structures of varying durability, the drug(s) may be released over a period of time to guarantee that it remains for a longer time in the range where it is pharmacologically active. Application of microencapsulation to fungicides may lead to successful field trials in the case of Dutch elm disease.

III. EXPLOITING THE CHITINASE GENE

A number of biological insecticides have been approved for use or are being considered for approval by the U.S. Environmental Protection Agency (EPA) (Table II). Our laboratory has been granted an approval

TABLE II

Biological Insecticides and Their Current Status Regarding Approval by the U.S. Environmental Protection Agency

Microorganism	Status
Bacterial insecticide	
Bacillus thuringiensis	Approved
Viral insecticides	
Nucleopolyhedrosis viruses infecting	
Cotton bollworm (*Heliothis zea*)	Approved
Douglas fir tussock moth (*Orgyia pseudotsugata*)	Approved
Gypsy moth (*Lymantria dispar*)	Approved
European pine sawfly (*Neodiprion sertifer*)	Approved
Alfalfa looper (*Autographa californica*)	Pending
Granulosis virus infecting	
Codling moth (*Cydia pomonella*)	Pending

by the EPA for use of two baculoviruses: the nucleopolyhedrosis virus (NPV) of the gypsy moth and the NPV of the European pine sawfly. Comparing the performance of these viral insecticides in the field, the NPV of the European pine sawfly is more effective than the NPV of the gypsy moth. However, the European pine sawfly is far less a pest than the gypsy moth. The situation is further confounded in that the biological insecticides registered for use are not very effective in controlling pest insect populations, with the exception of the NPV of the European pine sawfly (Podgwaite, 1985). This situation has been a major topic of concern, and genetic engineering of biological insecticides to increase their efficacy has been proposed (Miller *et al.*, 1983).

It has been demonstrated that the genome of a baculovirus can be genetically altered to express interferon β (Smith *et al.*, 1983) or β-galactosidase (Pennock *et al.*, 1984) in virus-infected insect cells maintained *in vitro*. In these experiments the NPV of the alfalfa looper (*Autographa californica*) was used as the vehicle for incorporating the genes in question into its DNA. Maeda *et al.* (1985) have carried such experimentation further. The NPV of the silkworm (*Bombyx mori*) was genetically engineered to produce interferon α during viral infection of silkworm larvae. By simply pricking the proleg of infected larvae, it was possible to collect interferon α in the hemolymph. This experiment demonstrated for the first time that living insects could produce a foreign product through a baculovirus vector.

In increasing the virulence of biological insecticides, the chitinase gene is of particular interest. Several laboratories in the United States and Japan are pursuing experimentation designed to break down chitin produced by marine invertebrates. In Japan especially the control of chitin waste is an acute problem, and in both countries it is realized that a number of useful products may be obtained through partial hydrolysis of chitin. These topics were discussed recently at a U.S.–Japan seminar held at the University of Delaware, at Newark (Zikakis, 1984). An area of genetic interest in controlling chitin deposits was represented in reports by Horwitz *et al.* (1984) and Soto-Gil and Zyskind (1984). These investigators are undertaking research designed to increase the level of chitinase in bacteria through genetic engineering, by cloning the gene that expresses chitinase. I believe that such research is especially applicable to the control of pest invertebrates.

The work of Horwitz *et al.* (1984) is directed to the bacterium *Serratia marcescens,* which naturally produces significant levels of extracellular chitinase. This observation was made known through screening programs for chitinase-producing microorganisms, conducted by Monreal and Reese (1969) and by Carroad and Tom (1978). Monreal and Reese screened 70 fungi and 30 bacteria for chitinase production. *Serratia marcescens,* strain QMB 1466, had the highest activity on milled chitin and the second highest on swollen chitin. Carroad and Tom (1978) screened 300 microorganisms from soil, insect, and marine residues and obtained *Bacillus cereus* as the most chitinolytic organism.

Horwitz *et al.* (1984) compared *S. marcescens* QMB 1466, *B. cereus,* and other high chitinolytic organisms. Based on NAG released during a 5-hr hydrolysis of shrimp shell or swollen chitin by cell-free culture filtrates, *S. marcescens* QMB 1466 produced the highest levels of chitinase. Another important factor in selecting *S. marcescens* for genetic improvement of chitinase production is that it is closely related to *Escherichia coli.* This is believed to increase the chances that *S. marcescens* genes would be expressed in *E. coli.* Many of the advanced genetic manipulations and recombinant DNA techniques demonstrated with *E. coli* possibly could be adapted to *S. marcescens.*

Horwitz *et al.* (1984) constructed DNA libraries from *S. marcescens* in an *E. coli* plasmid, pBR322, and a λ phage vector, Charon 4. Most chitinolytic organisms have an enzyme system consisting of two separate hydrolases (Jeuniax, 1966; Muzzarelli, 1977). Chitin is first hydrolyzed by an endochitinase, poly-β-1,4-(2-acetamido-2-deoxy)-D-glucoside glycanohydrolase (EC 3.2.1.14), to low-molecular-weight soluble multimers of NAG, with chitobiose (the dimer *N,N′*-diacetylchitobiose) being predominant. A chitobiase [chitobiose acetylamidodeoxyglucohydrolase (EC 3.2.1.29, now EC 3.2.1.30) then hydrolyzes chitobiose to NAG.

The libraries were screened for endochitinase and chitobiase gene products. Preliminary results indicated that several recombinant phages had both endochitinase and chitobiase activity. Restriction analysis revealed that the endochitinase and chitobiase genes were linked, sharing 1.8- and 1.2-kb *Eco*RI/*Sal* fragments. Research is in progress by these investigators to reintroduce such genes into *S. marcescens* QMB 1466 with the hope that this bacterial strain will secrete high levels of the chitinase enzymes.

Fuchs *et al.* (1986) observed that *S. marcescens* strain QMB 1466 produced five proteins with chitinase activity. From cloning experiments it was determined that the most abundant chitinase, with a subunit mass of 57 kDa, is apparently encoded by a 9.5-kb *Eco*RI fragment. Stable, constitutive, high-level expression vectors are being constructed for production of the 57-kDa chitinase in root-colonizing pseudomonads to determine the efficacy of these strains as biological control agents.

It has been reported that all luminous marine organisms produce an extracellular chitinase and are able to utilize the monomer of chitin, NAG, as a carbon source (Hastings and Nealson, 1977). Soto-Gil and Zyskind (1984) chose to work on the chitinolytic marine bacterium *Vibrio harveyi*. A clone library of hybrid plasmids containing DNA from this organism was constructed in *E. coli*. From the clones obtained, two were found which exhibited chitinase and chitobiase activity when cell extracts were prepared. Restriction analysis of the plasmids demonstrated that both chitinase and chitobiase activities are expressed from an insert 5.3 kb in length. It is believed that the genes coding for these enzymes are linked in *V. harveyi* (as was the case for these enzymes in the study of Horwitz *et al.*, 1984). Further analysis suggested that a chi operon exists in *V. harveyi*, coding for these enzymes and possibly a permease for chitobiose. Continuing studies are aimed at determining the order and localization of the genes in the chi operon.

In our research on inducing chitinase production in bacteria, we have shown that strains of the bacterial insecticide *B. thuringiensis* may be induced for chitinase activity. Such induced strains of *B. thuringiensis*, a *dendrolinus* variety and a *kurstaki* variety, were inhibitory to the DED fungus (Mazzone *et al.*, 1981). In this study, chitin was used as the sole carbon source in the growth medium. The surviving progeny of the bacteria would then be expected to metabolize chitin as a required carbon source, the facility for such lytic activity requiring induction within the bacteria of chitinase and related enzymes. The endurance of "chitinase" in such bacteria is suspect. In theory, the lytic capability should last for relatively long periods unless a carbon source more accessible than chitin, perhaps a carbohydrate, is available to the bacteria. In this situation the bacteria are likely to revert to their original state.

Induction studies involving the use of chitin have also been reported in the case of *S. marcescens* (Monreal and Reese, 1969; Ohtakara *et al.*, 1979). Reid and Ogydziak (1981) mutagenized *S. marcescens* QMB 1466 colonies with ultraviolet light and the survivors with ethylmethane sulfonate. The resulting isolate produced a much larger zone of clearing than either the original strain or the UV-mutagenized isolate. After induction by chitin, endochitinase and chitobiase activities appeared at similar times for both the mutant and the parent strain, suggesting coordinate control of these enzymes. Soto-Gil and Zyskind (1984) found that chitobiose was a strong inducer of both chitinase and chitobiase activities in *V. harveyi*. These enzymatic activities were detectable within minutes after the addition of chitobiose.

However promising such induction studies appear, the outlook is to recombinant DNA technology as offering greater potential for improvement in chitinase production by microorganisms. I am interested in extending this potential to obtain permanent bacterial and viral control agents that have been genetically altered to contain within their genomes the chitinase gene. These biological agents, possessing the property of chitinolysis in addition to their standard infectivity, should be more effective in the field against pest insects and disease-causing fungi. I believe this research approach would be more reliable than, for example, the procedure of adding chitinase to insecticidal spray formulations (Smirnoff, 1974, 1977), in which case the enzyme's activity in the field might be diminished by environmental factors. For microbial insecticides already approved for use, the strategy of inserting the chitinase gene within their genomes should not present a significant increase in the cost required to gain reapproval by the EPA for these genetically altered microbes.

REFERENCES

Alan, C. R., and Hadwiger, L. A. (1979). *Exp. Mycol.* **3**, 285–287.

Bailey, J. A. (1982). *In* "Phytoalexins" (J. A. Bailey and J. W. Mansfield, eds.), p. 289. Wiley, New York.

Belleau, B., and Burba, J. (1961). *Biochim. Biophys. Acta* **54**, 195–196.

Brillinger, G. U. (1979). *Arch. Microbiol.* **121**, 71–74.

Carroad, P. A., and Tom, R. A. (1978). *J. Food Sci.* **43**, 1158–1161.

Coxon, D. T. (1982). *In* "Phytoalexins" (J. A. Bailey and J. W. Mansfield, eds.), pp. 106–132. Wiley, New York.

Dähn, V., Hagenmeier, H., Höhne, H., König, W. A., Wolf, G. A., and Zähner, H. (1976). *Arch. Microbiol.* **107**, 143–160.

Endo, A., Kakiki, K., and Misato, T. (1970). *J. Bacteriol.* **104,** 189–196.
Fuchs, R. L., McPherson, S. A., and Drahos, D. J. (1986). *Appl. Environ. Microbiol.* **51,** 504–509.
Gooday, G. W. (1977). *J. Gen. Microbiol.* **99,** 1–11.
Hadwiger, L. A., and Beckman, J. M. (1980). *Plant Physiol.* **66,** 205–211.
Hadwiger, L. A., Fristensky, B., and Riggleman, R. C. (1984). *In* "Chitin, Chitosan, and Related Enzymes" (J. P. Zikakis, ed.), pp. 291–302. Academic Press, New York.
Hastings, J. W., and Nealson, K. H. (1977). *Annu. Rev. Microbiol.* **31,** 549–595.
Höhne, H. (1974). Dissertation, University of Tübingen.
Horwitz, M., Reid, J., and Ogrydziak, D. (1984). *In* "Chitin, Chitosan, and Related Enzymes" (J. P. Zikakis, ed.), pp. 191–208. Academic Press, New York.
Isono, K., Nagatsu, J., Kobinata, K., Sasaki, K., and Suzuki, S. (1967). *Agric. Biol. Chem.* **31,** 190–199.
Jeuniax, C. (1966). *In* "Methods in Enzymology" (E. F. Neufeld and V. Ginsburg, eds), Vol. 8, pp. 645–650. Academic Press, New York.
Lecadet, M. M., and de Barjac, H. (1981). *In* "Pathogenesis of Invertebrate Microbial Diseases" (E. W. Davidson, ed.), pp. 293–321. Allanheld, Osm, Totowa, New Jersey.
Leighton, T., Marks, E. and Leighton, F. (1981). *Science* **213,** 905–907.
Lim, F. (1984). "Biomedical Applications of Microencapsulation." CRC Press, Boca Raton, Florida.
Lindberg, G. D. (1981). *Plant Dis.* **65,** 680–683.
Lindberg, G. D., Whaley, H. A., and Larkin, J. M. (1980). *J. Nat. Prod.* **43,** 592–594.
McConnell, E., and Richards, A. G. (1959). *Can. J. Microbiol.* **5,** 161–168.
Maeda, S., Kawai, T., Obinata, M., Fujiwara, H., Horiuchi, T., Saeki, Y., Sato, Y., and Furusawa, M. (1985). *Nature (London)* **315,** 592–594.
Marks, E. P., and Sowa, B. A. (1976). *In* "The lnsect Integument" (H. R. Hepburn, ed.), pp. 339–357. Am. Elsevier, New York.
Marx, J. L. (1977). *Science* **197,** 1170–1172.
Mazzone, H. M. (1985). *Dev. Ind. Microbiol.* **26,** 471–477.
Mazzone, H. M., Kluck, J., Dubois, N. R., and Zerillo, R. (1981). *In* "Proceedings of the Dutch Elm Disease Symposium and Workshop" (E. S. Kondo, Y. Hiratsuka, and W. B. G. Denyer, eds.), pp. 36–45. Manitoba Dep. Nat. Resour., Winnipeg, Manitoba, Canada.
Miller, L. K., Lingg, A. J., and Bulla, L. A., Jr. (1983). *Science* **219,** 715–721.
Misato, T., and Kakiki, T. (1977). *In* "Antifungal Compounds" (M. R. Siegel and H. D. Sisler, eds.), Vol. 2, pp. 277–300. Dekker, New York.
Monreal, J., and Reese, E. T. (1969). *Can. J. Microbiol.* **15,** 689–696.
Mothes-Wagner, U., and Seitz, K.-A. (1984). *J. Invertebr. Pathol.* **43,** 218–225.
Muetterties, E. L. Roesky, H., and Wright, C. M. (1966). *J. Am. Chem. Soc.* **88,** 4856–4861.
Muzzarelli, R. A. A. (1973). "Natural Chelating Polymers." Pergamon, Oxford.
Muzzarelli, R. A. A. (1977). "Chitin." Pergamon, Oxford.
Neville, A. C. (1975). "Biology of the Arthropod Cuticle." Springer-Verlag, Berlin 2nd, New York.
Ohtakara, A., Mitsutomi, M., and Uchida, Y. (1979). *J. Ferment. Technol.* **57,** 169–177.
Pennock, G. D., Shoemaker, C., and Miller, L. K. (1984). *Mol. Cell. Biol.* **4,** 399–406.
Podgwaite, J. D. (1985). *In* "Viral Insecticides for Biological Control" (K. Maramorosch and K. E. Sherman, eds.), pp. 775–797. Academic Press, New York.
Post, L. C., DeJong, B. J., and Vincent, W. R. (1974). *Pestic. Biochem. Physiol.* **4,** 473–483.
Reid, J. D., and Ogrydziak, D. M. (1981). *Appl. Environ. Microbiol.* **41,** 664–669.
Šebesta, K., Farkas, J., Horska, K., and Vankova, J. (1981). *In* "Microbial Control of Pests and Plant Diseases 1970–1980" (H. D. Burges, ed.), pp. 249–281. Academic Press, New York.

Smirnoff, W. A. (1974). *Can. Entomol.* **106,** 429–432.
Smirnoff, W. A. (1977). *Can. Entomol.* **109,** 351–358.
Smith, D. A. (1982). *In,* "Phytoalexins" (J. A. Bailey and J. W. Mansfield, eds.), pp. 218–252. Wiley, New York.
Smith, G. E., Summers, M. D., and Fraser, M. J. (1983). *Mol. Cell. Biol.* **3,** 2156–2165.
Soto-Gil, R. W., and Zyskind, J. W. (1984). *In,* "Chitin, Chitosan, and Related Enzymes" (J. P. Zikakis, ed.), pp. 209–223. Academic Press, New York.
Vance, C. P., Kirk, T. K., and Sherwood, R. T. (1980). *Annu. Rev. Phytopathol.* **18,** 259–288.
Vincent, J. F. V. (1978). *Nature (London)* **273,** 339–340.
Wooley, D. W. (1952). "A Study of Antimetabolites," p. 48. Wiley, New York.
Yarmolinsky, M. B., and de la Haba, G. L. (1959). *Proc. Natl. Acad. Sci. U.S.A.* **45,** 1721–1729.
Zakakis, J. P., ed. (1984). "Chitin, Chitosan, and Related Enzymes." Academic Press, New York.

28

Improving the Effectiveness of Insect Pathogens for Pest Control

I. HARPAZ*

I. PHOTOSTABILIZATION

One of the major obstacles to expansion of the use of commercial baculovirus preparations for the control of lepidopterous pests is the sensitivity of the virus to photodegradation (Jaques, 1972; Payne, 1982; Miltenburger and Krieg, 1984). The problem is particularly severe in view of the fact that the climatic conditions prevailing in the countries with the greatest potential market for these viral insecticides provide long hours of intense sunlight, including the ultraviolet component. The exact mechanism by which UV irradiation inactivates the virus is not fully understood, although Ignoffo *et al.* (1977) speculated that hydrogen peroxide, produced by the near-ultraviolet irradiation of one or more amino acids, reduces both the vitality and pathogenicity of baculoviruses as well as other microbiological control agents.

The measures of UV protection so far adopted, whether in the manufacture of these bioinsecticides (see, for instance, Martignoni and Iwai, 1985; Shapiro, 1985), or in the field application techniques, are still unsatisfactory as regards preservation of infectivity after treatment.

*Professor I. Harpaz died during the production of this volume. His outstanding personality and contributions to the science of insect pathology will be sorely missed by his colleagues.

The term "application technique" in this context refers to the development of special application machinery, based on a variety of technological principles, all aiming to ensure that an adequate amount of the applied material adheres to the lower surface of the leaves. Thus, the leaves themselves will provide the protective shade against UV degradation of the pathogen.

A novel approach to the stabilization of photolabile insecticides, including baculoviruses, is currently being investigated by our group at Rehovot, Israel. It is based on utilization of the unique surface properties of certain clays, such as montmorillonite, to build systems in which photoexcited molecules can be deactivated by selected chromophores in an energy transfer mechanism before photodegradation starts. Early reports on this method, indicating its initial promise, have already been published (Margulies *et al.*, 1985, 1987).

II. BAIT FORMULATIONS

Another, though more conventional, technique for providing UV protection against photoinactivation is incorporation of the inclusion bodies of the baculovirus in a bait formulation, thereby utilizing the mass of the bait particle as a UV interceptor. However, baited pathogens, such as other bait–poison formulations, are of limited use in the control of lepidopterous larvae in agricultural crops. The reason is the underdevelopment of olfactory senses in these larvae, which are devoid of proper antennae, as compared to adult Lepidoptera, which are capable of sensing a pheromone-emitting source hundreds of meters away. The maximum distances from which caterpillars are able to respond to an olfactory stimulant are usually no more than a few centimeters. Hence nearly all the baits designed to lure caterpillars are at the best no more than phagostimulants, i.e., feeding enhancers. The larva in these cases responds positively to the stimulant by means of its gustatory rather than olfactory sense.

Our search for caterpillar chemoattractants that are capable of improving the performance of baits in the field has lately yielded some encouraging, though still preliminary, results. We are at present testing a substance which, besides being a powerful phagostimulant [superior to Coax, for instance (Sneh *et al.*, 1983)], also seems to attract noctuid larvae, though only for short distances. Notwithstanding these limitations, should this substance prove effective in the field, then a much more efficacious pathogen–bait formulation can be perfected. Such improved bait is also likely to reduce the cost of treatment considerably, since a much smaller amount of bait will be required to treat a unit area.

III. EXPANSION OF HOST RANGE

Another area in which the potential of baculovirus insecticides can be achieved is in expanding their host range. As a rule, baculoviruses are considered to be restricted in their host relations to one species of insect, or one genus at the most. The nucleopolyhedrosis virus of the alfalfa looper, *Autographa californica* Speyer, has attracted unusually wide attention because of its ability to cross-infect insect species outside the family Noctuidae, but still within the limits of the order Lepidoptera.

In the eyes of ecologists and environmentalists this narrow host range, or selectivity, of baculoviruses is a great advantage over chemical pesticides, which are notoriously nonselective and are thus likely to disrupt the environment by destroying many forms of life apart from the target pest species. On the other hand, industry would for obvious reasons prefer a microbial or viral insecticide with a much wider spectrum of killing than that of the monophagous or oligophagous baculoviruses. We are therefore faced here with an inherent, almost irreconcilable, dilemma.

Genetic engineering and biotechnology have opened up new prospects for the development of strains of baculoviruses with wider host ranges, plus other advantages for the purpose of pest control. Yet an old-fashioned screening of long-recognized strains of baculoviruses for possible "encroachments" on family or even order frontiers in their host relations may sometimes prove to be highly rewarding. Inspired by the spectacular discovery of *Bacillus thuringiensis israelensis* (BTI) in our laboratory in 1976 by Goldberg and Margalit (1977), we sometimes venture to bank on similar luck. In an exercise of this kind, we came across a baculovirus whose recognized host range has hitherto been restricted to one genus in the noctuid family. Much to our astonishment, we noticed that this virus could provoke a lethal viral disease in various species of locusts and grasshoppers. The latter belong to the hemimetabolous order of Orthoptera, which is part of an entirely different subclass of insects than the holometabolous Lepidoptera. Quite a leap for a genus-specific virus!

In fact, one should not be oversurprised by this skip of the virus over a couple of insect orders while still remaining within the class Insecta. It is now a well-recognized phenomenon, demonstrated for the first time by Fukushi (1935) in Japan with regard to the rice dwarf virus, of viruses that multiply both in a plant host and in their insect vector. The evolutionary distance between a rice plant and a leafhopper is obviously much greater than that between a caterpillar and a grasshopper. Regarding the latter cross-transmission a question is inevitably asked: Is it the same virus that was cross-transmitted from a moth to a locust, or did it merely activate a latent othopterous virus that had been present in the locust beforehand? An ELISA test showed that the baculovirus that spread systemically in the locust and killed it, sometimes within 4 days, was homologous to the virus extracted from the infected caterpillar. A dot blot hybridization test also confirmed the identity of the virus in both hosts. A reciprocal transmission of the virus from the locust back to the caterpillar produced the same syndrome and the same causal agent (Ben Simon *et al.*, in preparation).

Extra caution must be exercised while experimenting in these situations. It is commonly believed by entomologists that the peritrophic membrane in chewing insects separates the food bolus from the epithelial lining of the intestine, and the membrane is shed through the anus with the feces contained therein. Hence, soon after defecation, and in particular after a molt, the midgut of such insects should be virtually free from any food remnants. However, notwithstanding this common notion there is still a probability of traces of the perorally administered inoculum remaining in the alimentary tract of the grasshopper, particularly in the enteric ceca (blind sacs) of the midgut. These traces may persist in the gut even after molts and not be totally excreted with the peritrophic membrane (cf. Orihel, 1975; Adang and Spence, 1983). Thus, macerating entire grasshoppers for virus extraction and identification is likely to lead to mistaken etiological conclusions. A much safer procedure for this purpose is to look for the virus in carefully collected hemolymph or in cautiously excised fat body tissue of the infected grasshopper. Alternatively, intrahemocoelic administration of the viral inoculum, when possible, can obviate the difficulty discussed above. In fact, replication of the baculovirus in question has been positively demonstrated both in the hemolymph and in the fat body of grasshoppers that were inoculated either perorally or by intrahemocoelic injection.

The implications of this finding that lepidopterous baculoviruses are able to infect orthopterans are quite far-reaching as regards the two diametrically opposed requirements discussed above, namely (a) the need to widen the host range of the viruses in order to increase their prospects of commercialization and (b) the fear that they may at the same time become a threat to nontarget insects, in particular the beneficial ones like predators and parasites, which play an indispensible role in the natural control of both actual and potential pest species.

REFERENCES

Adang, M. J., and Spence, K. D. (1983). Permeability of the peritrophic membrane of the Douglas fir tussock moth (*Orgyia pseudotsugata*). *Comp. Biochem. Physiol. A* **75A,** 233–238.

Fukushi, T. (1935). Multiplication of virus in its insect vector. *Proc. Imp. Acad. Jpn.* **11,** 301–303.

Goldberg, L. J., and Margalit, J. (1977). A bacterial spore demonstrating rapid larvicidal activity against *Anopheles sergentii, Uranotaenia unguiculata, Culex univittatus, Aedes aegypti* and *Culex pipiens. Mosq. News* **37,** 355–358.

Ignoffo, C. M., Hostetter, D. L., Sikorowski, P. P., Sutter, G., and Brooks, W. M. (1977). Inactivation of representative species of entomopathogenic viruses, a bacterium, fungus and protozoan by an ultraviolet light source. *Environ. Entomol.* **6,** 411–415.

Jaques, R. P. (1972). The inactivation of foliar deposits of viruses of *Trichoplusia ni* and *Pieris rapae* and tests on protectant activities. *Can. Entomol.* **104,** 1985–1994.

Margulies, L., Rozen, H., and Cohen, E. (1985). Energy transfer at the surface of clays and protection of pesticides from photodegradation. *Nature (London)* **315,** 658–659.

Margulies, L., Cohen, E., and Rozen, H. (1987). Photostabilization of bioresmethrin by organic cations on a clay surface. *Pestic. Sci.* **18,** 79–87.

Martignoni, M. E., and Iwai, P. J. (1985). Laboratory evaluation of new ultraviolet absorbers for protection of Douglas-fir tussock moth (Lepidoptera: Lymantriidae) baculovirus. *J. Econ. Entomol.* **78,** 982–987.

Miltenburger, H. G., and Krieg, A. (1984). Bioinsecticides. II. Baculoviridae. *Adv. Biotechnol. Processes* **3,** 291–313.

Orihel, T. C. (1975). The peritrophic membrane: Its role as a barrier to infection of the arthropod host. *In* "Invertebrate Immunity: Mechanisms of Ivertebrate Vector–Parasite Relations" (K. Maramorosch and R. E. Shope, eds), pp. 65–73. Academic Press, New York.

Payne, C. C. (1982). Insect viruses as control agents. *Parasitology* **84,** 35–77.

Shapiro, M. (1985). Effectiveness of B vitamins as UV screens for the gypsy moth (Lepidoptera: Lymantriidae) nucleopolyhedrosis virus. *Environ. Entomol.* **14,** 705–708.

Sneh, B., Schuster, S., and Gross, S. (1983). Improvement of the insecticidal activity of *Bacillus thuringiensis* var. *entomocidus* on larvae of *Spodoptera littoralis* (Lepidoptera, Noctuidae) by addition of chitinolytic bacteria, a phagostimulant and a UV protectant. *Z. Angew. Entomol.* **96,** 77–83.

29

Entomogenous Nematodes and Their Prospects for Genetic Improvement

RANDY GAUGLER

I. INTRODUCTION

Problems associated with extensive chemical insecticide use have resulted in dramatically increased research efforts into alternative means of insect control. Biological control, because it offers the prospect of safe and effective reduction of pest populations with minimal concern for pest resistance, has been regarded by many as the alternative method of choice. The insect pathology component of biological control has surged strongly into the spotlight on the promise of a second generation of pathogens genetically modified by recombinant DNA technology.

Entomogenous nematodes are widely regarded as one of the major subdivisions of invertebrate pathology. Recognizing nematodes as met-

BIOTECHNOLOGY IN INVERTEBRATE
PATHOLOGY AND CELL CULTURE

azoans rather than microbes, recommendations have occasionally been made to accord nematodes status separate from pathogens and comparable in rank to parasitoids and predators (Nickle, 1974; Coppel and Mertins, 1977). These proposals have received little support, possibly because most insect nematologists identify themselves as pathologists by training and interest. Moreover, as we shall see in discussing entomogenous nematodes associated with microbes, the line between nematology and microbiology is not always distinct.

Despite being somewhat a "poor relation" to the rest of invertebrate pathology, many entomogenous nematodes possess the same favorable attributes for biological control as microbial agents, including safety, high virulence, culturability, high reproductive capacity, and life cycles well synchronized with their hosts. They share the same limitations as well, particularly sensitivity to environmental extremes and, in most cases, a limited commercial market because of their specificity. They also offer a few unique advantages over microbials, notably possession of chemoreceptors and mobility, enabling them to find hosts, mates, protected habitats, and other scattered resources by directed orientation rather than chance. Particularly advantageous, as insect nematologists have been quick to point out to government regulating agencies, as metazoans, nematodes are exempt from pesticide registration requirements.

Research into the bioinsecticidal potential of some entomogenous species has accelerated sharply in recent years (Gaugler and Kaya, 1983). This increased effort reflects major advances in technologies involving mass production of infective stages, advances followed by the formation of companies to produce and market some species for pest control. These firms are making strides in nematode standardization, culture, transport, and storage but, unlike many companies developing microbial agents, are not conducting research into biotechnology, at least if we take the narrow definition of biotechnology as being recombinant DNA technology. The reason is obvious: if lack of basic knowledge of the molecular biology and genetics of entomopathogenic protozoans and fungi has restricted their biotechnological development, it has virtually precluded that of entomogenous nematodes.

Although entomogenous nematodes have not shared in the current fervor for biotechnology, there are nonetheless prospects for their genetic improvement by genetic engineering as well as more conventional methods. A prerequisite to the implementation of genetic solutions, however, is a determination of what characters may be limiting control potential and what attributes are desirable. This chapter will focus on the current status of those nematodes accorded the greatest promise for biological control with emphasis on identifying targets of opportunity for genetic improvement.

II. NEMATODE FAMILIES OF BIOCONTROL IMPORTANCE

Insect–nematode interactions are usually benign or mildly debilitative to the host. Only species in 9 of the 27 families of entomogenous nematodes are considered to offer potential for biological control (Poinar, 1975; Petersen, 1982). These groups, exclusive of Rhabditidae, which possesses few legitimate attributes for biocontrol, are discussed here in ascending order of their perceived potential for biocontrol and genetic improvement.

A. Diplogasteridae

This large and diverse family comprised mostly free-living microbotrophs inhabiting soil or decomposing plant and animal remains. Associations with insects occupying the same habitat are common and almost always phoretic. It is a short step from phoresis to parasitism, however, and there is a clear evolutionary progression among the diplogasterids from external phoresis to internal phoresis to parasitism resulting in host mortality.

Species causing lethal host infections have similar life cycles. The oily dauer juveniles enter the host alimentary tract passively by ingestion or actively via the anus, where they feed on the normal gut flora, begin development to the adult stage, and establish reproducing colonies. When infections are heavy the nematodes eventually rupture the intestinal wall to enter the hemocoel, introducing coliform bacteria that kill the host by septicemia. Sandner *et al.* (1972) suggested that bacteria carried into the hemocoel share a symbiotic relationship with the nematodes, but Poinar (1979) argued persuasively that these bacteria were merely surface contaminants from the nematode cuticle.

Although some species cause lethal infections, the pathogenic nature of diplogasterids remains questionable because these nematodes are unable to invade the tissues of healthy insects (Poinar, 1969). Hemocoelic penetration is believed to be the result of gut weakening and pressure brought about by huge numbers of developing nematodes established in the gut. Thus, hemocoelic invasion appears to be inadvertent, a passive rather than active process. Once the gut barrier is breached it is the bacteria that cause death. Pathogenic effects possibly contributed by invading nematodes are obscured by the fast lethal action characteristic of bacterial septicemia. Poinar (1969) concludes that diplogasterid species causing host death should be considered as potential pathogens.

Despite the dubious pathogenicity of diplogasterids, Polish researchers report that at least one species, *Pristionchus uniformis*, should be accorded consideration for biological control. This nematode is widely distributed in Poland, where infection rates up to 83% have been reported from populations of the Colorado potato beetle, *Leptinotarsa decemlineata* (Fedorko, 1971). Preliminary tests suggest that this pest might be controlled when high nematode dosages are applied to soil. Further studies have not been reported.

As facultative parasites, diplogasterids retain the ability to develop and reproduce under free-living conditions. This critical distinction permits their culture xenically on artificial media in numbers enabling field manipulations. Their remarkably short generation time—4 days for *P. uniformis* (Fedorko and Stanuszek, 1971—greatly enhances mass rearing efforts. The infective juveniles produced are free-living resistant stages which may be stored in the laboratory for prolonged periods and show good survival in the soil. Host range has not been well studied, but the nonspecific manner in which they cause disease suggests that a wide spectrum of insect pests might be susceptible. There is, however, a serious lack of basic information on almost every aspect of parasite bioecology. This limitation, coupled with lingering questions regarding pathogenicity, has tended to deter investigations.

B. Allantonematidae

The family Allantonematidae contains 17 genera of insect parasites, attacking beetles, flies, fleas, bugs, thrips, ants, and even mites. Despite their diversity, only *Heterotylenchus autumnalis* has aroused significant interest among biological control workers. This nematode is reported to cause natural and experimental infections in several species of dung-breeding flies (Poinar, 1979), but its control potential appears limited to the face fly, *Musca autumnalis*.

The life cycle of *H. autumnalis* involves a complex alternation of gamogenetic and parthenogenetic generations. The free-living gamogenetic stages live and mate on cow dung, where the fertilized female penetrates host larvae and oviposits in the hemocoel. Hatching juveniles mature into parthenogenetic females, which in turn produce a large gamogenetic generation of small males and females that heavily parasitize the host ovaries and oviducts, causing parasitic castration. Infected females release the nematodes back into dung during a mock oviposition. Males are "dead-end" hosts since nematodes are unable to escape to reseed the dung and complete their life cycle.

The nematode is presumed to have been introduced into eastern North America from Europe together with the face fly. The host initially dispersed westward at a greater rate than its parasite, but inoculative releases were made in California and Montana to assist parasite dispersal (Nickle and Welch, 1984). Because allantonematids are obligate endoparasites unculturable on artificial media, these releases were accomplished through the use of nematode-infected face fly pupae. Both host and parasite are now widely distributed across the continent, and parasitism rates of 22% have been reported in Nebraska (Jones and Perdue, 1967), 30–40% in Missouri (Thomas and Puttler, 1970; Thomas *et al.*, 1972), 6–40% in California (Kaya and Moon, 1978), 4–37% in Mississippi (Robinson and Combs, 1976), and 7–10% in Iowa (Krafsur *et al.*, 1983).

These estimates of parasitism indicate that *H. autumnalis* exerts a substantial influence on natural face fly populations, as has been suggested by Nickle (1972, 1974) and Thomas *et al.* (1972). Treece and Miller (1968), however, found that host castration usually occurs after the first oviposition. In a calculation of the impact of parasitism, Krafsur *et al.* (1983) found that losses in the annual mean net reproductive rate of infected flies were small and concluded that losses would be greater only if more hosts were castrated before the first ovarial cycle. Even in California, where the prevalence of the nematode in host populations is higher and castration more likely to occur prior to the first oviposition, Kaya and Moon (1978) thought that the degree of natural biological control being exerted was probably insufficient to regulate host populations.

The lack of inundative or augmentative releases with *H. autumnalis* has not been a function of the argument as to whether this parasite promotes stability of host populations. Practical rather than theoretical considerations dictate that parasites (1) culturable by expensive *in vivo* means only, (2) having narrow host ranges, (3) attacking pests of modest economic importance, and (4) causing chronic instead of lethal infections will seldom reach the field trial stage of evaluation.

Howardula husseyi, an allantonematid parasite of the mushroom phorid *Megaselia halterata*, is strongly density-dependent but otherwise shares most of the same biocontrol limitations as *H. autumnalis:* it must be cultured *in vivo*, the single host species parasitized rarely induces direct yield loss (Rinker and Snetsinger, 1984), the principal effect is sterility or reduced fecundity, and it stores poorly. Its potential for phorid biological control has yet to be evaluated, although Richardson and Chanter (1979) proposed an augmentative strategy based on liberating parasitized female hosts into greenhouses to naturally release infective stages.

The allantonematids also include several bark beetle parasites of possible biocontrol interest, most notably in the genera *Contortylenchus, Neoparasitylenchus,* and *Parasitylenchus,* where infection causes sublethal debilitating effects. Their biocontrol prospects have been severely hampered by the impracticality of even laboratory maintenance, and Kaya (1984) concludes that despite Massey's (1974) optimism these nematodes offer little promise.

C. Sphaerulariidae

Sphaerulariid nematodes also rarely cause host mortality; infection is more commonly expressed in host sterility, reduced fecundity, delayed development, and/or altered behavior. Parasite-induced behavioral changes can be striking: bumblebee queens infected with *Sphaerularia bombi* are known as "the eternal seekers" because they fly continuously near the ground without ever constructing a nest (Poinar and van der Laan, 1972). Sphaerulariid parasitism in bark beetles distorts egg gallery shape and reduces gallery size (Kaya, 1984). The sphaerulariids are of considerable academic interest but as a group are not important regulators of pest density and appear to hold little potential for field application. Only *Tripius sciariae,* a parasite of sciarid flies, has garnered attention from biocontrol workers.

Tripius sciariae is an unusual sphaerulariid because infection is frequently fatal in host larvae. Infections are initiated by fertilized adult females ensheathed in the fourth-stage juvenile cuticle. The nematode's spearlike stylet and enzymatic action are used to gain entry into the larval hemocoel (Poinar and Doncaster, 1965). The exsheathed cuticle is left attached to the host, where it plugs the entry hole, providing protection from secondary invaders and fluid loss. The nematode uterus is then partially everted through the vulvar opening, where the uterine cells come into direct contact with the host hemolymph, enlarge, and begin to absorb nutrients. Eggs are deposited in the hemocoel, where they hatch and begin development to fourth-stage juveniles, which exit the larval host via the digestive tract. Juveniles carried into adult hosts are discharged back into the host environment through the ovipositor during mock oviposition.

Doubtlessly aided by the "nemapositing" habits of infected adult hosts, *T. sciariae* is widely distributed (Poinar, 1979), although the impact of this highly specific parasite on natural sciarid fly populations remains undetermined. There has been a single report of evaluation for biocontrol: Poinar (1965) inoculated several sciarid-infested greenhouse flats with infective-stage females and observed a steep drop in fly populations within 4 weeks, with 100% parasitism of the few adults recovered. He concluded that this nematode would be an ideal biocontrol agent for sciarids in greenhouses and mushroom houses (Poinar, 1979). Certainly *T. sciariae* benefits from an impressive list of attributes, particularly its apparent high effectiveness, excellent dispersal capacity, and ease of culture in *Bradysia paupera*. These advantages, however, are offset by the lack of a resistant dauer stage: infective-stage females survive a mere 2 weeks in soil and even shorter periods in water or agar (Poinar, 1965). Despite the advantages of this nematode, its short shelf-life seriously hinders development and may account for the lack of further study and evaluation over the past two decades.

D. Tetradonomatidae

The most extensively studied member of the six genera and seven species in the Tetradonomatidae is *Tetradonema plicans*. Like several other promising entomogenous nematodes, *T. plicans* is parasitic on greenhouse sciarid flies. Transmission is poorly understood but probably results from host ingestion of eggs or infective juveniles. Penetration of the larval host is followed by maturation to adult stages within a week and mating in the host body cavity. Gravid females exit the dead or dying larval host to deposit eggs in the host environment; although in long-lived hosts, eggs are sometimes released into the hemocoel, where a second generation of nematodes develops. Hosts infected as late instar larvae sometimes survive to the adult stage but are usually rendered sterile even from single female infections.

Originally isolated from *Sciara* (=*Bradysia*) *coprophila* by Cobb in 1919, the biological control potential of *T. plicans* remained unevaluated for more than 50 years, until Hudson (1972, 1974) initiated greenhouse studies against *B. paupera*. She achieved 100% larval parasitism in tubes containing 1 g of compost at a parasite : host ratio of 80 : 1. Reducing the ratio to 40 : 1 and increasing compost to 8 g decreased parasitism to 45%, but this rate remained constant even in 80-g pots of compost. Most encouraging, sciarid parasitism in pots at a 14 : 1 parasite–host ratio increased from 3–12% in the first host generation to 85–100% in the second.

Hudson (1974) provides an impressive list of attributes for *T. plicans*, citing the nematode's high pathogenicity, specificity to sciarid flies, high reproductive capacity, ease of *in vivo* rearing, storage of eggs for up to 1 year (although viability drops to 30% within 6 months), strong density dependence, and synchrony with the host life cycle. Regrettably, Hudson's positive but preliminary studies, completed more than 10 years ago, have not been pursued, so that the potential of this nematode for biocontrol remains unresolved.

E. Neotylenchidae

Most neotylenchids are free-living or plant parasitic. Two genera, *Deladenus* and *Fergusobia*, have entomogenous members although only *Deladenus* is of biocontrol significance. There are seven known entomogenous *Deladenus* species, primarily attacking siricid woodwasps; none have been evaluated so extensively as *D. siricidicola*, a parasite of *S. noctilio*. This insect pest was accidently introduced into Australia from New Zealand in 1951, with devastating consequences for coniferous forests. The subsequent discovery of *D. siricidicola* in New Zealand, where it is believed to have caused the collapse of *Sirex* populations (Zondag, 1962; from Bedding, 1984a), created a rare opportunity for classical biological control in entomogenous nematology.

The life history of *Deladenus* is remarkable, involving two separate life cycles, one parasitic in *Sirex* and the other a free-living mycetophagous cycle in the host gallery (Bedding, 1967, 1968, 1972, 1984a).

The parasitic cycle is initiated by allantonematid-like fertilized females, which use large tubular stylets to penetrate into the larval hemocoel. Parasitic development proceeds with enormous growth, but reproduction is inhibited until host pupation occurs, when juveniles are released ovoviviparously into the hemocoel and begin migration to the reproductive system, suppressing ovarian development and entering eggs. When infected females deposit their "eggs" the juveniles are transmitted to new trees. Male hosts are not sterilized because sperm are passed from the testes and into the vesiculae seminales prior to testicular invasion.

The parasitized adult oviposits nematode-filled eggs into fresh wood together with spores of the basidiomycete *Amylostereum areolatum*, a fungus symbiotically associated with siricid woodwasps. The fungus ordinarily grows throughout the tree, providing food for the woodwasp larvae. Ironically, it is now the juvenile nematodes which exit from the siricid eggs to feed on the developing fungus, maturing into neotylenchidlike adults reproducing oviparously. In contrast to the heavy stylet used by infective females to penetrate the thick host cuticle, mycetophagous stages have small syringelike stylets adapted for piercing fungal hyphae. The free-living phase ends only when mycetophagous juveniles develop near siricid larvae, triggering the formation of adult males and infective-stage females and the renewal of the parasitic cycle. The two cycles are tied together by the specificity of the mycetophagous nematode to the symbiotic fungus of the insect host (Bedding, 1972, 1984a).

The mycetophagous cycle can be maintained indefinitely without the intervention of a parasitic cycle (Bedding, 1984a) and is used for mass production purposes since the long life cycle of *Sirex* (1–3 years) virtually precludes *in vivo* mass culture of the parasitic stages. Bedding and Akhurst (1974) adapted Evans's (1970) method of rearing mycetophagous nematodes for *Deladenus* by eliminating contamination problems and increasing the wheat-to-water ratio, obtaining yields of 3–10 million nematodes per 500 ml flask. *Deladenus* nematodes lack a resistant stage, losing viability rapidly after 8 weeks of storage, so they must be utilized soon after production.

The nematode was released into *Sirex* populations by injecting inoculum into *Sirex*-infested trees, sometimes by allowing caged infected *Sirex* females to "nemaposit" directly into trees (Bedding and Akhurst, 1974; Bedding, 1984a). Establishment was most effective when trees or logs were inoculated at 1-m intervals with a wad hammer delivering a gelatin foam-based nematode inoculum. Field releases of *D. siridicola* for *Sirex* control were dramatically successful. Nematodes released from a single point in a Tasmanian forest in 1970 quickly became established and ended further tree mortality due to woodwasps (Bedding and Akhurst, 1974; Bedding, 1984a). Releases in a heavily *Sirex*-infested forest in 1972 resulted in a reduction from several thousand *Sirex*-killed trees annually to only 200 after two years, 5 the subsequent year, and none thereafter, with greater than 90% parasitism, and dispersal in other forests up to 13 km away (Bedding, 1984a). These spectacular results sparked a major program of parasite introductions, so that *D. siridicola* is now widely established in *S. noctilio*-infested areas of Australia.

Bedding's success in manipulating *D. siridicola* for the biological control of *Sirex* must be considered as the foremost accomplishment of applied insect nematology to date. *Deladenus* possesses a number of desirable biocontrol attributes, including high infectivity and high transmission efficiency (Kaya, 1986) and ease of mass production, but the key to its effectiveness is clearly the presence of a mycetophagous cycle capable of switching to the parasitic phase when insect hosts are present. This extraordinary biphasic life history permits mycetophagous stages not only to maintain themselves when insect hosts are unavailable but also to increase in number. Since parasitism does not reduce flight capacity (Bedding, 1984a) the nematodes are able to use the nemapositing habits of infected females for efficient dispersal to new host populations. The most formidable potential limitation, lack of a resistant stage, was resolved by introducing the nematodes shortly after production into their natural reservoir—the tree.

F. Mermithidae

This family of highly evolved nematodes contains no free-living, phoretic, or facultative members. Mermithids almost invariably cause lethal infections and several have been credited with causing massive epizootics in natural populations of insect pests (Phelps and DeFoliart, 1964; Welch and Rubtzov, 1965; Mongkolkiti and Hosford, 1971), points that have aroused strong interest in their biocontrol potential.

There are at least six distinct mermithid life cycles, but most species follow a similar plan with minor variations. Infections are initiated when the newly hatched preparasitic stage pierces the host body wall and enters the hemocoel, where nutrients are absorbed directly through the nematode cuticle. Mermithids generally lie free in the hemocoel, although a few species initially enter the brain to avoid the host defense response, moving back into the hemocoel only after pupation (Gaugler *et al.*, 1984). The stylet and anus of the preparasite disappear during parasitic development, while the intestine enlarges to form a food storage organ—the trophosome. During this period the parasite may increase in length from 1 to 16 mm (e.g., *Romanomermis culicivorax*) or even to 16 cm (e.g., *Mermis nigrescens*). When development is complete (a week to several months, depending on the species) the fully grown nematode emerges to renew a free-living existence. Host death rapidly ensues, presumably as a result of hemocoelic fluid loss and microbe entry through the large hole made by the exiting nematode. The postparasitic stage enters the soil, molts twice to the adult stage, mates, and oviposits to complete the life cycle.

The best known mermithid and among the most extensively studied of entomogenous nematodes is *R. culicivorax*. This nematode has a broad host range compared to most members of its highly specialized family, naturally infecting 17 mosquito species from six genera with more than 60 species reported to serve as hosts under laboratory conditions (Petersen and Chapman, 1979). Its life cycle closely follows the general mermithid plan and may be completed in 4–6 weeks. The attention this species has drawn is a function of its attributes for biological control: close adaptation to the host life cycle, host lethality, ease of *in vivo* mass production, high reproductive potential, and parasitism of a major insect pest.

The most significant factor in the development of *R. culicivorax* was the success of Petersen and Willis (1972a) in devising a means of mass culturing the parasite *in vivo* in quantities sufficient for field evaluations. Petersen and Willis' (1972b, 1974, 1975, 1976) subsequent series of field trials demonstrated that high levels of parasitism, sometimes 100%, could be achieved following inundative releases of preparasites. These tests culminated in an ambitious 7-week effort to control *Anopheles albimanus* in an El Salvador lake (Petersen *et al.*, 1978b); treatment of 13.2 ha during 11 releases resulted in a 94% reduction in larval density. Inoculative releases in California rice fields using postparasite stages have shown that a single early season introduction can result in increased nematode density and provide partial control for the entire season (Kerwin and Washino, 1983). Extensive field testing over the past decade and a half has shown that *R. culicivorax* can be effective in reducing mosquito populations (Petersen, 1982, 1984) but that the level of control is often variable and rarely, if ever, predictable.

Variability in field efficacy may be related to biological limitations of *R. culicivorax*. Parasitism is inhibited by low temperature (<15°C), mild salinity (0.04 *M* NaCl), and low oxygen tension (i.e., polluted water), restricting its practical usefulness to permanent and semipermanent freshwater pools with temperatures ranging from 15° to 35°C and electrical conductivity less than 1500 μS (Platzer, 1981; Petersen, 1984). Because it emerges from immature hosts, the nematode lacks an effective means of dispersal: Petersen *et al.* (1968) found *R. culicovorax* in only five of hundreds of mosquito pools sampled. Host susceptibility to infection declines sharply with increasing age (Petersen and Willis, 1970).

Enthusiasm for *R. culicivorax* is further tempered by problems inherent to the economics of *in vivo* production. Petersen's El Salvador trial used 14 million nematodes which required 600 man-hr to produce over a 6-week period (Petersen *et al.*, 1978s). Production economics coupled with difficulties in handling, storage, and shipment of mermithids on a commercial scale led Fairfax Biological Laboratories to abandon attempts to market *R. culicivorax* during the late 1970s (Petersen, 1982, 1984). However, most damaging to the prospects of this mermithid for inundative control was the development of *Bacillus thuringiensis* var. *israelensis* (Lacey and Undeen, 1986), which quickly displaced *R.*

culicivorax as the only effective biocontrol agent available against mosquitoes.

Although unable to compete with *B. thuringiensis* var. *israelensis* as an inundative agent, the development of inoculative release strategies for *R. culicivorax* remained an alternative based on persistent reports of mermithid establishment and recycling (Petersen and Willis, 1975; Brown-Westerdahl *et al.*, 1979; Walker *et al.*, 1985). The viability of this approach was delivered an abrupt blow with the recent publication of Hominick and Tingley's (1984) treatise on the vector control potential of mermithids. Focusing on *R. culicivorax*, these authors critically evaluated the interactions between mermithids and host populations and concluded that mermithids are unlikely to be useful for inoculative releases because their populations are "controlled by such tight density-dependent constraints that they can cause at most only moderate depressions in their host populations." Their model of mermithid–mosquito interactions predicts that a 90% reduction of the host population, regarded as minimal for vector control, would require an unrealistically high postparasite : host release ratio of nearly 13 : 1.

Blackfly mermithids suffer many of the same biocontrol limitations as mermithids parasitizing mosquitoes, not the least of which has been the development of *B. thuringiensis* var. *israelensis* for blackfly control (Gaugler and Finney, 1982). Moreover, the absence of procedures for mass culture have restricted studies. *In vivo* culture may now be within reach because of advances in colonizing blackflies (Simmons and Edman, 1981; Brenner *et al.*, 1981), although inherent difficulties of mermithid diapause, asynchronous egg hatch, and long life cycles will require attention. Mermithids have been credited with the eradication of blackflies from some streams (Phelps and DeFoliart, 1964; Welch and Rubtzov, 1965), but the theoretical assessment of mermithid population dynamics by Hominick and Tingley (1984) indicates that such epizootics are too localized and infrequent to have long-term effects on host populations.

Biological control manipulations with mermithids parasitizing terrestrial hosts have received scant attention, although many species (i.e., *M. nigrescens*) are regarded as important natural regulators of insect populations. Difficulties in rearing these long-lived nematodes have limited control efforts to a single species: *Filipjevimermis leipsandra*, a parasite of the banded cucumber beetle, *Diabrotica balteata*. Creighton and Fassuliotis (1981, 1982) devised methods for *in vivo* culture yielding 5 million eggs/week, or sufficient inoculum for conducting "micro field plot" (1.9-liter cans buried in soil) tests where 78% larval parasitism resulted (Creighton and Fassuliotis, 1983). Prevented from conducting larger trials by the labor and expense of *in vivo* culture, these researchers shifted emphasis into *in vitro* techniques and succeeded in rearing *F. leipsandra* through its entire life cycle using Schneider's *Drosophila* medium supplemented with fetal calf serum followed by preadult transfer to a solid substrate (Fassuliotis and Creighton, 1982). The first success in mermithid *in vitro* culture, this breakthrough nevertheless has not resulted in the anticipated increase in rearing efficiency. The *F. leipsandra* project has consequently been discontinued, and research emphasis for *Diabrotica* biocontrol now centers on steinernematid and heterorhabditid nematodes (C. S. Creighton, personal communication).

G. Steinernematidae/Heterorhabditidae

These two families of facultatively parasitic rhabditoids have generally similar life histories, pathologies, and biological control potentials and will be considered together. They are characterized by their association with bacteria in the genus *Xenorhabdus*.

The infective-stage juvenile carries its bacterial symbiont monoxenically in a specialized vesicle of the foregut. Once a suitable host has been found, and the hemocoel breached, the bacteria are released into the hemolymph, where they proliferate to kill the host by septicemia within 24–48 hr. The nematodes feed on the bacteria and liquefying host tissues, produce two or three generations, and emerge from the depleted host cadaver as infective-stage juveniles in search of new hosts. The life cycle is completed in 7–10 days. Steinernematid nematodes penetrate the host through natural body openings, while heterorhabditids are additionally capable of penetrating directly through the insect cuticle (Bedding and Molyneux, 1982). Heterorhabditid infective juveniles develop into hermaphroditic females, permitting host colonization even when only one juvenile finds the host (Khan *et al.*, 1976); subsequent generations are bisexual. Steinernematids are dioecious, so reproduction is dependent on host invasion by infectives of each sex.

It would be easy to consider the nematode as little more than a biological syringe for the bacterium, but the relationship between these two organisms is one of classical mutualism. The bacterium requires the nematode for protection from the environment, penetration into the host hemocoel, and inhibition of the host immune proteins. The nematode, in turn, is dependent on the bacterium for establishing nutrient conditions for reproduction and producing antibacterial compounds that repress microbial colonization of the cadaver by competing secondary invaders, delaying putrefaction.

Because the *Xenorhabdus* bacterium kills the host so quickly, its nematode partner need not adapt to any specific host life cycle; steinernematid and heterorhabditid nematodes act much like predators in that there is no intimate host–parasite relationship. Consequently, hundreds of insect species may be infected in the laboratory (Poinar, 1979), and doubtless many thousands are susceptible. Other invertebrates are also known to be susceptible, including some isopods (Poinar and Paff, 1985), symphylans (Swenson, 1966), and spiders (R. Gaugler, unpublished observations). Despite this impressive host range, the nematode–bacterium complex presents no hazard to mammals (Gaugler and Boush, 1979a), and the U.S. Environmental Protection Agency has exempted these organisms from registration and regulation requirements (Gorsuch, 1982), removing the greatest obstacle to their commercialization.

Rapid acceleration in the development of steinernematid and heterorhabditid nematodes over the past several years has largely been the result of Bedding's (1981, 1984b) advances in *in vitro* production. Because the associated bacterium can convert virtually any protein into a substrate suitable for nematode development, production on a commercial scale is feasible using the cheapest available protein source: chicken offal. By coating shredded polyurethane sponge with chicken offal homogenate (plus 10% beef fat for *Heterorhabditis* species), Bedding created a three-dimensional culture matrix with a large ratio of surface area to volume. This innovative approach has been geared up to an industrial level with aerated 3-kg plastic bags, producing average yields of 1000–2000 million infectives per day for less than $0.01/million (Bedding, 1984b).

Biologicals are not chemicals, but some degree of "shelf life" is necessary. Steinernematid infective stages are easily stored and shipped on damp sponge (Bedding, 1984b), but heterorhabditids are much less storage stable and sometimes are shipped in culture flasks.

Steinernematid and heterorhabditid nematodes are among the most extensively tested of insect pathogens. Field trials have established that the key to their successful use is choice of target habitat (Gaugler, 1981). Juvenile intolerance of environmental extremes, particularly desiccation (Kamionek *et al.*, 1974), solar radiation (Gaugler and Boush, 1978), and temperature (Schmiege, 1963), has strongly discouraged applications onto exposed surfaces such as foliage. Efforts to exploit the aquatic environment because it offers nearly ideal conditions for juvenile survival have proved even less successful: ill adapted to an aquatic habitat, infective juveniles are incapable of initiating active infections (Gaugler and Molloy, 1981; Gaugler *et al.*, 1983) and quickly settle out of the host feeding zone (Finney and Harding, 1981). The most promising field results have been obtained in habitats without ecological and behavioral barriers to infection. The soil is an attractive application site since it is the natural reservoir for steinernematid and heterorhabditid nematodes, provides a buffer against environmental extremes, and is the only target habitat where these nematodes offer any prospect of becoming established to provide long-term control. The greatest success has been recorded against the black vine weevil, *Otiorhynchus sulcatus*, where efficacy has been demonstrated in the United States (Stimman *et al.*, 1985) and Europe (Simons, 1981), and commercial use of *H. heliothidis* has been reported in Australia (Bedding, 1984b). Similarly positive results have been obtained from cryptic habitats (e.g., tree galleries) since these sites also offer a microclimate favorable for nematode survival (i.e., high humidity, shelter from solar radiation). For example, *Steinernema* (= *Neoaplectana*) *bibionis* is used commercially to control the black currant borer, *Synanthedon tipuliformis* (Miller and Bedding, 1982; Bedding, 1984b), and Lindegren and Barnett (1982) have used *S. feltiae* (= *N. carpocapsae*) to achieve total suppression of carpenterworms infesting commercial fig orchards. Although insects in cryptic habitats are often secure from conventional insecticides, infective nematodes may locate concealed pests by a directed klinotactic response to host-released chemostimulants (Gaugler *et al.*, 1980). Kaya (1985), Gaugler (1981), and Poinar (1979) provide detailed reviews and analyses of field trials with steinernematid and heterorhabditid nematodes.

Steinernematid and heterorhabditid commercialization followed quickly on the heels of the EPA exemption and Bedding's advances in mass rearing. There are presently companies producing and selling these nematodes for scientific and commercial use in the United States, Canada, Australia, and Europe. Several firms are touted as biotechnology companies (e.g., Biotechnology Australia, Biosis), which doubtless aids in recruiting venture capital, but none is known to have initiated investigations into genetic manipulations.

III. ANALYSIS

Efforts to genetically alter entomogenous nematodes must begin with the identification of candidate species. Although this phase remains largely an empirical process, nematodes selected for enhancement must (1) be easily mass producible, (2) have a rapid generation time, (3) have demonstrated their commercial appeal, and (4) have already been extensively studied and evaluated. Few entomogenous nematodes meet these criteria.

Diplogasterids are particularly unlikely subjects for improvement since it is uncertain whether they are even true pathogens. The greatest problem here is our poor understanding of diplogasterid biology and ecology; there is no foundation for initiating genetic improvement studies.

Low infectivity and low transmission capability are major drawbacks of allantonematid nematodes (Kaya, 1986). Even biotechnology offers little promise of overcoming the strong density-independent nature of *Heterotylenchus* nematodes. Furthermore, mass culture technology must be developed prior to genetic enhancement.

A key problem with the sphaeruliid *T. sciariae* is its lack of a resistant stage. This restricts commercial development, like that of insect predators and parasitoids, to cottage industries where infectives can be applied soon after production. Overall, too little has been published on this nematode to permit an assessment of biocontrol potential, a deficiency precluding its genetic enhancement.

Other than Hudson's preliminary tests, the tetradonematids have not been evaluated in the field. It would be senseless to attempt to propose genetic solutions to problems that are ill defined, if defined at all. The paucity of detailed work with nematodes parasitizing mushroom flies (e.g., the tetradonematid *T. plicans*, the sphaeruliid *T. sciariae*, the allantonematid *H. husseyi*) does not imply that these parasites lack biocontrol potential. Rather, it reflects a limited commercial market for highly specific parasites attacking pests of modest economic importance for which good control approaches are available. These are problems not amenable to biotechnological solutions.

Deladenus nematodes have proved remarkably successful in classical biological control efforts and appear to require little improvement. *Deladenus* was successful despite an understandable lack of interest from private industry in a nonlethal, highly host-specific agent possessing poor storage characteristics but excellent powers of establishment and recycling. Nevertheless, because of the ruinous impact of siricid woodwasps on Australian forests, the government developed, produced, and distributed the nematode for the common good. This example provides strong evidence that a parasite need not kill its host to regulate the host population and that lack of a resistant stage is not a drawback given sufficient incentive and absence of a profit motive. Perhaps the single limitation of *Deladenus* is its narrow host range: only siricid woodwasps are attacked. Increasing the host range may be a reasonable genetic target with microbials, but it scarcely seems realistic with an organism as highly evolved and specialized as *Deladenus* in view of the bewilderingly complex behavioral, physiological, and pathological modifications that would be required of a parasite dependent on both the host and the host's symbiotic fungus.

The mosquito mermithid *R. culicivorax* has been preempted as an inundative vector control agent, and the inoculative route is currently in serious question. Consequently, there is little or no commercial interest in developing this parasite for biocontrol and thus insufficient incentive to justify the expense of genetic improvement studies—all of which is disheartening considering the level of excitement of a few years ago. The loss of the prolific James J. Petersen to other areas of research and the dissolution of the University of Newfoundland's Research Unit on Vector Pathology underscore the loss of momentum with nematodes attacking vectors. The most fundamental hurdle to the development and improvement of other mermithid species is the absence of mass culture methodology. Fassuliotis and Creighton (1982) may have taken mermithid culture as far as it can go with the technology presently available. Biotechnology could make a major contribution here by providing information on the molecular basis for feeding and nutrition, speeding development of efficient *in vitro* rearing methods.

Six decades of study with steinernematids and one with heterorhabditids have demonstrated that field success with these parasites is largely dependent on determining what effects the physical environment in any particular target habitat will have on the nematodes. Thus, these parasites are not useful against all pests in all habitats, despite their theoretically broad host range. Moreover, laboratory success does not ensure field success, the field being infinitely more complex than a laboratory petri dish. Even with these limitations, steinernematid and heterorhabditid nematodes meet most of the attributes of a desirable biological control agent: high virulence, high reproductive capacity, broad host range, a host-search capability, and ease of mass culture using existing technology. These attributes, the extensive background information available, and the current high interest in their commercial development as biological insecticides make them obvious candidates for genetic improvement.

IV. GENETIC IMPROVEMENT

Once candidates for genetic manipulation have been identified, the question becomes what traits are desirable and amenable to enhancement. There is little logic in bolstering the host range of steinernematids and heterorhabditids, parasites already capable of causing lethal infections in hundreds or thousands of pest species. Pesticide resistance is a frequent target for improvement in some biocontrol agents (Hoy, 1979, 1985), but steinernematids, and presumably heterorhabditids, are unaffected by most agrichemicals (Dutky, 1974; Welch, 1971). Similarly, sex ratio and host preference are not factors limiting the field efficacy of steinernematid and heterorhabditid nematodes, while dispersal and overwintering capability are of questionable significance for inundative agents. Kaya (1985), however, has identified (1) virulence, (2) environmental persistence, and (3) heterorhabditid storage/shipping as traits needing improvement. Virulence is not a limitation; steinernematid and heterorhabditid nematodes are among the most potent of insect pathogens, but increases in mobility and sensitivity to host-released attractants could enhance searching capacity, providing *indirect* improvements in nematode virulence and efficacy. Increased environmental resistance deserves the highest priority, since previous field data has indicated that efficacy in environmentally unfavorable habitats is limited by nematode intolerance of physical extremes, particularly desiccation. This would serve a dual purpose in improving both field persistence and shipping/storage characteristics, the most significant limitations for steinernematid and heterorhabditid nematodes, respectively. An addendum to Kaya's list: (4) further increases in mass production efficiency could make the nematodes economically competitive with conventional insecticides, and (5) Akhurst (1983) has advocated selection to increase the proportion of infectives able to carry foreign *Xenorhabdus* as a means of creating a more effective nematode–bacterium association. Regardless of the change desired, traits controlled by the fewest number of genes would be the easiest to manipulate.

Subsequent to identifying characters worthy of improvement is determining methods for accomplishing the desired improvement. The best approach is to search for new isolates with desirable traits; it is always preferable to first exhaust the natural genetic variation available. Bedding (1984a) stresses that success with *Deladenus* was partially the result of screening numerous strains for those producing high levels of parasitism but not reducing host (and therefore nematode) dispersal. All indications point to enormous natural variation in steinernematid and heterorhabditid nematodes: they have been isolated from six continents and appear to be among the most ubiquitous of insect pathogens. Akhurst and Brooks (1984), for example, reported isolating these nematodes from 13 of 14 locations in North Carolina, while Mrácek (1980) recovered steinernematids from 21 of 57 soil samples taken from ecologically diverse habitats throughout Czechoslovakia. Bedding *et al.*, (1983) recognized that steinernematid/heterorhabditid strains can vary significantly in infectivity and recommend screening for the most effective strains. Isolations for useful strains would be most effective from extreme habitats where local natural selection pressure molds desired traits. For example, Finney-Crawley (1985) recovered cold-tolerant steinernematid strains by screening soils from Labrador and Newfoundland provinces in Canada. Similarly, pesticide-resistant strains would most likely be isolated from intensively managed agricultural soils.

Formulation is a nongenetic approach widely used to overcome problems of poor pathogen field persistence. Evaporetardants (Kaya and Reardon, 1982; MacVean *et al.*, 1982) and ultraviolet protectants (Gaugler and Boush, 1979b) have been used experimentally to improve steinernematid infective-stage survival, but the best example is provided by Kaya and Nelsen's (1985) encapsulation of steinernematid and heterorhabditid infectives with calcium alginate to provide desiccation protection. Although this innovative approach to achieving enhanced efficacy does not involve genetic manipulations, it represents biotechnology in the broad sense.

Genetic approaches to improving the usefulness of entomogenous nematodes have not been seriously investigated, possibly because of the controversial nature of genetically improving biocontrol agents in general. Roush (1979) outlines the arguments against the concept of genetic improvement as follows: (1) the features which must be selected cannot be satisfactorily identified, (2) natural selection in the field will act against artifically introduced attributes and the population will revert to its wild state, and (3) laboratory selection programs are inherently incapable of success because they reduce genetic variability and introduce correlated deleterious pleiotropic effects. The first objection is almost irrelevant to steinernematids and heterorhabditids, where field ineffectiveness of these extensively evaluated nematodes has been clearly linked to their environmental sensitivity. The second argument is equally irrelevant because, unlike most parasitoids and predators, steinernematid and heterorhabditid nematodes are being developed for inundative rather than inoculative releases. The bioinsecticide approach does not expect establishment and recycling, so natural selection is not a serious consideration. Argument three overlooks the *undirected* reduction of genetic variability characteristic of any rearing program, where genetic potential may be altered by sampling limitations, inbreeding, inadvertent selection, and genetic drift (Hoy, 1979). Whether directed selection for genetic improvement adversely affects overall fitness may be a question of minor significance in utilizing a bioinsecticide, where field effectiveness, not fitness, is the issue.

There are three approaches available to genetically alter entomogenous nematodes: (1) artificial selection, (2) hybridization, and (3) genetic engineering.

Artificial selection is a well-accepted means of improving domestic plants and animals and, more recently, insect predators (Hoy, 1985) and parasitoids (White *et al.*, 1970), although it has rarely been applied to insect pathogens. Steinernematid and heterorhabditid nematodes have several advantages as subjects for selection, including short generation time and ease of rearing and handling, but foremost is the apparently broad base of genetic variability available on which to impose selection. Variation might even be further increased by intraspecific hybridization of geographic strains or induced mutation. The only disadvantage is lack of information on the genetic basis of desired attributes. Still, enhanced virulence has been selected for in *S. feltiae* and a 2.5-fold increase obtained after two passages through gypsy moth larvae (Shapiro *et al.*, 1985). Subsequent serial passages did not produce any further increase in activity. Strains of *Steinernema* and *Heterorhabditis* have also reportedly been laboratory-selected for enhanced mobility (Deseö *et al.*, 1984). The possibility of developing highly fecund strains as a means of reducing rearing costs seems especially worthy of investigation and amenable to selection pressure. Care should be exercised in designing breeding programs to minimize undesirable effects of selection.

Hybridization of strains as a means of recruiting new genetic material would appear to hold considerable promise simply because there are many steinernematid and heterorhabditid geographic isolates with desirable traits. Various species have strains differing in virulence and infectivity (Bedding *et al.*, 1983), physiological tolerance to environmental extremes (Finney-Crawley, 1985), and mass rearing efficiency (Bedding, 1981). Laboratory hybridization might be used to increase genetic variability, followed by field releases where natural selection could shape a superior strain for inoculative releases. There is also the possibility that nematode crosses might result in heterosis, as has been demonstrated with some parasitoids (Legner, 1972). Some inferior hybrids might be produced as well, but these could easily be selected out. Hybridization remains unevaluated for entomogenous nematodes.

Genetic engineering offers unexpected prospects for improving steinernematid/heterorhabditid nematodes. It is not proposed that recombinant DNA techniques be immediately applied to these nematodes, since detailed insight into their fundamental genetics is needed before even considering gene manipulations. The *Xenorhabdus* bacteria mutualistically associated with these nematodes, however, are prime targets for improvement by the genetic engineer. The ease with which recombinant DNA technology can be done depends in large part on the size of the organism's nuclear genome: the smaller the genome, the less effort is required to create and screen recombinant DNA libraries and thus to isolate a particular gene. The *Xenorhabdus* genome is, of course, much smaller and easier to work with than that of its nematode partners, and lacks histones and other complicating proteins.

It is again essential to consider carefully which traits are to be improved: the virulence and host range of the bacterium are already nearly ideal, and environmental persistence and host finding are not pertinent because the nematode provides these functions. But the bacterium is the key to the culture of steinernematid and heterorhabditid nematodes. Akhurst (1980) has shown that *Xenorhabdus* occurs in two forms, primary and secondary, that differ sharply in their ability to support nematode growth and development. Only the primary form is usually isolated from infective-stage juveniles, but this form tends to be unstable and often converts to the secondary on artificial media. Nematode production on the secondary form is greatly reduced (Akhurst, 1980; Bedding, 1981). Engineering a bacterium unable to convert to the secondary form would clearly be advantageous, but first we must know what regulates the conversion. Our understanding of the biochemistry and physiology of *Xenorhabdus* spp. is minimal, and even less is known about its genetics. The lack of a sufficient base on which to impose biotechnological solutions is a recurrent theme in the present discussion of entomogenous nematodes.

Another intriguing if more distant possibility centers on the ability of infective-stage juveniles to produce an anti-immune factor which destroys the host immune proteins, allowing the bacterium to become established in the hemocoel without regard for a host defense response (Gotz *et al.*, 1981). Kirschbaum (1985) has suggested that such anti-immune substances might be genetically isolated, improved, and returned to the pathogen, or that the encoding genes might be transferred to another organism. This later approach could provide a means of achieving a long-standing goal in insect pathology: expanding the host spectrum of microbial pathogens.

Several laboratories have recently initiated research into the physiology and molecular biology of *Xenorhabdus,* and a number of genes are believed to have been isolated and cloned. Rapid progress should be expected in the near future, given the obvious commercial incentive.

Field releases of nematodes vectoring engineered bacteria will likely generate considerable controversy, judging from the furor caused by the proposed testing of engineered ice-nucleation bacteria, which are not even pathogens. Biotechnology is not just gene splicing; ecologists and others have important roles to play as well in resolving questions of safety, stability, and efficacy.

V. CONCLUSION

There are surprisingly good prospects for the genetic manipulation of some entomogenous nematode species, but progress is so hampered by an absence of information on their general and molecular genetics that even the most rudimentary studies would be valuable contributions. Major advances made with the free-living nematode *Caenorhabditis elegans* (Lewin, 1984; Karn *et al.*, 1985) have removed many technological barriers to such studies, providing a model for genetic and molecular biological analysis of entomogenous species. A sound foundation for genetic investigations must be developed by continuing studies into parasite biology, ecology, and behavior. Additional basic research directed, in particular, at steinernematid/heterorhabditid UV sensitivity, moisture requirements, host seeking, and culture would be helpful in suggesting lines of research for improvement of nematode efficacy.

ACKNOWLEDGMENTS

I thank Harry Kaya and Albert Pye for their valuable comments on the original manuscript. New Jersey Agricultural Experiment Station Publication No. F-08115-02-85, supported by state funds and U.S. Hatch Act.

REFERENCES

Akhurst, R. J. (1980). Morphological and functional dimorphism in *Xenorhabdus* spp., bacteria symbiotically associated with the insect pathogenic nematodes *Neoaplectana* and *Heterorhabditis*. *J. Gen. Microbiol.* **121,** 303–309.

Akhurst, R. J. (1983). *Neoaplectana* species: Specificity of association with bacteria of the genus *Xenorhabdus*. *Exp. Parasitol.* **55,** 258–263.

Akhurst, R. J., and Brooks, W. M. (1984). The distribution of entomophilic nematodes (Heterorhabditidae and Steinernematidae) in North Carolina. *J. Invertebr. Pathol.* **44,** 140–145.

Bedding, R. A. (1967). Parasitic and free-living cycles in the entomogenous nematodes of the genus *Deladenus*. *Nature (London)* **214,** 174–175.

Bedding, R. A. (1968). *Deladenus wilsoni* n. sp. and *D. siricidicola* n. sp. (Neotylenchidae), entomophagous–mycetophagous nematodes parasitic in siricid woodwasps. *Nematologica* **14,** 515–525.

Bedding, R. A. (1972). Biology of *Deladenus siricidicola* (Neotylenchidae) an entomophagous–mycetophagous nematode parasitic in siricid woodwasps. *Nematologica* **18,** 482–493.

Bedding, R. A. (1981). Low cost in vitro mass production of *Neoaplectana* and *Heterorhabditis* species (Nematoda) for field control of insect pests. *Nematologica* **27,** 109–114.

Bedding, R. A. (1984a). Nematode parasites of Hymenoptera. *In* "Plant and Insect Nematodes" (W. R. Nickle, ed.), pp. 755–795. Dekker, New York.

Bedding, R. A. (1984b). Large scale production, storage and transport of the insect-parasitic nematodes *Neoaplectana* spp. and *Heterorhabditis* spp. *Ann. Appl. Biol* **104,** 117–120.

Bedding, R. A., and Akhurst, R. J. (1974). Use of the nematode *Deladenus siricidicola* in the biological control of *Sirex noctilio* in Australia. *J. Aust. Entomol. Soc.* **13,** 129–137.

Bedding, R. A., and Molyneux, A. S. (1982). Penetration of insect cuticle by infective juveniles of *Heterorhabditis* spp. (Heterorhabditidae: Nematoda). *Nematologica* **28,** 354–359.

Bedding, R. A., Molyneux, A. S., and Akhurst, R. J. (1983). *Heterorhabditis* spp., *Neoaplectana* spp., and *Steinernema kraussei:* Interspecific and intraspecific differences in infectivity for insects. *Exp. Parasitol.* **35,** 249–257.

Brenner, R. J., Cupp, E. W., and Bernardo, M. J. (1981). Laboratory colonization and life table statistics for geographic strains of *Simulium decorum*. *Tropenmed. Parasitol* **31,** 487–497.

Brown-Westerdahl, B., Washino, R. K., and Platzer, E. G. (1979). Early season application of *Romanomermis culicivorax* provides continuous partial control of rice field mosquitoes. *Proc. Calif. Mosq. Vect. Cont. Assoc.* **47,** 55.

Cobb, N. A. (1919). *Tetradonema plicans* nov. gen. et spec., representing a new family, Tetradonematidae, as now found parasitic in larvae of the midge-insect *Sciara coprophilia* Lintner. *J. Parasitol.* **5,** 176–185.

Coppel, H. C., and Mertins, J. W. (1977). "Biological Insect Pest Suppression." Springer-Verlag, Berlin and New York.

Creighton, C. S., and Fassuliotis, G. (1981). A laboratory technique for culturing *Filipjevimermis leipsandra,* a nematode parasite of *Diabrotica balteata* larvae (Insecta: Coleoptera). *J. Nematol.* **13,** 226–227.

Creighton, C. S., and Fassuliotis, G. (1982). Mass rearing a mermithid nematode, *Filipjevimermis leipsandra* (Mermithida: Mermithidae) on the banded cucumber beetle (Coleoptera: Chrysomelidae). *J. Econ. Entomol.* **75,** 701–703.

Creighton, C. S., and Fassuliotis, G. (1983). Infectivity and suppression of the banded cucumber beetle (Coleoptera: Chrysomelidae) by the mermithid nematode *Filipjevimermis leipsandra* (Mermithida: Mermithidae). *J. Econ. Entomol.* **76,** 615–618.

Deseö, K. V., Grassi, S., Foschi, F., and Rovesti, L. (1984). Un sistema di lotta biologica contro il rodilegno giallo (*Zeuzera pyrina* L.; Lepidoptera, Cossidae). *Atti Giornate Fitopathol.* **2,** 403–414.

Dutky, S. R. (1974). Nematode parasites. *In* "Proceedings of the Summer Institute on Biological Control of Plant Insects and Diseases" (F. G. Maxwell and F. A. Harris, eds.), pp. 576–590. Univ. Press of Mississippi, Jackson.

Evans, A. A. F. (1970). Mass culture of mycetophagous nematodes. *J. Nematol.* **2,** 99–100.

Fassuliotis, G., and Creighton, C. S. (1982). In vitro cultivation of the entomogenous nematode *Filipjevimermis leipsandra*. *J. Nematol.* **14,** 126–131.

Fedorko, A. (1971). Nematodes as factors reducing the populations of Colorado beetle, *Leptinotarsa decemlineata* Say. *Acta Phytopathol. Acad. Sci. Hung.* **6,** 175–181.

Fedorko, A., and Stanuszek, S. (1971). *Pristionchus uniformis* sp. n. (Nematoda, Rhabditida, Diplogasteridae), a facultative parasite of *Leptinotarsa decemlineata* Say and *Melolontha melolontha* L. in Poland. Morphology and biology. *Acta Parasitol. Pol.* **19,** 95–112.

Finney, J. R., and Harding, J. B. (1981). Some factors affecting the use of *Neoaplectana* sp. for mosquito control. *Mosq. News* **41,** 798–800.

Finney-Crawley, J. R. (1985). Isolation of cold tolerant steinernematid nematodes in Canada. *J. Nematol.* **17,** 496 (abstr.)

Gaugler, R. (1981). Biological control potential of neoaplectanid nematodes. *J. Nematol.* **13,** 241–249.

Gaugler, R., and Boush, G. M. (1978). Effects of ultraviolet radiation and sunlight on the entomogenous nematode, *Neoaplectana carpocapsae*. *J. Invertebr. Pathol.* **32,** 291–296.

Gaugler, R., and Boush, G. M. (1979a). Nonsusceptibility of rats to the entomogenous nematode, *Neoaplectana carpocapsae*. *Environ. Entomol.* **8,** 658–660.

Gaugler, R., and Boush, G. M. (1979b). Tests on materials as ultraviolet protectants of an entomogenous nematode. *Environ. Entomol.* **8,** 810–813.

Gaugler, R., and Finney, J. (1982). A review of *Bacillus thuringiensis* var. *israelensis* as a biological control agent of black flies (Diptera: Simuliidae). *Misc. Publ. Entomol. Soc. Am.* **12,** 1–17.

Gaugler, R., and Kaya, H. K. (1983). A bibliography of the entomogenous nematode family Steinernematidae. *Bibl. Entomol. Soc. Am.* **1,** 43–64.

Gaugler, R., and Molloy, D. (1981). Field evaluation of the entomogenous nematode *Neoaplectana carpocapsae*. *J. Nematol.* **13,** 1–5.

Gaugler, R., LeBeck, L., Nakagaki, B., and Boush, G. M. (1980). Orientation of the entomogenous nematode *Neoaplectana carpocapsae* to carbon dioxide. *Environ. Entomol.* **9,** 649–652.

Gaugler, R., Kaplan, B., Alvarado, C., and Montoyo, J. (1983). Assessment of *Bacillus thuringiensis* serotype 14 and *Steinernema feltiae* for control of the *Simulium* vectors of onchocerciasis in Mexico. *Entomophaga* **28,** 309–315.

Gaugler, R., Wraight, S., and Molloy, R. (1984). Bionomics of a mermithid parasitizing snowpool *Aedes* spp. mosquitoes. *Can. J. Zool.* **62,** 670–674.

Gorsuch, A. M. (1982). Regulations for the enforcement of the Federal Insecticide, Fungicide, and Rodenticide Act exemption from regulation of certain biological control agents. *Fed. Regist.* **47,** 23928–23930.

Gotz, P., Boman, A., and Boman, H. G. (1981). Interactions between insect immunity and an insect-pathogenic nematode with symbiotic bacteria. *Proc. R. Soc. London, Ser. B* **212,** 333–350.

Hominick, W. M., and Tingley, G. A. (1984). Mermithid nematodes and the control of insect vectors of human disease. *Biocontr. News Inf.* **5,** 7–20.

Hoy, M. A. (1979). The potential for genetic improvement of predators for pest management programs. *In* "Genetics in Relation to Insect Management" (M. A. Hoy and J. J. McKelvey, Jr., eds.), pp. 106–115. Rockefeller Found. Press, New York.

Hoy, M. A. (1985). Recent advances in genetics and genetic improvement of the Phytoseiidae. *Annu. Rev. Entomol.* **30,** 345–370.

Hudson, K. E. (1972). Nematodes as biological control agents: Their possible application in controlling insect pests of mushroom crops. *Mushroom Sci.* **8,** 193–197.

Hudson, K. E. (1974). Regulation of greenhouse sciarid fly populations using *Tetradonema plicans* (Nematoda: Mermithoidea). *J. Invertebr. Pathol.* **23,** 85–91.

Jones, C. M., and Perdue J. M. (1967). *Heterotylenchus autumnalis*, a parasite of the face fly. *J. Econ. Entomol.* **60,** 1393–1395.

Kamionek, M., Maslana, I., and Sandner, H. (1974). The survival of invasive larvae of *Neoaplectana carpocapsae* Weiser in a waterless environment under various conditions of temperature and humidity. *Zesz. Probl. Postepow Nauk Roln.* **154,** 409–412.

Karn, J., Dibb, N. J., and Miller, D. M. (1985). Cloning nematode myosin genes. *Cell Muscle Motil.* **6,** 185–237.

Kaya, H. K. (1984). Nematode parasites of bark beetles. *In* "Plant and Insect Nematodes" (W. R. Nickle, ed.), pp. 727–754. Dekker, New York.

Kaya, H. K. (1985). Entomogenous nematodes for insect control in IPM systems. *In* "Biological Control in Agricultural IPM Systems" (M. A. Hoy and D. C. Herzog, eds.), pp. 283–302. Academic Press, New York.

Kaya, H. K. (1987). Epizootiology of nematode diseases. *In* "Epizootiology of Insect Diseases" (J. Fuxa and Y. Tanada, eds.). Wiley, New York (in press).

Kaya, H. K., and Moon, R. D. (1978). The nematode *Heterotylenchus autumnalis* and face fly *Musca autumnalis:* A field study in northern California. *J. Nematol.* **10,** 333–341.

Kaya, H. K., and Nelsen, C. E. (1985). Encapsulation of steinernematid and heterorhabditid nematodes with calcium alginate: A new approach for insect control and other applications. *Environ. Entomol.* **14,** 572–574.

Kaya, H. K., and Reardon, R. C. (1982). Evaluation of *Neoaplectana carpocapsae* for biological control of the western spruce budworm, *Choristoneura occidentalis:* Ineffectiveness and persistence of tank mixes. *J. Nematol.* **14,** 595–597.

Kerwin, J. L., and Washino, R. K. (1983). Field evaluation of *Bacillus thuringiensis* var. *israelensis, Lagenidium giganteum,* and *Romanomermis culicivorax* in California rice fields. *Proc. Annu. Conf. Calif. Mosq. Vect. Contr. Assoc.*, pp. 19–25.

Khan, A., Brooks, W. M., and Hirschmann, H. (1976). *Chromonema heliothidis* n. gen., n. sp. (Steinernematidae, Nematoda), a parasite of *Heliothis zea* (Noctuidae, Lepidoptera), and other insects. *J. Nematol.* **8,** 159–168.

Kirschbaum, J. B. (1985). Potential implication of genetic engineering and other biotechnologies to insect control. *Annu. Rev. Entomol.* **30,** 51–70.

Krafsur, E. S., Church, C. J., Elvin, M. K., and Ernst, C. M. (1983). Epizootiology of *Heterotylenchus autumnalis* (Nematoda) among face flies (Diptera: Muscidae) in central Iowa, USA. *J. Med. Entomol.* **20,** 318–324.

Lacey, L., and Undeen, A. H. (1986). Microbial control of black flies and mosquitoes. *Annu. Rev. Entomol.* **31,** 265–296.

Legner, E. F. (1972). Observations on hybridization and heterosis in parasitoids of synanthropic flies. *Ann. Entomol. Soc. Am.* **65,** 254–263.

Lewin, R. (1984). Why is development so illogical? *Science* **224,** 1327–1329.

Lindegren, J. E., and Barnett, W. W. (1982). Applying parasitic nematodes to control carpenterworms. *Calif. Agric.* **36,** 7–8.

MacVean, C. M., Brewer, J. M., and Capinera, J. L. (1982). Field tests of antidesiccants to extend the infection period of an entomogenous nematode, *Neoaplectana carpocapsae,* against the Colorado potato beetle. *J. Econ. Entomol.* **75,** 97–101.

Massey, C. L. (1974). Biology and taxonomy of nematode parasites and associates of bark beetles in the United States. *U.S., Dep. Agric., Agric. Handb.* **446.**

Miller, L. A., and Bedding, R. A. (1982). Field testing of the insect parasitic nematode, *Neoaplectana bibionis* (Nematode: Steinernematidae) against the currant borer

moth, *Synanthedon tipuliformis* (Lep.: Sesiidae) in black currants. *Entomophaga* **27,** 109–114.
Mongkolkiti, S., and Hosford, R. M., Jr. (1971). Biological control of the grasshopper *Hesperotettix viridis pratensis* by the nematode *Mermis nigrescens*. *J. Nematol.* **3,** 356–363.
Mrácek, Z. (1980). The use of "*Galleria* traps" for obtaining nematode parasites of insects in Czechoslovakia (Lepidoptera: Nematoda, Steinernematidae). *Acta Entomol. Bohemoslov.* **77,** 378–382.
Nickle, W. R. (1972). Nematode parasites of insects. *Proc. Annu. Tall Timbers Conf., 1972,* pp. 145–163.
Nickle, W. R. (1974). Nematode infections. *In* "Insect Diseases" (G. E. Cantwell, ed.), Vol. 2, pp. 327–376. Dekker, New York.
Nickle, W. R., and Welch, H. E. (1984). History, development, and importance of insect nematology. *In* "Plant and Insect Nematodes" (W. R. Nickle, ed.), pp. 627–653. Dekker, New York.
Petersen, J. J. (1982). Current status of nematodes for the biological control of insects. *Parasitology* **84,** 177–204.
Petersen, J. J. (1984). Nematode parasites of mosquitoes. *In* "Plant and Insect Nematodes" (W. R. Nickle, ed.), pp. 797–820. Dekker, New York.
Petersen, J. J., and Chapman, H. C. (1979). Checklist of mosquito species tested against the nematode parasite *Romanomermis culicivorax*. *J. Med. Entomol.* **15,** 468–471.
Petersen, J. J., and Willis, O. R. (1970). Some factors affecting parasitism by mermithid nematodes in southern house mosquito larvae. *J. Econ. Entomol.* **63,** 175–178.
Petersen, J. J., and Willis, O. R. (1972a). Procedures for the mass rearing of a mermithid parasite of mosquitoes. *Mosq. News* **32,** 226–230.
Petersen, J. J., and Willis, O. R. (1972b). Results of preliminary field applications of *Reesimermis nielseni* (Mermithidae: Nematoda) to control mosquito larvae. *Mosq. News* **32,** 312–316.
Petersen, J. J., and Willis, O. R. (1974). Experimental release of a mermithid nematode to control *Anopheles* mosquitoes in Louisiana. *Mosq. News* **34,** 316–319.
Petersen, J. J., and Willis, O. R. (1975). Establishment and recycling of a mermithid nematode for the control of mosquito larvae. *Mosq. News* **35,** 526—532.
Petersen, J. J., and Willis, O. R. (1976). Experimental release of a mermithid nematode to control floodwater mosquitoes in Louisiana. *Mosq. News* **36,** 339–342.
Petersen, J. J., Chapman, H. C., and Woodard, D. B. (1968). The bionomics of a mermithid nematode of larval mosquitoes in southwestern Louisiana. *Mosq. News* **28,** 346–352.
Petersen, J. J., Willis, O. R., and Chapman, H. C. (1978a). Release of *Romanomermis culicivorax* for the control of *Anopheles albimanus* in El Salvador. I. Mass production of the nematode. *Am. J. Trop. Med. Hyg.* **27,** 1265–1276.
Petersen, J. J., Chapman, H. C., Willis, O. R., and Fukuda, T. (1978b). Release of *Romanomermis culicivorax* for the control of *Anopheles albimanus* in El Salvador. II. Application of the nematode. *Am. J. Trop. Med. Hyg.* **27,** 1268–1273.
Phelps, R. J., and DeFoliart, G. R. (1964). Nematode parasites of Simuliidae. *Univ. Wis. Res. Bull.* **245,** 1–78.
Platzer, E. G. (1981). Biological control of mosquitoes with mermithids. *J. Nematol.* **13,** 257–262.
Poinar, G. O., Jr. (1965). The bionomics and parasite development of *Tripius sciarae* (Bovien) (Sphaerulariidae: Aphelenchoidia), a nematode parasite of sciarid flies (Sciaridae: Diptera). *Parasitology* **55,** 559–569.

Poinar, G. O., Jr. (1969). Diplogasterid nematodes (Diplogasteridae: Rhabditida) and their relationship to insect disease. *J. Invertebr. Pathol.* **13,** 447–454.

Poinar, G. O., Jr. (1975). "Entomogenous Nematodes." Brill, Leiden, The Netherlands.

Poinar, G. O., Jr. (1979). "Nematodes for Biological Control of Insects." CRC Press, Boca Raton, Florida.

Poinar, G. O., Jr., and Doncaster, C. C. (1965). The penetration of *Tripius sciarae* (Bovien) (Sphaerulariidae: Aphelenchoidea) into its insect host, *Bradysia paupera* Thom. (Mycetophilidae: Diptera). *Nematologica* **11,** 73–78.

Poinar, G. O., Jr., and Paff, M. (1985). Laboratory infection of terrestrial isopods (Crustacea: Isopoda) with neoaplectanid and heterorhabditid nematodes (Rhabditida: Nematoda). *J. Invertebr. Pathol.* **45,** 24–27.

Poinar, G. O., Jr., and van der Laan, P. (1972). Morphology and life history of *Sphaerularia bombi. Nematologica* **18,** 239–252.

Richardson, P. N., and Chanter, D. O. (1979). Phorid fly (Phoridae: *Megaselia halterata*) longevity and the dissemination of nematodes (Allantonematidae: *Howardula husseyi*) by parasitized females. *Ann. Appl. Biol.* **93,** 1–11.

Rinker, D. L., and Snetsinger, R. J. (1984). Damage threshold to a commercial mushroom by mushroom-infesting phorid (Diptera: Phoridae). *J. Econ. Entomol.* **77,** 449–453.

Robinson, J. V., and Combs, R. L., Jr. (1976). Incidence and effect of *Heterotylenchus autumnalis* on the longevity of face flies in Mississippi. *J. Econ. Entomol.* **69,** 722–724.

Roush, R. T. (1979). Genetic improvement of parasites. *In* "Genetics in Relation to Insect Management" (M. A. Hoy and J. J. McKelvey, Jr., eds.), pp. 97–105. Rockefeller Found. Press, New York.

Sandner, H., Seryczynska, H., and Kamionek, M. (1972). Preliminary microbiological and ultrastructural investigations of bacteria isolated from *Pristionchus uniformis* Fedorko and Stanuszek. *Bull. Acad. Pol. Sci.* **20,** 567–569.

Schmiege, D. C. (1963). The feasibility of using a neoaplectanid nematode for control of some forest insect pests. *J. Econ. Entomol.* **56,** 427–431.

Shapiro, M., Poinar, G. O., Jr., and Lindegren, J. E. (1985). Suitability of *Lymantria dispar* (Lepidoptera: Lymantriidae) as a host for the entomogenous nematode, *Steinernema feltiae* (Rhabditida: Steinernematidae). *J. Econ. Entomol.* **78,** 342–345.

Simmons, K. R., and Edman, J. D. (1981). Sustained colonization of the black fly *Simulium decorum* Walker (Diptera: Simuliidae). *Can. J. Zool.* **59,** 1–7.

Simons, W. R. (1981). Biological control of *Otiorhynchus sulcatus* with heterorhabditid nematodes in greenhouses. *Neth. J. Plant Pathol.* **87,** 149–158.

Stimman, M. W., Kaya, H. K., Burlando, T. M., and Studdert, J. P. (1985). Black vine weevil management in nursery plants. *Calif. Agric.* **39,** 25–26.

Swenson, K. G. (1966). Infection of the garden symphylan, *Scutigerella immaculata*, with the DD-136 nematode. *J. Invertebr. Pathol.* **2,** 133–134.

Thomas, G. D., and Puttler, B. (1970). Seasonal parasitism of the face fly by the nematode *Heterotylenchus autumnalis* in central Missouri, 1968. *J. Econ. Entomol.* **63,** 1922–1923.

Thomas, G. D., Puttler, B., and Morgan, C. D. (1972). Further studies of field parasitism of the face fly by the nematode *Heterotylenchus autumnalis* in central Missouri with notes on the gonadotrophic cycles of the face fly. *Environ. Entomol.* **1,** 759–763.

Treece, R. E., and Miller, T. A. (1968). Observations on *Heterotylenchus autumnalis* in relation to the face fly. *J. Econ. Entomol.* **61,** 454–456.

Walker, T. W., Meek, T. W., and Wright, V. L. (1985). Establishment and recycling of

Romanomermis culicivorax (Nematoda: Mermithidae) in Louisiana ricelands. *J. Am. Mosq. Control Assoc.* **1,** 468–473.

Welch, H. E. (1971). Various target species: Attempts with DD-136. *In* Biological control programmes against insects and weeds in Canada, 1959–1968. *Tech. Commun., CIBC* **4,** 147–154.

Welch, H. E., and Rubtzov, I. A. (1965). Mermithids (Nematoda: Mermithidae) parasitic in blackflies (Insecta: Simuliidae). I. Taxonomy and bionomics of *Gastromermis boophthorae* sp. n. *Can. Entomol.* **97,** 581–596.

White, E. B., DeBach, R., and Garber, M. J. (1970). Artificial selection for genetic adaptation to temperature extremes in *Aphytis lingnanensis* Compere (Hymenoptera: Aphelinidae). *Hilgardia* **40,** 161–192.

Zondag, R. (1962). A nematode disease of *Sirex noctilio*. *N. Z. For. Serv. Interm Res. Release.*

30

Genetically Engineered Microbial and Viral Insecticides: Safety Considerations

KARL MARAMOROSCH

Severe effects from repeated use of toxic chemical insecticides prompted the development of alternative, biological insecticides. Genetically engineered microbial and viral organisms are now being obtained and new mass production technology will make available cost-effective products for combating insect pests and vectors of pathogens. The manipulation of invertebrate pathogens and cells by recombinant DNA and cell fusion technology to improve crop production and human health hold considerable promise for the future. Genetic engineering permits us to study DNA, RNA, and protein structure in a rapid and new way. This genetic engineering will yield basic information about biological systems as well as provide practical results.

Within the past 10 years it became clear that genetic engineering and cell fusion represent the biggest single advance in biology in the 20th century. The potential of genetic engineering must be exploited, without endangering the environment. Biological control of pests and vectors is a very important area where true progress has already been made (Maramorosch and Sherman, 1985). We can expect that the application of biotechnology to insect pathology will result in increased agricultural productivity. By producing more food and fiber on currently

BIOTECHNOLOGY IN INVERTEBRATE
PATHOLOGY AND CELL CULTURE

cultivated land, it may be possible to leave as yet uncultivated areas unaltered, without disturbing the ecology.

The genetic engineering for the creation of *Bacillus thuringiensis* of greater efficacy is now in the forefront of research. Novel and unexpected results, in addition to the expected ones, can turn up in the near future. Vaccines against malaria and vaccines against virus diseases will be manufactured soon by recombinant DNA technology. Even a birth control vaccine is being contemplated.

However, there are very serious constraints that scientists are encountering. One is the funding problem. Vaccine research is very expensive, and profits from a conventionally available vaccine are always lower than profits from drugs that must be used by patients for long periods. Human vaccines are less profitable investments for pharmaceutical companies than drugs. Even more important constraints might be imposed by the regulatory agencies. Requirements for safety are very stringent. Finally, suitable and acceptable adjuvants will have to be developed for the genetically engineered vaccines. Then we have to face the substantial difficulties of using any living, genetically engineered organism as a carrier for antigens and the release of such organisms in the environment. Not only *Escherichia coli* but also vaccinia virus, yeast, and baculoviruses as well as nontoxic pathogens of cholera and typhoid can be engineered to provide vaccines. Today this is a very active and exciting area of research. Vaccines eventually produced by such means promise to be comparatively inexpensive. The problems with such vaccines are the purification of antigens from all other products made by baculoviruses or other carriers and the question of their antigenic strength.

Vaccines are the most cost-efficient means of preventing diseases. A malarian sporozoite vaccine through recombinant DNA technology will soon be tested. A merozoite vaccine based on recombinant DNA technology is also being planned. Should both prove effective, their eventual combination into a single compound vaccine would be feasible and this vaccine could save millions of lives (Maramorosch, 1985). Currently antimalarial vaccines are planned against *Plasmodium falciparum*, but if this most dangerous of the malarial parasites can be controlled, *Plasmodium vivax* and other plasmodia will eventually be tackled by the same techniques. The number of malaria cases is now estimated at 200 million per year, and in Africa and some parts of Southeast Asia 50 of every 1000 children die from malaria before the age of five. Over 1 million die annually in Africa alone. The resurgence of malaria during the past two decades is due to the resistance of mosquitoes to chemical insecticides and the emergence of drug-resistant strains of malarial parasites. Advances in biotechnology will provide modified, genetically engineered and highly efficient *B. thuringiensis* H-14 strains that will curtail the spread of malaria by mosquito vectors, further reducing the impact of this greatest killer of humans today (Nossal, 1985).

Biological control of pests and vectors is an area where great progress has already been achieved and where genetically engineered pesticides and vector control agents can have a real impact. The Genex Corporation has predicted that by the year 2000 total sales of products obtained through recombinant DNA technology will reach 40 billion per year (Schneiderman, 1984). Various aspects should be discussed in this context. One is the threat of secrecy in the area of new developments. Molecular biology grew rapidly because of openness and free exchange of information. This tradition is now being threatened and it may even be destroyed by the rapidly expanding genetic engineering. There have been several meetings at which scientists who once eagerly discussed their procedures and results have refused to reveal technical data. This is deleterious, and the trend to discover, patent, call a press conference to boost the stock, and only later publish is seen fairly often. Let us hope that in insect pathology the motivation of scientists to present their complete results at seminars, lectures, and conferences and in peer-refereed journals will prevail.

Now the problem of safety and legitimate environmental concern about the release of genetically engineered microorganisms and viruses on a large scale must be considered. We are facing opposition to genetic engineering by some parts of society. This is not surprising. There have always been those who are opposed to vaccination, as well as to other activities undertaken in the common interest of societies, and there are now many more who are frightened by the new powers that can give humans the possibility of creating life forms in ways not intended by nature. There is a tendency of people to resist the unknown, to resist change, to preserve the status quo, and to resent innovations. Disillusionment with technology and science is deeply ingrained in some parts of society. This perception of science as antihuman and destructive must be refuted through education. We scientists have to explain to the public that conventional genetic techniques that have been used in plant and animal breeding for thousands of years have provided our domestic animals and staple food crops and have had a profound impact on the ecosystem. It is very difficult therefore to distinguish a rust-resistant wheat or a photoperiod-independent wheat obtained by conventional breeding before 1975 from one that might be created by DNA technology. Certain regulatory agencies have tried to prevent the release into the environment of varieties newly created by DNA technology, while at the same time permitting the release of similar, if not identical, varieties created in the traditional, old manner. Introducing DNA from one cell into another is classified as genetic engineering, but cell fusion, which achieves genetic hybridization, is not. If one country enacted restrictive regulations to stop the use of genetically engineered microorganisms or viruses, another country would go ahead and research would continue.

As scientists, we are primarily interested in the practical end results of scientific development. There are very few scientists in high positions in governments. Fortunately, this number is now increasing in high-technology industry management positions and this is very important in the decision-making process. We have an obligation to inform the public what we are doing and what practical results can be expected from our work.

Addressing the safety aspects, we have to face those who fear a willful or accidental release of new, inadequately tested and highly pathogenic species into the biosphere and those who expect that in the near future a golden age of prosperity will be created by DNA technology. Each extreme view is exaggerated. The dangers of creating disasters through genetic engineering will not materialize. Before 1975 new genes were introduced into cells by viruses and plasmids, genes jumped within cells, and through sexual reproduction genes were exchanged in novel and often unpredictable ways. These "natural" events produced epidemics of bubonic plague and immense grasshopper populations that devastated crops in America and parts of Asia. Nevertheless, civilization was not destroyed by bubonic plague nor crops eliminated by grasshoppers.

We now know how to alter the genetic bases underlying baculovirus capabilities of infection, reproduction, and maintenance. The outcomes of such genetic manipulations are complex, and the potential impacts beyond those intended are not always immediately apparent. There are no good guidelines to assess the potential risks posed by genetically altered baculoviruses (Miltenburger, 1980). This newly emerging area of research faces new regulations and the impact on the hosts, on nontarget organisms, and on the environment will have to be assessed very carefully (Quaraishi, 1985).

As far as the safety of biologically engineered baculoviruses is concerned, the question arises: Is the insertion of DNA into cells of other species as restricted as generally believed? Can DNA be inserted artificially to become innocuous? (Doller, 1985; McIntosh and Shamy, 1980; Kurstak *et al.*, 1978). Constant vigilance will certainly be needed in monitoring the results. New developments in cell fusion, in the development of serum-free inexpensive media for cell cultivation, and in genetic engineering are forthcoming and international cooperation will undoubtedly help to speed up progress in this exciting and important area. It is not easy to define the range of impacts of genetically engineered organisms in the environment. Speculations and conclusions vary greatly, with alarming predictions at one end and reassuring generalizations at the other.

Field tests of genetically engineered microorganisms and viruses undoubtedly will be approved, although at this time in the United States regulatory agencies and ecologists are raising questions about genetically engineered biological insecticides and thus are delaying approval.

It is easier to answer the question of whether there is potential harm to mammals, humans, and nontarget organisms than to assess the potential environmental impact. Baculoviruses do not infect nontarget organisms and microbial insecticides can be selected to affect predominantly target organisms. Still, it is very difficult to decide whether there is a potential risk when spraying baculoviruses as biological control agents in the field or releasing *B. thuringiensis*, irrespective of whether the control agents were produced by conventional or by DNA technology.

At present, a regulatory impasse affects those who are developing new strains of baculoviruses of microbial insecticides (Sun, 1986). Regulatory agencies and ecologists are attempting to forestall approval, requiring additional information before any releases into the environment can be made. The Environmental Protection Agency, one of several U.S. government agencies involved in the regulatory process, has formidable difficulties in formulating, implementing, and obtaining acceptance for its regulatory decisions (Fox, 1985). The agency is expected to evaluate proposals on the basis of their scientific, environmental, economic, and public policy merits and consequences. Many scientists are convinced that there is nothing unsafe about using recombinant DNA technology. From a scientific point of view, recombinant DNA technology allows researchers to make more profound genetic changes than were possible by using older genetic techniques. Ecologists who oppose the release into the environment of viruses and microbes altered by DNA technology point out the inherent difficulties of retrieving or eliminating such altered entities once they are released. We cannot predict how long genetically altered microbes or viruses will survive in the environment. Mere survival does not imply problems for the native organisms. There exist numerous alien species of insects in the United States, but only a few of them have had deleterious effects. Ecologists realize this and they are not against the introduction of certain beneficial parasites into new areas, a practice which proved very successful in the past. Yet the novel techniques have been greeted with mixed feelings. There have been indications that a single plasmid carrying a few antibiotic resistance genes can become widely distributed in different organisms and in different environments and maintained in microorganisms despite selective pressures. One can only wonder whether a transferred gene could cause problems in a setting that is different from one that was intended (Fox, 1985). Is it correct to state that the existing naturally occurring exchanges have already exhausted all possible recombinations?

A uniform, well-formulated policy on the release of new genetically engineered microorganisms or viruses is urgently needed. There is now disagreement among government agencies, scientists at university and industrial laboratories, and the concerned public as to what will constitute efficient, consistent, and safe environmental release. Attempts are being made to draft a bill that would specify initial, expanded, and commercial use of genetically engineered biological insecticides. If such a bill were to be introduced, it would permit different levels of risk assessment under different conditions. Eventually both the U.S. Department of Agriculture and the Environmental Protection Agency would probably be involved in issuing permits for environmental release.

As has been pointed out by Pramer (1983), the intense interest in biotechnology has been the consequence of many complex interacting economic and scientific events and not, as seemed at first, the result of the new recombinant DNA techniques. The decision that genetically engineered new strains of viruses and microorganisms could be patented, concerns of industry that world leadership might move to other countries, and the realization by university faculty and students that this new area is one of great promise all had an effect on the rapid growth of biotechnology.

It became apparent that the establishment of a biogenetically modified pathogen—essentially a new species of a bacterium, fungus, nematode, or virus—in an ecosystem where it formerly did not occur, can be followed by unforeseeable consequences. Some of the consequences could be very slight, while other might be profound. Therefore, every care should be taken to apply newly developed biocontrol agents with the least possible risk to health of humans and other vertebrate and invertebrate animals and plants. Evaluations of cost–risk benefits are necessary and such evaluations should be made at the time of preliminary testing of novel bioengineered pathogens. The difficulties encountered with recently developed genetically engineered bacteria are fresh in our memory. In 1986 the Environmental Protection Agency suspended the permit to field-test genetically altered frost-free bacteria developed by Advanced Genetic Sciences (AGS) because the open-air test in California has not been approved (Anonymous, 1986).

Opposition to biotechnology is not limited to uninformed people who fear the unknown and object to it for emotional, rather than rational, reasons. Some people object to it for more abstract or for religious reasons. A genuine threat is perceived by Pope John Paul II, who, in an address to UNESCO, stated that "the future of man and mankind is threatened, radically threatened, . . . by men of science. . . . This can be verified as well in the realm of genetic manipulations and biological experiments as well as in those of chemical, bacteriological or nuclear armaments" (Nossal, 1985). At present such extreme views about biotechnology have wide support, unfortunately, and it is up to scientists to refute them as simply not true. Biotechnology appears to be the most successful enterprise in which human beings have ever been engaged.

Under the term "safety" we must include not only safety to humans, domestic animals, fish, birds, beneficial insects, and cultivated plants and forests but also broader possible environmental changes. Some effects can be predicted, while others are unforeseen. This should not stop us from developing and testing bioengineered microbial and viral pesticides, but it should encourage constant vigilance.

In conclusion, there are obvious advantages in biotechnology applications for biological pest control in agriculture. At the same time, environmental implications are extremely important and these have to be carefully studied and considered before new forms produced by biotechnology are released in the environment. The praise of biotechnology is certainly justified, but the environmental and economic implications must not be overlooked.

REFERENCES

Anonymous (1986). *Genet. Eng. News* **6**(4).

Doller, (1985). The safety of insect viruses as biological control agents. *In* "Viral Insecticides for Biological Control" (K. Maramorosch and K. E. Sherman, eds.), pp. 399–439. Academic Press, New York.

Fox, J. L. (1985). Fixed up in Philadelphia: Genetic engineers meet with ecologists. *ASM News* **51**, 832–836.

Kurstak, E., Tijssen, P., and Maramorosch, K. (1978). Safety considerations and development problems make an ecological approach of biocontrol by viral insecticides imperative. *In* "Viruses and Environment" (E. Kurstak and K. Maramorosch, eds.), pp. 571–592. Academic Press, New York.

McIntosh, A. H., and Shamy, R. (1980). Biological studies of a baculovirus in a mammalian cell line. *Intervirology* **13,** 331–341.

Maramorosch, K. (1985). The future of mosquito-borne diseases in the world. *J. Am. Mosq. Control Assoc.* **1,** 419–422.

Maramorosch, K., and Sherman, K. E., eds. (1985). "Viral Insecticides for Biological Control." Academic Press, New York.

Miltenburger, H. G., ed. (1980). "Safety Aspects of Baculoviruses as Biological Insecticides." Bundesminist. Forschungsber. & Technol., Bonn, Germany,

Nossal, G. J. V. (1985). "Reshaping Life: Key Issues in Genetic Engineering." Cambridge Univ. Press, London and New York.

Pramer, D. (1983). Interdependence in biotechnology. *BioScience* **33,** 357.

Quraishi, R. (1985). The government grapples with biotechnology. *Science* **231,** 667–668.

Schneiderman, H. A. (1984). What entomology has in store for biotechnology. *Bull. Entomol. Soc. Am.* **30,** 54–61.

Sun, M. (1986). Local opposition halts biotechnology tests. *Science* **231,** 667–668.

Index

E

F

R

S